BIOINFORMATICS OF GENOME REGULATION AND STRUCTURE II

BIOINFORMATICS OF GENOME REGULATION AND STRUCTURE II

Edited by

NIKOLAY KOLCHANOV

Institute of Cytology & Genetics, Siberian Branch of the Russian Academy of Sciences, Novosibirsk, Russia

RALF HOFESTAEDT

Bielefeld University, Bielefeld, Germany

LUCIANO MILANESI

CNR-ITB Institute of Biomedical Technologies, Segrate (Milano), Italy

Sponsored by the Institute of Cytology and Genetics, Siberian Branch of the Russian Academy of Sciences (Russia) and the Institute for Biomedical Technologies ITB-CNR (Italy)

Springer

Library of Congress Control Number: 2005933710

ISBN-10: 0-387-29450-3 e-ISBN-10: 0-387-29455-4
ISBN-13: 978-0-387-29540-6

Printed on acid-free paper.

Printed in the United States of America.

9 8 7 6 5 4 3 2 1 SPIN 11411925

springeronline.com

Contents

Contributing Authors

I. Abnizova
D.A. Afonnikov
A. Alexeevski
E. Ananko
P. Arrigo
A.S. Arseniev
T.V. Astakhova
G. Bazykin
S.M. Bendre
E. Beresikov
P.M. Beskaravainy
H. Binder
A. Blinov
R. te Boekhorst
A.S. Brok-Volchanski
T. Busygina
S. Cerutti
M.B. Chaley
C. Chaouiya
P. Chavatte
M. Chen
A. Chugunov
V.A. Debelov
A.A. Deev
P.S. Demenkov
G.V. Demidenko
R.G. Efremov
A. Ershova
S.I. Fadeev
M. Fattore
E.Ya. Frisman
D. Furman
M. Fursov
Yu.A. Gaidov
M.S. Gelfand
A.V. Gerasimova
W.R. Gilks
T. Gojobori
V.P. Golubyatnikov
V. Gor
A. Gorban
K. Gorbunov
D.A. Grigorovich
M. Heisler
R. Hofestaedt
E. Ignatieva
K. Ikeo
V.A. Ivanisenko
P. Jain
H. Jönsson
S.G. Kamzolova
A.A. Kanapin

A. Karyagina
A. Katokhin
A. Katyshev
T.M. Khlebodarova
J. Ko
A.V. Kochetov
V.V. Kogai
N.A. Kolchanov
A. Kondrashov
Yu. Konstantinov
E.V. Korotkov
E.G. Kostyanicina
E.A. Kotelnikova
M.A. Krestyanova
N.A. Kudryashov
K. Kumar
V.A. Kuznetsov
O.N. Laikova
A.A. Laskin
V.G. Levitsky
V.A. Likhoshvai
N. Limova
E.K. Litvenko
I. Lokhova
V.I. Lukyanov
V. Lyubetsky
V.Y. Makeev
I.S. Masulis
I. Merelli
V. Merkulov
T. Merkulova
E.M. Meyerowitz
D. Miginsky
L. Milanesi
A.A. Mironov
V. Mironova
C.K. Mitra
E. Mjolsness
L.F. Murga
E. Myasnikova
E. Nedosekina
Y. Nishio
O. Novikova
A. Ogurtsov
N. Omelyanchuk
M.J. Ondrechen
Yu.L. Orlov
D. Oshchepkov
A.F. Osipov
A.A. Osypov
O.N. Ozoline
A. Palyanov
E.M. Panina
L. Pattini
S.V. Petrova
A.G. Pichueva
V.V. Pickalov
S.S. Pintus
O.A. Podkolodnaya
N. Podkolodny
A.A. Polyansky
A. Poplavsky
T. Popova
M.A. Pozdnyakov
A.L. Proscura
Yu.A. Purtov
S. Ramachandran
D.A. Ravcheev
G.V. Reddy
J. Reinitz
D.A. Rodionov
I.B. Rogozin
A. Romashchenko
M.A. Roytberg
L. Rusin
G. Sachdeva
A. Samsonova
M. Samsonova
L. Sánchez
K.V. Shaitan
B.E. Shapiro
K.S. Shavkunov
A.A. Shelenkov
T.I. Shipilov

K.G. Skryabin
O.G. Smirnova
A.A. Sorokin
S. Spirin
I. Stepanenko
S. Surkova
Y. Surya pavan
R.N. Tchuraev
K.B. Tereshkina
D. Thieffry
I. Titov
V. Trifonov
I.I. Tsitovich
V.P. Turutina
Y. Usuda
V. V'yugin
G. Vasiliev
O.V. Vishnevsky
A.G. Vitreschak
E.E. Vityaev
E.P. Volokitin
P.E. Volynsky
D. Vorobiev
K. Walter
Y. Wei
X. Xia
O.L. Zhdanova
A. Zinovyev

Preface

The last 15 years in development of biology were marked with accumulation of unprecedentedly huge arrays of experimental data. The information was amassed with exclusively high rates due to the advent of highly efficient experimental technologies that provided for high throughput genomic sequencing; of functional genomics technologies allowing investigation of expression dynamics of large groups of genes using expression DNA chips; of proteomics methods giving the possibility to analyze protein compositions of cells, tissues, and organs, assess the dynamics of the cell proteome, and reconstruct the networks of protein–protein interactions; and of metabolomics, in particular, high resolution mass spectrometry study of cell metabolites, and distribution of metabolic fluxes in the cells with a concurrent investigation of the dynamics of thousands metabolites in an individual cell.

Analysis, comprehension, and use of the tremendous volumes of experimental data reflecting the intricate processes underlying the functioning of molecular genetic systems are unfeasible in principle without the systems approach and involvement of the state-of-the-art information and computer technologies and efficient mathematical methods for data analysis and simulation of biological systems and processes.

The need in solving these problems initiated the birth of a new science—postgenomic bioinformatics or systems biology *in silico.* The following problems embody the key directions of the systems computer biology: (1) Integration of the heterogeneous experimental data that are obtained by various methods of structural and functional genomics, transcriptomics, proteomics, metabolomics, and other approaches of modern biology in databases; (2) Integrated computer analysis aimed at detection of the patterns in functioning of molecular genetic systems of the living organisms from microorganism to humans; (3) Construction of mathematical models of molecular biological and molecular genetic systems and processes that would form the background for analyzing the mechanisms involved in realization of genetic information; and (4) Investigation of the principles underlying organization and function of gene networks, which provide for formation of phenotypic characteristics of the living organisms basing on the information encoded in their genomes.

These issues of the utmost importance were discussed at the International Conference on Bioinformatics of Genome Regulation and Structure (BGRS), organized biennially by the Laboratory of Theoretical Genetics with the Institute of Cytology and Genetics, Siberian Branch of the Russian Academy of Sciences, Novosibirsk, Russia. BGRS'2004, hold in July 2004, was the fourth event in the series (http://www.bionet.nsc.ru/meeting/bgrs2004/).

The program of the Fourth International Conference comprised five main sections, covering most topical aspects of theoretical, experimental, and applied directions of bioinformatics research, namely: (i) computer structural and functional genomics; (ii) computer structural and functional proteomics; (iii) computer evolutionary biology; (iv) computer systems biology, and (v) new approaches to analysis of data and processes in molecular biology.

We are glad to offer to the readers the second book *Bioinformatics of Genome Regulation and Structure,* which summarizes results of the works presented at BGRS'2004. The book contains selected reviewed papers of the Conference participants from many countries all over the world including Canada, China, France, Germany, India, Italy, Japan, The Netherlands, Russia, Singapore, Sweden, United Kingdom, USA.

We hope that this book will be useful for the scientists involved in basic and applied research in the field of experimental and theoretical studies of structure–function organization of genomes, on the one hand, and teachers and students of universities and colleges, mathematicians and biologists, on the other, that is, to a wide range of readers who are interested in the modern state, problems, possibilities, and prospects of bioinformatics.

Professor Nikolay Kolchanov
Head of the Laboratory of Theoretical Genetics
Institute of Cytology and Genetics, Siberian Branch
of the Russian Academy of Sciences, Novosibirsk, Russia
Chairman of the Conference

Professor Ralf Hofestaedt
Faculty of Technology, Bioinformatics Department
University of Bielefeld, Germany
Co-Chairman of the Conference

PART 1

COMPUTATIONAL STRUCTURAL, FUNCTIONAL AND EVOLUTIONARY GENOMICS

RECOGNITION OF CODING REGIONS IN GENOME ALIGNMENT

T.V. Astakhova[1], S.V. Petrova[1], I.I. Tsitovich[2], M.A. Roytberg[1*]

[1] *Institute of Mathematical Problems in Biology, Russian Academy of Sciences, ul. Institutskaia, 4, Pushchino, Russia, 142290, e-mail: Roytberg@impb.psn.ru;* [2] *Institute of Information Transmission Problems, Russian Academy of Sciences, Bol'shoi Karetnyi per., 19, Moscow, Russia, 127994*

[*] *Corresponding author*

Abstract: Gene recognition is an old and important problem. Statistical and homology-based methods work relatively well, if one tries to find long exons or full genes but are unable to recognize relatively short coding fragments. Genome alignments and study of synonymous and non-synonymous substitutions give a chance to overcome this drawback. Our aim is to propose a criterion to distinguish short coding and non-coding fragments of genome alignment and to create an algorithm to locate aligned coding regions. We have developed a method to locate aligned exons in a given alignment. First, we scan the alignment with a window of a fixed size (~ 40 bp) and assign a score to each window position. The score reflects if numbers K_S of synonymous substitutions, K_N of non-synonymous substitutions, and D of deleted symbols look like those for coding regions. Second, we mark the 'qualified exon-like' regions, QELRs, i.e., sequences of consecutive high-scoring windows. Presumably, each QELR contains one exon. Third, we point out an exon within every QELR. All the steps have to be performed twice, for the direct and reverse complement chains independently. Finally, we compare the predictions for two chains to exclude any possible predictions of 'exon shadows' on complementary chain instead of real exons. Tests have shown that ~ 93 % of the marked QELRs have intersections with real exons and ~ 93 % of the aligned annotated exons intersect the marked QELRs. The total length of marked QELRs is ~ 1.30 of the total length of annotated exons. About 85 % of the total length of predicted exons belongs to annotated exons. The runtime of the algorithm is proportional to the length of a genome alignment.

Key words: coding region; gene recognition; genome alignment; synonymous and non-synonymous substitution

1. INTRODUCTION

Existence of powerful genome alignment methods (Roytberg et al., 2002; Brudno et al., 2003;) and availability of many complete genomes, including several eukaryotic ones, lead to new formulations of classic problems of sequence analysis. Indeed, we can analyze pairwise (or, if possible, multiple) sequence alignment instead of one genome sequence. In case of gene recognition, the problem of genome alignments allows one to exploit two ideas. First, coding regions are, in general, more conservative than the non-coding ones. Thus, one can try to recognize genes as sequence of well-aligned genome fragments (Bafna and Huson, 2000; Batzoglou et al., 2000; Novichkov et al., 2001; Taher et al., 2003).

Such methods are efficient for relatively distant species, but some genes can be unrecognizable because of a low interspecies similarity. From the other hand, alignment of close genomes often gives many false positive exons because of existence of conservative non-coding regions (Shabalina and Kondrashov, 1999). Second, one can additionally pay attention to the difference between substitution patterns in coding and non-coding regions; the former tend to be synonymous, i.e., preserve a coded residue.

The methods using alignment-based HMMs or pair HMMs (Meyer and Durbin, 2002; Pedersen and Hein, 2003) take into account the differences between various parts of a genome alignment implicitly, in course of HMM training. Despite the promising results shown by these methods, we think that it is worth learning explicitly what benefit one can get from the differences in substitution patterns.

The explicit usage of the differences is implemented by Nekrutenko et al. (2001); the abilities of this approach were demonstrated by Nekrutenko et al. (2002). However, the goal of the paper by Nekrutenko et al. (2001) was mainly to study the ability of the proposed criterion to recognize relatively long exons as a whole; authors did not try to recognize the exon borders or short coding regions.

We propose a two-stage procedure combining prediction techniques of traditional identification of exons in DNA sequence and methods based on information about genome alignment. First, using investigation of substitution patterns, we perform an alignment filtration, i.e., locate 'exon-like regions' (ELR) in the alignment. Then, the putative exon within ELR can be found with classic statistical approach. Below, we will demonstrate advantages and drawbacks of the approach and will discuss possible ways to improve it.

2. METHODS AND ALGORITHMS

General description of the approach. The algorithm works in four steps. Three first steps have to be performed independently for the direct and reverse complement chains. At the last step, we compare the results obtained for two chains and prepare the final prediction. We start (first step) with scanning of the alignment with a window of a fixed size w and a given shift s. For each considered window, we make a decision if it is exon-like or not. Then (second step), we reveal the 'exon-like' regions, ELRs. An ELR is a set of consecutive window positions (see details below). Any two ELRs marked on a chain do not intersect each other. Presumably, each ELR contains one exon. During this step, we work only with exon/non-exon marks of window positions, the marks were assigned at previous step. At the third step, we reveal a putative exon for each ELR and ascribe the exon with a score. If ELR does not contain a pair of aligned exons of high enough score, the ELR is to be rejected. Finally, we compare ELRs found on the direct and inverse chains. If two ELRs from different chains intersect each other, we keep only one of them, the ELR having an exon of higher score.

Table -1. The values $Score_E(K_N, K_S)$

K_S \ K_N	0	1	2	3	4	5
0	2.6	3.8	5.1	7.4	9.2	11.0
1	2.1	3.4	4.7	6.7	8.2	10.2
2	1.7	3.0	4.2	6.0	7.7	9.4
3	1.4	2.6	3.8	5.3	7.0	8.6
4	1.0	2.2	3.4	4.6	6.2	7.8
5	0.6	1.8	3.0	3.9	5.5	7.0
6	0.2	1.4	2.3	3.2	4.7	6.2
7	–0.2	1.0	1.8	2.7	4.0	5.4
8	–0.6	0.6	1.2	1.9	3.3	4.6
9	–1.0	0.2	0.7	1.1	2.6	3.8
10	–1.4	–0.2	0.3	0.8	1.8	3.0
11	–1.7	–0.6	–0.2	0.4	1.1	2.2
12	–2.1	–1.0	–0.5	–0.1	0.3	1.4
13	–2.5	–1.4	–0.9	–0.5	–0.1	0.6
14	–2.9	–1.8	–1.4	–1.0	–0.5	–0.2
15	–3.3	–2.5	–2.0	–1.5	–0.8	–1.0

Here, we set $Score_E(K_N, K_S) = 20$ if $K_S > 5$, and $Score_E(K_N, K_S) = -20$ if $K_S \leq 5$ and $K_N > 15$. The values are obtained from the statistics of windows of length 45.

Analysis of window position. Let w be a size of the window. For a window at the position P, i.e., for the fragment of alignment from position P

to position $P + w - 1$, the program calculates its score $H(P)$. The score characterizes the presence of stabilizing selection at the protein level. We have tested two approaches to define the score $H(P)$, the 'theoretical' approach and the 'empiric' one. Within the *theoretical* approach, the basic characteristics of the window at a given position of alignment are (1) *FMatch*—the fraction of match alignment positions, i.e., superposition of identical nucleotides and (2) the probability $Pr(K_T, K_S, D)$ to obtain K_S or more synonymous substitutions if K_T random independent substitutions were performed and D codons are deleted. We calculate $Pr(K_T, K_S, D)$ for three possible frames. The score $H_T(P)$ is a negative binary logarithm of the minimum of the three probabilities. We say that the window position P is 'exonic', if both *FMatch*(P) and the score $H_T(P)$ exceed the threshold.

Within the *empiric* approach, the score H_E is computed based on the statistics of the appearance of the windows with the given number of non-synonymous substitutions K_N and synonymous substitutions K_S in coding and non-coding regions. Table 1 shows the pre-computed *Scores* assigned to the different pairs (K_N, K_S); the values in Table 1 are obtained as log–likelihood ratios of the corresponding empiric frequencies (only windows without deletions were taken into account). Table 1 confirms significant difference between the two-dimensional distribution of (K_N, K_S) of the windows without deletions in coding and non-coding regions. If the window at the position P does not contain deletions ($D = 0$), then we get the value $H_E(P)$ from the pre-computed table (Table 1); the values in the table are obtained as log–likelihood ratios of the corresponding empiric frequencies. If $D > 0$ but the value *FMatch* exceeds a proper threshold, the value $H_E(P)$ is computed from the same table, but with recalculated values K_N and K_S.

Exon-like regions (ELRs). A region is a set of consecutive windows, i.e., the windows at positions P, $P + s$, $P + 2s$, ..., where s is a given shift. A region starts in the beginning of the first window and ends at the end of the last window. An exon-like region (ELR) is a region meeting the following conditions:

(1) the first window of the region contains a putative acceptor site or START codon; the last window of the region contains a putative donor site or STOP codon (see below);

(2) a region is 'exonic-dense', i.e., a difference between the numbers of non-exonic and exonic windows within a consecutive part of a region cannot exceed a threshold *InnerCut*;

(3) the number of non-exonic windows at the beginning and at the end of a region cannot exceed a threshold *EdgeCut*; and

(4) a region is not a part of another fragment meeting conditions (1)–(3).

Putative exons and qualified ELRs (QELRs). Our algorithm first finds all ELRs in both chains and then reveals among them the 'qualified' ELRs (QELRs). The definition of QELR is based on the notion of a *putative exon.*

Putative exon is a part of an exon-like region starting with a putative acceptor site or START codon and ending with a putative donor site or a STOP codon. A putative acceptor (donor) site is aligned, i.e., present in both sequences, dinucleotide 'AC' ('GT'), its neighborhood has Berg–von Hippel score (Berg and Hippel, 1987) exceeding a given cut-off. Putative START- and STOP-codons also have to be present in both sequences and to be aligned. If the exon starts with a START-codon and/or ends with a STOP-codon, then it should not contain STOP-codons in the corresponding frame.

We assign each putative exon E with a statistical score $S(E)$ and an alignment score $A(E)$. The score $S(E)$ is calculated by the method described by Gelfand et al. (1996). The value $S(E)$ depends on the scores of splicing sites, codon potential, and exon length. Alignment score $A(E)$ reflects the difference between the ratios $K_S/\max(K_N, 1)$ for the exon calculated for the considered chain and the inverse chain.

We ascribe each exon-like region R with the score $G(R)$ that is a sum of the maximal values of $S(E)$ and $A(E)$ for putative exons belonging to the region. We say that the ELR R is a qualified ELR (QELR) if R meets the following conditions:

(1) the value $G(R)$ of the region exceeds the cut-off *ELRScoreCut* and

(2) the region does not intersect an ELR on the opposite chain or the intersecting ELR on the opposite chain has a lower value ELR_Score.

The result of the algorithm's work is lists of QELR for both chains. The exon E corresponding to the maximal score $S(E)$ among all putative exons within a QELR R is considered as a predicted exon for the region R.

Genome alignments. We used two sets of genome alignments. The first set is the alignment of syntenic regions of the *Homo sapience* chromosome 6 (GenBank ACCESSION NT_007592) and the *Mus musculus* chromosome 17 (GenBank ACCESSION NT_002588) of ~700 000 nucleotides long. The human sequence contains 55 annotated genes and the mouse sequence contains 58 annotated genes.

Alternative splicing variants are given for 17 human genes and for only 1 mouse gene. Mouse genes contain 567 annotated exons, 476 of them are aligned correctly with the corresponding human exons. Incorrect alignment of other genes mostly can be explained by inconsistency of exon annotation in human and mouse genome. The total length of all the annotated mouse exons is 93 162; the average length is 165. The alignment was obtained by OWEN program (Ogurtsov et al., 2002).

The second set is the set of 117 orthologous mouse and human genes from Batzoglou et al. (2000). The genes were also aligned with the OWEN program. The mouse genes contain 476 exons; 397 of them are aligned correctly, while the other exons have incorrectly aligned ends. The total length of mouse genes is 105 450; the average length is 222.

The alignment of syntenic regions of the *Homo sapience* chromosome 6 and the *Mus musculus* chromosome 17 was used as the training data for the algorithm; the set of Batzoglou et al. (2000) was used as testing sets.

Finally, we have analyzed the four pairs of orthologous mouse and rat mRNA with atypical ratio of the numbers of synonymous and non-synonymous substitutions; the set was proposed by G. Bazykin.

Testing parameters. We used the following values of parameters (see above): (1)window size $w = 45$, window offset $s = 15$ bp; (2) *FMatch* cut-off for 'exonic' window *FMatchMin* = 0.65, $H(P)$ cut-off for 'exonic' window: $H_T\_Min = 1.2$ (for 'theoretical' score H_T), $H_E\_Min = 3.0$ (for 'empiric' score H_E); (3) ELR cut-offs *InnerCut* = 6, *EdgeCut* = 6; (4) minimal score of an acceptor splicing sites *ACC_Score* = –17, minimal score of a donor splicing sites *DON_Score* >= –7; and (5) the cutoff for the ELR score $G(R)$ is 2.5.

3. RESULTS AND DISCUSSION

3.1 Results

The algorithm produces two types of objects (see Materials and Methods): qualified exon-like regions (QELR) and putative exons. The results on QELR prediction are given in Table 2; the results on exon prediction in Table 3. All results are given for the mouse chromosome; the results for the human chromosome are very similar. Results for training and testing sets are in good agreement. For the testing set, we have not reported 40 QELR predicted on the inverse chain, because we have no information about exons on this chain (Batzoglou et al., 2000, also do not consider predictions on inverse chain). If we take into account these extra QELR, the percent of QELR (line '% Inters QELR') will fall to 87 %.

The goal of the presented algorithm is to locate the aligned exons, not to give their precise borders; we consider the latter problem as a separate task and now continue to work on it; the results to be reported later. For example, we will propose the method to process correctly the QELRs containing more than one exon; this is a common situation for genes with short introns.

To check the applicability of the method to genes with nonstandard relation between K_N and K_S, we have considered four pairs of orthologous mouse and rat mRNA (Table 4).

Table -2. Qualified exon-like regions (QELR) predicted for the alignment of syntenic regions of the *Homo sapience* chromosome 6 and *Mus musculus* chromosome 17 (training set) and the set of 117 orthologous mouse and human genes from Batzoglou et al., 2000 (testing set)

	Training		Testing	
	Theor.	Empiric	Theor.	Empiric
N QELR	441	425	334	310
Tot L QELR	116591	116209	127827	144652
% Tot L Exon	125	125	121	137
Ave L QELR	264	273	339	414
% Ave L Exon	161	166	153	187
N Inters QELR	400	396	324	302
% Inters QELR	91	93	97	97
Covered Exon %	99	99	98	99
N Lost Exon	23	15	43	34
% Lost Exon	4	3	9	7

The data are given for both theoretical (columns 'Theor' and empiric versions of the scoring function $H(P)$. We use the following notation: 'N QELR', number of revealed QELRs; 'Tot L QELR', total length of revealed QELRs; '% Tot L Exon', ratio of the total length of revealed QELRs and the total length of annotated exons; 'Ave L QELR', average length of ELR; '% Ave L Exon', ratio of the average length of revealed QELRs and the average length of annotated exons; 'N Inters QELR', number of QELRs having intersection with an annotated exon; '% Inters QELR', percent of revealed QELRs having intersection with annotated exons; 'Covered Exon (%)', average part of the exon covered by intersecting QELR; 'N Lost Exon', number of 'lost exons', i.e., correctly aligned exons that do not intersect QELRs; and '% Lost Exon', percent of the lost exons among all correctly aligned exons.

Table -3. Correspondence between the predicted putative exons and annotated exons

	Training		Testing	
	Theor.	Empiric	Theor.	Empiric
% Lost Exon	27	31	30	36
Covered Exon (%)	87	90	88	88
Exactly Recogn (%)	41	39	43	39

'% Lost Exon', percent of correctly aligned annotated exons having no intersection with the predicted exons; 'Covered Exon (%)', average part of the correctly aligned annotated exon covered by intersecting predicted exon; and 'Exactly Recogn (%)', percent of the correctly aligned annotated exons that coincide with a predicted exon.

Table -4. Orthologous mouse and rat genes with nonstandard K_N and K_S ratio

Mouse GI	Rat GI	K_N	K_S	K_N/K_S
6678712	19705461	0.10	0.09	1.15
21312956	20806163	0.164	0.156	1.05
8394248	9507069	0.18	0.17	1.05
6753828	6978833	0.19	0.186	1.02

In all pairs, the program detected one QELR that contained the desired segment. In two cases (6 678 712 vs. 19 705 461 and 8 394 248 vs. 9 507 069), the coding region was predicted exactly. For the pair 6 753 828 vs. 6 978 833, the correct exon has rank 2 among all putative exons; the predicted exon has correct donor site and covers 77 % of the correct coding region. In the last case, the predicted QELR coincides with the correct coding region, but the predicted exon is significantly shorter.

3.2 Discussion

The algorithm addresses two problems. First, it approximately locates the area where it is reasonable to look for exons (generation of qualified exon-like regions, QELRs). Second, it points out the putative exons within QELRs. The problem is relatively independent, i.e., we can use arbitrary gene recognition algorithm to solve the second problem, when the first problem is already solved. We have studied whether the problems can be solved based on the difference of the substitution patterns in coding and non-coding regions.

Our main efforts were directed to the first problem, and the algorithm effectively solves it. Taking into account its linear runtime, the algorithm can serve as a useful filtration tool for any exon-recognition algorithm working with genome alignments. We have demonstrated that statistics of the possible values of pairs (K_N, K_S) in coding and non-coding regions can serve as the background to distinguish between the coding and non-coding fragments.

Putative exons show up worse correlation with the annotated exons than ELRs as well as the predictions made by the programs that use more sophisticated training technique (see Introduction). We plan to improve significantly this part of our algorithm. For example, we plan to generate for a given ELR several putative exons having different frames and link them to predict the whole gene. Another possible development of the project is to realign genomes in the vicinity of putative exon borders. General genome alignment algorithms often misalign conservative positions of splicing sites.

AKNOWLEDGMENTS

Authors are thankful to A. Kondrashov, A. Ogurtsov, S. Shabalina, G. Bazykin, S. Sunyaev for helpful discussions. The work was supported by the Russian Foundation for Basic Research (grants Nos. 03-04-49369 and 02-07-90412); the Ministry for Industry, Science, and Technologies of the Russian Federation (grants Nos. 20/2002 and 5/2003); NIH (grant TW005899, co-PIs V. Tumanyan and M. Borodovsky), and NWO and ECO-NET grants. The work was partly performed during the visit of M.A. Roytberg to NCBI (January-March, 2004). We thank anonymous referees for their helpful comments.

PREDICTING sRNA GENES IN THE GENOME OF *E. COLI* BY THE PROMOTER-SEARCH ALGORITHM PlatProm

A.S. Brok-Volchanski[1], I.S. Masulis[1], K.S. Shavkunov[1], V.I. Lukyanov[1], Yu.A. Purtov[1], E.G. Kostyanicina[1], A.A. Deev[2], O.N. Ozoline[1*]

[1] *Institute of Cell Biophysics, Russian Academy of Sciences, Pushchino, Moscow Region, 142290, Russia; e-mail: ozoline@icb.psn.ru;* [2] *Institute of Theoretical and Experimental Biophysics, Russian Academy of Sciences, Pushchino, Moscow Region, 142292, Russia*

[*] *Corresponding author*

Abstract: The potentially transcribed regions in the genome of *E. coli* were searched for on a systematic basis using the novel pattern recognition software PlatProm. PlatProm takes into consideration both the sequence-specific and structure-specific features in the genetic environment of the promoter sites and identifies transcription start points with a very high accuracy. The whole genome scanning by PlatProm along with the expected promoters upstream of the annotated genes identified several hundred of very similar signals in other intergenic regions and in many coding sequences. Most of them are expected as start points for independent RNA transcripts, providing a unique opportunity to reveal genes encoding antisense and/or alternative RNAs. The potential PlatProm as a tool revealing sRNA genes is discussed.

Key words: promoters; genome regulatory regions; gene expression; promoter-search software; transcription; untranslated RNAs; genome annotation

1. INTRODUCTION

The current annotation of bacterial genomes relies on computational methods identifying coding sequences on the basis of certain specific features in base composition, codon usage, the database of *N*-terminal peptide sequences, matches for the ribosome binding site, transcription and translation terminators, homology with known genes in other bacterial species, expression efficiencies measured by microarrays, and other available information (Blattner et al., 1997; Lukashin and Borodovsky, 1998; Delcher et al., 1999; Besemer et al., 2001; Walker et al., 2002; Azad and Borodovsky, 2004). Although these methods are not perfectly precise, they allow identifying most protein-encoding genes, and the longest possible *N*-termini were generally selected in cases when multiple in-frame start codons were found. A high level of conservation and the specific features of the three-dimensional structures of rRNAs and tRNAs were employed to identify genes of these RNA species. However, the detection of sRNA genes (small untranslated RNAs) in bacterial genome is still problematic. These RNAs regulate diverse cellular functions, such as RNA processing, mRNA stability, translation, protein stability, and secretion. The first 13 were discovered fortuitously on the basis of high abundance or functions related to protein synthesis or activity. After genome-wide identification of these gene species became a focus of attention, 49 novel sRNA genes with yet unknown functions were found.

The largest contribution to the set of novel sRNAs was given by the approach primarily based on the sequence conservation within intergenic regions and exploiting microarrays as well as other techniques for experimental verification (Wassarman et al., 2001). Overall, 17 novel sRNAs and 6 ORFs were suggested by this combinatorial approach; however, sRNA genes that had evolutionary conserved secondary structures rather than nucleotide sequences might be overlooked as well as any sRNA genes overlapping the neighboring coding sequences. The former limitation was surmounted by Rivas et al. (2001), who distinguished conserved RNA secondary structures from a background of other conserved sequences using probabilistic models of expected mutational patterns in pairwise sequence alignments. Sequence-based structure comparison among genomes of closely related bacteria allowed detecting eight novel sRNAs. Carter et al. (2001) analyzed intergenic regions in terms of nucleotide and dinucleotide composition, occurrence frequency of sequence motifs typical of RNA structural elements, free energy of folding, and some other considerations, which led to the prediction of 562 sRNAs. Approaches relying mainly on the biological features of sRNAs were suggested by Vogel et al. (2003) and Zhang et al. (2003). In the former case, 50–400 nt RNA products were picked out from the total fraction of RNAs and their sequences were estimated. Most of them were derived from within the coding regions including the small fraction (~ 5 %) matching to the antisense strand. Fragments from intergenic regions comprise 18 % of all samples, and seven novel sRNA genes were identified among them. In the latter case (Zhang et al., 2003), a set of new sRNAs was chosen from the fraction of RNAs that co-immunoprecipitated with Hfq. Although only ~ 30 % of the known sRNAs interact with Hfq, this method allowed revealing five additional sRNAs.

The predictive potential of transcription signals was exploited by Argaman et al. (2001) and Chen et al. (2001). The first team mostly relied on the accuracy of terminator prediction. Promoters were searched for upstream of terminators by approaches combining the consensus and weight matrix considerations. All the sequences 50–400 bp long located between the promoter and terminator within empty intergenic regions were compared to the genomes of other bacteria, and conserved sites were assayed experimentally. Expressions of 14 out of 24 predicted genes were detected. Chen et al. (2001) used RNAMotif and the thermodynamic scoring system to detect the set of terminators in the bacterial genome. Several hundreds were suggested as potential stop signals for genes yet to be discovered. Combining these data with a set of promoters predicted by profile-based software (Bucher et al., 1996) and requiring the presence of both signals on the same strand 45–350 bp away from each other, 227 candidates were selected. Expression of eight candidates was tested and confirmed in seven cases. Thus, both approaches employing transcription signals as a primary search criterion demonstrated a very good proportion between the predicted and the confirmed candidates.

What part of the overall population of sRNAs do the 62 verified genes comprise? More than 1000 other potential sites for sRNA synthesis were predicted in intergenic regions (Hersberg et al., 2003), and approximately the same number of transcripts generated from these regions was detected by the expression analysis (Tjaden et al., 2002). This essentially exceeds the estimates made for the total number of sRNAs in *E. coli* (50–200 genes) (Eddy, 1999; Wassarman et al., 2001). Many of the predicted genes may therefore be false positives. On the other hand, the discovered noncoding RNAs range from 45 to 370 nt in length. Considering that shorter RNAs (21–25 nt) in eukaryotic organisms are involved in RNA interference, processing, chemical modification, and stabilization, while gene silencing is controlled by longer RNAs, some untranslated RNA species may be overlooked due to commonly used size limitation. Moreover, most methods employed so far are focused on intergenic regions, excluding a possibility to find new transcripts generated from sequences encoding proteins, although experimental approaches demonstrate the possibility of both antisense and alternative (shortened) transcription from these sequences (Selinger et al., 2000; Vogel et al., 2003). This means that scanning techniques capable of predicting genes with parallel transcriptional output are required. Katz and Burge (2003) and Pedersen et al. (2004), who proposed methods revealing local RNA secondary structures within bacterial genes, made the first step in this direction. However, folding propensity may not be a general feature of intrinsic transcripts. Since the methods of comparative genomics cannot be used to detect them within ORF, approaches searching for transcription signals may have the highest potential here.

Promoter search algorithms proposed so far (Alexandrov and Mironov, 1990; Horton and Kanehisa, 1992; Mahadevan and Ghosh, 1994; Pedersen and Engelbrecht, 1995; Hertz and Stormo, 1996; Yada et al., 1999; Vanet et al., 1999; Leung et al., 2001; Gordon et al., 2003; Huerta and Collado-Vides, 2003) are usually based on the sequence preferences in the regions of specific contacts with RNA polymerase. Most of them can identify more than 80 % of promoters from testing compilations but even the best protocols at this level recognize a large portion (0.8–3.4 %) of non-promoter DNA as promoter-like signals (Horton and Kanehisa, 1992; Gordon et al., 2003). Within the genome size sequences, the background noise is, therefore, more than one order of magnitude greater than the required signal. That is why promoters are usually searched for as the most probable candidates of several promoter-like signals found within a limited region upstream of a particular gene (or terminator, as discussed above).

We tried to increase the performance of computational prediction by taking into account structural features in the genetic environment of the promoter sites, considering them as a generalized platform for transcription complex formation. The resultant algorithm PlatProm was used for promoter prediction within the entire genome of *E. coli*. The disposition of promoter-like signals according to the gene borders is discussed.

2. METHODS AND ALGORITHMS

Compilations. The training set contained 400 σ^{70}-promoters with single start point. The testing set contained 290 known promoters with single or multiple transcription start points. Overlapping and homologous promoters were removed from both sets. All sequences were 411 bp long (–255/+155 according to the start point, nominated as 0). The control set contained 400 sequences taken from the coding regions of convergently transcribed genes.

Weight matrices. Three types of weight matrices were used to formalize structural organization of the promoter sites. The matrices of the first type reflect distribution of nucleotides in the conservative elements (–35 and –10) and dinucleotides near the start point (position –1) and in the 5′-flanking region of element –10 ('extended –10 element'). These matrices are designed exactly as described by Hertz and Stormo (1996) and contain 6×4 (or 1×16) scores equal to natural logarithms of normalized frequencies of the appearance of each nucleotide (dinucleotide) at each position of the preliminary aligned promoters. Occurrence frequency of particular nucleotides (dinucleotides) in genome was used for normalization. Allowed variations of the spacer and the distance between the start point and the element –10 were 14–21 and 2–9 bp, respectively. This means that 64 alignments were tested for each sequence to find the maximal score. Any deviations from the optimal spacer (17 bp) and optimal distance (6 bp) were penalized based on their frequencies. Optimal matrices were generated by the procedure of expectation–maximization.

Table -1. Weight matrix for $(T)_n$ in the region –83/–76

n	Scores	FS
< 4 (penalty)	–0.02	–0.01
4	0.27	0.14
5	0.74	0.38
≥6	1.32	0.69

Table -2. Weight matrix for TA in the region –49/–47

Order	Position	FS
1 check	–48	0.56
2 check	–47	0.52
3 check	–49	0.39
Penalty	No	–0.13

The $(T)_n$ or $(A)_n$ ($n \geq 4$) tracts interacting with RNA polymerase α-subunits (Wada et al., 2000) or stabilizing the transcription complex by a properly induced bend (Hivzer et al., 2001) were accounted by the set of 26 simplified matrices scoring the presence of these elements in 26 regions from –20 to +34 (distributed with a periodicity of ~ 1 helix turn). An example of such matrices, accounting the presence of $(T)_n$ in the region –83/–76 is shown in Table 1. Positive scores represent log probabilities to find 5′-end of $(T)_n$ in the region. The penalty is estimated as a probability of $(T)_n$ absence. Final scores (FS) were reduced by the coefficient estimated as a ratio of the average information content in the region to the information content at the sixth position of aligned element – 35 (the least significant). This reduction was aimed to balance the endowments of structure-specific elements with the contributions of conservative base pairs.

Flexible dinucleotides TA, supporting adaptive isomerization of the DNA on the protein surface (Ozoline et al., 1999a; Masulis et al., 2002), were accounted by 20 cascade matrices exemplified in Table 2. The presence of TA is first checked at the position where probability to find it is maximal and than at the adjacent points. An absence of TA is penalized. The whole set of such matrices reflects a regular distribution of TA in the region of – 98 to + 24.

A large contribution to the sensitivity of PlatProm is made by mixed A/T-tracts, putatively involved in the polymerase sliding along the DNA (Ozoline et al., 1999b). These elements are distributed with periodicities of

~ 1 and ~ 1.5 helix turns. To take into account this regularity, we scored the presence of paired rather than single A/T-tracts: www(*n*)www (w = A = T; where *n* is 7, 8, 13, or 14 random base pairs) in the region of – 139 to + 11. Overall 15 cascade matrices (similar to that shown in Table 2) were used for this purpose.

Ideal direct and inverted repeats (5–11 bp long separated by 5 or 6 bp) were considered as putative targets for interaction with transcription regulators. Their presence was scored as natural logarithms of the lengths. Contributions of all elements were summarized, giving the total score. An average score for non-promoter sequences amounted to – 3.8. The value of Std was equal to 3.0. An average score of promoters from the testing set was 7.29.

The distribution of promoters predicted by PlatProm within the genome of *E. coli*. Only highly reliable signals scoring ≥ 8.2 ($p < 0.00005$) were used. The genome map of *E. coli* K12 (NCBI, GenBank entry U00096) was basically used for this purpose, but ~ 300 genes additionally annotated in the Colibri DataBase were added. The allowed distances between the transcription start points and the coding sequences were deduced based on positional coördinates typical of the known promoters (Figure 1). Although ~ 90 % of them are < 250 bp far from coding sequences, some genes have leaders as long as 600 bp and more. That is why we considered 750 bp as the distance at which the predicted promoter may be ascribed to the downstream gene. Other promoter-like points were sorted according to their location relative to gene borders. The map of known and predicted transcription start points is available by request (ozoline@icb.psn.ru).

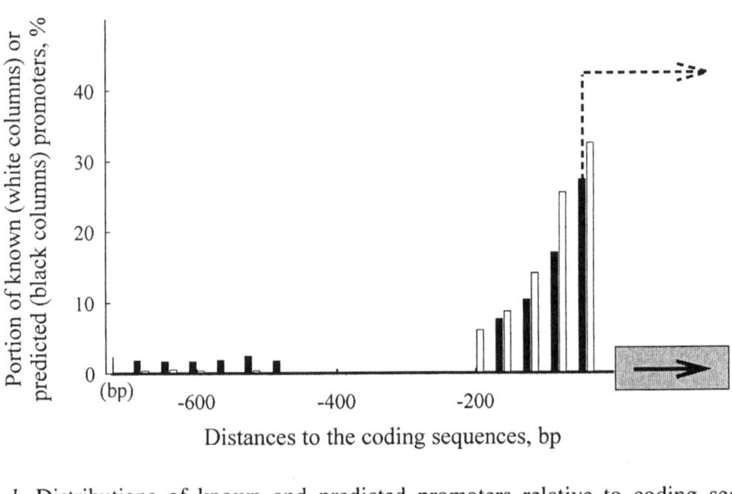

Figure -1. Distributions of known and predicted promoters relative to coding sequences. Arrows indicate orientation of the gene (rectangle) and the direction of transcription.

3. RESULTS AND DISCUSSION

The current version of PlatProm identifies 84.8 % promoters of testing compilation at the level when zero false positives were found in the control set (scores > 3.0). This means that the combination of sensitivity (((true positives (TP))/(TP + false negatives (FN))) × 100 = 84.8 %) and specificity (((TN/(TN + false positives (FP))) × 100 = 100.0 %) of our approach is better than in the case of algorithms based on neural networks (80.6 % and 99.14 %, respectively; Horton and Kanehisa, 1992); logic grammar formalism (68.7 % and 82.23 %; Leung et al., 2001); and sequence alignment kernel (82 % and 84 %; Gordon et al., 2003). Another three parameters characterizing performance of promoter search algorithms are AE (average error) = (((FN + FP)/(N + P)) × 100), CC (correlation coefficient) = $(TP \times TN - FP \times FN)/((TP + FP) \times (TN + FN) \times (TP + FN) \times TN + FP))^{1/2}$, and accuracy of the transcription start point prediction among others promoter-like signals detected in the same region. AE and CC for PlatProm are 6.3 and 0.87, respectively, which is better than reported originally (16.5 and 0.67) (Gordon et al., 2003). Positions with maximal scores coincided with experimentally estimated start points or were located at the nearest two positions in ~ 80 % cases, while this value is usually lower than 50 % (Hertz and Stormo, 1996; Huerta and Collado-Vides, 2003).

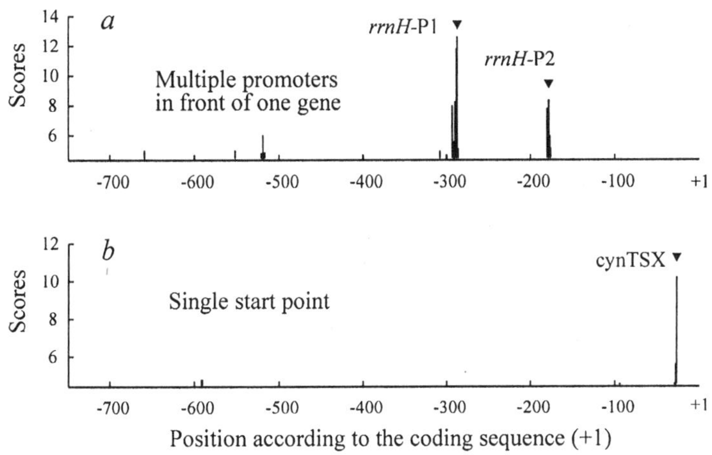

Figure -2. Distributions of promoter-like signals (columns) within 750-bp regions upstream of (*a*) rrnH and (*b*) cynT genes. The positions of known promoters are indicated by triangles.

The percentage of recognized promoters increases up to 90 % if ± 5-bp variations in the positioning of the start point were allowed. For the training set, this value is 94 %, while the percentage of known promoters recognized in the genomic DNA is 91.4 %. Figure 2 exemplifies distribution of promoter-like signals within the regulatory regions of two genes. In the case of gene encoding 16S RNA, both known promoters (*rrnH*-P1 and *rrnH*-P2) are surrounded by other promoter-like signals forming two clusters (Figure 2*a*). Huerta and Collado-Vides (2003) already mentioned this phenomenon and assumed that additional sites might be used for polymerase trapping.

Putative promoters were found upstream of 2229 genes. Most of them form only one cluster or appear as single promoter-like point, like the known promoter *cynTSX* (Figure 2*b*). The rest form several clusters, providing a possibility to predict multiple promoters with potentially different regulation. Similarly to known promoters, the predicted sites are usually located less than 250 bp away from coding sequences (Figure 1, black columns). The information given by this set of predicted promoters facilitates their experimental identification and sometimes gives a chance to find functional promoters in the places where they otherwise may be overlooked.

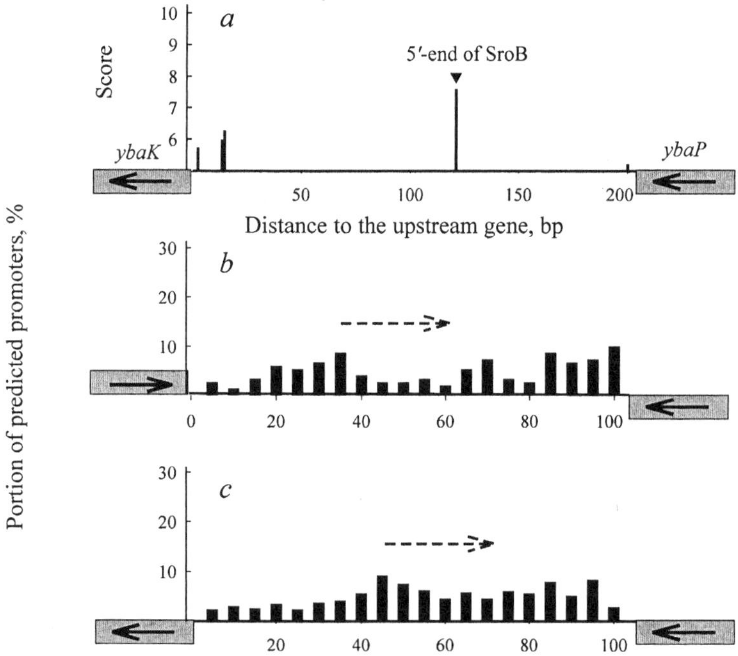

Figure -3. Distribution of promoter-like points between *yba*K and *yba*P genes. The 5′-end of the SroB sRNA is indicated (*a*); relative positioning of the predicted promoters relative to adjacent genes (rectangles) (*b* and *c*). Solid arrows show directions of transcription; dashed arrows, orientation of the predicted promoters.

3.1 Putative promoters in intergenic regions

More than 600 promoters were predicted between the genes transcribed convergently or from the opposite DNA strand. Their positioning does not show any preference (Figures 3*b* and 3*c*). The average distance between non-overlapping genes in the genome of *E. coli* is 148 bp. The average length of intergenic regions containing potential promoters is larger (440 and 298 for the sets in Figures 3*b* and 3*c*, respectively), thus suggesting the existence of additional genes. Several hundreds of sRNA genes are predicted in such regions (see above). Thus, the presence of functional promoters for their expression is also expected. Figure 3*a* shows the potential promoter for the predicted SroB sRNA (Vogel et al., 2003).

3.2 Potential promoters for antisense transcription

The 709 promoter-like signals detected on the opposite strand of protein-encoding sequences are perhaps of the highest significance (Figure 4*b*). the antisense RNAs produced from such promoters may block translation by base pairing with mRNAs or regulate their processing and stability. Such RNAs control expression of many plasmid and transposon genes, but they were not considered so far as typical of the bacterial genes (Wagner et al., 2002). All the genes bearing promoter-like signals were tested previously for the synthesis of antisense RNAs (Selinger et al., 2000), and in all but one cases such products were found. Although the data from the expression analysis cannot be considered as strong evidence, at least some of the found signals may be true positives. Assuming antisense transcription, Vogel et al. (2003), who analyzed short RNA products and detected 21 RNAs generated from the antisense strand, obtained other data. Eight of the RNAs may be produced as run-through transcripts from the neighboring genes, while appearance of the remaining 13 requires a different explanation, and antisense transcription may be the most evident. At least five of them may be produced from promoters predicted by PlatProm.

3.3 Potential promoters for alternative transcription

A genome-wide promoter screening unexpectedly detected 379 genes having promoters with a propensity to produce shortened RNA products from the sense strand (Figure 4*a*). At least 46 of the 275 previously detected short RNAs (Vogel et al., 2003) may be initiated at such promoters, thus supporting the suggestion that bacterial genes may have a parallel transcriptional output. The role of such RNAs remains vague. Some of them may encode alternative proteins, but the appearance of functional promoters in coding sequences may also have other reasons. Thus, additional promoters may be involved in polymerase trapping or may intensify the transcription of properly oriented downstream genes even if they are > 750 bp away from these promoters. On the other hand, the preferred location of promoter-like sites at the beginning of genes allows a speculation that coordinates of some genes require correction, here, our data may be used to pick them out.

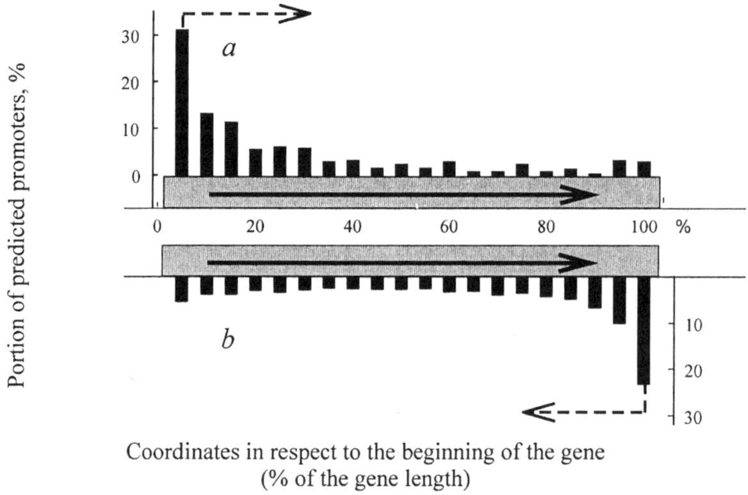

Figure -4. Relative positioning of the predicted promoters within the coding sequences: (*a*) potential promoters for alternative transcription and (*b*) promoters for antisense transcription.

In any case, PlatProm provides a unique opportunity of predicting independently transcribed regions within coding sequences.

ACKNOWLEDGMENTS

The work was supported by RFBR grant No. 03-04-48339 and RFBR–Ministry of Industry and Science (Moscow oblast) grant No. 04-04-97280.

CONTENT SENSORS BASED ON CODON STRUCTURE AND DNA METHYLATION FOR GENE FINDING IN VERTEBRATE GENOMES

X. Xia

University of Ottawa, 150 Louis Pasteur, Ottawa, Canada, e-mail: xxia@uottawa.ca

Abstract: All vertebrate genomes are heavily methylated at CpG dinucleotide sites, and methylated CpG dinucleotides are prone to CpG→TpG mutations through spontaneous deamination. This leaves different footprints on coding and non-coding sequences. We capture these different fingerprints by five indices that can be used to discriminate between coding and non-coding (intron) sequences. We also show that a linear discriminant function derived from a training set of coding and intron sequences from human chromosome 22 can be successfully used in gene-finding of the zebrafish genome.

Key words: content sensor; DNA methylation; gene finding; vertebrate genome; codon structure

1. INTRODUCTION

There are two major categories of gene-finding methods. The first is based on known genes in molecular databases and uses homology search by FASTA (Pearson and Lipman, 1988) and BLAST (Altschul et al., 1990; 1997). The second is based on known gene structures and represented by GENSCAN (Burge and Karlin, 1997). Existing software for gene-finding often combine both approaches, e.g., GenMark (Hayes and Borodovsky, 1998), GLIMMER (Salzberg et al., 1998), Orpheus (Frishman et al., 1998), Projector (Meyer and Durbin, 2004) and YACOP (Tech and Merkl, 2003).

For structure-based prediction, the hidden Markov model (HMM) is frequently used in combination with the Viterbi algorithm (Pevzner, 2000; Baldi and Brunak, 2001). Whether HMM is effective depends much on whether the hidden states (e.g., exon, intron, tRNA, rRNA, intergenic sequence, etc.) emit different symbols (i.e., nucleotide combinations). Take for illustration the classical HMM example of the dishonest casino dealer who switches between a fair die and a loaded die (i.e., two hidden states). If the loaded die differ much from the fair die, e.g., if the probability of having six is nearly unity for the loaded die, then a short stretch of sixes is sufficient to identify the point of switching. If the loaded die differs only slightly from the fair die, then the switching event will be difficult to identify unless the dealer throws the loaded die consecutively for a long time. Similarly, if exons and introns differ dramatically in nucleotide combinations, then we would be able to distinguish between short exon and intron sequences. If they differ little, then we may not be able to tell them apart even with long sequences.

In vertebrate genomes, exons and introns are often quite short. Thus, it is important to find the features that differ substantially among exons, introns, RNA genes, etc. For this reason, extensive studies have been carried out to characterize the distinctions between different sequence states leading to a variety of signal sensors, such as the relatively uniform splicing sites (Gelfand, 1989; Tenney et al., 2004; Foissac and Schiex, 2005) and the much less uniform exon–exon junctions (Gelfand, 1992), and content sensors, such as unusual frequency distributions of words (Pevzner et al., 1989; Gelfand et al., 1992; Borodovsky and McIninch, 1993), that can be potentially used in gene-finding.

This paper focuses on the content sensors that can discriminate between exons and introns based on the sequence pattern created by DNA methylation. DNA methylation is a ubiquitous biochemical process, particularly pronounced in vertebrate genomes (Rideout et al., 1990; Sved and Bird, 1990; Bestor and Coxon, 1993). A typical representative of the vertebrate methyltransferase is the mammalian DNMT1 with five domains, of which the NlsD, ZnD, and CatD domains bind specifically to unmethylated CpG, methylated CpG, and hemimethylated CpG sites, respectively (Fatemi et al., 2001). Methylation of C in the CpG dinucleotide elevates greatly the mutation rate of C to T through spontaneous deamination of the resultant m^5C (Brauch et al., 2000; Tomatsu et al., 2004), generating strong footprints in both prokaryotic and vertebrate genomes (Xia, 2003; 2004). Here, we develop indices to capture the differential methylation-mediated substitution patterns and nucleotide triplet structures between exons and introns for gene detection.

1.1 Nucleotide and dinucleotide frequencies by triplet sites

Designate the nucleotide frequencies of a sequence as, p_C, p_G, p_A, and p_T, and the sequence length as L. Consider both coding and non-coding sequences as a linear sequence of consecutive triplets and the nucleotide frequencies at the three sites of the triplets designated as p_{i1}, p_{i2}, and p_{i3}, where i = 1, 2, 3, and 4 corresponding to nucleotides A, C, G, and T, respectively.

For non-exon (e.g., intron) sequences, there is no codon structure. Consequently, we expect $p_{i1} \approx p_{i2} \approx p_{i3} \approx p_i$, where p_i is the average of p_{i1}, p_{i2}, and p_{i3}. For coding sequences, methylation will create heterogeneity in nucleotide frequencies among the three sites. Take NCG codon as an example, where N stands for any of the four nucleotides. DNA methylation and spontaneous deamination tend to change these NCG codons to NTG and NCA codons (note that CpG→TpG mutations in one DNA strand lead to CpG→CpA mutations in the complementary strand), with the former change being nonsynonymous and the latter synonymous. Because nonsynonymous substitution is generally much rarer than the synonymous substitutions in a large number of protein-coding genes in many organisms studied (Xia et al., 1996; Xia, 1998; Xia and Li, 1998), we should expect NCG→NCA mutations more often than NCG→NTG mutations. This tends to increase the frequency of A at the third codon position. Similarly, dicodons, such as 'NNC GNN', tend to mutate synonymously to 'NNT GNN' with DNA methylation, increasing the frequency of T at the third codon position. Thus, in contrast to non-coding sequences where we expect $p_{i1} \approx p_{i2} \approx p_{i3} \approx p_i$, we should expect $p_{i1} \neq p_{i2} \neq p_{i3} \neq p_i$ in coding sequences. This suggests that the deviation of p_{ij} (where j = 1, 2, and 3 corresponding to the three triplet positions) from p_i can contribute to the discrimination between coding and non-coding sequences. A measure of this deviation that is independent of L is as follows:

$$\varphi_{\text{Nuc}} = \frac{\sum_{i=1}^{4} \sqrt{\dfrac{\sum_{j=1}^{3} \dfrac{(f_{ij} - f_i)^2}{f_i}}{N_i}}}{4}, \tag{1}$$

where f_{ij} stands for the number of nucleotide i at codon position j, f_i is the mean number of nucleotide i averaged over the three codon (triplet) positions, and N_i is the sum of nucleotide i in the sequence. We expect φ_{Nuc} to be greater for coding sequences than for non-coding sequences.

Following a similar line of reasoning, we expect the dinucleotide frequencies at triplet positions (1, 2), (2, 3), and (3, 1) to be similar to each other in non-coding sequences but different in coding sequences. Designate the number of dinucleotides as $f_{ij.k}$, where ij = AA, AC, ..., TT, respectively, and k = 1, 2, 3 corresponding to the triplet positions (1, 2), (2, 3), and (3, 1), respectively. The deviation of $f_{ij.k}$ from f_{ij}, which is the number of dinucleotide i averaged over the three triplet positions, should also contribute to the discrimination between coding and non-coding sequences. A measure of this deviation, which is independent of L is:

$$\varphi_{\text{DiNuc}}=\frac{\sum_{i=1}^{4}\sum_{j=1}^{4}\sqrt{\dfrac{\sum_{k=1}^{3}\dfrac{(f_{ij.k}-f_{ij})^2}{f_{ij}}}{N_{ij}}}}{16}. \tag{2}$$

For short sequences, N_{ij} may be zero, in which case φ_{DiNuc} is not defined, or very small, in which case φ_{DiNuc} would fluctuate widely. To avoid this problem, the computation can be done by setting valid $N_{ij} \geq 6$, and the denominator will be the number of valid N_{ij} values.

1.2 Differential methylation intensity

DNA methylation and deamination decrease the CG-containing triplets and increase the UG- and CA-containing triplets. However, their effect is stronger on introns than on coding sequences because of a weaker selection constrains on introns than on coding sequences, e.g., all CGN→TGN, CGN→CAN, and NCG→NTG mutations are nonsynonymous and should be selected against in coding sequences but not in non-coding sequences. For this reason, the intensity of methylation (designated I_m) should be greater in introns than in coding sequences:

$$I_m=\frac{(f_{\text{NUG,UGN,NCA,CAN}}-f'_{\text{NUG,UGN,NCA,CAN}})-(f_{\text{NCG,CGN}}-f'_{\text{NCG,CGN}})}{f_{\text{NUG,UGN,NCA,CAN}}+f_{\text{NCG,CGN}}}, \tag{3}$$

where f is the sum of frequencies of those subscripted codons and f' is the corresponding expectation computed simply by

$$f'_{ij\kappa}=N_{\text{triplet}}P_iP_jP_k, \tag{4}$$

where N_{triplet} is the total number of non-overlapping triplets in the sequence. A more reasonable expectation would be (by taking AAA and AAG for illustration):

$$f'_{\text{AAA}}=N_{\text{Lys}}\cdot\frac{f_{\text{AAA}}}{f_{\text{AAA}}+f_{\text{AAG}}}$$

$$f'_{\text{AAA}}=N_{\text{Lys}}\cdot\frac{f_{\text{AAG}}}{f_{\text{AAA}}+f_{\text{AAG}}}, \tag{5}$$

where N_{Lys} is the number of triplets identical to lysine codons. However, such a formulation is not equally applicable to non-coding sequences.

1.3 Codon avoidance

Among UG- and CA-containing codons that tend to be increased by DNA methylation of CpG dinucleotides, five (AUG, CAA, CAC, CAU, and UUG) are generally avoided in coding sequences in vertebrate genomes, either caused by reduced amino acid usage or other unknown factors. Designating these avoided UG- and CA-containing triplets as f_1 and the other UG- and CA-containing triplets as f_2, we define the triplet avoidance index as

$$I_{ta} = \frac{(f_2 - f_2') - (f_1 - f_1')}{f_1 + f_2}. \qquad (6)$$

1.4 Index of polypurine and polypyrimidine formation

Polypurine and polypyrimidine stretches are ubiquitous among eukaryotic genomes (Birnboim et al., 1979; Mills et al., 2002; Ohno et al., 2002), but their frequencies in coding sequences are constrained by the necessity of codons with mixed purines and pyrimidines. For this and perhaps other reasons, the polypurine and polypyrimidine triplets tend to be more frequent in non-coding sequences than in coding sequences. We define the following index to measure the tendency of polypurine and polypyrimidine triplets:

$$I_{pp} = \frac{(f_{\mathrm{RRR,YYY}} - f'_{\mathrm{RRR,YYY}}) - (f_{\mathrm{Mixed}} - f'_{\mathrm{Mixed}})}{f_{\mathrm{RRR,YYY}} + f_{\mathrm{Mixed}}}. \qquad (7)$$

In this paper, we demonstrate the utility of these indices in discriminating between introns and coding sequences.

2. MATERIALS AND METHODS

The ten annotated contigs from human chromosome 22 (ref_chr22.gbk), perhaps the best annotated human chromosome sequence, was retrieved from the FTP site of GenBank. The CDSs, exons, and introns were extracted according to the sequence annotation in the FEATURES table, and their triplet/codon frequencies were computed, by using DAMBE (Xia, 2001; Xia and Xie, 2001). The indices shown in Eqs. (1)–(7) were also computed by DAMBE for introns and CDSs.

For intron sequences, the indices differ little for the six different triplet frames (i.e., three on each strand), and the numerical results are presented only for the triplet frame starting with the first intron site.

The sequences are grouped into length categories. The indices from sequences with $L \geq 2000$, referred to hereafter as the training set, were used in a linear discriminant analysis performed with the DISCR procedure in SAS (SAS Institute Inc., 1989).

The normal-theory methods (METHOD=NORMAL) was used, equal variance (POOL=YES) of the variables (indices) in the two groups (CDSs and intron) was assumed, and the prior probability was left as the default value of 0.5. Multivariate analysis of variance (MANOVA) was performed to test the significance of the difference in these indices between the two groups. The fitted discriminant function derived from the training set was used to classify shorter coding and intron sequences grouped in length categories 1000–1999, 500–999, and 200–499 to investigate how the discriminating power would change with the sequence length.

To check the general utility of these indices in discriminating between coding and non-coding sequences, we downloaded the Refseq file (zebrafish-rna.gbff), containing 6668 zebrafish (*Danio rerio*) protein-coding genes. The gene sequences contain only the CDS and its 5′-end and 3′-end flanking sequences. The coding sequences were again extracted by using DAMBE (Xia, 2001; Xia and Xie, 2001), and their indices were similarly computed. The discriminant function derived from the training set was then applied to the classification of these sequences.

3. RESULTS AND DISCUSSION

The annotated human chromosome 22 contains 111 CDSs and 1824 introns with $L \geq 2000$. In this set of sequences, the MANOVA test shows that five indices differ significantly between the CDS and intron sequences, with $F = 1770.72$, $DF_{Numerator} = 5$, $DF_{Denominator} = 1929$, and $p < 0.0001$. Univariate significance tests (Table 1) show that all five indices can contribute significantly to the discrimination between CDS and intron sequences.

Table -1. Results of univariate significance tests

Index	STD_T[1]	STD_P[1]	STD_B[1]	F	p
φ_{Nuc}	0.0565	0.0245	0.072	8340.66	<.0001
φ_{DiNuc}	0.0811	0.0392	0.1005	6364.42	<.0001
I_{pp}	0.0578	0.0526	0.0341	407.07	<.0001
I_m	0.0848	0.0838	0.0184	46.41	<.0001
I_{ta}	0.1079	0.0896	0.0849	868.1	<.0001

[1] Subscripts *T*, *P*, and *B* stand for total, pooled, and between, respectively.

The estimated parameters of the linear discriminant function for the training set are shown in Table 2, obtained from the linear discriminant analysis. The discriminant function can be used to classify unknown sequences. In short, the five indices are calculated for each sequence and Y_{CDS} and Y_{Intron} are then computed according to the estimated parameters in Table 2. A sequence with $Y_{CDS} > Y_{Intron}$ is classified as a CDS and otherwise, as an intron. In this limited study, we used only CDS and intron sequences to demonstrate the discriminating power of the indices between these two classes of sequences. A practical study for gene finding involving these content sensors would include many other classes of sequence states. For example, the GENSCAN program (Burge and Karlin, 1997; 1998) uses 17 different sequence states.

Table -2. Linear discriminant function

Index	Y_{CDS}	Y_{Intron}
Constant	−66.030880	−19.142580
φ_{Nuc}	334.618640	−53.342820
φ_{DiNuc}	65.782790	87.759840
I_{pp}	60.572250	61.689980
I_m	42.675790	43.121830
I_{ta}	−17.072500	6.354970

For these 1935 sequences in the training set, 6 coding sequences out of a total of 111 were misclassified as intron and two introns out of a total of 1824 were misclassified as coding sequences, with the error rate of the classification being 0.0276 (Table 3). The discriminant function (Table 2) derived from this training set can be used successfully in discriminating between the CDS and the intron sequences, but the power of discrimination offered by these five indices decreases with decreasing sequence length (Table 3). Note that the 'misclassified' sequences with $L \geq 500$ are nearly always annotated in the GenBank file as hypothetical and may have wrong annotation in the first place.

Table -3. Results of classification with the discriminant function

L (bp)	From	Classified to		Error
		CDS	Intron	
> 2000	CDS	105	6	0.0276
	Intron	2	1822	
1000–1999	CDS	225	18	0.0376
	Intron	1	876	
500–999	CDS	155	23	0.0717
	Intron	10	696	
200–499	CDS	80	29	0.2494
	Intron	156	514	

I would like to illustrate the discriminating power of these indices by a particular 'intron' that has its index values similar to coding sequences and is classified by the linear discriminant function (Table 2) as a coding sequence with a posterior probability of nearly unity. The 'intron' belongs to a gene annotated as 'LOC284861' in the ref.chr22.gbk file, starts with GT and ends with AG, and is 'derived by automated computational analysis' according to the FEATURES table in the GenBank file. However, it is annotated as a part of the coding sequence in other cloned homologous human cDNA sequences (GenBank accession: AL117481, AL122069, and AL133561).

There are three lines of evidence to suggest that this 'intron' is not an intron. First, when the intron and its two flanking exons are treated as a single exon, there is no embedded stop codon. Second, it has at least four indels when aligned with the GenBank sequence XM_375042, and all the indels are inframe triplets. Such indel events are typical of coding sequences. Third and perhaps the most important, aligning the 'intron' with other homologous human cDNA genes shows that its starting GT and ending AG are not conserved, which is not what we would expect if the starting GT and ending AG represent true donor and acceptor sites. All these suggest that the 'misclassification' of the intron as a coding sequence by the discriminant function may in fact represent a correct identification. When hypothetical sequences are excluded, then the error rate of the classification decreases by nearly one order of magnitude.

As a test of the general utility of these five indices in discriminating coding and non-coding sequences, we extracted the coding sequences from 6668 protein-coding genes in the zebrafish-rna.gbff file retrieved from GenBank. These coding sequences range in length between 117 and 18 600 bp. Also extracted are the sequences upstream of the initiation AUG codon and downstream of the termination codon.

The application of the discrimination function (Table 2) to the classification of these three classes of zebrafish sequences (Table 4) shows that only a small fraction of the sequences were misclassified, i.e., when CDS sequences were not classified as CDSs and when upstream and downstream sequences were classified as CDS sequences. This suggests the similarity between the intron and the upstream and downstream sequences. In other words, the discriminant function, which is based only on the difference between CDS and intron sequences, can not only pick up CDS sequences from a mixture of CDS and intron sequences, as shown in Table 4, but can also separate coding sequences from a variety of non-coding sequences.

In short, the five indices we develop in this paper may be used for detecting protein-coding genes across all vertebrates. The test results are better, if the discriminant function is applied to DNA sequences from other mammalian species (e.g., mouse and rat) instead of sequences from zebrafish.

Table -4. Classification of upstream and downstream sequences

Sequence	*L* (bp)	*N*	To CDS	Error, %[1]
CDS	≥ 500	5948	5454	8.305
	< 500	720	667	7.361
Upstream	≥ 500	71	10	14.085
	< 500	124	28	22.581
Downstream	≥ 500	3198	74	2.314
	< 500	602	76	12.625

[1] Percentage of misclassification.

We should finally mention four shortcomings of this study. First, the indices in Eqs. (1)–(7) are still rather crude and can be improved essentially. For example, φ_{Nuc} and φ_{DiNuc} may have better statistical properties, if they are based on trinomial distributions. Second, there are other significant content sensors that can be derived from methylation patterns. For example, the ratio of UG-containing triplets to CA-containing is nearly constant across the three triplet sites, i.e., (1, 2), (2, 3), and (3, 1), in non-coding sequences but differs dramatically in coding sequences. Third, the footprint of DNA methylation on non-intron and non-CDS sequences has not been thoroughly explored. The forth shortcoming is inherent in content sensors in that the indices are for detecting genes, but not for predicting the exact exon–intron boundaries. The latter would require information on signal sensors, such as splicing sites (Gelfand et al., 1996; Tenney et al., 2004; Foissac and Schiex, 2005).

In summary, given the fact that even relatively crude formulation of these indices can allow us to discriminate between coding and non-coding sequences, we believe that the differential footprints on coding and non-coding sequences left by the methylation-mediated substitutions can serve as powerful content sensors in gene detection in vertebrate genomes.

ACKNOWLEDGMENTS

This study is supported by grants from the University of Ottawa and NSERC discovery and strategic grants. I thank many of my Russian colleagues who have provided many helpful comments and suggestions. Two anonymous reviewers corrected errors in an early draft presented at the BGRS'04 conference. Two other reviewers corrected more errors.

THE SITEGA TOOL FOR RECOGNITION AND CONTEXT ANALYSIS OF TRANSCRIPTION FACTOR BINDING SITES: SIGNIFICANT DINUCLEOTIDE FEATURES BESIDES THE CANONICAL CONSENSUS EXEMPLIFIED BY SF-1 BINDING SITE

V. Levitsky[1, 2*], E. Ignatieva[1, 2], G. Vasiliev[1], N. Limova[1], T. Busygina[1], T. Merkulova[1], N. Kolchanov[1, 2]

[1] *Institute of Cytology and Genetics, Siberian Branch of the Russian Academy of Sciences, prosp. Lavrentieva 10, Novosibirsk, 630090, Russia, e-mail: levitsky@bionet.nsc.ru;*
[2] *Novosibirsk State University, ul. Pirogova 2, Novosibirsk, 630090, Russia*
[*] *Corresponding author*

Abstract: Development of computational methods to search for transcription factor binding sites (TFBSs) is important in investigation of regulatory regions of eukaryotic genes and in genome annotation. We propose a SiteGA method for recognition of TFBSs, providing the search of SF-1 binding sites as an example. The SiteGA method was implemented using a genetic algorithm (GA) involving an iterative discriminant analysis of local dinucleotide context characteristics. These characteristics were compiled not only over the core binding site (BS) region, but over its flanks as well. The major advancement of this approach is an improvement in accuracy by a large window capturing the meaningful context features besides the canonical consensus. The experimental verification confirmed the majority of predicted sites. The program SiteGA is available at http://wwwmgs2.bionet.nsc.ru/mgs/programs/sitega/.

Key words: steroidogenic genes; SF-1 BS recognition; local dinucleotide context; discriminant analysis; genetic algorithm

1. INTRODUCTION

Recognition of TFBSs by computer methods is an effective approach to the search and analysis of the gene regulatory regions. A widely used weight matrix (WM) models for TF–DNA binding imply that there is some contribution from each base at each position and that the sum of all the contributions is above a certain threshold (Stormo, 2000). These well-known approaches were applied in the Tfsitescan (Ghosh, 2000) and MatInspector systems (Quandt et al., 1995). Despite the fact that the additivity assumption does not fit the data perfectly, in most cases it provides a very good approximation of the true nature of the specific protein–DNA interactions (Benos et al., 2002). Nevertheless, the WM representation is severely limited by the assumption that positions in a site contribute independently to the total score. As a result, the major drawback of using WM for genome-wide TFBS prediction is its high false-positive rate (Long et al., 2004). The high-scoring false positives can be filtered out by an additional criterion: the condition that most true binding sites co-occur with a second binding site, for either the same transcription factor or a different one (Bulyk et al., 2004; Long et al., 2004). The context information found in flanking regions of TFBSs is not often used for development of recognition methods. The approach trying to reveal additional features in the flanking regions of sites was developed for the recognition of GREs (Glucocorticoid Regulatory Elements) (Seledtsov et al., 1991). It may be concluded that TF–DNA interactions depend on multiple factors, including the presence of other *cis* regulatory elements.

The SF-1 is a member of the nuclear receptor superfamily. In contrast to most of its members interacting with DNA as a homodimer or a heterodimer, SF-1 activates gene expression by binding to DNA in a monomeric form (Val et al., 2003). SF-1 is a transcription factor known as a key regulator of the steroidogenic gene expression in gonads and adrenals (Val et al., 2003). Moreover, SF-1 is required for development and differentiation at all the levels of the hypothalamic–pituitary–gonadal and adrenal axis (Val et al., 2003). There is experimental evidence for the presence of the SF-1 binding sites in the regulatory regions of many genes functioning within this axis (Busygina et al., 2003). Nevertheless, an overall pattern of the SF-1 mediated regulation is far from complete, and no reliable computer methods for the SF-1 BS recognition for genome-wide analysis are yet available.

The main aim of our research is to develop and validate a TFBS recognition approach accurate enough for a genome-wide application. Our approach, SiteGA, takes advantage of observing relations between the nearest and distant dinucleotide positions located within central and both flanking regions of BS. These dependencies were identified by a genetic algorithm based on iterative discriminant analysis. The SF-1 BS recognition is provided here as an example of how the SiteGA method was implemented. Finally, the capability of the predicted SF-1 BSs of interacting with this protein was experimentally validated.

2. METHODS AND ALGORITHMS

2.1 Sequences used for analysis

To develop the recognition method, we used the SITES sample containing 54 nucleotide sequences with a centrally localized SF-1 BS and flanks. The total length of the SF-1 BS and its flanking regions was 93 bp. The SITES sample was derived from the TRRD (Kolchanov et al., 2002) and contained almost all the known sites for which the interaction with the SF-1 factor was experimentally ascertained by (i) electrophoretic mobility shift assay (EMSA) with purified SF-1 protein, (ii) DNase I footprinting with purified SF-1, and (iii) EMSA with nuclear extracts and antibodies against SF-1.

We used for analysis two samples of sequences. The EMBL(STER) sample of 54 sequences served as the control set for development of the recognition methods. This sample was composed of sequences of genes whose assignment to the steroidogenic system is generally recognized. Hence, they likely contain SF-1 BSs, but experimental evidences for the presence of these sites are still absent. The EMBL(STER) sample was built up directly from the EMBL nucleotide sequences database using the literature sources. In addition, for searching putative SF-1 BSs, we used the TRRD(STER–) sample of 1274 sequences of genes for which the presence of the SF-1 BSs was not experimentally demonstrated. This sample was derived from the TRRD (Kolchanov et al., 2002).

2.2 Experimental verification of the predicted SF-1 BSs

SF-1 binding to 32 bp labeled double-stranded oligonucleotides corresponding to the predicted SF-1 BS was studied by EMSA. The nuclear extracts of testes dissected from 14-day-old male Wistar rats were prepared (Gorski et al., 1986) with the described modifications (Shapiro et al., 1988). The binding reactions were carried out in a 10-μl reaction volume containing 3 μg of nuclear extract preincubated with 0.4 μg calf thymus DNA at 4 °C for 10 min, 1 ng (^{32}P)-labeled DNA probe in a binding buffer (25 mM HEPES pH 7.6), 100 mM KCl, 0.1 mM EDTA, 1 mM dithiothreitol, and 10 % glycerol). For antibody-containing reactions, the extracts were preincubated with 1 μl of antibodies to SF-1 (The Upstate Biotechnology) at 4 °C for 10 min. After further incubation for 10 min at a room temperature, these samples were analyzed by electrophoresis in a 5 % nondenaturing polyacrylamide gel in 50 mM Tris–borate and 1 mM EDTA at 180 V at 4 °C. After electrophoresis, gels were dried and the protein–DNA complexes were visualized by autoradiography.

2.3 Algorithm

The SiteGA method was implemented using a genetic algorithm (GA) involving a discriminant function of the local dinucleotide context characteristics. The method divides the entire analyzed region (93 bp) into three overlapping segments (the left, central, and right) each 36 bp long. First, a partition of each segment into local fragments is searched for; then, the most significant frequencies of dinucleotides within the fragments obtained are selected. The GA utilizing iterative discriminant analysis of the distribution of dinucleotide frequencies over the fragments of a current partition is used at both steps (Levitsky and Katokhin, 2003). Let us consider one of the segments and the corresponding (1) real and (2) random nucleotide sequences sets (obtained by shuffling of the real sequences). The partition $\Omega(b_1, b_2, \ldots, b_{p-1})$ of a segment (a, b) is defined as a set of P fragments (a_p, b_p), where $(p = 1, \ldots, P)$ meets the following conditions: (1) $a_1 = a$, (2) $a_p = b_{p-1}$ for $p = 2, \ldots, P$, and (3) $b_P = b$. The fitness of partition Ω was assessed using the Mahalanobis distance $R^2(\Omega)$ between the two sets of sequences in the space of $N = 16 \times P$ values of dinucleotide frequencies for P local fragments (the value P was optimized).

$$R^2(\Omega) = \sum_{k=1}^{N}\sum_{n=1}^{N}\left\{\left[f_n^{(2)} - f_n^{(1)}\right] * S_{n,k}^{-1} * [f_k^{(2)} - f_k^{(1)}]\right\}. \tag{1}$$

Here, $f_n^{(1)} = f_{i,p}^{(1)}$ is the mean frequency of the ith dinucleotide in the pth partition fragment for the set of the real sequences; $f_n^{(2)}$, the respective frequency for the random sequences set ($n = (p-1) \times 16 + i$, $p = 1, \ldots P$, $i = 1, \ldots 16$, $n = 1, \ldots, N$); and $S_{n,k}^{-1}$ is an element of the matrix $|S^{-1}|$ inverse to the matrix $|S| = |S^{(1)}| + |S^{(2)}|$. These two matrices are the covariance matrices of the vectors of dinucleotide frequencies over sets 1 and 2, respectively.

The search for optimal partition starts with the random assignment of a certain partition. The GA is constructed for the partition using operations of two types: mutation (changes in the positions of borders between regions of the same partition) and recombination (exchange of fragments between two partitions). The selection of the most significant frequencies of dinucleotides within the fragments obtained define the full set of N context variables $\{ f_{i,p} \}$ for the chosen partition. This set is divided into two subsets: one, $\{ f_{i,p}^{m+} \}$, of M_{k+} variables used for the recognition and the other, $\{ f_{i,p}^{m-} \}$, of M_{k-}, the discarded variables ($M_{k+} + M_{k-} = N$). This division specifies one individual S_k. The final subset of the most significant context characteristics is searched with the GA. The initial GA population consists of individuals of arbitrarily chosen variables used and discarded. Mutation is defined as one exchange for the subset S_k between its used and discarded characteristics; recombination means an exchange between the used characteristics of the two subsets S_{k1} and S_{k2}. The fitness $R(S_k)$ is defined in the same way as in Eq. (1) by summing over M_{k+} used variables. Finally, the recognition function value for each segment is calculated for an arbitrary nucleotide sequence at each position of the window of 36 bp (the fragment X_r):

$$\varphi_r(X_r) = \frac{1}{R^2} \times \sum_{n=1}^{M} \sum_{k=1}^{M} \{[f_n(X_r) - (\tfrac{1}{2}) \times [f_n^{(2)} + f_n^{(1)}] \times S_{n,k}^{-1} \times [f_k^{(2)} - f_k^{(1)}]\}. \quad (2)$$

The distance R^2 is calculated using Eq. (1) by summing over M selected frequencies. The higher reliability of recognition corresponds to the values of the function $\varphi_r(X_r)$ closer to +1. A specified Z-score Z_r (threshold) transforms the function $\varphi_r(X_r)$ as follows:

$$\varphi_r(X_r, Z_r) = \begin{cases} 1 - |1 - \varphi_r(X_r)|, & \text{if } |1 - \varphi_r(X_r)| < Z_r \times \sigma_\varphi \\ 0, \text{otherwise.} \end{cases}, \quad (3)$$

Here, σ_φ is the standard deviation of the recognition function φ_r values for the corresponding train sample. To recognize a BS in a nucleotide sequence, a sliding window (X) and the division $\{X_1, X_2, X_3\}$ were used. Finally, the recognition function value φ (X, Z_1, Z_2, Z_3) was calculated from the equation:

$$\varphi(X, Z_1, Z_2, Z_3) = \begin{cases} 0, \text{ if } \varphi_r(X_r, Z_r) = 0, \text{ for one of } r,\ 1 \le r \le 3 \\ (\tfrac{1}{3}) * \sum_{r=1}^{3} \varphi_r(X_r, Z_r), \text{ otherwise.} \end{cases} \quad (4)$$

3. RESULTS AND DISCUSSION

3.1 Accuracy comparison of the SiteGA and weight matrix methods

To compare the weight matrix (WM) method with the ours, we calculated the accuracy estimations using the standard jack-knife technique. For the SITES set, we sampled many times a new training subset, which contained all but one sequence from the full set. A new method trained on the basis of this subset was applied to the sequence that was not used for the training process. The results of all produced training were averaged, and the false positive rate was estimated. The log-odds method (Stormo, 2000) was used to obtain WM elements:

$$w_{i,j} = \ln\left(\frac{F_{i,j}+b_j}{N+b}\right) - \ln p_j \,,\; b = \sum_j b_j \,. \quad (5)$$

Here, N is the number of sites; p_i is an *a priori* probability for the nucleotide I; $F_{i,j}$, the number of sites for which nucleotide j appears in position I; and all pseudocounts b_j are equal to 1 (Berg and von Hippel, 1987).

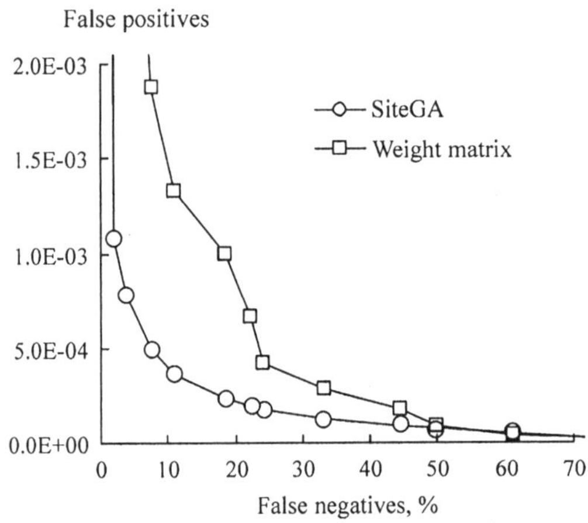

Figure -1. Comparison of the recognition performance of the SiteGA and weight matrix methods.

As illustrated in Figure 1, SiteGA approach has proved successful. When compared with WM approach, the SiteGA produced a 3–5-fold reduction in the false positive rate at any false negatives percentage below 50 %. In contrast to the WM method, the SiteGA application resulted in notable accuracy growth, when the sequence length increased from 36 to 93 nt.

Finally, the optimal Z-score sets (thresholds) for the recognition functions of the three segments were searched for. These sets were found using the control samples of random sequences. Three optimal sets of Z-scores for the recognition functions were chosen (Table 1). For example, for the set II, the percentage of false negatives calculated for the SITES sample was 37 %, while false positives rate for the control sample of random sequences were 7.9E-06 (1/127296).

Table -1. The optimal Z-score sets for the SiteGA SF-1 BS recognition function

Z score set name	Z-score sets (central, right, and left)	False negatives, %	False positives
I	2, 2.2, 2.4	5.6	5.4E-05
II	1.2, 1.4, 1.6	37.0	7.9E-06
III	1, 1, 1	64.8	2.1E-06

3.2 Context analysis of the SF-1 BSs and their flanks

The analysis of significant correlations ($p < 0.05$) between dinucleotide frequencies located within three segments (see section 2.3, Algorithm) demonstrated that most of the correlations were positive (114 from 177; 64 %); their great majority resided in the central segment (147; 83 %); and the right flank was found to be more informative than the left (21 correlations against 9).

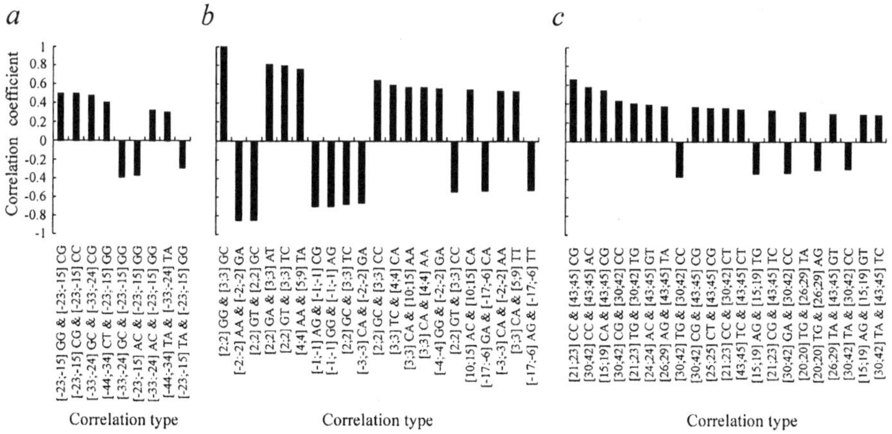

Figure -2. The diagrams of significant correlations used in the recognition method construction for the left flank (*a*) $p < 0.05$, central segment (*b*) $p <$ 1E-04, and right flank (*c*) $p < 0.05$. The dinucleotide position +1 corresponds to the first 'G' of the most conserved GG dinucleotide in consensus.

Figure 2 presents diagrams of significant correlations used in the recognition method construction for the left (*a*, $p < 0.05$), central (*b*, $p <$ 1E-4), and right segments (*c*, $p < 0.05$). For example (Figure 2*b*), the GT frequency in the [2; 2] region positively correlated with the TC frequency in the [3; 3] region and negatively with the CC frequency in the [3; 3] region. The pattern beyond the canonical SF-1 footprint (not longer than 20 bp) reflects the genome nucleotide context around the site whose consideration helps to increase the recognition accuracy. The significant context feature besides the footprint may be related to the presence of other still unknown specific to SF-1 BS *cis* regulatory elements. Our approach used this hidden context information; thereby a higher recognition performance is achieved.

Figure 3 presents the diagram of significant correlation ($p < 0.01$ and $p < 0.05$) density per one dinucleotide position used in the recognition method construction for the central segment (36 bp long). For example, the correlation of 0.52 between the TT frequency in the region [5; 9] and the CA frequency in the region [3; 3] contributes equally to all dinucleotide positions of both regions (i.e., the total contribution at positions 3 and 5–9 is unity). One region [–5; 4] of the maximal correlation density (Figure 3) is clearly observed. It coincided with the location of consensus sequence GTCAAGGTCA (Busygina et al., 2003).

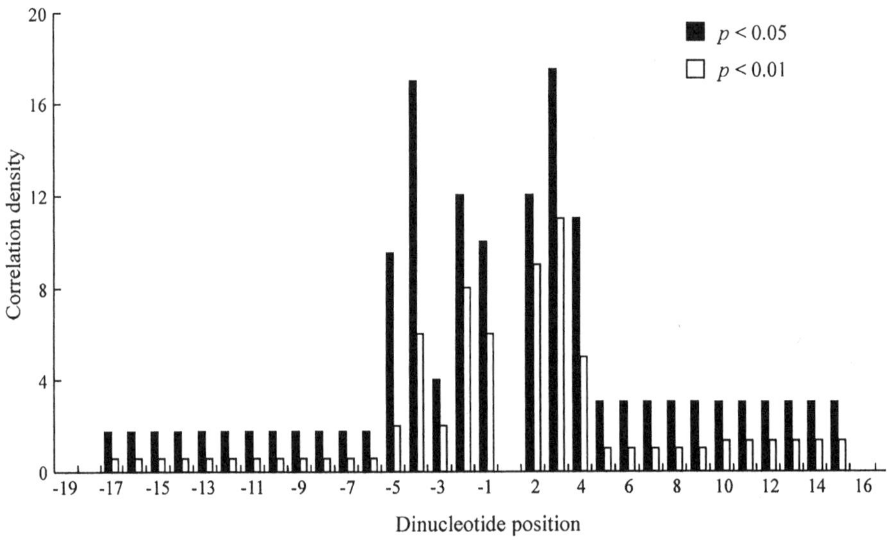

Figure -3. The number of significant correlations ($p < 0.01$ and $p < 0.05$) per one dinucleotide position (correlation density) for central segment of SF-1 BS.

3.3 Experimental verification of the predicted SF-1 BSs

We have experimentally tested 29 potential SF-1 BSs (Table 2). We tested BSs identified in the regulatory regions of the genes of the EMBL(STER) sample and a number of BSs in genes we identified as the probable SF-1 targets in the (TRRD(STER–) sample.

Table -2. The SF-1 binding sites predicted by the SiteGA method and experimental support

	Gene	Gene region	Position[1]	Direct or reverse strand	SiteGA prediction Z-score set[2]			Experi-mental support[3]
					I	II	III	
1	*HSD3b* (Mouse)	Promoter region	–112	←	+	+	–	+
2	*Ad4BP/SF-1* (Mouse)		–225	→	+	+	–	+
3	*Ad4BP/SF-1* (Mouse)		–208	←	+	+	–	+
4	*StAR* (Sheep)		–104	←	+	–	–	+
5	*StAR* (Macaque)		–228	←	+	+	+	+
6	*StAR* (Mouse)		–1356	←	+	+	–	+
7	*CYP17* (Porcine)		–52	→	+	+	–	+
8	*CYP17* (Porcine)		–139	←	+	–	–	–
9	*HSD17BI* (Rat)		–83	←	+	+	+	+
10	*LH beta* (Porcine)		–113	←	+	+	–	+
11	*CRBP2* (Mouse)		–566	←	+	+	–	–
12	*CRBP2* (Mouse)		–229	→	+	–	–	+
13	*D2* (Rat)		–686	→	+	+	–	+
14	*D2* (Rat)		–160	→	+	+	+	+
15	*iNOS* (Rat)		–220	←	+	+	–	–
16	*IRBP* (Human)		–253	→	+	–	–	+
17	*IRBP* (Human)		–1204	←	+	+	+	+
18	*SPRR1A* (Human)		–677	→	+	–	–	–
19	*SPRR1A* (Human)		–50	←	+	–	–	+
20	*HSD17BII* (Human)	Non-coding exon	+440	←	+	+	–	+
21	*iNOS* (Rat)		+63	←	+	+	+	+
22	*IRBP* (Human)		+5	←	+	+	–	+
23	*Slp* (Mouse)	Intron	+798	←	+	+	+	+
24	*HSD3b* (Human)		+5847	→	+	+	–	–
25	*Slp* (Mouse)		+5299	→	+	+	–	–
26	*Slp* (Mouse)		+12460	←	+	–	–	+
27	*Slp* (Mouse)		+12692	←	+	+	+	+
28	*StAR* (Bovine)		+2370	←	+	+	+	+
29	*StAR* (Bovine)		+5442	←	+	–	–	+
The percentage of predicted sites confirmed experimentally, %					79.3	81.9	100	

[1] The position relative to the transcription start; [2] see Table 1 for Z-scores setting; [3] '+' indicates that the capability of interacting with SF-1 of predicted site was confirmed in gel shift experiments using nuclear extract and specific antibodies to SF-1; otherwise, was not confirmed.

The ability of the predicted SF-1 BSs to interact with the protein was confirmed for 79.3 % (23 of 29), 81.9 % (17 of 21), and 100 % (8 of 8) of the predicted sites within Z-score sets I, II, and III, respectively (Table 2). A number of results provided by EMSA are shown in Figure 4. The most stringent threshold (Z-score set III) assured the perfect experimental verification of the predicted sites. The experimentally supported sites were located in the regulatory regions of 11 genes relevant to the steroidogenic system and of 5 other genes (rat *iNOS* and *D2*, mouse *CRBP2*, and human *IRBP* and *SPRR1A*; Table 2). The experimental verification demonstrated a high prediction accuracy of the SiteGA method.

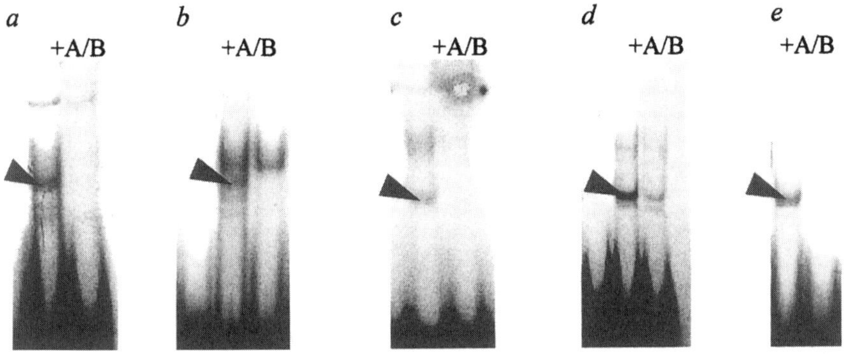

Figure -4. Experimental support for a number of SF-1 BSs predicted by the SiteGA method. Gene names and site positions relative to the transcription start are the following: (*a*) sheep *StAR*, –105; (*b*) porcine *Cyp17*, –51; (*c*) mouse *Ad4BP/SF-1*, –225; (*d*) mouse *Slp*, +798; and (*e*) mouse *HSD3b*, –113. The shifted SF-1/DNA complex is indicated by arrow. Disappeared or fainter bands due to antibodies against SF-1 (A/B) in the right lane confirmed SF-1 binding to the site.

3.4 Conclusion

The discriminant analysis we applied, allowed us to reveal the context characteristics not only of the core region of the SF-1 BS, but also of their flanking regions (Figures 2, 3). The number of significant correlations between dinucleotide frequencies are found both within the region corresponding to the consensus sequence and outside it. By taking into account the local positioning of the dinucleotide context, we achieved a considerable increase in the recognition accuracy of SiteGA when compared to that for the WM method (Figure 1).

It will be recalled that the WM approach suffers from the underlying assumption that the positions in the site may contribute additively to the total score (Benos et al., 2002). The consideration of relations between the nearest and distant dinucleotide positions located within three separate segments (central and both flanking regions of the site) allowed us to reveal subtle context features and to improve the recognition accuracy. The experimental tests of the SF-1 BSs predicted by the SiteGA method supported its high efficiency. The number of correctly predicted SF-1 BSs ranged from 79.3 % to 100 % (Table 2) depending on threshold setting (Table 1). We achieved a relatively low false positive rate in comparison with widely used weight matrix method. This high accuracy (Figure 1) and experimental support (Table 2) demonstrate the advantages of the SiteGA approach application for genome-wide TFBS search.

ACKNOWLEDGMENTS

The work was supported by the Russian Foundation for Basic Research (grants Nos. 03-04-48469, 05-04-49111, and 03-07-90181); the Ministry of Industry, Science, and Technologies of the Russian Federation (grant No. 43.073.1.1.1501); Siberian Branch of the Russian Academy of Sciences (integration project No. 119); MCB RAS (grant No. 10.4); NATO (grant No. LST.CLG.979816); and US Civilian Research and Development Foundation for the Independent States of the Former Soviet Union (CRDF) and the Ministry of Education of Russian Federation within the Basic Research and Higher Education Program (award No. REC-008 and grant Y2-B-08-02). The authors are grateful to A.S. Poplavsky for technical support and A.V. Osadchuk and A.V. Katokhin for helpful discussions.

TRANSCRIPTION REGULATORY REGIONS DATABASE (TRRD): A SOURCE OF EXPERIMENTALLY CONFIRMED DATA ON TRANSCRIPTION REGULATORY REGIONS OF EUKARYOTIC GENES

N. Kolchanov[1, 2*], E. Ignatieva[1, 2], O. Podkolodnaya[1], E. Ananko[1], I. Stepanenko[1, 2], T. Merkulova[1, 2], T. Khlebodarova[1], V. Merkulov[1], N. Podkolodny[1, 2, 3], D. Grigorovich[1], A. Poplavsky[1], A. Romashchenko[1]

[1] *Institute of Cytology and Genetics, Siberian Branch of the Russian Academy of Sciences, prosp. Lavrentieva 10, Novosibirsk, 630090, Russia, e-mail: kol@bionet.nsc.ru;* [2] *Novosibirsk State University, ul. Pirogova 2, Novosibirsk, 630090, Russia;* [3] *Institute of Computational Mathematics and Mathematical Geophysics, Siberian Branch of the Russian Academy of Sciences, Lavrentieva 6, Novosibirsk, 630090, Russia*

[*] *Corresponding author*

Abstract: The goal of creation of the Transcription Regulatory Regions Database (TRRD) was to provide a complete and adequate description of the structure–function organization of transcription regulatory regions in eukaryotic genes. TRRD contains only experimentally confirmed data about (i) transcription factor binding sites; (ii) regulatory units (promoter regions, enhancers, and silencers); and (iii) locus control regions. The main tool for searching TRRD and navigation in it is SRS. TRRD has hierarchically organized vocabularies and thesauruses used for developing specialized data retrieval tools. The current TRRD release contains information about 2308 eukaryotic genes (of them, 34 % are human genes) inputted basing on annotation of 7565 scientific papers. For these genes, the largest in the world sets of experimentally confirmed regulatory units (3439) and transcription factor binding sites (10 045) are collected in TRRD. Of them, 37 % of regulatory units and 38 % of binding sites are related to human genes. This paper characterizes groups of experiments basing on which regulatory units and binding sites are annotated. Examples of TRRD entries are given. The database is available at http://www.bionet.nsc.ru/trrd/.

Key words: database; transcription regulation; TRRD: Transcription Regulatory Regions Database

1. INTRODUCTION

Transcription of RNA polymerase II (Pol II) transcribed eukaryotic genes involves regulatory units (promoter regions, enhancers, and silencers) that are localized predominantly in the noncoding regions of genes. They function via specific interactions with transcription factors. In certain cases, transcription of genes or whole gene loci is regulated by locus control regions (LCRs; Kielman et al., 1994; Jones et al., 1995). In experiments with transgenic animals, LCRs enable a coordinated copy number–dependent chromosomal position–independent gene expression (Festenstein and Kioussis, 2000).

At present, a number of specialized databases exist that accumulate data on individual aspects of the structure–function organization of regulatory regions. As a rule, these databases contain data obtained by computer analysis of genomic sequences and the information extracted from other databases along with the experimentally confirmed data extracted by analyzing scientific papers. For example, the Eukaryotic Promoter Database (EPD) includes sequences of promoters of eukaryotic genes and data on positions of transcription start sites. Over a half of the transcription start sites are identified by computer analysis involving 'in silico primer extension' technique (Schmid et al., 2004).

The DataBase of Transcriptional Start Sites (DBTSS) accumulates the information on the positions of transcription start sites of genes obtained by analyzing full-length cDNA sequences constructed by oligo-capping method (Suzuki, et al., 2004).

The Mammalian Promoter Database (MPromDb; Sun et al., 2003) contains promoter sequences from only three species (human, mouse, and rat) with indication of positions of transcription factor binding sites; the information about these positions is taken from both scientific papers and the field FEATURES in the corresponding entries of GenBank. PlantCARE is a database of plant cis-acting regulatory elements. The number of entries in this database is not large; it describes approximately 160 individual promoters from higher plant genes (Lescot et al., 2002). The object-oriented Transcription Factors Database (ooTFD) is aimed at capturing information regarding the polypeptide interactions that comprise and define the properties of transcription factors and contains the data about transcription factor binding sites (Ghosh, 2000). TRANSFAC compiles the data on transcription factors of both prokaryotes and eukaryotes and on nucleotide sequences of their binding sites (Matys et al., 2003). COMPEL accumulates the information about composite regulatory elements—pairs of closely located binding sites for various transcription factors (Kel-Margoulis et al., 2002).

Box 1. Glossary

Regulatory unit: extended DNA sequence (promoter region, enhancer, silencer, etc.), that alter the levels of gene transcription and can determine spatial patterns of gene expression.

Promoter region: a specific region usually upstream of the transcription start site that binds and directs RNA polymerase to the correct transcriptional start site and contains binding sites for transcription factors that regulate the rate of transcription of the adjacent gene.

Enhancer: a cis-acting sequence that increases the utilization of (some) eukaryotic promoters and can function in either orientation and in any location (upstream or downstream) relative to the promoter.

Silencer: a cis-regulatory sequence that can reduce the levels of transcription from an adjacent promoter. Silencers can be 5' or 3' to the promoter they regulate.

Transcription factor binding site: a short DNA sequence that is recognized by the transcription factor and binds them in a specific manner.

Transcription Regulatory Regions Database (TRRD), which we are presenting in this paper, is a unique information resource developed to provide an integrated description of transcription regulation of Pol II transcribed eukaryotic genes. TRRD contains only experimentally confirmed data about the structure–function organization of the following hierarchical levels of transcription regulation: (i) transcription factor binding sites; (ii) regulatory units (promoter regions, enhancers, and silencers); and (iii) LCRs. TRRD compiles also patterns of gene expression with references to the regulatory units and binding sites that provide realization of these patterns. Data are inputted into TRRD by experts–biologists basing on analysis and annotation of scientific literature.

Two versions of TRRD were designed. The relational TRRD version was developed in ORACLE8i environment. Its scheme is available at http://www.bionet.nsc.ru/trrd/RelScheme/. Unlike the relational version, the SRS version is available via the Internet (http://www.bionet.nsc.ru/trrd/). The Internet-accessible TRRD comprises seven databases integrated with one another and with external databases by Sequence Retrieval System (SRS) v. 6. In addition to conventional options for data retrieval provided by SRS, a specialized search tool based on hierarchically organized vocabularies and thesauruses was designed. TRRD is linked to over 20 world information resources, including the databases of nucleotide sequences EMBL and GenBank as well as to Ensembl, UniGene, GeneCards, OMIM, EPD, Swiss-Prot, TRANSFAC, and COMPEL.

2. TRRD STRUCTURE AND FORMAT

The unit entry of TRRD is gene. The information in the database is distributed in interlinked tables. The main table is TRRDGENES. In addition to the data identifying a gene, this table contains links to external information resources as well as cross-links to the records in the other TRRD tables related to this gene. These tables comprise information about (1) transcription factor binding sites (TRRDSITES); (2) regulatory units—promoter regions, enhancers, and silencers (TRRDUNITS); (3) LCRs (TRRDLCR); (4) transcription factors interacting with sites (TRRDFACTORS); (5) qualitative specific features of gene expression depending on developmental stage of organisms, stage of the cell cycle, types of cells and degree of their differentiation, etc. (TRRDEXP); and (6) references to original publications (TRRDBIB). The types of experiments where particular data were obtained are indicated in the corresponding fields of the tables TRRDSITES and TRRDUNITS as a digital code. Information fields of TRRD were described earlier (Kolchanov et al., 1999; 2000; 2002). The format of TRRDSITES in the current TRRD version (as of February 11, 2004) is expanded. A new field **PreferredName** is added; this field contains the standard site name. The field PreferredName is filled automatically basing on the data in the field FactorName (TRRDFACTORS) using a specialized vocabulary of transcription factors; its organization is detailed in Section 5.

3. PRINCIPLES OF DESCRIPTION OF TRANSCRIPTION REGULATION IN TRRD

A distinctive feature of TRRD is that it contains only the information confirmed by special experiments. Examples of experiments are listed in Tables 1 and 2. Experimental methods that were used to obtain the information are fixed in the database as digital codes. For example, attachment of a DNA fragment under study to reporter gene (assay code 6.8), deletion analysis of the attached DNA (assay code 6.1.1), and comparative analysis of expression of plasmid constructs under various conditions (assay codes 6.3.1 and 6.5) are used for delineation and initial analysis of extended regulatory regions (promoter regions, enhancers, silencers, etc.; Table 1). Identification of transcription start sites involves experiments on extension of radioactively labeled primer to 5′-end of mRNA template (assay code 5) and analysis of the DNA fragments protected from nucleases (RNase T1, RNase A, and S1 nuclease) in a complex with RNA (assay code 5.5).

Table -1. Examples of assays providing the information about regulatory units inputted into TRRD*

Type of experiment	Assay code in TRRD
Determination of transcription start sites	
Extension of radioactively labeled primer to 5′-end of mRNA template	5
Analysis of the DNA fragments protected from nucleases (RNase T1, RNase A, and S1 nuclease) in a complex with RNA	5.5
Delineation and initial analysis of large regulatory regions	
Insertion of the promoter region upstream of reporter gene	6.8
Attachment of DNA fragment of interest to homologous or heterologous promoter and reporter gene	6.3.1
Deletion analysis	6.1.1
Assessing appropriate regulation by different agents in transient transfection assay	6.5

* The complete list of experimental assays providing the data on regulatory units and transcription factor binding sites inputted in TRRD is available at http://wwwmgs.bionet.nsc.ru/mgs/gnw/trrd/digcodes.shtml.

Figure 1*a* shows the regulatory units of human PECAM1 (platelet/endothelial cell adhesion molecule-1) gene, whose description is included into TRRD basing on experimental data. PECAM1 gene has two tissue-specific promoter regions: one is endothelial-specific and the other, myeloid-specific. They are localized at a distance of 300 bp from one another. Correspondingly, two groups of multiple transcription start sites were detected in endothelial and myeloid cells.

The transcription in myeloid cells is regulated by transcription factors Egr1 and Sp1, which interact with the corresponding binding sites within the first promoter region. The transcription in endothelial cells is controlled by NF-κB and GATA2, whose binding sites are discovered in the second promoter region. A fragment of entry P01766 from TRRDUNITS table, describing the endothelium-specific promoter region of human PECAM1 gene, is shown in Figure 1*b*.

As is evident from the record in the field ExperimentCodes, the transcription start sites were identified by primer extension method (assay code 5) in HUVEC and DAMI cell lines. The functional characteristics of this promoter region were studied in DAMI, U937, and K562 cell lines by transformation with plasmid constructs where a fragment of this gene containing the promoter region (assay code 6.8) and its variants with successive deletions (assay code 6.1.1) were attached to reporter gene. To determine inducibility of the promoter region, the effects of PMA and TNF-alpha on expression of plasmid constructs (assay code 6.5) were studied.

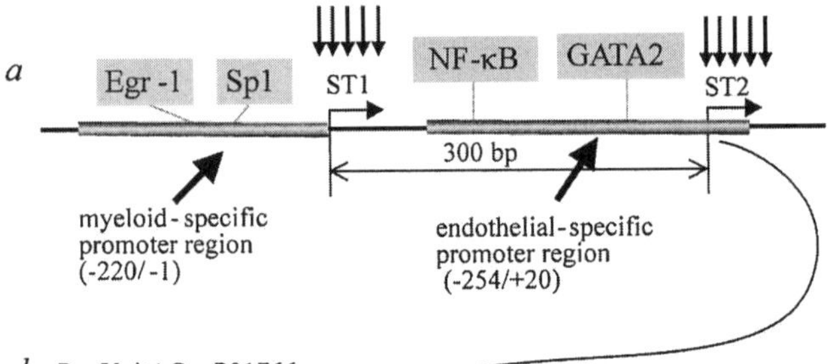

b RegUnitAC P01766
GeneID Hs:PECAM1
RegRegion 5'region
RegUnit promoter region; ST2; 254 to +20; S5996, S5997
DNA BankLink EMBL; HSDNAPP; X96848; 989 to 1263
SeqLength 275
Sequence

```
gatctttctg  gatgtcctgg  gagagcagcc  ctttgggtgg  tgggtggagg
ctggaggcag  ggaggaatcc  cctcacagtg  ccatgagaag  ggcccccaaa
cccaggcgag  acagagggag  ggtcaagaac  gccaaggcaa  atgtcacttg
tgccttgttt  tttccctaaa  gaaactaaac  aaagcggccg  cgttcggtgg
cccctcagga  aggccggtca  tttcctgagg  agatatcagg  ccagcccagg
ccccattgtt  cccggtttcc  agcca
```

PromotTisSp TS
PromotInd I
MultipleStarts 4, 7, 9, 10, 22, 26, 27 [Gumina R.J. et al., 1997]
ExperimentCodes HUVEC: 5 [Gumina R.J. et al., 1997]
DAMI: 5, 6.1.1, 6.8 [Gumina R.J. et al., 1997]
U937: 6.5 (PMA, TNF-alpha), 6.8 [Botella L.M. et al., 2000]
K562: 6.5 (TNF-alpha), 6.8 [Botella L.M. et al., 2000]
Comments promoter is an endothelial-specific [Gumina R.J. et al., 1997]
Site:(S5996) -199 to -176; NF-κB bs ;
Site:(S5997) -30 to -13; GATA bs ;
//

Figure -1. Presentation of human PECAM1 (platelet/endothelial cell adhesion molecule-1) gene in TRRD. (*a*) Regulatory units of the human PECAM1 gene and (*b*) description of the endothelial-specific promoter region of PECAM1 gene in TRRDUNITS table: RegUnitAC, GeneID, RegRegion, and RegUnit contain general description of the promoter region; DNA_BankLink, reference to the database of nucleotide sequences; SeqLength, length of the promoter sequence; Sequence, nucleotide sequence of the promoter region; PromotTisSp, tissue specificity; PromotInd, inducibility of the promoter region; MultipleStarts, positions of multiple transcription start sites, and ExperimentCodes, cells and codes of assays that were used to obtain the information.

The experimental methods that provided the information about transcription factor binding sites inputted into TRRD (Table 2) are directed to (1) initial detection of transcription factor binding sites; (2) identification of the transcription factor that interacts with a particular binding site; and (3) determination of the site's functionality.

Shown in Figure 2 is the rat gene of 7-dehydrocholesterol reductase (DHCR7). It is known that 25-hydroxycholesterol decreases the expression of DHCR7. This fact is reflected in TRRDEXP table in a form of expression pattern (Figure 2*a*). The information fields RegUnitLink and SiteLink of TRRDEXP contain the links to regulatory unit of the gene (Acc. number P02936) and the transcription factor binding sites (Acc. numbers S9059, S9060, and S9061) mediating the influence of 25-hydroxycholesterol (Figure 2*b*). The regulatory unit of DHCR7 gene is a sterol responsive region located between –287 and –1. It contains the binding sites for Spl and NF-Y and E-box, interacting with SREBP1. A fragment of SREBP1 binding site description from TRRDSITES table is shown in Figure 2*c*. The record in the field ExperimentCodes indicates that the site position was determined by DNase I footprinting with purified SREBP1 protein (assay code 1.1.5); the functionality and important positions of the site were studied by analysis of point mutations in the site (assay code 6.2) during cotransfection of the vector carrying SREBP1 gene (assay code 6.6.1.1) and by study of the effect of 25-hydroxycholesterol on expression of plasmid constructs (assay code 6.5).

Table -2. Examples of assays providing the information about transcription factor binding sites inputted into TRRD

Type of experiment	Assay code in TRRD
Detection of transcription factor binding sites	
DNase I footprinting with nuclear extract	1.1.1
DNase I footprinting with purified or recombinant protein	1.1.5
Genomic footprinting	1.5
Methylation protection assay	4.1
Methylation interference assay	4.2
Electrophoretic mobility shift assay (EMSA) with nuclear extract	3.1
EMSA performed in the presence of competitive oligonucleotides	3.2
EMSA performed with mutant probes or competitors	3.3
Identification of DNA-binding proteins	
DNase I footprinting with purified or recombinant protein	1.1.5
DNase I footprinting with nuclear extract and specific antibodies	1.1.6
EMSA with purified or recombinant protein	3.5
EMSA with nuclear extract and specific antibodies	3.6
Confirming the functional importance of the site	
Insertion of isolated site 5′ of homologous or heterologous promoter	6.3.2
Comprehensive mutant analysis	6.2
Trans-activation of a reporter gene by overexpression of a distinct transcription factor	6.6
Genomic footprinting	1.5
Insertion of isolated site 5′ of homologous or heterologous promoter	6.3.2
Comprehensive mutant analysis	6.2

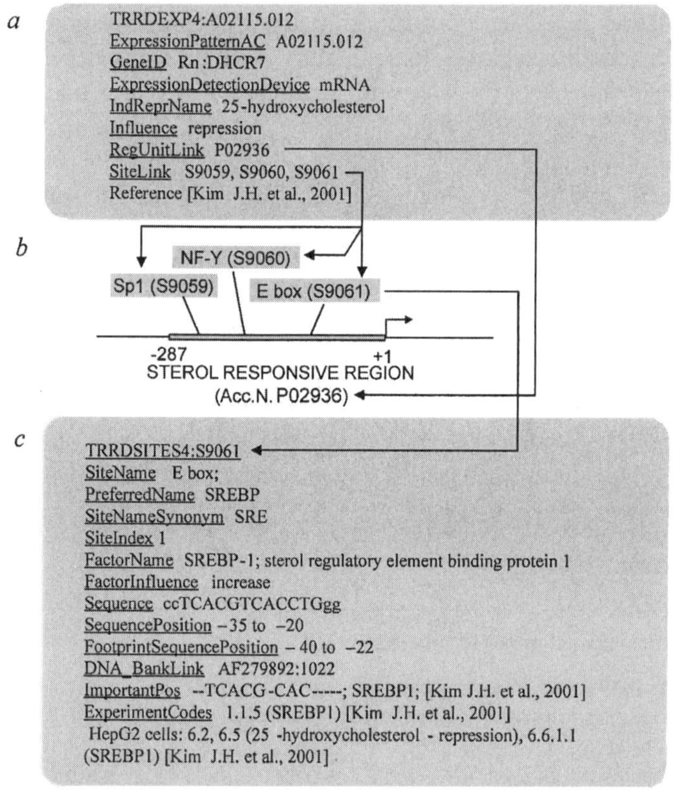

Figure -2. Regulation of rat 7-dehydrocholesterol reductase (DHCR7) gene under the effect of 25-hydroxycholesterol: (*a*) a fragment of description of expression pattern from TRRDEXP table; (*b*) sterol responsive region and the transcription factor binding sites localized to it; and (*c*) a fragment of description of SREBP1 binding site in TRRDSITES. SiteName, PreferredName, SiteNameSynonym, and SiteIndex give a general information about the site; FactorName, name of the factor interacting with the site; FactorInfluence, the effect of transcription factor on expression of the gene; Sequence and FootprintSequencePosition, site's sequence; DNA_BankLink, reference to the database of nucleotide sequences; ImportantPos, the nucleotides within the site whose mutations impair binding of the factor; and ExperimentCodes, codes of experiments wherefrom the data and cell names used were taken.

4. INFORMATION CONTENT OF TRRD

TRRD is the largest in the world informational resource that compiles experimentally confirmed data on the structure–function organization of regulatory regions of eukaryotic genes and natural transcription factor binding sites. The largest amount of information accumulated in TRRD pertains to human, mouse, and rat genes (Table 3). Topical sections including genes united by various functional characteristics (http://wwwmgs.bionet.nsc.ru/

mgs/gnw/trrd/sections1.shtml) were developed within TRRD. Overall, 15 sections were described earlier (Kolchanov et al., 2002; Ignatieva et al., 2004). Recently, two new sections were formed, namely, Genes Expressed in B cells (B-TRRD) and Hepatitis C Virus–Induced Genes (HCV-TRRD).

Table -3. Informational content of TRRD (as of February 2005)

	Totally in TRRD	Of the below species, %			
		Human	Mouse	Rat	Other species
Genes	2308	34	25	14	27
Regulatory units	3439	37	22	14	27
Transcription factor binding sites	10045	38	19	15	28
Expression patterns	14231	38	29	18	15
Scientific papers	7565	40	23	16	21

5. HIERARCHICALLY ORGANIZED THESAURUSES AND VOCABULARIES

Thesauruses on mammalian tissues and organs are developed within TRRD; these thesauruses provide the user with the possibility of obtaining supplemental information about cell composition, localization, and origin of tissues as well as about the functions of various organs and their parts. Thesauruses are represented by html pages with cross-references (http://wwwmgs.bionet.nsc.ru/mgs/gnw/trrd/thesaurus/). An example of one entry of the thesaurus for organs was described earlier (Ignatieva et al., 2004). Basing on the technology of development and support of controlled vocabularies (Ananko et al., 1998), hierarchically organized vocabularies on tissues, cells, organs, developmental stages of organisms, external stimuli, and transcription factors were composed as well as dictionaries of synonymic terms.

Let the organization of the vocabulary on transcription factors be an example. In addition to names of transcription factors and their synonyms, the vocabulary contains names of closely related factors, which are used for automatic marking of binding sites (the field PreferredName from TRRDSITES table). For example, the preferred name COUP-TF is assigned to the binding sites containing 13 name variants of transcription factors in the FactorName field (Figure 3). It is significant that the preferred name COUP-TF is ascribed not only to the sites with groups of synonymic names in their FactorName field (for example, COUP-TF, COUP-TF1, Ear3/COUP-TF, COUP-alpha, and v-erbA-related protein-3), but also to the sites interacting with closely related factors (COUP-TFI, COUP-TFII, and Ear-2).

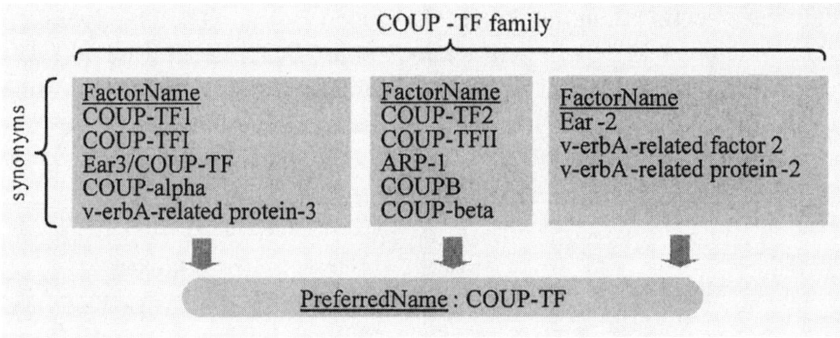

Figure -3. Records in the fields FactorName from which PreferredName of site is determined as COUP-TF.

6. DATA RETRIEVAL POSSIBILITIES OF TRRD

The data retrieval tools of TRRD were characterized earlier (Kolchanov et al., 2002; Ignatieva et al., 2004). SRS (Sequence Retrieval System) is the main tool for searching TRRD by keywords within 132 indexed fields. Browsers (of species and gene names) are supported in TRRD as well as topical sections.

The program BLAST (http://wwwmgs.bionet.nsc.ru/mgs/systems/fastprot/units_blast.html) allows sequences of regulatory regions homologous to an analyzed DNA sequence to be retrieved from TRRD. The regions homologous to transcription factor binding sites in a DNA sequence are searched for by the program BinomSite (http://wwwmgs.bionet.nsc.ru/mgs/programs/mmsite/).

The specialized search systems developed within TRRD (http://wwwmgs.bionet.nsc.ru/mgs/gnw/trrd/thesaurus/search.html and http://wwwmgs.bionet.nsc.ru/mgs/gnw/trrd/thesaurus/search_hidden.html) provide the following options: (1) search for genes induced (or repressed) by an external stimulus; (2) search for genes expressed in a specified organ, tissue, or cells or at a certain developmental stage of the organism; (3) combined search for genes expressed in a specified tissue, organ, or cells under induction by a specified external stimulus (simultaneously); and (4) search for genes or sites regulated by a defined transcription factor.

The system is realized basing on queries to SRS version of TRRD and on hierarchical vocabularies (of tissues, cells, organs, developmental stages of organisms, external stimuli, and transcription factors) and thesauruses on mammalian organs and tissues.

ACKNOWLEDGMENTS

The work was supported by the RFBR (grants Nos. 03-07-90181, 03-07-90181, 03-04-48469, and 03-04-48506); PCB RAS (grant No. 10.4); Siberian Branch of the Russian Academy of Sciences (integration project No. 119); Federal Agency of Science and Innovations (grant No. 43.073.1.1.1501); Russian Federal Research and Development Program Research and Development in Priority Directions of Science and Technology (contract No. 38/2004); US Civilian Research and Development Foundation (CRDF); and the Ministry of Education of the Russian Federation under The Basic Research and Higher Education Program (grant No. NO-008-X1). The authors are grateful to I.V. Lokhova for bibliographical support and to O.G. Smirnova, S.S. Ibragimova, T.V. Busygina, A.L. Proscura, E.L. Mishchenko, V.V. Suslov, S.A. Fedorova, I.I. Turnaev, A.L. Zakharenko, S.G. Udalieva, and E.S. Voronich for annotating efforts.

ARTSITE DATABASE: COMPARISON OF *IN VITRO* SELECTED AND NATURAL BINDING SITES OF EUKARYOTIC TRANSCRIPTION FACTORS

T. Khlebodarova*, O. Podkolodnaya, D. Oshchepkov, D. Miginsky, E. Ananko, E. Ignatieva

Institute of Cytology and Genetics, Siberian Branch of the Russian Academy of Sciences, prosp. Lavrentieva 10, Novosibirsk, 630090, Russia, e-mail: tamara@bionet.nsc.ru

* *Corresponding author*

Abstract: The ArtSite database was developed; the database compiles the information on the structures of eukaryotic transcription factor binding sites and/or their DNA-binding domains obtained from *in vitro* selected sequences. Current release of the database comprises 420 matrices describing specific features of binding sites or their DNA-binding domains for over 200 transcription factors. The matrices were constructed basing on alignments of representative samples of transcription factor binding sites, totally containing over 10 thousand sequences.

The information compiled in ArtSite was used to compare the structures of natural and *in vitro* selected binding sites for USF1, SP1, YY1, RAR/RXR, and E2F/DP1 transcription factors. The structures of the natural and *in vitro* selected binding sites for each transcription factor analyzed were found similar, suggesting that at least for the factors in question the structures of binding sites correlated with the affinities of the corresponding factors. Insignificant differences in detected frequencies of certain nucleotides reflect the general trend, namely, a higher occurrence of moderate affinity sites in the natural population of sequences compared with the sequences obtained *in vitro*.

Key words: databases; transcription factor; binding sites; transcription regulation

1. INTRODUCTION

Recently, development of new technologies, in particular, SELEX (Systematic Evolution of Ligands by EXponential enrichment), SAAB (Selected And Amplified Binding site imprint assay), REPSA (Restriction Endonuclease Protection Selection and Amplification), CASTing (Cyclical Amplification and Selection of Targets), and other *in vitro* selection procedures (Kinzler and Vogelstein, 1989; Blackwell and Weintraub, 1990; Wright et al., 1991; Hardenbol et al., 1997), yielded numerous data on the structures of binding sites for various transcription factors, both eukaryotic and prokaryotic. However, the questions on whether these data reflect the genuine structures of natural binding sites and what are the potential of applying these data to search for and prediction of natural sites are yet to be answered.

The opinions on this issue are inconsistent and ambiguous (Robison et al., 1998; Shultzaberger and Schneider, 1999; Roulet et al., 2000; Ehret et al., 2001) and unfavorable for at least several prokaryotic transcription factors (Robison et al., 1998). Thus, it is no wonder that having a considerable volume of information on the structures of natural binding sites complied with the database TRRD (Kolchanov et al., 2002), we decided to create a database that would integrate these pieces of information.

We developed the database ArtSite, whose contents allowed us to make a comparative analysis of the structures of natural and artificial binding sites. Note that several other databases also compile this information, namely, TRANSFAC (Matys et al., 2003), SELEX_DB (Ponomarenko et al., 2000), and JASPAR (Sandelin et al., 2004); however, either the databases contain data on only artificial sites and their resources are small or the information is beyond the public domain. This work present the ArtSite database and its content as well as comparison of the structures of binding sites for USF1, SP1, YY1, RAR/RXR, and E2F/DP1, extracted from the ArtSite and TRRD.

2. RESULTS AND DISCUSSION

2.1 Description of the ArtSite database

ArtSite is a database accumulating information about the structures of sequences that specifically interact with the DNA-binding domains of eukaryotic transcription factors (TF). Characteristics of these sequences are described in ArtSite by means of frequency matrices, which are constructed based on alignments of representative samples of TF binding sites.

The samples are constructed of both the genomic and *in vitro* synthesized DNA sequences binding TF in a specific manner, which are described in the literature and discovered by various selection methods. The ArtSite web interface allows for various queries (by TF name, its synonyms, structure of DNA-binding domain, origin of factor, and/or literature source) and to output a list of the corresponding entries (Figure 1).

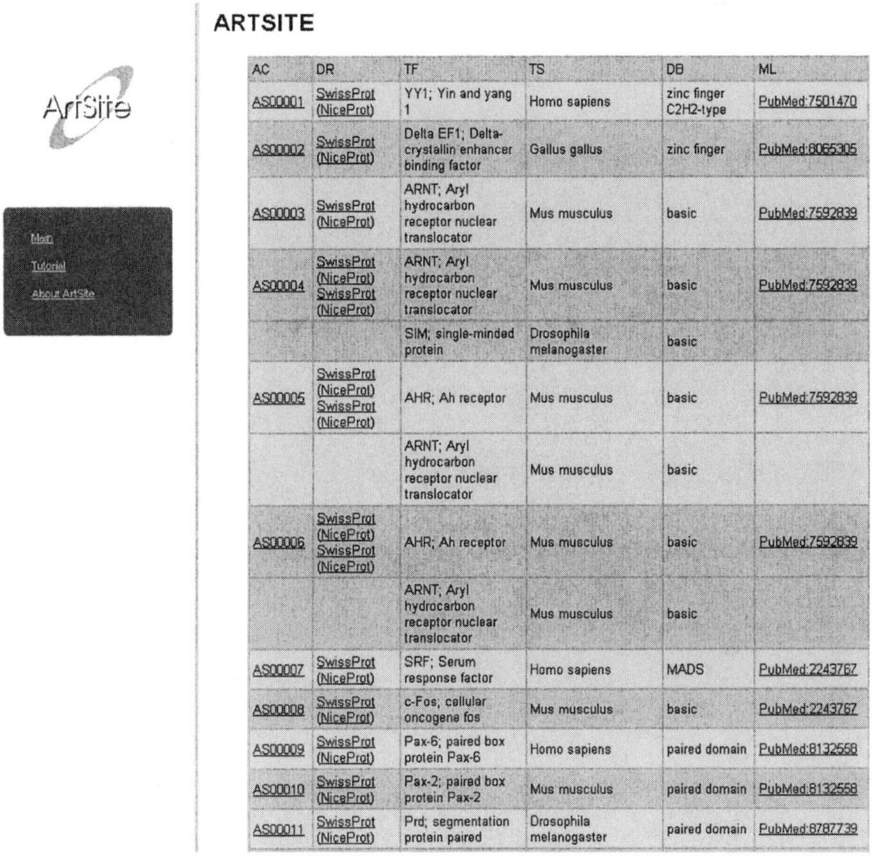

AC	DR	TF	TS	DB	ML
AS00001	SwissProt (NiceProt)	YY1; Yin and yang 1	Homo sapiens	zinc finger C2H2-type	PubMed:7501470
AS00002	SwissProt (NiceProt)	Delta EF1; Delta-crystallin enhancer binding factor	Gallus gallus	zinc finger	PubMed:8065305
AS00003	SwissProt (NiceProt)	ARNT; Aryl hydrocarbon receptor nuclear translocator	Mus musculus	basic	PubMed:7592839
AS00004	SwissProt (NiceProt) SwissProt (NiceProt)	ARNT; Aryl hydrocarbon receptor nuclear translocator	Mus musculus	basic	PubMed:7592839
		SIM; single-minded protein	Drosophila melanogaster	basic	
AS00005	SwissProt (NiceProt) SwissProt (NiceProt)	AHR; Ah receptor	Mus musculus	basic	PubMed:7592839
		ARNT; Aryl hydrocarbon receptor nuclear translocator	Mus musculus	basic	
AS00006	SwissProt (NiceProt) SwissProt (NiceProt)	AHR; Ah receptor	Mus musculus	basic	PubMed:7592839
		ARNT; Aryl hydrocarbon receptor nuclear translocator	Mus musculus	basic	
AS00007	SwissProt (NiceProt)	SRF; Serum response factor	Homo sapiens	MADS	PubMed:2243767
AS00008	SwissProt (NiceProt)	c-Fos; cellular oncogene fos	Mus musculus	basic	PubMed:2243767
AS00009	SwissProt (NiceProt)	Pax-6; paired box protein Pax-6	Homo sapiens	paired domain	PubMed:8132558
AS00010	SwissProt (NiceProt)	Pax-2; paired box protein Pax-2	Mus musculus	paired domain	PubMed:8132558
AS00011	SwissProt (NiceProt)	Prd; segmentation protein paired	Drosophila melanogaster	paired domain	PubMed:8787739

Figure -1. ArtSite web interface: view of search results.

2.2 Format of the ArtSite database

An entry of ArtSite corresponds to one selection experiment wherefrom the matrix was generated that describes the binding site for a TF or one of its domains if the factor–DNA interaction is complex. This format also allows for describing binding sites for heterodimeric proteins and intricate complexes of transcription factors. Description of such entry is shown in Figure 2.

Accession number	AS00117
Creation date	19/03/03
Annotator	Khlebodarova T.M.
Reference	SwissProt (NiceProt)

DNA

Number of sequences	40
Selection rounds	8
The synthetic template used for selection experiment	5'-TCCGAATTCCACAG-N18-TGCAATGGATCCGTCT-3'
Methods	DNA selection and amplification EMSA with purified recombinant protein Methylation interference

Proteins

	protein 1
Transcription factor name	RXRA; Retinoid X receptor alpha
Synonyms	Retinoic acid receptor RXR-alpha NR2B1
Origin of factor	Mus musculus
Binding form	homodimer
Domain	nuclear receptor-type
Organ	
Tissue	
TD	
Cell line	

Binding site recognition tool

Matrix

A	0	0	0	0	1	0	39	31	33	3	2	0	3	33	11	8	5	0
G	19	28	31	38	8	4	1	2	5	29	27	2	4	3	5	26	30	21
C	3	1	1	1	5	33	0	5	2	1	4	3	28	3	20	4	3	9
A	0	0	0	1	26	3	0	2	0	7	7	35	5	1	4	2	1	4
Consensus	G	G	G	G	T	C	A	A	A	G	G	T	C	A	C	G	G	G

Koenig R.J., Subauste J.S., Yang Y.Z. (1995) Retinoid X receptor alpha binds with the highest affinity to an imperfect direct repeat response element.. SO 7:136, 2896-903

PubMed:7789315

Figure -2. An example of the entry of the ArtSite database describing sites for binding of the transcription factor RXRA to DNA detected in *in vitro* experiments.

An entry comprises 32 fields; of them, 21 fields are obligatory for filling. An entry includes ArtSite accession number; date of entry creation; name of the annotator who created the entry; references to SWISSPROT and TrEMBL; short, full, and synonymous names of the TF; organ, tissue, and cell line wherefrom the TF was isolated in the case it is of endogenous origin; the form of TF binding to DNA; name of the DNA-binding domain of the TF described; the synthetic template used in selection experiment; number of selection rounds; number of sequences binding this TF and detected in the experiment; brief description of the methods used for selecting sequences; matrix shown as the number/frequencies of nucleotides at a certain position within the sequence aligned with respect to most frequently met nucleotides; consensus; and detailed information concerning the publication annotated.

The field 'Comments' contains textual comments on the specific features of experiment performed and construction of the weight matrix for the binding site described that cannot be input into the format developed but are important from the annotator's standpoint for a correct understanding of the data stored.

2.3 Specificity of the ArtSite database format for natural sites

This database also contains the entries describing matrices for natural binding sites obtained by selection of cloned genomic DNA fragments, and in this case, the format is virtually similar to that described above except that the synthetic template used for selection experiment is absent.

2.4 Content of the ArtSite database

The ArtSite database is a natural extension of the database TRRD. The first release of the former database contains 420 matrices describing the binding sites for over 200 transcription factors and their DNA-binding domains. These matrices were constructed basing on alignments of more than 10 000 sequences detected using various variants for selecting transcription factor binding sites described in 215 original publications. The database content is shown in Tables 1 and 2.

Table -1. Content of the ArtSite database

Organism	Number of matrices
Yeast	16
C. elegans	1
Plants	24
Insect (Drosophila)	15
Vertebrate	348
Clawed frog	6
Chicken	24
Mammals	318
Mammalian viruses	16
Total	420

Table -2. The ArtSite database content with reference to the structure of transcription factor DNA-bindind domains

DNA-binding domain (class according classification of Wingender, 1997)	Number of matrices
Basic domain (1.1, 1.2, 1.3, 1.6)	100
Nuclear receptor type (2.1)	30
Zinc finger (2.2, 2.3, 2.4)	97
Homeodomain (3.1)	55
POU domain (3.1.2)	16
Paired domain (3.2)	7
Fork-head domain (3.3)	10
Myb domain (3.5.1)	20
Ets domain (3.5.2)	19
MADS (4.3)	18
HMG box (4.6)	14
Others (3.4, 3.6, 4.1, 4.2, 4.5)	34
Total	420

Table 1 lists the data on species origin of transcription factors used for selection of binding sites by *in vitro* technologies. As clearly seen, more than 80 % of the matrices are related to the structures of animal transcription factor binding sites with 70 % of them referring to mammals. Thus, only 10 % of the matrices refer to plants and insects. The data given in Table 2 illustrate the grouping of matrices according to the structure of DNA-binding domains. As shown, almost half of the matrices (197) accumulated in ArtSite describe the structures of TF binding sites containing basic or zinc finger domains in their structure. Note also that a considerable number of matrices stored in the database, i.e., 85 matrices, describe the structure of binding sites of various homeodomain proteins and nuclear receptors. The other groups of transcription factors are less represented.

2.5 Comparison of *in vitro* selected and natural binding sites

Thus, what are the potential of applying the data obtained by *in vitro* selection for recognition of *in vivo* sites in genomes of various organisms? To clarify this issue, we compared the matrices constructed using the sequences selected *in vitro* with the matrices constructed using the sequences extracted from TRRD. Table 3 shows the data obtained upon such comparison for three TFs having different DNA-binding domains and different types of binding to DNA.

As is evident from Table 3, the matrices constructed using different sources are nonetheless virtually similar with reference to most frequently occurring nucleotides within the detected cores for USF1 and SP1 and differ inessentially in significant nucleotides at positions –2, –3, and +4 in the case of RAR/RXR.

Table -3. Matrices describing the natural and *in vitro* selected binding sites for USF1, SP1, and RAR/RXR transcription factors

Factor, binding type, and DNA-binding domain	Nucleotide	Nucleotide position									
		-5	-4	-3	-2	-1	0	+1	+2	+3	+4
	A	**22**	6	0	**48**	2	4	2	1	**19**	4
USF1	C	10	**23**	**51**	1	**36**	10	1	0	8	**23**
	T	2	**17**	0	2	3	0	**48**	4	5	9
Homodimer	G	**18**	9	1	1	11	**37**	1	**47**	**20**	**16**
	Consensus	**R**	**Y**	**C**	**A**	**C**	**G**	**T**	**G**	**R**	**S**
bHLH domain	(natural)										
	A	**8**	1	1	**30**	1	1	0	0	**9**	3
	C	2	**16**	**30**	1	**28**	1	2	0	3	**12**
	T	0	**11**	0	0	2	0	**28**	0	5	**13**
	G	**7**	2	0	0	0	**29**	1	**31**	**14**	1
	Consensus	**R**	**Y**	**C**	**A**	**C**	**G**	**T**	**G**	**R**	**Y**
	(selected)	(AS00239, AC ArtSite_DB; Bendall, Molloy, 1994)									
	A		41	42	3	5	55	5	8	41	27
SP1	C		18	10	9	7	**144**	8	4	19	39
	T		39	6	8	5	34	14	21	14	16
Monomer	G		**147**	**186**	**224**	**227**	11	**216**	**210**	**169**	**161**
	Consensus		**G**	**G**	**G**	**G**	**C**	**G**	**G**	**G**	**G**
Zinc fingers C2H2	(natural)										
type	A		2	3	0	0	0	3	0	1	1
	C		1	1	0	0	**8**	0	1	0	2
	T		2	1	0	0	1	2	3	2	1
	G		**6**	**6**	**11**	**11**	2	**6**	**7**	**8**	**7**
	Consensus		**G**	**G**	**G**	**G**	**C**	**G**	**G**	**G**	**G**
	(selected)	(AS00239 - AC ArtSite_DB; Thiesen, Bach, 1990)									
	A		6	**13**	6	**12**	2	2	2	1	**29**
RAR/RXR	C		**18**	7	5	3	0	0	3	**28**	2
	T		2	**10**	5	2	0	7	**27**	1	1
Heterodimer	G		8	4	**18**	**17**	**32**	**25**	2	4	2
	Consensus		**C**	**W**	**G**	**R**	**G**	**G**	**T**	**C**	**A**
Nuclear receptor	(natural)										
zinc fingers C4-type	A		9	**13**	**7**	**9**	1	0	1	1	**6**
	C		4	2	1	0	0	0	1	**11**	2
	T		0	1	1	2	0	2	**11**	0	2
	G		3	0	**7**	**5**	**15**	**14**	3	4	6
	Consensus		**V**	**A**	**R**	**R**	**G**	**G**	**T**	**C**	**R**
	(selected)	(AS00305, AC ArtSite_DB; Kurokawa et al., 1993)									

In the last case, G occurs most frequently in the natural sites at position –2, whereas G or A in the *in vitro* selected sequences; A and T are met with equal probabilities at position –3 in natural sites, whereas A is typical of *in vitro* selected sites; and A is most frequent at position +4 in natural sites, whereas A and G are equiprobable in the *in vitro* selected sequences. Thus, at least one of the significant nucleotides is necessarily present at all the three positions in both matrices, suggesting that the distinctions detected are not crucial and may stem from a moderate size of one of the matrices. Note in this connection that a similar fact played no negative role when comparing the *in vitro* selected and natural binding sites for SP1. The matrix for this factor constructed basing on natural sequences and containing 244 functional sites did not differ in significant nucleotides from the matrix containing only 11 sequences detected by EMSA (Thiesen and Bach, 1990). We believe that these data indicate that the functionality of SP1 binding site correlates directly with the affinity of this TF for the site in question. However, this correlation in the case of RAR/RXR may be not so clearly pronounced, and certain nucleotides that are important for binding of this factor to DNA may impair the fine expression regulation of the genes controlled by the factor in question and, consequently, lose their significance during natural selection.

At present, the resource of our database allows for assessment of probable functionality of the sites detected basing on comparison of natural and artificially selected sequences for 17 transcription factors. Find below two examples confirming this possibility. Shown in Table 4 are the matrices describing the binding sites for two transcription factors, E2F/DP1 and YY1. Comparison of the matrices constructed based on the natural functional binding sites for YY1 and the *in vitro* selected sequences suggests that the flanking nucleotides do not play a significant role in the function of these sites. Moreover, data on the affinity of the sequences detected in *in vitro* experiments (Hyde-DeRuyscher et al., 1995) and displaying significant flanking nucleotides (C, G, and G at positions –4, –3, and +5, respectively) demonstrate that these nucleotides also have no effect on the level of binding, as they are met in all the types of sequences—with high, medium, and low affinities (Table 5).

Analysis of these sequences demonstrates that the detected nucleotides (at positions –4 and –3) are contained in the primer used for selection and thus, were selected randomly. As for the G nucleotide at position +5, its appearance is not so evident. It cannot be excluded that the last nucleotide is necessary for the site function; however, a small sample of the natural sites prevented from detection of its significance. Nonetheless, note a trend of increase in its occurrence in the natural population and its rather similar frequencies in natural and *in vitro* selected populations—41 and 53 %, respectively. Note also that the *in vitro* selected sequences were tested for their ability to activate a reporter gene (Hyde-DeRuyscher et al., 1995). Thus, no wonder that both matrices have the same consensuses—CCATNTT. Moreover, analysis of these matrices and the data listed in Table 4 allowed us to detect within the site the nucleotides responsible for the affinity of the factor for DNA. All the four nucleotides are met at position +2 with equal probabilities; however, only the presence of nucleotide A decreases drastically the affinity of the site (Table 5).

Similar analysis of the binding site for E2F/DP1 transcription factor gives the following results. The matrices from Table 3 exhibit very close structures, namely, stringently fixed C and G nucleotides at positions 0 and +1 and inessential differences in significant nucleotides at positions +3 and +4. In natural sites, nucleotide A occurs most frequently at position +4, whereas in the *in vitro* selected, T and A are found. As for position +3, C and G are equiprobable in the natural sites, whereas C is most frequent in the *in vitro* selected sequences. Thus, at least one of the significant nucleotides is present at these positions in both matrices, making the distinctions detected not principal. Presumably, these distinctions are related to a higher frequency of moderate affinity sites in the population of natural sequences. Data shown in Tables 4 and 5 give grounds for such inference.

Table -4. Matrices describing the natural and *in vitro* selected binding sites for E2F/DP1 and YY1 transcription factors

Factor	Nucleotide	Nucleotide position											
		-5	-4	-3	-2	-1	0	+1	+2	+3	+4	+5	
	A		4	5	2	4	**17**	2	4	3	3	4	
	C		9	7	**19**	**17**	1	0	7	3	5	6	
	T		4	5	1	1	4	**19**	9	**14**	**12**	3	
	G		5	5	0	0	0	1	2	2	2	9	
YY1	Consensus (natural)		N	N	C	C	A	T	N	T	T	N	
	A		2	0	0	0	**56**	0	9	4	0	9	
	C		**39**	0	**56**	**56**	0	0	12	1	2	6	
	T		0	15	0	0	0	**56**	25	**50**	**50**	9	
	G		15	**41**	0	0	0	0	10	1	3	**30**	
	Consensus (selected)		C	G	C	C	A	T	N	T	T	G	
		(AS00001, AC ArtSite_DB; Hyde-DeRuyscher et al., 1995)											
	A	3	4	0	1	0	0	0	0	2	**21**	**17**	**16**
	C	1	0	0	**22**	**12**	**36**	2	**26**	**16**	7	4	3
	T	**29**	**32**	**37**	1	0	0	0	0	2	4	**11**	**11**
E2F/DP1	G	5	2	1	**14**	**26**	2	**36**	**12**	**18**	6	6	8
	Consensus (natural)	T	T	T	S	S	C	G	S	S	A	W	W
	A	4	2	0	0	0	0	3	0	0	**9**	**5**	**5**
	C	2	1	1	**17**	**14**	**24**	1	**19**	**18**	4	2	1
	T	**18**	**22**	**24**	0	0	0	0	0	4	**10**	**16**	**17**
	G	1	0	0	**8**	**11**	1	**21**	**6**	3	1	1	0
	Consensus (selected)	T	T	T	S	S	C	G	S	C	W	W	W
		(AS00123, AC ArtSite_DB; Tao et al., 1997)											

Presumably, the presence of C nucleotide in the core sequence (CCCGCC) is characteristic of high affinity sites (Table 5), whose occurrence rate is higher among the *in vitro* selected sequences (Table 4); however, G at position +1 is the most significant nucleotide in the site structure. Other conditions being equal, substitution of G with A decreases drastically the affinity of the factor for DNA, and only a high resolution of the method used allowed such sites to be detected among the *in vitro* selected sequences. A virtually complete absence of the natural sites with any other nucleotides except for G at this position also confirms the significance of this position.

Table -5. Binding sites for the transcription factors YY1 (Hyde-DeRuyscher et al., 1995) and E2F/DP1 (Tao et al., 1997) detected in *in vitro* experiments and their degree of affinity for the corresponding factors

Factor	Sequence	Affinity
YY1	cgCCATTTTaag	High
	gtCCATTTTtgt	Medium
	atCCATCTTgac	Medium
	cgCCATGTTgcg	Medium
	cgCCATTTGccg	Medium
	cgCCAT**A**TTcct	Low
	cgCCAT**A**TTgtc	Low
	gtCCAT**A**TTgta	Low
E2F1/DP1	ttattTTTCCCGCCTTT	High
	tCTTCCCGCCTTAttc	High
	tgatTTTGGCGGGATTc	Medium
	ttGTTCCC**A**GCCACtc	Very low

Table -6. Kullback-Leiber distances between matrices for natural and *in vitro* selected binding sites, respectively

Matrix type	USF	SP1	YY1	RXR/RAR	E2F1/DP1
Distance	0.1265	0.1676	0.1801	0.1799	0.1930
Positions considered	10	9	10	7	12

In general, these results give evidence about a high level of similarity between the natural and *in vitro* selected binding sites of USF, SP1, YY1, RXR/RAR, and E2F1/DP1 transcription factors. For verifying our conclusion, we measured the similarity between matrices described above and the Kullback–Leiber distance according to Aerts et al. (2003). The authors discriminate three levels of similarity values of matrices: 0.2 corresponds to a high stringency; 0.3, to a moderate stringency; and 0.4, to a low stringency (Aerts et al., 2003). Following our estimates (Table 6), the distances between natural and *in vitro* selected matrices considered are less than 0.2, thus, showing high stringency and supporting our conclusion.

Thus, the above examples demonstrate that analysis and comparison of the matrices constructed basing on the *in vitro* selected and natural sites expand considerably our knowledge about the structure of the site as well as suggest applying more optimistically various methods for detecting sites in unstudied genes and assessing theoretically their functionality. A comprehensive analysis of the structure of CNF/NF1 binding sites using the *in vitro* selected and natural site sequences (Roulet et al., 2000) gave similar results. This allowed the authors to use the results obtained for developing a highly sensitive method for prediction and recognition of DNA-binding sites in eukaryotic genomes. Analysis of the corresponding matrices for STAT proteins also discovered the similarity between the structures of natural and *in vitro* selected binding sites (Ehret et al., 2001). However, an analogous analysis of Lrp binding sites in *E. coli* genome and Lrp-binding sequences produced by *in vitro* experiments resulted in an opposite conclusion—the matrices constructed using the synthesized sequences selected *in vitro* differed essentially from those obtained with the natural sites (Shultzaberger and Schneider, 1999) and were inappropriate for site recognition. Moreover, this result was confirmed by a comprehensive study of DNA-binding sequences in *E. coli* genome and their comparison with the corresponding sequences obtained *in vitro* (Robison et al., 1998). This demonstrates that the specific features discovered in the structure of transcription factor binding sites of a certain type are in no way applicable to sites of another type, as formation of the binding sites is determined not only by the structure of the transcription factor DNA-binding domain, but also by specific mechanisms involved in fine regulation of the genes controlled by particular transcription factors. These mechanisms have been formed during the evolution and fixed in the structure of the binding site for a particular transcription factor in various genes it controls. In a similar way, the conditions used for selecting the synthesized sequences, which differ from the natural mechanisms involved in site formation, also influence the structure of *in vitro* selected sites. All this requires a caution in data interpretation, on the one hand, and comprehensive approach in studies, on the other.

ACKNOWLEDGMENTS

The work was supported in part by the Russian Foundation for Basic Research (grants Nos. 03-07-90181 and 03-04-48469), Russian Academy of Sciences (grant No. 10.4), Siberian Branch of the Russian Academy of Sciences (integration project No. 119), and governmental contract with the Federal Agency for Science and Technology 'Identification of potential targets for novel medicinal drugs based on reconstructed gene networks' with the priority direction 'Living system'.

COMPARATIVE ANALYSIS OF ELECTROSTATIC PATTERNS FOR PROMOTER AND NONPROMOTER DNA IN *E. COLI* GENOME

S.G. Kamzolova*, A.A. Sorokin, P.M. Beskaravainy, A.A. Osypov
Institute of Cell Biophysics, Russian Academy of Sciences, Pushchino, Moscow oblast, Russia, e-mail: kamzolova@icb.psn.ru
* *Corresponding author*

Abstract: Distribution of electrostatic potential of the complete sequence of *E. coli* genome was calculated. It is found that DNA is not a uniformly charged molecule. There are some local inhomogeneities in its electrostatic profile, which correlate with promoter sequences. Electrostatic patterns of promoter DNAs can be specified due to the presence of some distinctive motifs that may be involved as promoter signal elements in RNA–polymerase–promoter recognition.

Key words: *E. coli* genome; promoters; coding region; electrostatic potential distribution

1. INTRODUCTION

There are about 4000 promoters in the genomes of *Escherichia coli* and related bacteriophages that are recognized by RNA polymerase ($E\sigma^{70}$). It is shown that promoters vary considerably in their nucleotide sequences. Statistical analysis of nucleotide sequences for all known promoters detected two homologous hexamer motifs, 5′TATAAT3′ and 5′TTGACA3′, centered around −10 and −35 positions, respectively (Harley and Reynolds, 1987). The functional role of these hexanucleotides as universal promoter determinants involved in recognition of RNA polymerase has been proved. However, even in these conserved regions, nucleotide sequences differ essentially in individual promoters containing, as a rule, only 7.9 canonical nucleotide pairs from 12. Therefore, the precise identification of promoters based on the canonical consensus sequences appeared to be impossible. A large set of promoter search algorithms taking account two canonical hexamers and some preferential nucleotide sequences in the regions flanking these elements and around transcription start point has failed in correct prediction of promoter sites in *E. coli* genome (Horton and Kanehisha, 1992; Hetz and Stromo, 1996; Huerta and Collado-Vides, 2003).

Taking into account the sequence diversity of the promoters, a wide range of their functional activities, and differential response to the same physiological signals, it was suggested that some noncanonical specific determinants should be involved in the process of RNA polymerase interaction with different promoters and their groups. Some new attempts to reveal noncanonical promoter determinants on the basis of nucleotide sequence led to identification of many new hexanucleotides preferred in promoter sites as compared with the total genome structure (Kamzolova et al., 2004). Overall, 542 different hexanucleotides were shown to be twofold more frequent in the promoters than in the chromosome. These hexanucleotides are very large in number to consider all of them as sequence-specific promoter determinants. It is natural to suppose that some of them can be involved in the interaction with RNA polymerase due to their specific physical properties. Some physicochemical characteristics of promoter DNA, such as overall geometry, deformability, thermal instability, and dynamical features were shown to play an important role in differential interaction of the enzyme with various promoters (Kamzolova and Postnikova, 1981; Margalit et al., 1988; Travers, 1989; Leirmo and Gourse, 1991; Perez-Martin et al., 1994; deHaseth and Helmann, 1995; Kamzolova et al., 1999).

DNA is a highly charged polyelectrolyte and, therefore, its electrostatic potential may be one of the main features recognized by DNA-binding proteins. Really, the important role of DNA electrostatic features in the multistep process of protein–DNA recognition was shown for some DNA-binding proteins like transcription factors of eukaryotic bZIP family (Strauss-Soukup and Maher, 1998), homeodomain (Labeots and Weiss, 1997), or lambda cI repressor and EcoRI endonuclease (Misra et al., 1994). However, until recent years, such study for RNA polymerase ($E\sigma^{70}$) was hampered due to considerable difficulties in theoretical calculations of electrostatic potential distribution for long DNA fragments of the size not less than promoter length (> 150 nucleotide pairs). In 1999, we proposed a simplified method for calculation of electrostatic potential distribution for long DNA fragments as large as 1000 bp (Polozov et al., 1999), which was later modified to provide a means for calculating electrostatic profiles for DNA sequences of millions base pairs including complete genomes (Kamzolova et al., 2000; Sorokin, 2001). Development of this method opened a field for studying electrostatic properties of promoter DNAs (Kamzolova et al., 2000; Sorokin, 2001; Sorokin et al., 2001; Dzhelyadin et al., 2001a, b). Electrostatic interactions between promoter DNA and RNA polymerase ($E\sigma^{70}$) were shown to be of considerable importance in regulating promoter function for 'early' T4 phage promoters (Polozov et al., 1999). Electrostatic characteristics of promoter DNA were suggested to be a new promoter determinant marked by its relative independence from promoter nucleotide sequence (Dzhelyadin et al., 2001a, b; Sorokin et al., 2001).

Here, distribution of electrostatic potential of the complete sequence of *E. coli* genome was calculated. As an example to illustrate a functional meaning of the information hidden in the electrostatic map of the genome, a comparative analysis of electrostatic patterns of promoter and nonpromoter DNA sites was made. It was found that genome DNA is not a uniformly charged molecule. There are some local inhomogeneities in its electrostatic profile, which correlate with promoter sequences. Electrostatic patterns of promoter DNAs differ for different promoters by the presence of some specific elements. No direct correlation between the nucleotide sequence of promoters and their electrostatic characteristics is observed: (a) on the one hand, there are promoters possessing a high 'homology score' of nucleotide sequences but differing by their electrostatic patterns and (b) on the other, there are promoters differing by their sequences but showing similar electrostatic potential distribution.

2. METHODS

Nucleotide sequence data were extracted from the following databases: the complete sequence of *E. coli* K-12 genome (GenBank accession number U00096), Regulon (http://www.cifn.unam.mx/Computational_Genomics/), and Promec (http://bioinfo.md.huji.ac.il/marg/promec/). Overall, 359 experimentally confirmed promoters were extracted from *E. coli* genome nucleotide sequence. When studying electrostatic profiles, promoter nucleotide sequences (–250 to +150 bp) were aligned according to the transcription start point. The 359 coding DNA fragments (each 400 bp long) flanking promoter sites were used as nonpromoter DNA sequences.

Electrostatic potential distribution around double-helical DNA molecule was calculated by the Coulombic method (Polozov et al., 1999; Kamzolova et al., 2000; Sorokin, 2001) using the computer program of A.A. Sorokin (lptolik@icb.psn.ru). A full atom model of DNA molecule was used with atom coordinates taken from (Landolt-Bornstein, 1989). The dependence of helix geometry on nucleotide sequences was allowed for according to (Ponomarenko et al., 1997). The electrostatic potential $V(\vec{R})$ of DNA fragments were calculated in accordance with Coulomb's law:

$$V(\vec{R}) = \sum_i \frac{Q_i}{\varepsilon(\vec{R})R_i},$$

where R_i is the distance from the atom with the charge Q_i to the point of observation and $\varepsilon(\vec{R})$, distance-dependent dielectric constant. The charges that were summed over σ- and π-electron clouds were placed at the centers of atoms according to Zhurkin et al. (1980). Potential was calculated on the surface of a cylinder with 15-Å radius, the cylinder axis coinciding with the DNA double-helix axis with 1Å step along the cylinder axis and 1° step over the azimuthal angle. Two-dimensional distribution of the potential was averaged with 31 Å sliding window along the helix axis completely surrounding the cylinder circumference. For more details, see (Polozov et al., 1999; Kamzolova et al., 2000).

3. RESULTS AND DISCUSSION

E. coli genome contains 4288 genes combined into 2580 operons that are controlled by ~ 4000 promoters. Most of them are specifically recognized by RNA polymerase $E\sigma^{70}$ (σ^{70}-specific promoters of *E. coli*). So far, 359 promoters interacting with σ^{70}-RNA polymerase were identified and characterized biochemically. Since their localizations on the genome map were known, the corresponding functionally specified DNA fragments were chosen from *E. coli* complete nucleotide sequence for comparative analysis.

Using the original method (Kamzolova et al., 2000), electrostatic potential distribution was calculated for the complete nucleotide sequence of *E. coli* genome, containing 4 639 221 base pairs. The results, representing the profile of electrostatic potential distribution around the complete genome, are stored in our database (kamzolova@icb.psn.ru). A huge body of information hidden in the electrostatic map of *E. coli* DNA may be useful in studying manifold problems of the genome organization and functioning.

The possibility of extracting some functional information from the electrostatic map of *E. coli* genome can be demonstrated by the example provided by a large-scale analysis of electrostatic patterns of promoter and nonpromoter DNA sites. Electrostatic profiles of 359 promoters and their nearby coding sequences were analyzed according to the presence of peaks and valleys as well as their arrangement and values. Figure 1 shows some representative examples of electrostatic patterns for (*a*) promoters and (*b*) DNA coding regions.

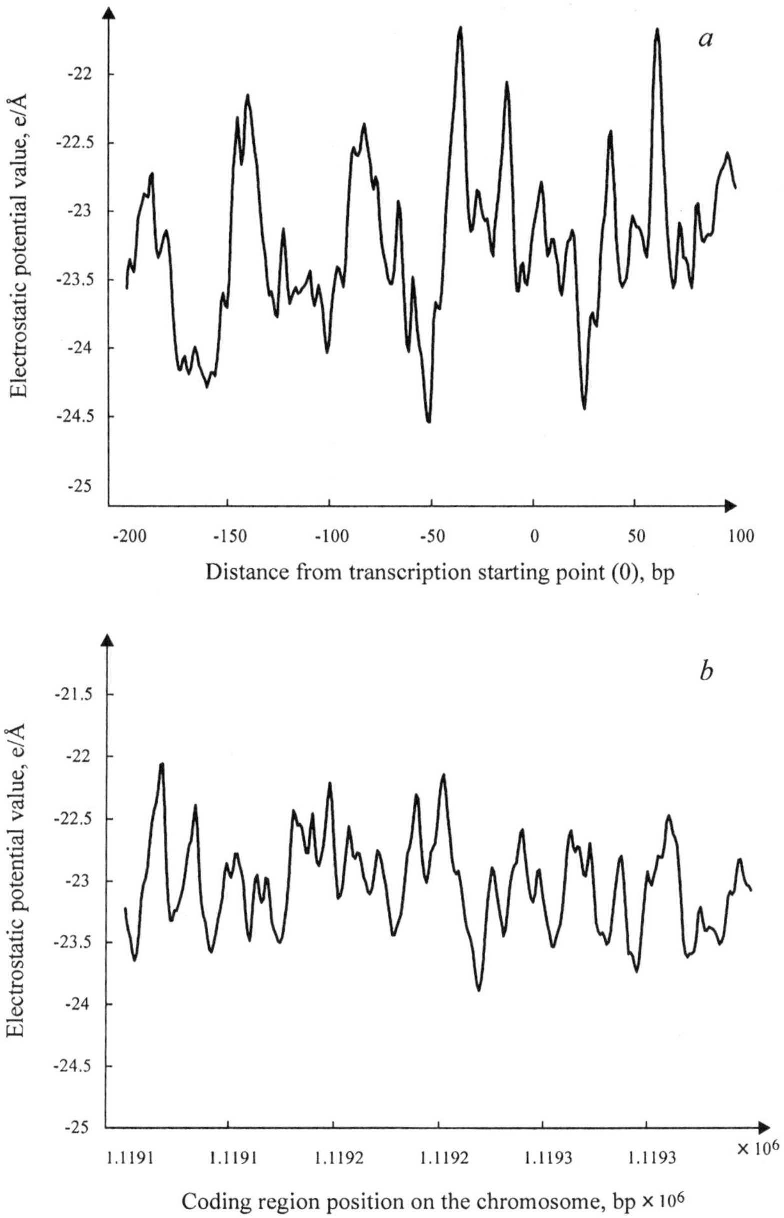

Figure -1. Electrostatic potential distribution for (*a*) promoter and (*b*) coding region in *E. coli* genome.

It is found that nonpromoter DNA regions are characterized by more homogeneous distribution of electrostatic potential, whereas local inhomogeneities, with the greatest and smallest values of potential corresponding to promoter sites. There are no common specific elements in electrostatic profiles of the DNA fragments containing different nonpromoter regions.

Electrostatic profiles of promoters are of a complex-shaped design with alternating areas of peaks and valleys. Individual promoters differ in their electrostatic patterns, but they can be clustered on the basis of certain common electrostatic motifs.These results are in agreement with our data obtained previously for 15 σ^{70}-specific promoters of T4 phage 'early genes' (Kamzolova et al., 2000; Dzhelyadin et al., 2001a). T4 early promoters were shown to be grouped due to the presence of some specific elements in far upstream region of their electrostatic profiles.

There was a good correlation between the patterns of electrostatic potential distribution in far upstream regions of the promoters and their functional behavior in response to ADP-ribosylation of RNA polymerase α-subunit, thus indicating the role of the specific electrostatic elements as promoter determinants in electrostatic interaction with the enzyme (Kamzolova et al., 2000; Dzhelyadin et al., 2001a). It is interesting that electrostatic profiles of some *E. coli* promoters display similar motifs in far upstream region. It can be illustrated by the example of three ribosomal promoters: two individual rrnABP1 promoters and one rrnG-P1 promoter, located in *E. coli* genome at 3 939 139, 4 032 828 and 2 729 469 base positions, respectively.

Nucleotide sequences of the promoters are given in Figure 2. All promoters are identical in their nucleotide sequences in the region from –41 to +20 bp; the two rrnABP1 promoters share sequence identity in the region from –75 to +20 bp.

```
          -90       -80       -70       -60       -50       -40       -30       -20
1 AAAGACTATATTTAAGATGTTTTGCCTGAAAAGTGAGCGAACGATAAAGTTTTATATTTTTCGCTTGTCAGGCCGGAAT
2 ACCGACGCTGAAATAAGCATAAAGAATAAAAAATGCGCGGTCAGAAAATTATTTTAAATTTCCTCTTGTCAGGCCGGAAT
3 TATGGCACATTAACGGGGCTTTTGCTGAAAAAATGCGCGGTCAGAAAATTATTTTAAATTTCCTCTTGTCAGGCCGGAAT

  -40       -30       -20       -10       +1        10        20
1 TTCGCTTGTCAGGCCGGAATAACTCCCTATAATGCGCCACCACTGACACGGAACAACGGCA
2 TCCTCTTGTCAGGCCGGAATAACTCCCTATAATGCGCCACCACTGACACGGAACAACGGCA
3 TCCTCTTGTCAGGCCGGAATAACTCCCTATAATGCGCCACCACTGACACGGAACAACGGCA
```

Figure -2. Nucleotide sequence of ribosomal promoters (1, rrnGP1-2 728 496; 2, rrnABP1-4 032 828; and 3, rrnABP1-3 939 139): +1, transcription start point; –10 and –35 consensus hexanucleotides are indicated. Differing nucleotide sequences are underlined.

In spite of their considerable sequence similarity, the promoters differ in electrostatic patterns in far upstream regions (Figure 3). The far upstream regions, corresponding to the positions from –75 to –100 bp of the promoters, are indicated by vertical lines.

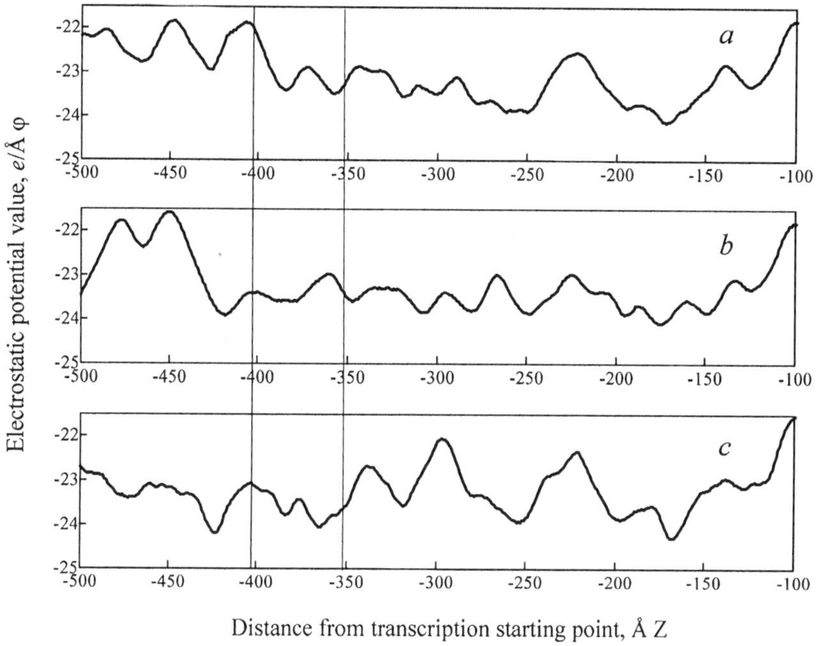

Figure -3. Electrostatic potential distribution around double-helix DNA containing ribosomal promoters: (*a*) rrnABP1-4 032 828; (*b*) rrnABP1-3 939 139; and (*c*) rrnGP1-2 728 496.

The data obtained indicate that two ribosomal promoters (rrnABP1-4 032 828 and rrnG-P1-2 729 469) have electrostatic patterns similar in design with electrostatic profile of T4 promoter P164.5, whose activity is inhibited due to ADP-ribosylation of RNA polymerase α-subunit (Kamzolova et al., 2000; Dzhelyadin et al., 2001a). Such behavior was shown to be determined by the presence of a specific element containing two most negatively charged sites at –75 and –90 bp in the far upstream region of promoter DNA involved in the interaction with α-subunit (Kamzolova et al., 2000). The most negatively charged character of this region should prevent its electrostatic interaction with ADP-ribosylated α-subunit carrying some additional negative charge due to the modification, thus resulting in a decrease in promoter activity in the case of the modified enzyme (Kamzolova et al., 2000; Dzhelyadin et al., 2001a).

A distinctly different electrostatic element is found in the far upstream region of rrnABP1-3 939 139 promoter (Figure 3). There is an extended peak at about –75 bp and a more negative flanking site at –60 bp. The promoter shares these specific electrostatic features with T4 promoter P131.7, which is activated in response to ADP-ribosylation of the enzyme (Kamzolova et al., 2000; Dzhelyadin et al., 2001a).

Although the ribosomal promoters are characterized by a high level of homology in core regions, some difference in far upstream sequences influence the whole profile of electrostatic potential distribution, including both core and far upstream sites of the promoters. The difference in electrostatic properties of various ribosomal promoters possessing extended sequence homology can result in a difference in their functional behavior, allowing a high level of ribosomal RNA synthesis to be maintained in bacterial cells in different conditions. This also can explain some conflicting data on synthesis of ribosomal RNAs in infected *E. coli* containing ADP-ribosylated form of RNA polymerase.

Thus, the data obtained indicate that electrostatic patterns of promoter DNAs can be specified due to the presence of some distinctive motifs, which differ in different promoter groups and may be involved as signal elements in differential recognition of various promoters by the enzyme.

Interestingly, the two rrnABP1 ribosomal promoters, which are characterized by the most extended sequence similarity, belong to different groups according to their electrostatic properties. However, promoters rrnABP1-4 032 828 and rrnG-P1-2 729 469, possessing a lower level of sequence homology, may be assigned to the same group according to the common electrostatic element that they share in the functionally important region of promoter DNA.

Summarizing, it is reasonable to suggest that characteristic variations in electrostatic potential of DNA may contribute to RNA polymerase–DNA recognition by specifying promoter sites as electrostatic traps or barriers for the enzyme. In addition, alternating areas of negative and positive potential in promoter sites may enforce charged RNA polymerase molecules to orient properly relative to the transcription start point. Thus, DNA electrostatic component may be one of the signals allowing RNA polymerase to identify promoter sites in genomes.

ACKNOWLEDGMENTS

The authors are grateful to E.G. Saveljeva for technical support.

The work was supported by the Russian Foundation for Basic Research (grant RFBR-naukograd No. 04-04-97275).

ANALYSIS OF NUCLEOSOME FORMATION POTENTIAL AND CONFORMATIONAL PROPERTIES OF HUMAN J1-J2 AND D2-D1 TYPE ALPHA SATELLITE DNA

A. Katokhin[1*], V. Levitsky[1, 2], D. Oshchepkov[1], A. Poplavsky[1], V. Trifonov[1], D. Furman[1, 2]

[1] *Institute of Cytology and Genetics, Siberian Branch of the Russian Academy of Sciences, prosp. Lavrentieva 10, Novosibirsk, 630090, Russia, e-mail: katokhin@bionet.nsc.ru;*

[2] *Novosibirsk State University, ul. Pirogova 2, Novosibirsk, 630090, Russia*

[*] *Corresponding author*

Abstract: The structure-forming function of alpha satellites in arrangement of the centromeric heterochromatin implies the presence of certain contextual and conformational signals (codes) for compacting DNA of these regions into nucleosomes and chromatin structures of higher level. However, this aspect of informational content of alpha satellite primary DNA sequences yet requires further studies. Computer analysis of nucleosome formation potential (NFP) was performed using a sample of J1–J2 and D2–D1 type alpha satellites from the human genome. A number of regions with the context favorable for several variants of nucleosome positioning were detected. Statistical analysis of DNA conformational profiles, in particular, concerning the property Wedge, demonstrates a superposition of the context-dependent and CENP-B-dependent nucleosome positionings. The corresponding software packages are available at http://wwwmgs.bionet.nsc.ru/mgs/programs/recon/ (RECON method) and http://wwwmgs.bionet.nsc.ru/mgs/programs/sitecon/ (SITECON method).

Key words: alpha satellite DNA; nucleosome positioning; DNA conformational properties

1. INTRODUCTION

Alphoid or alpha satellite DNA is an essential component of the primate genome, comprising up to 5 % in the human genome. Alphoid DNA clusters in the centromeric and pericentromeric chromosome regions, forming blocks with a length of 200 kbp to 5 Mbp. A monomer with a length of ~ 171 bp is the elementary unit of alphoid DNA. Ten types of the monomers somewhat varying in their nucleotide composition are known. The homology between the consensus sequence of a monomer of certain type and individual sequences of monomers belonging to this particular type amounts usually to approximately 80–90 % (Alexandrov et al., 2001; Black et al., 2004; Paar et al., 2005).

The second level in organization of centromeric alpha satellites is repeated units composed of two (J1–J2 or D1–D2) or five (W1–W5) monomers of various types. These units are named suprachromosomal families. The monomers within each suprachromosomal family are subdivided into two classes according to the presence of B box, a specific binding site for the centromeric protein CENP-B (Yoda et al., 1998). The monomers J2, D1, W1, W2, and W3 carry the B box (Alexandrov et al., 2001).

Finally, the higher organization level is the so-called high order repeats (HORs), composed of members of one or other suprachromosomal families. HORs are specific of each chromosome. The homology between the units of the same HOR is 95–96 % (Alexandrov et al., 2001; Paar et al., 2005). The repeated units form tandem repeats at each level of the hierarchical organization.

The centromeric heterochromatin is formed through packaging of alpha satellites into nucleosomes. In this process, each monomer having a length of ~ 171 bp permits positioning of only one nucleosome (146 bp) with a short linker region, resulting in a supercompact DNA packaging in these regions (Gilbert and Allan, 2001). The mechanisms involved in formation of nucleosome arrays with an increased density in the centromeric regions are yet vague.

Presumably, interaction of the centromeric DNA with specific proteins, designated as CENP (CENtromere Proteins) may represent one of such mechanisms. In particular, it is known that the nucleosomes of centromeric heterochromatin differ from the conventional nucleosome by that they contain the histone-like CENP-A protein, capable of binding to DNA in a nonspecific manner, instead of histone H3. Several other CENP proteins, which are constitutive components of the centromeric chromatin, were detected, in particular, CENP-C, displaying a nonspecific binding to DNA (Politi et al., 2002). Among the proteins of this group, only CENP-B is the protein capable of binding specifically to alpha satellites, namely, to the sequence CTTCGTTGGAAACGGGA of the B box, mentioned above (Mitchell, 1996; Enukashvili et al., 2003; Masumoto et al., 2004). However, the role of CENP-B in formation of the centromeric heterochromatin is likely to be rather limited, since the Y chromosome centromeric heterochromatin also formed of alphoid sequences but without involvement of CENP-B (Enukashvili et al., 2003; Masumoto et al., 2004).

The event of CENP-B binding to B box in the monomers of J2 and D1 types determines an unambiguous nucleosome positioning within the J1–J2 and D1–D2 dimers, as was demonstrated in *in vitro* and *in vivo* experiments (Yoda et al., 1998; Ando et al., 2002). In the absence of CENP-B protein, several equivalent translational positions for the entire chain of nucleosomes with retained fixed internucleosome interval were realized in experiments on *in vitro* reconstitution of nucleosomes (Yoda et al., 1998). This fact suggests that the nucleosome positioning depends rather on the alphoid contexts than on CENP-B binding to B box; however, the nature of this dependence is yet vague.

In this work, computer analysis of distribution of contextual and conformational nucleosome positioning signals in J1–J2 and D2–D1 type dimers was performed. The contextual signals were analyzed by RECON method for calculating nucleosome formation potential (NFP; Levitsky et al., 2001; Levitsky, 2004). The conformational signals were studied by SITECON method allowing DNA conformational and physicochemical properties to be analyzed (Oshchepkov et al., 2004). The results obtained agree well with the data on experimental nucleosome mapping.

2. METHODS AND ALGORITHMS

Samples of actual J1–J2 and D1–D2 dimers differing contextually from the consensus no more than by 5 % were used in the work. The J1 and J2 sequences were extracted from the following genomic contigs: BX284928 (chromosome I), AC069355 (chromosome III), AC135046 (chromosome V), AB005791 (chromosome VI), NC_000007 (chromosome VII), BX322613 (chromosome X), AADD01123003 (chromosome XII), M58446 (chromosome XVI), and AC135053 (chromosome XIX). The D1 and D2 sequences were extracted from the genomic contigs J04773 (chromosome II), M81229 (chromosome II), AC0079999 (chromosome IV), AC009609 (chromosome IV), AC027263 (chromosome IV), M64779 (chromosome VIII), AL603712 (chromosome XIV), M65181 (chromosome XVIII), D29750 (chromosome XXI), and BX294002 (chromosome XXII).

The sequence characteristics essential for nucleosome positioning, such as NFP, were assessed by RECON method (Levitsky et al., 2001; Levitsky, 2004). To construct NFP, the method considers two samples of sequences—nucleosome formation sites with a length of 160 bp and random sequences with equal nucleotide frequencies. As the sample of nucleosome formation sites, we used 141 DNA sequences from the Nucleosomal DNA database, which were centered at their dyad axes according to the positions indicated

(Ioshikhes and Trifonov, 1993). First, a partition of the studied region of nucleosome formation site into fragments is searched for. We define the partition $\Omega(b_1, b_2, \ldots, b_{p-1})$ of the site $[a, b]$ as a set P of the nonoverlapping local fragments $[a_p, b_p]$ ($p = 1, \ldots, P$) meeting the following conditions: $a_1 = a$; $a_{p+1} = b_p$, for $p = 1, \ldots, P-1$; $b_P = b$. The number of regions P is was taken equal to 13. The search for an optimal partition is intended to provide minimal errors during recognition. The Mahalanobis distance R^2 (Mahalanobis, 1936) between distributions over two samples is used as the parameter for assessing the quality of partition. The value R^2 depends on $N = 16 \times P$ variables, dinucleotide frequencies in the partition fragments (16 is the number of dinucleotides). Growth in R^2 corresponds to mutual distancing of the centers of distributions over the samples (1) and (2).

When analyzing a random DNA sequence, the value of function $\varphi(X)$ was calculated at each position of the sliding window (fragment X, 160 bp):

$$\varphi(X) = \frac{1}{R^2} \times \sum_{n=1}^{N} \sum_{k=1}^{N} \{[f_n(X) - (½) \times [f_n^{(2)} + f_n^{(1)}] \times S_{n,k}^{-1} \times [f_k^{(2)} - f_k^{(1)}]\}. \quad (1)$$

Here, $f_n(X)$ is the vector of dinucleotide frequencies constructed with account of the partition of the fragment X into local fragments. The NFP $\varphi(X)$ [2] is constructed so that its mean value over the sample of nucleosome formation site sequences equals +1; over the sample of random sequences, –1. This means that a higher probability of nucleosome formation corresponds to the values of NFP $\varphi(X)$ [2] close to +1.

Finally, to find the values of NFP $\varphi(X)$ with a specified significance level α, it was transformed as follows:

$$\varphi_\alpha(X) = \begin{cases} \dfrac{|1 - \varphi(X)|}{P_\alpha \times \sigma_\varphi} & \text{if } |1-\varphi(X)| < P_\alpha \times \sigma_\varphi, \\ 0, & \text{otherwise.} \end{cases} \quad (2)$$

In our analysis the significance level α is selected equaling 0.95.

Representation of the NFP as [2] is used to bring into correlation larger values of NFP for a dinucleotide position to larger probability of a nucleosome formation with its dyad axis located at the position. Thus, the values $\varphi_\alpha(X) > 0$ [2] correspond to reliable prediction of nucleosome formation sites.

The conformational characteristics essential for nucleosome positioning were assessed by SITECON method (Oshchepkov et al., 2004), which takes into account distribution of 38 statistically significant conservative context-dependent DNA conformational and physicochemical properties. In the SITECON analysis, a training sample of aligned functional sequences is used for detection of significantly conservative properties at the positions of alignment.

Successively recoding the sequences into one of the 38 properties, we calculate the mean square deviation of the property at the positions and the mean value. Thus, a low variance of a particular property indicates its conservation at a particular position. The significance of the mean square deviation is estimated using chi-square test.

3. RESULTS AND DISCUSSION

Shown in Figure 1 are NFP profiles of J1, J2, D1, and D2 type alphoids constructed using the tool RECON (Levitsky et al., 2001). NFP value at each point of a DNA sequence corresponds to the probability of a nucleosome formation with its dyad axis located at the position, e.g., of positioning of the nucleosome at this point.

The maximal NFP values within J2 and D1 type monomers, interpreted as the signals for translational nucleosome positioning, are observed at three to four positions, namely, 378, 414, and 449 for J2 and 378, 392, 415, and 441 for D1. The J1 and D2 type monomers contain four–five such signals.

This result complies well with the data of *in vitro* experimental studies on nucleosome positioning along J1–J2 dimers in the absence of CENP-B (Yoda et al., 1998). These MNase footprint experiments showed several positions for nucleosome edges along the monomers. Most of them were suppressed upon CENP-B addition; however, those flanking the B box were enhanced.

These nucleosome edge positions allowed authors to draw an overall scheme of nucleosome positioning along J1–J2 dimers (Figure 1*a*; the layout above the plot). Nucleosome mapping along the J1–J2 and D2–D1 dimers was performed in *in vivo* experiments with similar results (Ando et al., 2002).

To investigate in more detail the contribution of specific dinucleotide blocks of each monomer to alphoid DNA folding into nucleosome, we analyzed the distribution of context-dependent DNA conformational and physicochemical properties along the monomers. To avoid a chromosome-specific bias in assessment of dinucleotide frequencies, the samples of J1, J2, D1, and D2 sequences for SITECON analysis contained one sequence of each corresponding type from each chromosome.

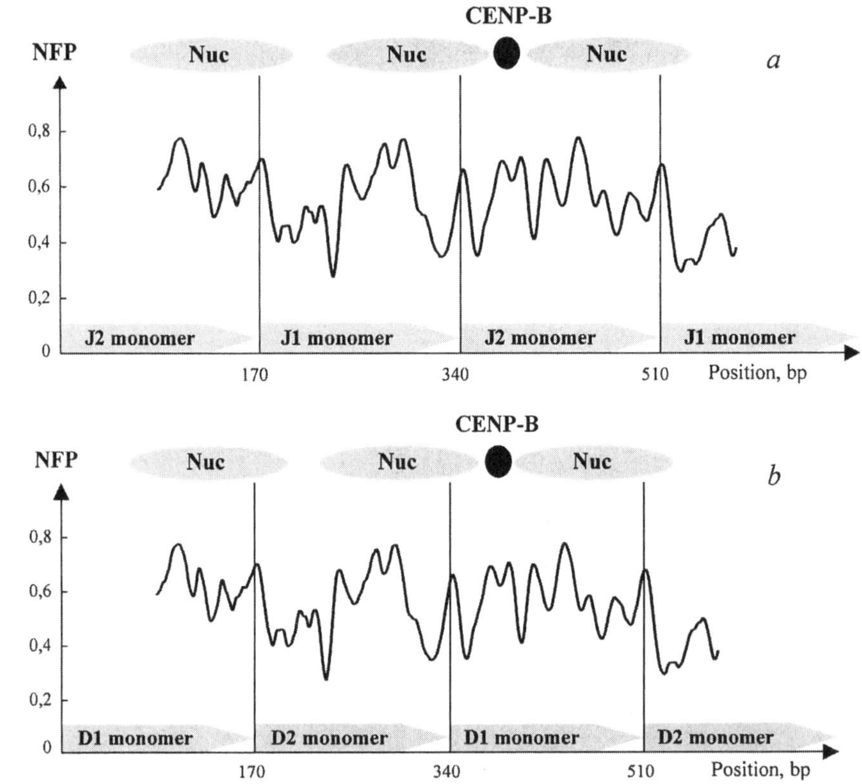

Figure -1. NFP profile along (*a*) J2, J1 and (*b*) D1 and D2 type alpha satellite sequences. The layout of nucleosome positioning within each monomer according to experimental data (Yoda et al., 1998; Ando et al., 2002) is shown above as well as the location of CENP-B protein over the B boxes in J2 type monomers.

Analysis of the sample of J1, J2, D1, and D2 type alpha satellites using the tool SITECON (Oshchepkov et al., 2004) detected existence of several conserved blocks when considering several sets of analyzed properties (data not shown). As the analyzed set of properties was redundant, we focused on one most illustrative property. The Wedge property summarizes the data on roll and tilt angles and characterizes a total curvature of a free B-DNA (Ulanovsky and Trifonov, 1987). On the one hand, Drew and Travers (1985) suggested that DNA curvature might account for sequence-specific positioning of nucleosomes. On the other hand, it was hypothesized that the sequence-dependent curvature of satellite DNA was an important feature of centromeric heterochromatin condensation (Radic et al., 1987; Lobov et al., 2001). Thus, we used the property Wedge in our analysis (Figure 2). The property Wedge illustrates well the distribution pattern of conservative properties related to DNA curvature along the J1, J2, D1, and D2 monomers.

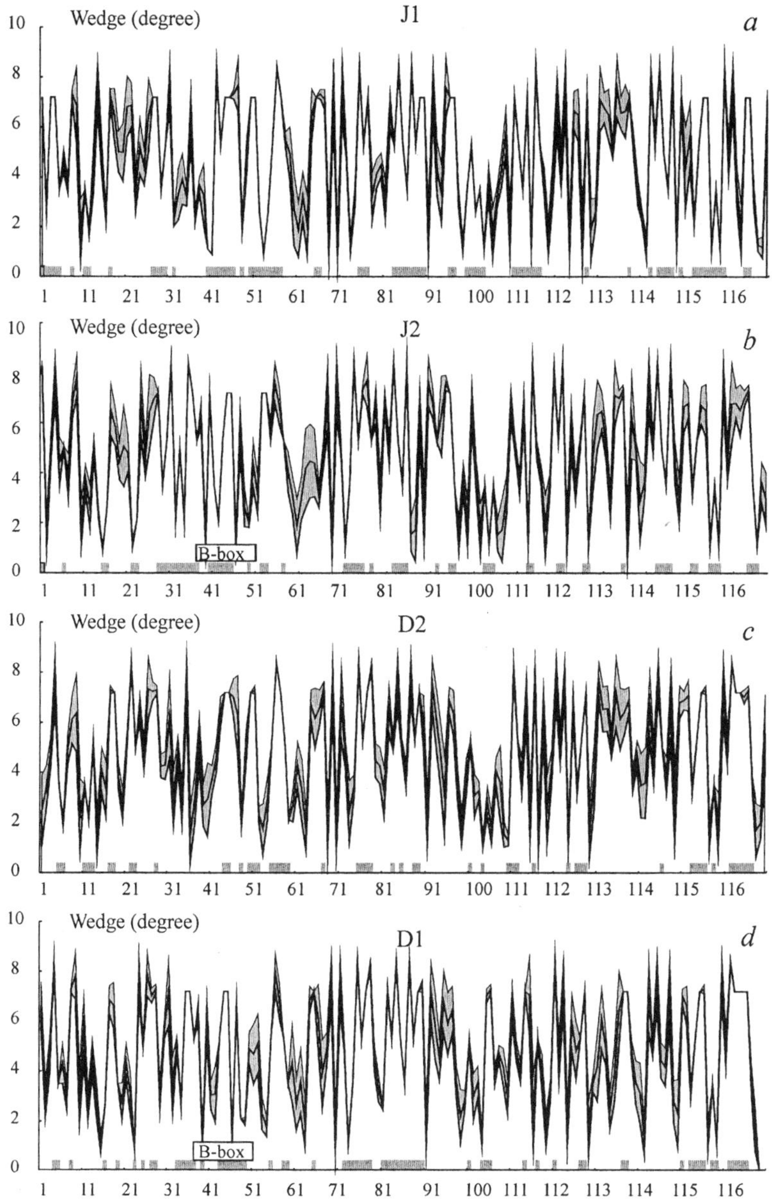

Figure -2. Profiles of the property Wedge along (*a*) J1, (*b*) J2, (*c*) D2, and (*d*) D1 type monomers: firm line indicates the profile of mean values for the sample; gray area around the firm line, the range of standard deviation; the abscissa, nucleotide positions; and the ordinate, value of the property in degrees. Position of B box is shown. Regions with statistically significant conservation of the property Wedge (99.9 %) are indicated with gray rectangles.

The block of Wedge conservation in the J2 type monomer is localized to the region of B box, reflecting a high conservation of its context as a site for binding CENP-B protein (Yoda et al., 1998; Ando et al., 2002; Figure 2*b*). The similar pattern is seen for the region of B box in the D1 type monomer (Figure 2*d*). However, the Wedge profiles are insufficiently coordinated and hence, not additive. This means that the intrinsic curvature of the B box free of CENP-B protein is insignificant. Indeed, X ray structure analysis data demonstrate that B box as a DNA fragment with a length of 21 bp is bent to 60° only in the complex with CENP-B protein (Tanaka et al., 2001).

In the sample of J1 type monomers, the region 36–52, which corresponds to the B box region in J2 type monomers, also displays a conservation of the property Wedge. However, unlike the genuine B box, increased Wedge values within this region are observed in the 5-bp region that corresponds to a half-turn of the DNA helix (Figure 2*a*). The similar pattern is seen for the region 36–52 in the D2 type monomer (Figure 2*c*). Presumably, this is connected with a prominent intrinsic curvature of DNA within this region and reflects the linker function of these regions, which, as a rule, are located between two nucleosomes positioned by the complex B box–CENP-B (see the layout in Figure 1).

Monomers of all the four types display conservation of the property Wedge over the region 97–105 as well. As is evident from Figure 2, typical of this region are decreased values of the property in question, suggesting that DNA here is almost straight. It was demonstrated earlier that this particular conformation characterized the region of nucleosome dyad axis in mammalian satellites including alpha satellites of Primates (Fitzgerald et al., 1994; Fitzgerald, Anderson, 1999); hence, this region may be assumed the center of nucleosome site, i.e., one of the preferable positions of the nucleosome dyad.

Thus, the NFP profile and profile of the property Wedge along J1, J2 D1, and D2 type alpha satellites suggest a superposition of the context-dependent and CENP-B-dependent nucleosome positionings. First, the context of these alpha satellite dimers enhances formation of compact nucleosome arrays due to the presence of several signals for nucleosome positioning, as detected by RECON analysis. Second, a regular pattern of regions with a pronounced intrinsic curvature, characteristic of linker regions, in J1 and D2 monomers as well as straight regions in all the four monomers, as demonstrated by SITECON analysis, results in formation of regular compact nucleosome arrays. In addition, such pattern of conformational properties of the B box–containing alpha satellites presumably provides a characteristic helical curvature of their axis in the 3D space (Lobov et al., 2001).

Finally, CENP-B bends the DNA in the region of the cognate box (rather straight according to the data of SITECON analysis) when interacting with the box, thereby additionally compacting the nucleosome rows and fixing the heterochromatin structure formed. Interestingly, the latest data report that the centromeric heterochromatin in cell culture displays two spatially separated domains: in one domain the B box–containing alpha satellites are associated with CENP-B; in the other, the same B box–containing alpha satellites are associated with CENP-C (Politi et al., 2002). Presumably, the conformational distinctions between the B box–containing alpha satellites with CENP-B and without it are of great importance for the function of centromeres.

ACKNOWLEDGMENTS

The work was supported in part by the Russian Foundation for Basic Research (grants Nos. 03-04-48555, 05-04-49111 and 03-07-96833); CRDF (Y2-B-08-02); Ministry of Industry, Science, and Technologies of the Russian Federation (grant No. 43.073.1.1.1501); Russian Federal R and D Program Research and Development in Priority Directions of Science and Technology (contract No. 28/2002); and Program for Basic Research of the Presidium of the Russian Academy of Sciences (contract No. 10002-251/P-25/155-270/200404-082).

VMM: A VARIABLE MEMORY MARKOV MODEL PREDICTION OF NUCLEOSOME FORMATION SITES

Yu.L. Orlov[1*], V.G. Levitsky[1, 2], O.G. Smirnova[1], O.A. Podkolodnaya[1], T.M. Khlebodarova[1], N.A. Kolchanov[1, 2]

[1] *Institute of Cytology and Genetics, Siberian Branch of the Russian Academy of Sciences, prosp. Lavrentieva 10, Novosibirsk, 630090, Russia, e-mail: orlov@bionet.nsc.ru;*

[2] *Novosibirsk State University, ul. Pirogova 2, Novosibirsk, 630090, Russia*

[*] *Corresponding author*

Abstract: Prediction of the DNA capacity to form nucleosome structure based on sequence statistics is of importance in analysis of gene expression regulation in eukaryotes. Context analysis of nucleotide sequences of experimentally defined nucleosome formation sites allows the determination of the sequence preference for nucleosome formation relying on statistical information. However, context analysis does not allow identifying the clear-cut consensus making feasible site prediction. One has to make recourse to more general context sequence features, such as dinucleotide frequencies. Markov model is a common approach to the prediction of the functional regions in DNA sequences that disregards positional information. Here, we use an improved version of the Markov model to predict the preference of DNA sequences to be within a nucleosome structure. The developed VMM program (the Variable Memory Markov model) computes the nucleosome formation potential for genomic DNA sequences of arbitrary lengths, including the short transcription factor binding sites. The VMM is publicly available at <http://wwwmgs.bionet.nsc.ru/programs/VMM/>.

Key words: gene expression regulation; nucleosome formation site

1. INTRODUCTION

The compaction degree of DNA within chromatin in the eukaryotic cells is different. The main structural element of chromatin, the nucleosome, is formed by a DNA fragment of about 147 bp packaged around a histone octamer (Khorasanizadeh, 2004). Experimentally defined binding sites of DNA to histone octamer underlie statistical analysis of nucleosome formation sites (Ioshikhes and Trifonov, 1993; Levitsky et al., 2005). The mechanisms of sequence-directed nucleosome positioning have been studied in numerous *in vivo* and *in vitro* experiments that suggested the existence of a specialized nucleosome code determining this positioning as a result of multiple histone–DNA interactions (Kiyama and Trifonov, 2002).

The questions of the context specificity of nucleosome DNA in relation to the regulatory regions stir interest. It is feasible, in principle, to develop algorithms for identification of nucleosome sites (Levitsky et al., 2001). Software tools for estimation of the nucleosome formation potential (the tendency to bind histone octamer and to form nucleosome structure) of arbitrary nucleotide sequences have been developed in the RECON program by Levitsky (2004). The RECON method is based on discriminant analysis. It takes into account the frequencies of dinucleotides in the local regions of nucleosome sites. The program detects the block structure of nucleosome formation site during its partition into local regions with a specific dinucleotide context.

The use of different mathematical approaches to the statistics-based prediction of a sequence preference to be within a nucleosome structure broadens our tools for analysis of the genome structure and the regulatory mechanisms behind gene expression. We describe here an Internet-available program, VMM, that allows calculation of the probability of a DNA region to be within a nucleosome structure.

A distinguishing feature of VMM is that it provides analysis of set of sequences and is capable of predicting the preference to be in a nucleosome structure for relatively short sequences of length less than 147 bp, including the transcription factor binding sites. The division into local regions used in RECON is inapplicable to analysis of short sequences; therefore, we use the more common Markov model. We implemented an expansion to the standard Markov model for the appearance of letters in a text depending on the preceding context. The expansion we called the variable memory Markov (VMM) model. This model uses the nucleotide correlations more exhaustively than standard Markov models do and it proved to be efficient in analysis of protein sequences (Bejerano, 2004).

Construction of such a Markov model is feasible for a given set of DNA sequences using the TreeComplexity program (Orlov et al., 2002). The VMM program implements method for estimation of the fit of a DNA sequence to the predefined model of nucleosome formation site <http://wwwmgs.bionet.nsc.ru/programs/VMM/>.

2. METHODS AND ALGORITHMS

We used DNA sequences from the NPRD databases as sources of the data on the nucleosome formation site (Levitsky et al., 2005). More than 400 sequences of 200-bp long were phased relative to the center of the experimentally defined nucleosome formation site.

To estimate the fit of the queried sequence to the nucleosome site, we took advantage of the context tree source model (Orlov et al., 2002). Otherwise defined, it is a variable memory Markov model (also called variable length Markov models (Rissanen, 1986; Buhlmann and Wyner, 1999). Like the standard Markov model, the newly developed model is stationary: the probability of a nucleotide occurrence is not dependent on the position in a sequence. It is dependent on the local preceding (left) context only. The $X = X_1X_2 \dots X_n$ sequence is generated with the probability

$$P(X) = P(X_1|S_1)P(X_2|S_2)\dots P(X_n|S_n). \tag{1}$$

Every S context is not longer than a certain fixed length d that determines the probability distribution of a letter occurrence in a sequence immediately to the right. Contrary to the fixed order Markov models, these models are not restricted to a predefined order. Rather, by examining the training data, a model is constructed that fits higher order dependencies where it exists, while using lower order dependencies if not thus reducing the number of parameters. Indeed, there are often insufficient data available to estimate the exponentially increasing number of parameters in full Markov models. As shown, VMM models are capable of capturing rich signals from a modest amount of training data without using hidden states (Bejerano, 2004).

Ron et al. (1996) suggested the approach expressed as a subclass of probabilistic finite automata. They have demonstrated that an optimum model can be developed using the prediction suffix tree construction. Rissanen had shown an algorithm for construction of such suffix tree that could be used for effective data compression (Rissanen, 1986; Barron et al., 1998).

It is convenient to arrange all the possible contexts of fixed lengths into a tree-like structure. Figure 1 provides an example of a context tree for a four-letter DNA alphabet.

Figure 1 gives an example of a complete context tree that contains all the 16 possible contexts of length 2. The contexts are read from bottom to top, from the leaves of the tree.

The tree can be incomplete, i.e., not all the contexts can be represented. An example of such a context tree is shown in Figure 2.

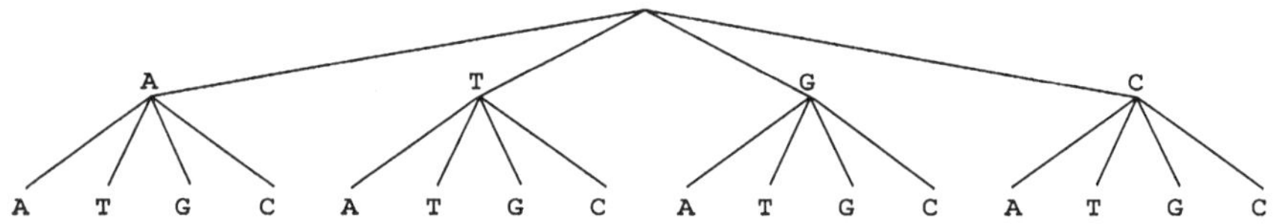

Figure -1. A tree-like structure containing all the contexts of length 2.

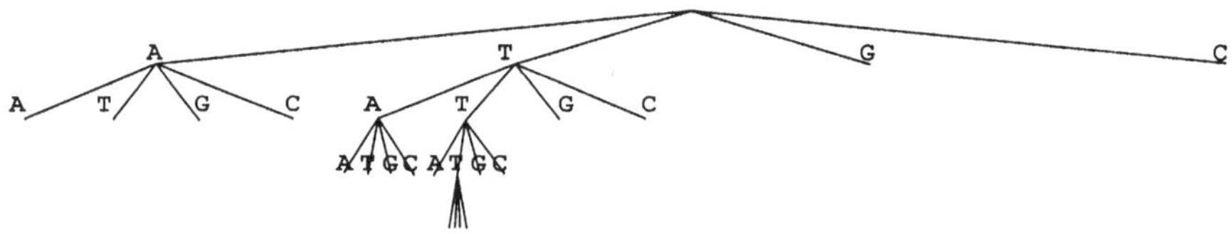

Figure -2. An example of a generating tree source for nucleosome formation sites. The tree is constructed using the TreeComplexity program (http://wwwmgs.bionet.nsc.ru/mgs/programs/complexity/).

The probability $P(Z_i|S_j)$ of the letter occurrence $Z_i \in \{A,T,G,C\}$, $i = 1, 2, 3, 4$ in every one of the S_j contexts is defined by the respective θ parameter as

$$P(Z_i|S_j) = \theta_j^i, \tag{2}$$

where $\sum_{i=1}^{4} \theta_j^i = 1$, $j = 1, 2, ...|T|$, $|T|$ is the number of all the contexts in T model (the number of leaves in the tree), $1 \le |T| \le 4^d$, and d defines the maximum context length. To calculate a reasonable statistics, the number of all possible contexts (that always less than the total length of the training sequence) should be significantly less than 4^d. We select $d = 6$ for nucleosome sequence sample.

Contexts may vary in length, but no context is the end of some other context in the tree. Every path from leaves to root corresponds to the word in a DNA sequence. Every context has its own distribution for generation of the next letter in a sequence.

To illustrate, for the tree T depicted in Figure 2, there are six contexts of two nucleotides in length (AA,TA,GA,CA,GT,CT), seven contexts of three nucleotides ({A,T,G,C}AT, {A,G,C}TT), two contexts of one nucleotide (G and C), and four contexts of four nucleotides ({A,T,G,C}TTT). In all, there are 19 preceding contexts, thus, the size $|T|$ of the tree is 19. Every context defines four numbers, the occurrence probability to the right of it. To define these numbers, we use the frequencies of the corresponding oligonucleotides, in total 76 (4×19).

An optimum model of generating tree source was built on the starting data using the Context algorithm (Barron et al., 1998). The algorithm for building the model is related to a mathematical estimation of stochastic complexity, and the Internet-available program TreeComplexity implements it (Orlov et al., 2002). The algorithm constructs complete context tree up to the depth *d* and then prunes insignificant leaves (rare contexts). Thus, not all the contexts from the training sequence are presented in the final context tree.

It is interesting to note that asymmetric trees of this type with the context extended only at A or T nodes and never at C or G nodes can be found in poly-W tract. Such a structure of the tree extends information on context rules in nucleosome site. As an example of such rules reported earlier, periodic distribution of AA and TT dinucleotides is important for nucleosome formation (Bolshoy et al., 1997).

The nucleosome formation potential was preformed by estimation of the correspondence probability of a sequence fit to the Markov model trained on the database. Fit function was constructed as the logarithm of the probability to obtain a sequence X in a sliding window of the fixed length *n*:

$$F(X) = \log (P(X)) = \sum_{j=1}^{n} \log (P(X_j|S_j)).$$

The program outputs this profile together with the level expected by random for a DNA sequence with equal nucleotide frequencies. A window may be of any size, but it is expedient to have it smaller than 147 bp.

VMM provides the calculation of the fit function of a sequence not only to the nucleosome formation sites, but also to any DNA sequence set whose tree model was constructed using the TreeComplexity program (http://wwwmgs.bionet.nsc.ru/mgs/programs/complexity/). With the VMM program, users can analyze a single up to 1 Mb sequence and obtain a profile of the preference for nucleosome formation. The program enables the calculation of the average profile for a set of phased (i.e., of the same length) DNA sequences. The latter possibility is of great interest for defining the preference for the formation of nucleosomes for different sequence sets, the gene promoters, for example.

To calculate the nucleosome potential by the given method, a homogenous model of a sequence is used. In contrast, in the previous method (Levitsky, 2004), the queried region of the nucleosome formation site was considered as a whole and the minimum length of the queried sequence was 160 bp. Thus, the model we propose allow the estimation of the nucleosome potential for sequences as short as 20 bp. Comparison of nucleosome potential function based on variable memory Markov models (the VMM program) and function based on dinucleotide frequencies and discriminant analysis (the RECON program) revealed a correlation of these fit functions on test genomic data. The proposed program for nucleosome site prediction is novel in that it allows the calculation of the nucleosome potential for short sequences, such as the transcription factor binding sites.

3. RESULTS AND DISCUSSION

3.1 Prediction of nucleosome location in single gene sequence

Let us consider how the VMM program is applied to a single gene for which nucleosome location was tested experimentally. The human *c-myc* gene (AC X00364) is provided as an example (Pullner et al., 1996). Figure 3 contains the fit profile calculated by the VMM software and schematic representation of nucleosome location defined experimentally (Pullner et al., 1996). The gray circles indicate the nucleosomes observed in all cell lines, the hatched indicate only those that have not been identified in all cell lines investigated or only in some cell lines and in minor amounts. The straight line indicates the level expected by random. The region of the promoter P1/P2, where no nucleosome formation was observed experimentally, corresponds to a profile minimum. The sequence stretches where nucleosomes were observed not in all cell lines are neighboring with the other local minima. Another 'narrow' profile minimum located around 3400 corresponds to the linker region between nucleosomes.

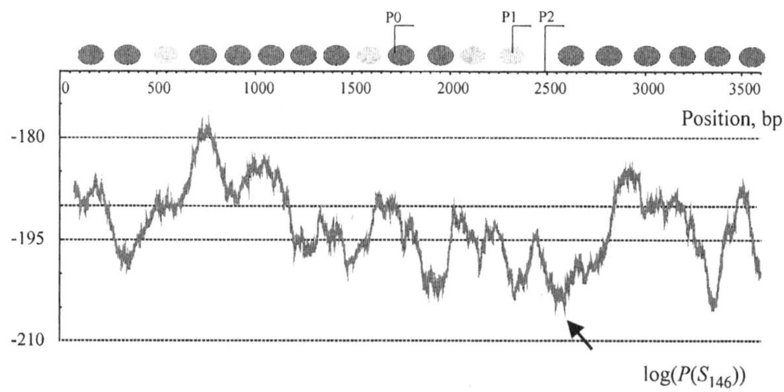

Figure -3. The nucleosome formation profile for human c-myc gene sequence. The relevant experimental data on nucleosome positioning are shown schematically by circles (Pullner et al., 1996). The arrow below indicates the profile minimum.

The specificity of the queried nucleosomal sites makes impossible an accurate prediction for these sites. In fact, the nucleosome formation is determined not only by the DNA context features, but also by external influences, in particular, the disposition of the neighboring nucleosomes. The accuracy of the most of experimentally mapped data available in the literature is limited to about 10–20 nucleotides. Even the very existence of a nucleosomal context code defined as degenerate periodic signals of particular nucleotides was debatable. Despite the questions raised, the program we propose nevertheless allows the delineation of regions where nucleosome formation is hindered. The important point is that only statistical data for the nucleosome formation sites from the NPRD database (Levitsky et al., 2005) were used for the prediction.

3.2 Statistical comparison of nucleosome potential in gene regions

The query of the sequence set allows statistical inferences about its average preference for nucleosome formation to be made. We analyzed two sets of gene promoters phased relative to the transcription start as [–300; +100]. The samples were taken from the TRRD database (http://wwwmgs.bionet.nsc.ru/mgs/gnw/trrd) by the gene expression level: the promoters of housekeeping genes and the promoters of tissue-specific genes (32 and 86 sequences, respectively). Figure 4 presents averaged profiles for these samples.

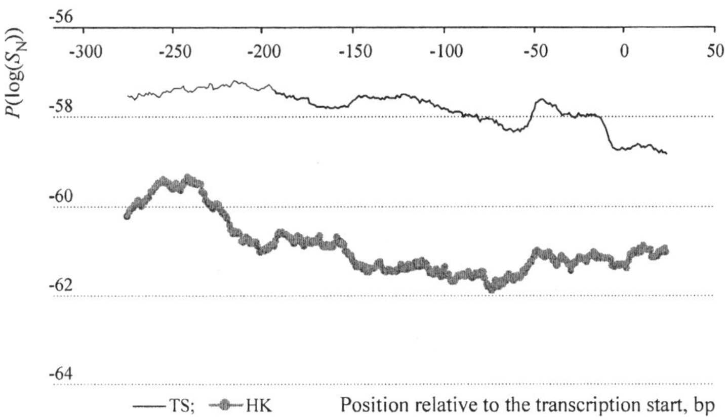

Figure -4. Nucleosome formation potential profiles in a sliding window of 50 bp for sets of eukaryotic gene promoters phased relative to the transcription start [–300; +50]. Averaged profile for the set of promoters of the housekeeping genes and tissue-specific genes are indicated by solid black and light gray lines, respectively.

The promoters of the housekeeping genes have smaller fit to nucleosome formation sites than promoters of tissue-specific genes. Thus, the results suggest that the genes expressed all the time in the nucleus should be devoid of tight chromatin packing unlike the tissue-specific genes, whose expression is inducible by other factors. These results are in good agreement with our previous data (Levitsky et al., 2001).

We have calculated the average nucleosome fit function for sets of exons and introns extracted from EID database (Saxonov et al., 2000). The results show that introns and 5'-untranslated gene regions have a higher nucleosome formation potential than exons (Figure 5).

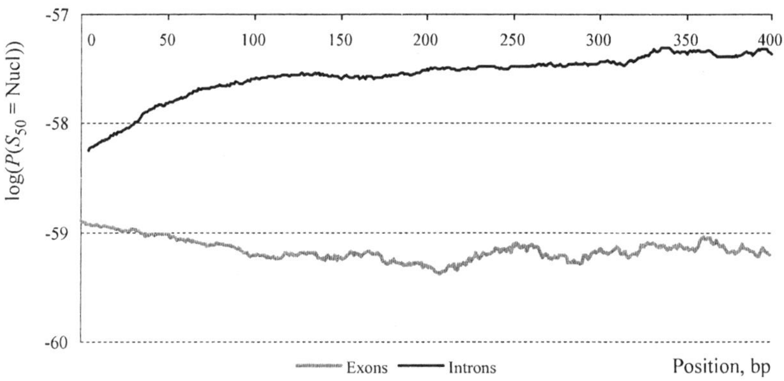

Figure -5. Averaged nucleosome formation potential profiles in a sliding window of 50 bp for sets of exons and introns extracted from EID database. The sequences were phased to 5′-end.

Moreover, introns have similar values of nucleosome formation potential as regulatory regions (promoters). See for comparison the profile for tissue-specific promoters in Figure 4. One can see an edge effect at the start point of both profiles in Figure 5. This is because the sequences were phased to 5′-end.

3.3 Nucleosome formation and nucleotide bias

Let us consider some factors that could be connected with the context code of nucleosome in genome. An interpolation of genetic codes—triplet code of amino acids, code of RNA structure, etc. (Trifonov, 1997)—is of great importance. The presence of different codes is connected with the text complexity, defined via nucleotide bias (entropy of letters in the text) or as a number of words (strings) in the text (linguistic complexity and other measures). We can define quantitatively the text complexity of a DNA sequence and compare these measures.

Complexity profiles for long genomic DNA sequences including complete human chromosomes were constructed using the Complexity program (Orlov and Potapov, 2004). Nucleosome formation potential of the same genomic sequences was estimated by the VMM program. The fit function $F(X)$ for nucleosome potential of sequence X is normalized as

$$F'(X) = \log(P(X))/\log(P(S_{\text{random}})),$$

where $P(S_{\text{random}})$ stays for probability by random for sequence of the same size. Figure 6 contains these sliding window profiles for the human globin gene cluster. A significant correlation of these parameters was established. For example, for the human globin gene cluster, the nucleosome fit function in a sliding window correlates with entropy of nucleotides (Figure 6). Correlation coefficient $r = 0.66$. However, correlation of the nucleosome fit function and text complexity by modified Lempel–Ziv method is also significant and equals 0.35.

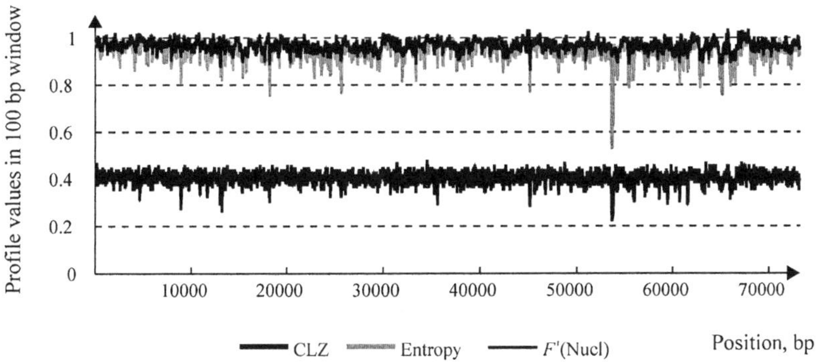

Figure -6. Profiles of nucleosome formation potential F' (normalized to the level expected for random sequences), text complexity CLZ (by Lempel–Ziv method), and entropy of nucleotides (normalized to [0;1] scale for human globin gene cluster (73 308 bp).

3.4 Transcription factor binding site analysis

Another field of research is analysis of transcription factor binding site (TFBS) sequences. Using the VMM program, we calculate the fit profile for sequences containing TFBS with flanks of a total length of 100 bp. The sequences were extracted from the TRRD database. Every set contains 15–50 sequences, which bind by a transcription factor.

We demonstrated that nucleotide sequences containing binding sites of HMG1, NFATP, and Oct transcription factors have the greatest nucleosome formation potential (Table 1).

Table -1. Averaged complexity and nucleosome formation potential for sets of transcription factor binding sites

TFBS	CLZ/N	Nuc	TFBS	CLZ/N	Nuc
HMG1	0.491	-55.34	RAR	0.492	-58.90
NFATp	0.500	-56.58	TRE	0.481	-58.93
Oct	0.504	-57.04	Pax	0.492	-59.01
DBP	0.508	-57.07	MyoD	0.494	-59.01
GATA3	0.495	-57.30	NFE2	0.499	-59.04
GRE	0.504	-57.53	NF-Y	0.499	-59.22
HNF1	0.511	-57.53	JunFos	0.502	-59.23
PPRE	0.506	-57.61	EKLF	0.478	-59.36
HNF3	0.510	-57.78	ATF1	0.499	-59.37
GATA2	0.492	-57.79	CREB	0.500	-59.38
Pu1	0.494	-57.80	p53	0.500	-59.42
CRX	0.512	-57.81	CRE	0.498	-59.44
cMyb	0.508	-57.83	Ets	0.494	-59.53
STAT1	0.500	-57.90	USF	0.495	-59.76
CEBP	0.502	-57.94	SRF	0.504	-59.78
Pdx1	0.519	-58.00	CLOCK/BMAL1	0.498	-60.01
NF1	0.501	-58.01	SRE	0.500	-60.16
GATA4	0.496	-58.02	AP2	0.481	-60.17
ER	0.494	-58.05	SP3	0.480	-60.33
STAT1,3,5	0.496	-58.10	E2F	0.486	-60.34
Ftz	0.516	-58.20	cKrox	0.464	-60.37
COUP	0.495	-58.25	SRE	0.500	-60.39
NF-κB	0.499	-58.27	EGR1	0.475	-60.43
SF1	0.500	-58.31	SP1	0.479	-60.58
HNF4	0.496	-58.46	Ebox_sre	0.502	-60.76
GATA1	0.481	-58.58	ARNT	0.494	-61.26
ATF2	0.499	-58.61	cMyc	0.491	-61.70
YY1	0.494	-58.68	HIF	0.491	-61.79
TTF1	0.505	-58.71	WLTF1	0.445	-61.89

CLZ/N: complexity according to the method by Lempel and Ziv expressed as a number of repeated fragments in sliding window. Nuc: logarithm of the probability of a sequence in sliding window to fit the nucleosome site model -log($P(S_{50})$). Sliding window size N is 50 bp.

Indeed, the capacity of these protein transcription factors to bind DNA packaged in a nucleosome structure has been demonstrated (Bagga et al., 2000; Belikov et al., 2004; Johnson et al., 2004). The TFBS are sorted by averaged nucleosome fit function. NF-κB represents the level expected by random (marked by gray in Table 1). The sites with a lower capacity to be within nucleosome structure could be characterized by the presence in proximal promoters and the corresponding transcription factors bind DNA free of nucleosome packing.

Thus, even for short DNA sequences with a length less than 146 bp, the method developed could predict nucleosome formation capacity. It is interesting to note that the text complexity of TFBS correlates with the nucleosome fit function (Table 1). This supports and extends the idea that a low text complexity of DNA sequences enhances nucleosome formation (Bolshoy et al., 1997).

Thus, the VMM software allows the advantageous use of context properties of a DNA sequence as a basis for predictions of nucleosome formation potential. It should be noted that context approach for nucleosome site prediction is not complete. The propensity of nucleosome formation could depend on the CpG methylation, DNA flexibility, non-specific DNA–histone interactions, and other factors. We will develop other computer methods for nucleosome positioning code analysis, including physicochemical properties of the DNA, 3D models, and histone octamer models.

ACKNOWLEDGMENTS

The authors are grateful to V.N. Potapov, S.V. Lavryushev, A.V. Katokhin, and D.P. Furman for helpful discussion and technical support. This work was supported in part by the RFBR (grants Nos. 03-04-48506, 05-04-49111, 05-07-90185, and 05-07-98012), CRDF (Y2-B-08-02), Siberian Branch of the Russian Academy of Sciences (Integration Project No. 119), and Presidium of the RAS Program on molecular physicochemical biology (grant No. 10.4).

DNA NUCLEOSOME ORGANIZATION OF THE PROMOTER GENE REGIONS

V.G. Levitsky[1*], A.G. Pichueva[1], A.V. Kochetov[1], L. Milanesi[2]
[1] *Institute of Cytology and Genetics, Siberian Branch of the Russian Academy of Sciences, prosp. Lavrentieva 10, Novosibirsk, 630090, Russia, e-mail: levitsky@bionet.nsc.ru;* [2] *Istituto di Tecnologie Biomediche Avanzate, via F. cervi 93, 20090 Segrate, Milano, Italy*
[*] *Corresponding author*

Abstract: DNA nucleosome organization is an important factor in gene expression patterning. The nature of the context signals determining nucleosome formation sites are not completely understood. Given these considerations, the relation between the nucleosome positioning and the regulation of gene expression appears of great interest. Taxon-specific nucleosome organization of the yeast and mammalian core promoters was identified. The context parameters of the DNA nucleosome organization differ by the distribution pattern in the mammalian and yeast promoters. We suppose that a common nucleosome positioning pattern may exist in promoters regions of mammalian genes. The programs used for nucleosome positioning analysis are available at http://wwwmgs.bionet.nsc.ru/mgs/programs/recon/, http://wwwmgs.bionet.nsc.ru/mgs/programs/phase/.

Key words: gene expression pattern; nucleosome positioning; periodic dinucleotide density; nucleosome formation potential

1. INTRODUCTION

An essential feature of eukaryotic DNA is its packaging in chromatin structure. Today, nucleosomes are recognized as highly dynamic units through which the eukaryotic genome can be regulated (Khorasanizadeh, 2004). The open chromatin represents regions that are poised for gene activity, perhaps as a precursor to remodeling chromatin to allow transcription (Gilbert et al., 2004). Strong experimental evidence is accumulating for nucleosome positioning in the promoter regions of the eukaryotic genes being of great importance in their regulation of transcription (Goriely et al., 2003).

The problem of contextual specificity of nucleosomal DNA (Trifonov, 1997) is of particular interest during the last 25 years. It is now known that nucleosome arrangement in the genomic DNA is determined by the interaction of different regulatory and structural proteins with their cognate sites and context nucleosome positioning signals. Taken together, the interactions form the code of chromatin nucleosome organization (Lowary and Widom, 1997; Kiyama and Trifonov, 2002).

We have previously shown that the nucleosome formation potential is greater in tissue-specific gene promoters than in the genes expressed in many tissues and housekeeping genes (Levitsky et al., 2001). Hence, the capability of nucleosome positioning in the promoter region may serve as a regulatory factor of the gene expression pattern.

A periodic occurrence of the dinucleotides AA and TT rendering DNA able to bend was discovered (Trifonov and Sussman, 1980). The presence of a pattern with consensus non-T(A/T)G, found in human exons and introns with periodicity of roughly 10 nucleotides (Baldi et al., 1996), was related with phased bending potential and nucleosome positioning.

Here, we present the results of a computational analysis of the characteristics potentially related to nucleosome positioning in the core promoters of the yeast and mammalian genes. The analysis allowed us to reveal the common pattern of nucleosome positioning characteristics in promoters of mammals.

2. MATERIALS AND METHODS

Saccharomyces cerevisiae mRNAs were extracted from the EMBL nucleotide sequence databank. Full-size 5′UTRs were selected from the entries containing a description of the transcription start sites (TSS) and complete coding regions. Only the mRNAs with the full size 5′UTRs (i.e., containing reference to an experimentally mapped TSS) were used. This resulted in a set of 5′UTRs of 240 yeast genes. To avoid bias due to redundant sequence data in statistical analysis, redundant sequences (coding sequence homology higher than 70 %) were removed. Finally, the set comprised 5′UTRs of 98 yeast genes with a single TSS and a complete coding sequence. A sample of promoter sequences spanning 150 nucleotides upstream of the major TSS was also compiled and combined with the corresponding 5′UTRs; 271 promoter sequences with a length of 1500 bp ([−1000; +500] with respect to the TSS) of the eukaryotic genes (mostly mammalian) were selected from the TRRD database. All the promoter sequences were aligned with the TSS.

The PHASE and RECON computer programs were used in comparative analysis of the nucleosome formation ability. The RECON program calculates the nucleosome formation potential (NFP; Levitsky et al., 2001; Levitsky, 2004). The program RECON was developed using discriminant analysis basing on the method of genetic algorithm utilizing the statistics of dinucleotide location within local regions of nucleosome formation sites.

The PHASE method is based on an estimate of the extent to which the occurrence frequencies of phased dinucleotides in real sequence are different from those expected in random.

Let the dinucleotide of j-type ($1 \le j \le 16$) be at the ith position of a sequence of sample $\{R\}$.

For each dinucleotide position in a sequence, the dinucleotides at distances of one or two helical turns (one turn is of 10.5 bp in average) are taken into consideration. Let us examine a sequence S which is L long and a set of 10 distances $\{D_n\} = \{\pm 10, \pm 11, \pm 20, \pm 21, \pm 22 \text{ bp}\}$ (five distances in the opposite direction from any sequence position 22 bp from the edges). Let us calculate the complete set of dinucleotide frequencies of the sequence S and then generate a sample of random sequences $\{R\}$ of the same dinucleotide content. For this sample, let us count the number of dinucleotides of the j type that have exactly k dinucleotides at distances $\{D_n\}$ k ($0 \le k \le N$). We evaluate the frequency of this occurrence as $p_j(k)$. Note that $p_j(0)$ probability means the absence of any periodic dinucleotides and $p_j(N)$ means the highest permissible by the sample $\{D_n\}$ periodicity. Then, the dinucleotide frequency f_j calculated only for the positions that are not located too close for both edges of sequence S (the distance is greater than 22 bp) may be presented as follows:

$$f_j = \sum_{k=0}^{N} p_j(k). \quad (1)$$

Next, we determine the function $w_j(k)$, which is the logarithm of the relative probability of periodic dinucleotide of the j type:

$$w_j(k) = -\lg \frac{p_j(k)}{f_j}. \quad (2)$$

Note that the function may be considered as the logarithm of a probability to observe a periodic *j*th dinucleotide of *k* type. Next for this function, we count the expected average by all possible *k* values:

$$w_j = -\sum_{k=0}^{N} \left(\frac{p_j(k)}{f_j} \lg \frac{p_j(k)}{f_j} \right). \tag{3}$$

Let the maximum value of $w_j(k)$ be $w_{\max}$ (most probably, it equals $w_j(N)$). Finally, we calculate the periodic dinucleotide density function (PDD_j) as follows:

$$PDD_j(k) = \frac{w_j(k) - w_j}{w_{\max} - w_j}. \tag{4}$$

Let us calculate the integral function *PDD* for the arbitrary dinucleotide set $\{J\}$ as follows:

$$PDD = \sum_{j \in \{J\}} PDD_j \,. \tag{5}$$

From Eq. (4) and Eq. (5) it follows that the maximum value of the functions PDD and PDD_j is 1, while the value 0 denotes the average value for random sequences, i.e., dinucleotides do not tend to be phased.

The PHASE program estimates the density of dinucleotides phased with helical turn periodicity in DNA sequences. i.e., first, for each dinucleotide of *j* type found in the analyzed sequence, a *k* number of phased dinucleotides are counted. Than, the score is calculated according to Eq. (4). The PHASE program yields the function PDD (periodic dinucleotide density). In the current study, we applied PDD only to AA and TT dinucleotides; the smoothing window size was 145 bp.

3. RESULTS

The profiles of the PDD and NFP for the promoter and 5′UTR regions of the yeast gene sequences aligned with the TSS are shown in Figure 1. Clearly, the PDD is maximal in the wide [–100; +50] region overlapping the TSS. The NFP has, as a rule, constant values in the upstream region and slightly increased, in the downstream.

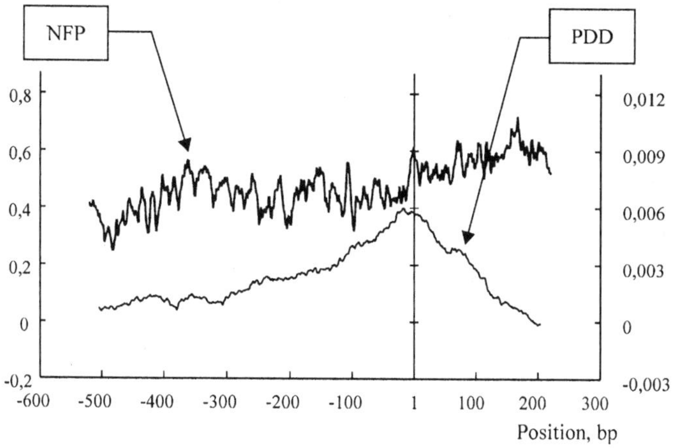

Figure -1. Profile of periodic dinucleotide density (PDD) of the AA and TT dinucleotides and nucleosome formation potential (NFP) for the promoter and 5'UTR regions of yeast genes.

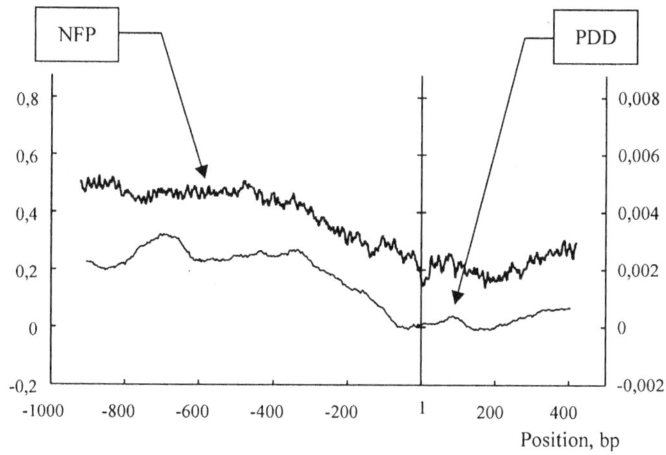

Figure -2. Profile of the periodic dinucleotide density (PDD) and the nucleosome formation potential (NFP) of the TRRD gene sequences (*Mammalia*).

Unlike yeast, the core promoters and downstream regions of *Mammalia* show other PDD and NFP profiles (Figure 2): both profiles start to decrease about 400 bp upstream of the TSS and remain low within the downstream region. Similar behavior of PDD and NFP profiles indicates that a common nucleosome positioning pattern may exist in promoters regions of *Mammalia* genes. The main feature of this pattern is a negative trend located from –400 to +1 bp relative to TSS (Figure 2). The density of nucleosome packaging is low in the proximal region [–100; +1] in comparison with the distant one [–1000; –400]; the region between these proximal and distant regions have transient behavior (Figure 2).

4. DISCUSSION

It is known that periodicities of the dinucleotides AA and TT are most important in nucleosome site formation. Particular periodicities of these dinucleotides are related to the DNA curvature, thereby facilitating nucleosome formation (Ioshikhes et al., 1996). The periodicities of AA and TT dinucleotides near the TSS may be related to DNA conformation. This DNA conformation is very important in transcription initiation.

The NFP data for *Mammalia* are in good accordance with our previous analysis. The negative trends within the proximal promoter region [–400;+1] for nucleosome positioning ability were deduced from RECON approach (Levitsky et al., 2001a) and on the basis of conformational and physicochemical properties (Levitsky et al., 1999). It may be supposed that there are common pattern of nucleosome positioning characteristics in promoters of higher eukaryotes. It should be noted that the NFP values in distant region are similar to intron NFP values (Levitsky et al., 2001b), which most likely characterize the typical noncoding genomic DNA. The NFP and PDD low values within the downstream region may be related with the presence of coding regions. We showed earlier that exons in comparison with introns had a low nucleosome positioning ability (Levitsky et al., 1999; 2001b).

The absence of common pattern for mammals and yeast *Saccharomyces cerevisiae* promoters may be related to some very known peculiarities of the yeast genome organization. There are several remarks about yeast *Saccharomyces cerevisiae* that may explain this: the absence of histone H1; short cell cycle; and very small genome size (quite similar to prokaryotic). Besides that, only a very small proportion of yeast genes (5 %) are interrupted by an intron (Spingola et al., 1999). Yeast introns are preferentially located at the 5′-end of the open reading frames (Spingola et al., 1999). Moreover, there are ~ 10 times more genes in mammals than in simple eukaryotes like yeast. The number of different tissue type follows the same rule. It may be concluded that for higher eukaryotes, unlike the yeast most of the time, the great majority of genes are silenced. The transcription processes and chromatin organization of the yeast *Saccharomyces cerevisiae* are quite different from those of the higher eukaryotes. We suppose that these specific features could influence the yeast genome organization and nucleosome formation potential.

ACKNOWLEDGMENTS

The work was supported by the RFBR (grants Nos. 03-07-90181, 05-04-49111, and 03-04-48469); Ministry of Industry, Science, and Technologies of the Russian Federation (grant No. 43.073.1.1.1501); Program of the Russian Academy of Sciences (Dynamics of Plant, Animal, and Human Gene Pools); Siberian Branch of the Russian Academy of Sciences (integration project No. 119); US Civilian Research and Development Foundation for the Independent States of the Former Soviet Union (CRDF); and the Ministry of Education of the Russian Federation under the Basic Research and Higher Education Program (grants Nos. REC-008 and Y2-B-08-02); MIUR (Laboratory of Interdisciplinary Technologies-LITBIO–RBLA0332RH); MCB RAS (grant No. 10.4); and NATO (grant No. LST.CLG.979816).

A TOOLBOX FOR ANALYSIS OF RNA SECONDARY STRUCTURE BASED ON GENETIC ALGORITHM

I. Titov[1, 2*], D. Vorobiev[1], A. Palyanov[1, 2]
[1] *Institute of Cytology and Genetics, Siberian Branch of the Russian Academy of Sciences, prosp. Lavrentieva 10, Novosibirsk, 630090, Russia, e-mail: titov@bionet.nsc.ru;*
[2] *Novosibirsk State University, ul. Pirogova 2, Novosibirsk, 630090, Russia*
[*] *Corresponding author*

Abstract: Versatile and rapid web-servers are timely for applied tasks requiring calculations of RNA secondary structure. We describe a web-server that integrates programs for analysis of RNA secondary structure. The server currently offers: solution of tasks of RNA folding and inverse folding based on the genetic algorithm; estimation of a sequence capability to form secondary structure; and an editor for interactive RNA folding with imposed constraints. Our server is publicly available at http://wwwmgs2.bionet.nsc.ru/mgs/systems/garna/.

Key words: RNA secondary structure; genetic algorithm; inverse folding

1. INTRODUCTION

Analysis of the RNA secondary structure is not a pure theoretical occupation. Its applied extensions are manifold including the search for non-coding RNAs, RNA design, search for effective hybridization oligonucleotides, etc. Today's most popular web-servers for calculations of RNA secondary structure are the mfold site (Zuker, 2003) and Vienna package site (Hofacker, 2003), which rely on dynamical algorithms. Genetic algorithm has been previously used for calculations of RNA secondary structure (Gultyaev et al., 1995; Currey and Shapiro, 1997; Titov et al., 2000a) intending to search for kinetic intermediates and to account of tertiary interactions. The genetic algorithms based on available programs were not fast enough to calculate the RNA structures via Internet. GArna was the computer program that provided the lacking rapidity.

We have developed a program set for analysis of RNA secondary structure on the base of a genetic algorithm (Titov et al., 2002). Here, we describe this integrated toolbox. Our web-server provides an online access to four programs: GAfold, MatrixSS, GAinverse, and GAedit. The programs can be used for (i) fast search of RNAs in genomic sequences; (ii) prediction of RNA secondary structure using predefined constrains; and (iii) design of antisense oligonucleotide or RNA sequence with a given structure. The interfaces for GAfold and MatrixSS programs are static html-pages. The GAinverse and GAedit programs were provided with the Java-interface. The Java-applets use Java Virtual Machine 4.1 and are supported by most browser versions. All the programs may be reached by the links from the entrance page http://wwwmgs2.bionet.nsc.ru/mgs/systems/garna, where the applications are briefly described and the user manual is provided.

2. GAfold

GAfold is a program designed for the calculation of RNA secondary structure on the base of a genetic algorithm. This unit is used as auxiliary by the GAinverse and GAedit programs. RNA sequence is the input. By default, GAfold calculates one of the low-energy secondary structures for a sequence as long as 250 nucleotides. Turner's free energy parameters (Jaeger et al., 1989) are used to calculate RNA secondary structure on the server. The other structures can be calculated by changing the algorithm parameters (see below). GAfold finally yields RNA secondary structure as a conventional secondary structure graph and in the ct-format.

RNA alternative stable structures or kinetic intermediates may be frequently of interest. Calculations for such structures can be done by changing parameters, such as initial random number, calculation time, and selection strength of evolutionary simulations. Random numbers set the trajectory of stochastic optimization; when random number is changed, the algorithm can converge to another RNA structure with similar energy. When calculation time (defined by stop criterion) or selection strength of the algorithm is reduced, the algorithm can find less stable structures.

A feature of GAfold is that it enables the estimation of nonrandomness of the input sequence relative to its potential of formation of RNA secondary structure. The nucleotides should be properly arranged in a sequence to provide the stability of RNA structure. This ordering is estimated by comparing the energy of the secondary structure of a given sequence with that of the sequences resulting from its random shuffling. This estimate removes the G/C composition effect on the secondary structure energy. The relative deviation (the Z-score) of the structural RNA energy was estimated as about −1.8 (Rivas and Eddy, 2000; Titov et al., 2002). GAfold outputs the Z-score value for an input sequence and its distribution parameters of the energy of random sequences with the same nucleotide composition and length. These distributions were previously calculated (Titov et al., 2002) and tabulated in the current GAfold program.

Table -1. Z-score value for certain RNA classes (Rivas and Eddy, 2000; Titov et al., 2000b; 2002)

RNA class	Random sequences	tRNAs	5SRNAs	5′UTRs
Average Z-score	0	−1.84	−1.81	−0.52

GAfold allows for incorporation of the simplest constraints on the secondary structure, prohibiting a given set of nucleotides from pairing with any other nucleotides. Preventing a string of consecutive nucleotides from pairing is a taken advantage for calculation of the secondary structure of RNA hybridized with a short oligonucleotide. The constraints of other types are itemized below in Section GAedit.

3. MatrixSS

Much has been achieved in deciphering genomic sequences. This made timely the search for non-coding RNAs in genomic sequences without explicitly incorporating calculation of their secondary structures into the algorithms. Despite the improved algorithms for prediction of RNA secondary structures, calculations of long RNAs are still unacceptably time-consuming for real-time computations via Internet. The MatrixSS program was designed for a less-specific purpose, namely, the search in sequences longer than 250 nt of subsequences that are capable of forming a stable secondary structure. The MatrixSS computes the nucleotide frequency characteristics, E-score, expressed as

$$\text{E-score} = 9G\bullet C + 3A\bullet U + 2G\bullet U,$$

where G, C, A, U < 1 are the frequency of a corresponding nucleotide in the sequence (Titov et al., 2002).

The correlation between the E-score and the secondary structure energy is considerably higher than between G+C, (G-C)/(G+C) or other known scores (Table 2).

Table -2. Correlation between secondary structure energy of random RNAs and nucleotide scores (Titov et al., 2002)

Nucleotide score	E-score	G•C	G+C
Linear correlation, r^2	0.89	0.82	0.37

The sequence input in MatrixSS is similar to that in GAfold. The MatrixSS outputs a symmetrical dot-like matrix of potential complementary interactions. Summing by the matrix columns gives the E-score profile that expresses the potential involvement of a subsequence in a secondary structure. Then, the secondary structure of a candidate subsequence can be calculated using the program GAfold.

4. GAinverse

RNA offers promising material for nanobiotechnology due to specificity of its complementary interactions. GAinverse makes possible computation of the RNA sequences that fold into a given secondary structure, thus becoming suitable for RNA design.

The GAinverse program extends the adaptive walk procedure implemented by the inverse folding server of the Vienna package. GAinverse uses the genetic algorithm operating over a sequence population. Our algorithm optimizes the thermodynamic probability of the target structure by maximizing the energy gap between the competitive structures and the desired target structure. The algorithm details are described elsewhere (Titov and Palyanov, 2004). The algorithm interface uses similar Java-applets as GAedit program (Figure 1, Section 5). The user predefines first the length of a yet unknown sequence, and the constraints on the secondary structure are imposed as in the case of RNAfold. Then by clicking 'Calculate sequence', the calculating module of the inverse folding program is started. The module yields the sought nucleotide sequence, which is displayed along with its secondary structure. This helps to redesign the sequence.

5. GAedit

GAedit is designed for interactive RNA folding. The program is 'a point-and-click' graphical interface for setting constrains on the secondary structure and the output of calculations. The user can apply the program for inputting experimental data to increase the accuracy of structure predictions. The data sources may be RNAs either subjected to enzymatic digestion or homologous on the basis of computer analysis.

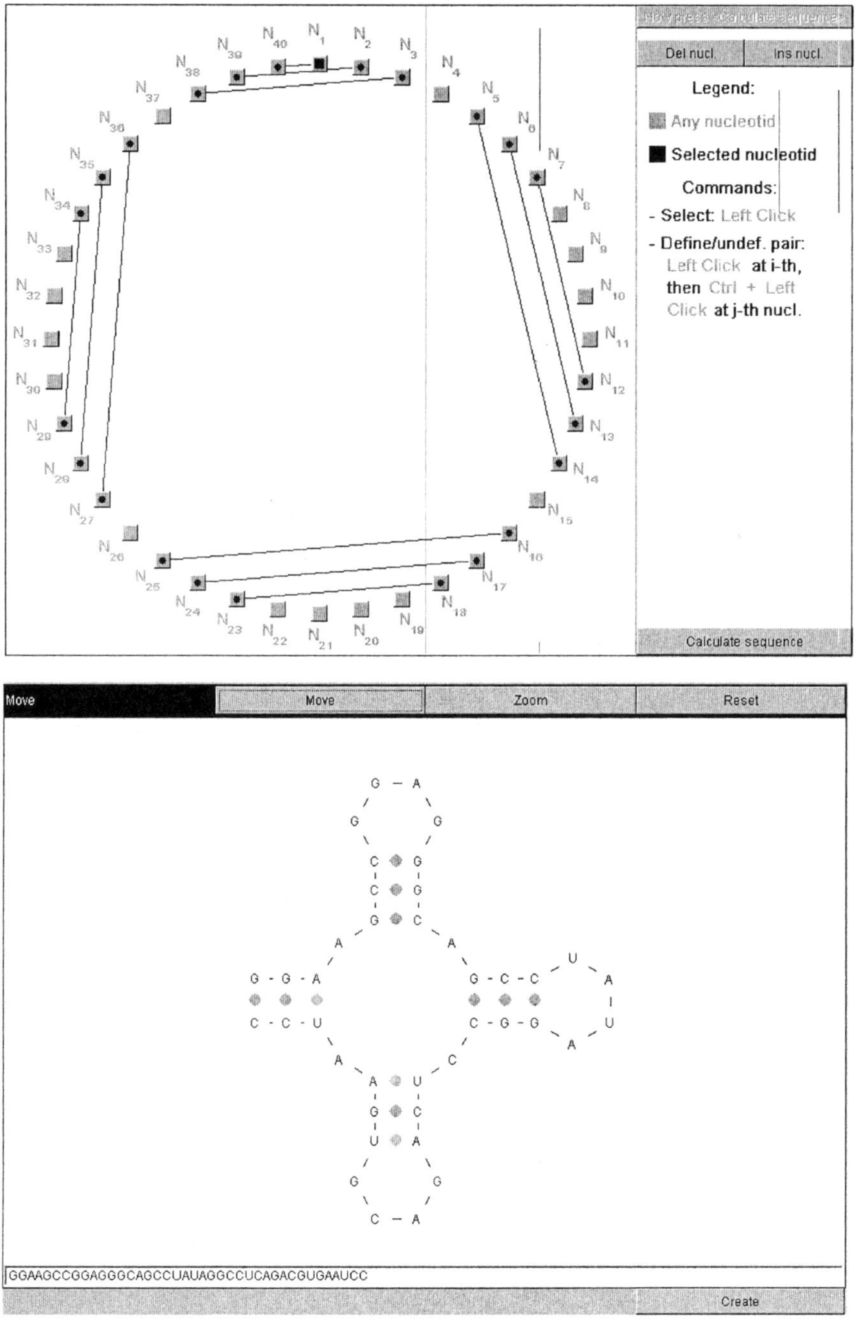

Figure -1. The GAedit interface.

The experimental data are taken into account by imposing the following constraints on the secondary structure: a) forcing the pairing of the complementary nucleotides *i* and *j*; b) forcing the pairing of a given nucleotide with an unknown partner; c) a given nucleotide will not pair with any other nucleotide.

The user inputs an *a priory* information about RNA structure as constraints of three kinds. Secondary RNA structure is calculated by minimizing the structure free energy using the genetic algorithm. Experimental data do not consistently provide unequivocal information about the state of a nucleotide. Using iterative calculations, the GAedit program allows verification of hypotheses of a nucleotide state for a query.

The editor window has a field for input of RNA sequence, controlling elements, and it displays RNA as an encircling nucleotide sequence (Figure 1). A click on the 'calculate structure' will provide calculation of the secondary structure and its display in the applet window. One can return to the editing mode to introduce changes by clicking 'create' and to reiterate the procedure until the desired results are obtained.

6. CONCLUSION

The task of analysis of RNA secondary structure is a milestone in the history of theoretical molecular biology. In the past decades, numerous computer programs have been developed to analyze RNA secondary structure and servers were created to predict structures via Internet. The distinguishing features of our GArna server are as follows: a) it is based on the genetic algorithm that can be tuned for searching both suboptimal and intermediate structures; b) it enables to perform a rapid estimate of RNA structure potential and then, to make the estimate more accurate by comparing the query with the random sequences; c) it offers a user friendly interface for interactive RNA folding and RNA design; and d) it incorporates the programs that solve the problems of RNA structure calculation, RNA design, and analysis at one site.

ACKNOWLEDGMENTS

The work is supported by the US Civilian Research and Development Foundation for the Independent States of the Former Soviet Union (CRDF), Ministry of Education of the Russian Federation within the Basic Research and Higher Education Program (Award No. REC-008), and Governmental contract with the Federal Agency for Science and Technology 'Identification of potential targets for novel medicinal drugs based on reconstructed gene networks' within the priority direction 'Living system'.

COMPARATIVE GENOMICS AND EVOLUTION OF BACTERIAL REGULATORY SYSTEMS

M.S. Gelfand[1, 2, 3*], A.V. Gerasimova[2], E.A. Kotelnikova[2], O.N. Laikova[2], V.Y. Makeev[2], A.A. Mironov[2, 3], E.M. Panina[1], D.A. Ravcheev[1, 3], D.A. Rodionov[1], A.G. Vitreschak[1]

[1] *Institute for Information Transmission Problems, Russian Academy of Sciences, Bolshoy Karetny pereulok 19, Moscow, 127994, Russia, e-mail: gelfand@iitp.ru;* [2] *State Scientific Center GosNIIGenetika, 1-j Dorozhny proezd 1, Moscow, 117545, Russia;* [3] *Department of Bioengineering and Bioinformatics, Moscow State University, Vorobievy Gory, 1-73, Moscow, 119992, Russia*

[*] *Corresponding author*

Abstract: Recent advances in genome sequencing and development of comparative genomics techniques allow one to study evolution of regulation in prokaryotes at several different levels: microevolution of orthologous regulatory sites, changes in regulon content, evolution of interacting regulatory systems, and co-evolution of transcription factors and their binding signals. Regulatory interactions appear to be very dynamic in some cases and surprisingly stable in others. The review presents several examples where comparative analysis uncovered plausible scenarios of evolution of regulatory systems.

Key words: regulation; evolution; transcription; binding site; riboswitch

1. INTRODUCTION

Comparative analysis of regulatory signals is a powerful tool for functional annotation of genomes (Gelfand, 1999). Based on the assumption of conservation of regulatory interactions, it uses two related but somewhat different approaches.

Phylogenetic footprinting assumes that regulatory sites evolve slower than the surrounding non-coding sequences and thus are seen as conservation islands in alignments of intergenic regions. The term was introduced in analysis of eukaryotes (Gumucio et al., 1992), where it is a much-used technique (Frazer et al., 2004) that sometimes even provides motivation for sequencing of genomes (Boffelli et al., 2003; Cliften et al., 2003; Kellis et al., 2003; Thomas et al., 2003). At the same time, until lately it had not been applied to the analysis of bacterial genomes, as no genomes at the suitable evolutionary distances were available. For several years, the only group allowing for such analysis was enterobacteria (Florea et al., 2003).

A related approach is *consistency filtering* of candidate sites that has been successfully applied to many bacterial regulatory systems (Gelfand et al., 2000). Despite the fact that in most cases it is impossible to construct a reliable recognition rule for transcription factor binding sites, simultaneous analysis of multiple genomes allows one to retain only the sites occurring upstream of orthologous genes (and thus, likely to be true). The false positives are scattered at random and thus can be ignored. This approach was applied to the analysis of many diverse systems, and allowed us to make a number of functional predictions that were subsequently confirmed in experiment (Rodionov et al., 2000, 2002 a, b; Makarova et al., 2001; Panina et al., 2001, 2003a, b; etc.).

However, this approach allows one to find only the conserved regulon cores retained at relatively large evolutionary distances. Sequencing of numerous genomes uniformly spanning the evolutionary space made it possible to study the taxon-specific regulation and evolution of regulatory systems (Gelfand and Laikova, 2003). In particular, availability of many closely related genomes allowed for the use of *phylogenetic shadowing* (Boffelli et al., 2003) for identification of prokaryotic regulatory sites that look like conservation islands in multiple alignments (Figure 1).

Evolution of regulatory sites has several aspects:

1. Evolution of sites regulating expression of orthologous genes;
2. Co-evolution of transcription factors and their binding signals;
3. Evolution of *regulons*, that is, sets of co-regulated genes; and
4. Evolution of interacting systems.

We cannot yet suggest a uniform theory, or even drafts of a theory; however, there exist a number of non-trivial observations that can serve as a raw material for creating such a theory.

Orthologous sites: unexpected conservation of non-consensus nucleotides. The traditional view on non-consensus nucleotides in transcription factor binding sites is that they represent random noise tolerated while the deviations from the consensus pass some threshold (Berg and von Hippel, 1988). A more complicated theory is that deviations from the consensus allow for activation or repression to occur at a fixed, gene-dependent level.

```
EC    AAA-GAGAAAAAAGCAGCAAACTTCGGTTGAAAAAGCCGCTATGATCGCCGGATAATCGTTTGCTTTTTTTA---
ST    AAA-GCATAAAAAGCGGCAAAGTTCAGTTGAAAAAGCGTTGATGATCGCTGGATAATCGTTTGCTTTTTTTTG--
YP    AAATGTATTAAATGTCGCATTCGGGTGTTGATTAGTCACCACTGATGGCTAGATAATCGTTTGCCTTAAATGACA
      *** *    *** *  ***       *****  *  *     **** **  ************* **   *

EC    -CCACCC--------GTTTTGT--------ATGCGCG----GAGCTAAACGTTTGCTTTTTTGCGACGCAGCA-A
ST    -CCACCC--------GTTTTGT--------ATACGTG----GAGCTAAACGTTTGCTTTTTTGCGGCGCCCCG-G
YP    TCTGCCCTAAACTTCGATTTTTTTTCAGTCATGCGTTCTCCCAGCTAATCGTTTGCTATTTTTCCCCGCTCTATG
       *  ***        * *** *        ** **       ****** ******** **** *  ***

EC    ATTGTCGCAAACCTGGA----------GCAGGAA-GATAACGTTTCGCTGGCAGGGGATTGTCCGCCACGCATCT
ST    -TTGTCAGTAATGTAGC----------ACAAGGA-GATAACGTTGCGCTGTTAGTGGATTACCTCCCACGTATAC
YP    AGTCAGGGAGAGTTAGTGAGTTCATCGACAGGAACGGAAACGATTACGTAGAGAAGGGCGCTTGGCTTGGCATGA
        *       *  * *            ** * * *  **** *    *      **        *   * **

EC    TGACGAAAATTAAACTCTCAGGGGATGTTTTCTTATGTCT------ACGCCATCAGCGCGTACCGGCGGTTCACT
ST    CGACGAATAATAAATTCTCAGGGGATGTTTTCT-ATGTCT------ACGCCTTCAGCGCGTACCGGCGGTTCACT
YP    CTATTTTAAATGA-CACACAGGGGACATCACC--ATGTCTAGCAGCAACCCTCAAGCACAGCCAAAGGGCACGCT
        *     * *   *******  *   *  ******      *  **   *** *   *    **  * ** *
```

```
                                                            -35box
                       [------FNR-------]              [---===FNR----[===]-NrdR---
EC    CCGTACGCTCTGCTTTTTACTTTGAGCTACATCAAAAAAAGCTCAAACATCCTTGATGCAAAGCACTATATATAG
ST    CTGTACGCTCTGATTTTTACCTTGTTCTACATCAATAAAATTGCAAACATCCTTGATGCAAATCACTACATATAG
KP    CCGTACTCTCACCTTTTTACCTTGTTCTGGGTCAATAAAATCGCAAACATCTTTGATGCAAATCACTACATATAG
      * **** ***   ******* ***  **   **** ****   ******** ********** ***** ******
      --] -10box       >[-----NrdR-----]
EC    ACTTTAAAATGCGTCCCAACCCAATATGTTGTATTAATCGACTATAATTGCTACTACAGCTCCCCACG--AAAAA
ST    ACTTTAAAATGCACGCCGACCCAATATGTTGTATTAATTGACTACAATTGCTACAACACCTGTTCACT--CGACA
KP    AACTTAAAATGCGCCTCGGCCCAACATATTGTATTAATCGTCTATTAT-GTCACCATATCTTGTCGATGTCTGGC
      *  *********    *  ***** ** ********** * *** ** *  ** * * **    *
                [-DnaA--]
EC    GGTGCGGCGTTGTGGATAAGC-GGATGGCGATTGCGGA-AAGCACCGGAAAACGAAACGAAAAAACCGGAAAACG
ST    CAAGGTGAATTGTGGATAACCTGGGTCAGGATTGCGGG-AAGTCATTGGAAAAGAGATGAATAAACCTGTTA-TG
KP    GGTGATGAGATGTGGATAAAACGGGCCGGATCCGAAGGTAAACAGCACGAGCCGTAGCGTGCAGCGCCTTCG-GG
         *  *   *********   **    *    *  *  **        *   *    *   *   *       *
                    [-DnaA--]
EC    CCTTTCCCAATTTCTGTGGATAACCTGTTCTTAAAAATATGGAGCGATCATGACACCGCATGTGATGAAACGAGA
ST    GCTTCCCCGGCCTCTGTGGATAACCTGTTCTTACAAATATGGAGTGATCATGACACCGCATGTGATGAAACGAGA
KP    ATAACCTCCGCCTCTGTGGATAACCTGTTCT---ATATATGGAGTGATCATGACACCGCATGTGATGAAACGTGA
           * *    *******************   * ******** *************************** **
```

Figure -1. Phylogenetic shadowing. Binding sites are set in boldface; promoter boxes, Shine–Dalgarno boxes; and genes, in italics. EC: *E. coli*, ST: *Salmonella typhimurium*, KP: *Klebsiella pneumoniae*, YP: *Yersinia pestis*. (Top) PurR binding sites upstream of *yjcD* genes in enterobacteria look like conserved islands. (Bottom) Multiple overlapping FNR, DnaA, and NrdR binding sites in the regulatory region of *nrdR* gene of *E. coli* and close relatives. Overlapping sites are shown by '='. Transcription start is marked by '>'.

However, analysis of binding sites regulating expression of orthologous genes demonstrated unexpectedly high conservation of non-consensus nucleotides (examples are shown in Table 1). Note that deviations from the consensus occur at different positions and thus cannot be explained by erroneous assignment of consensus nucleotides.

The simplest explanation for this phenomenon could be that insufficient time has passed for mutations that would revert a non-consensus position to the consensus state or change a non-consensus nucleotide to another non-consensus one. Indeed, if one considers very close genomes, e.g., different strains of the same species, coincidence of non-consensus nucleotides would be absolutely natural. However, statistical analysis demonstrated that the observed degree of conservation is much higher than the one expected under a neutral model (Kotelnikova et al., 2005).

This phenomenon was analyzed in two ways. Firstly, the degree of conservation in non-consensus positions was shown to be much higher than conservation in synonymous codon positions assumed the best available approximation to the neutrally evolving DNA. Secondly, ANOVA analysis demonstrated that dependence of the non-consensus nucleotide on the orthologous row of genes is higher than the dependence on the genome.

Table -1. Orthologous sites with conserved non-consensus nucleotides

Genome	Binding site	
	PurR site upstream of purL	PurR site upstream of purM
Escherichia coli	ACGCAAACGgTTtCGT	tCGCAAACGTTTGCtT
Salmonella typhi	ACGCAAACGgTTtCGT	tCGCAAACGTTTGCtT
Yersinia pestis	ACGCAAACGgTTtCGT	tCGCAAACGTTTGCcT
Haemophilus influenzae	AtGCAAACGTTTGCtT	tCGCAAACGTTTGCtT
Pasteurella multocida	ACGCAAACGTTTtCGT	tCGCAAACGTTTGCtT
Vibrio cholerae	ACGCAAACGgTTGCtT	ACGCAAACGTTTtCcT

Non-consensus nucleotides are shown by lower case boldface symbols; conserved non-consensus positions are underlined.

One possible explanation for this phenomenon is the following. Recent experimental studies demonstrated that the speed of gene activation depends on gene position in a metabolic pathway (Zaslaver et al., 2004).

Thus, the binding sites perform a fine-tuning of the regulation level by maintaining the gene-specific binding constant of the transcription factor. The latter depends on the site sequence and, in particular, on nucleotides in non-consensus positions. Thus, these positions are not neutral, and evolution of each particular site follows a rather narrow path dependent on the gene position in the metabolic pathway.

Regulons: plasticity of content. Comparison of even very close genomes demonstrates that point mutations can destroy a site and thus release a gene from regulation (Figure 2). Analysis of bacterial regulatory systems demonstrates that, beside the conserved core, many regulons contain taxon-specific members. A regulated gene may be genome-specific and absent in related genomes, or be released from regulation by a given factor.

An example is provided by the NadR regulon in enterobacteria. It is well studied in *E. coli*, where it includes the main NAD-synthesis genes—*nadA*, *nadB*, and *pncB*. However, even in very close genomes of *Yersinia* and *Erwinia* spp., the candidate binding sites are observed only upstream of *nadA*, but not other genes. On the other hand, these genomes have a conserved NadR-binding site upstream of the *nadR* gene itself, so that the latter is autoregulated. Thus, even relatively simple regulons covering essential metabolic pathways may be quite flexible.

In taxonomic groups evenly covered by sequenced genomes, one can study evolution of regulons in detail. Other interesting examples are the fructose, ribose, and purine regulons.

Consensus	ttGtACAagttaactaGTacaa
Escherichia coli	gtcgccga**ATGTACTAGAGAACTAGTGCAT**tagcttat
Salmonella typhimurium	accgcagg**ATGTACTAGTAAACTAGTTTAA**tggattgg
Yersinia pestis	gtcgtcggATGTTTTAACTAAATATTTTCAtgagtgat
Erwinia chrysanthemi	ctcgccgcATGTACTGATGGGTAACCGGCGctgaactg
Conserved positions	•++•••+ ++++•••+• •• •+ • • • •

Figure -2. Degeneration of TrpR binding site upstream of the trpH gene. The site region is set in capitals; functional sites, in boldface; and non-consensus nucleotides are underlined.

The fructose repressor FruR is a global regulator of the *E. coli* metabolism (Ramseier et al., 1995). However, in Vibrionaceae and Pasteurellaceae, it regulates only transport and metabolism of fructose. Preliminary analysis shows that expansion of the regulon occurred in the *E. coli* lineage.

A slightly more complicated story is that of purine and ribose repressors. The common ancestor of gamma-proteobacteria contained a ribose repressor that regulated the ribose catabolism operon; this state was retained in Pseudomonadaceae. Somewhere along the branch leading to Enterobacteriaceae, Vibrionaceae, and Pasteurellaceae, this repressor was duplicated. One copy (RbsR) retained the specificity towards ribose, but its DNA binding signal has changed. The other copy retained the signal, but changed the specificity, becoming the repressor of purine biosynthesis genes, PurR. Analysis of genomes from the latter three families demonstrated a gradual sliding of the regulon on the metabolic map (Ravcheev et al., 2002).

Interacting regulatory systems. Although regulation in bacteria is simpler than in eukaryotes, many genes are regulated by several factors and thus belong to several regulons simultaneously. In particular, this is a common feature of genes encoding enzymes belonging to several metabolic pathways (we do not discuss here a very non-trivial question of what set of reactions may constitute a pathway).

A somewhat more interesting situation occurs when one functional system is controlled by several regulators. Sometimes these regulators act independently, e.g., tryptophan attenuator and repressor TrpR of *E. coli.* In other cases, a complex functional system uses several regulators responding to different external stimuli.

One of examples of the latter kind is regulation of respiration in *E. coli* involving aerobic/anaerobic switch FNR, two-component regulator ArcAB, and nitrate/nitrite regulators NarPQ/NarLX. These regulators form different cascades in different genomes; orthologous and non-homologous isofunctional operons are regulated by different factors in different genomes (Table 2; Gerasimova et al., 2004). Similar observations were made for heat shock regulators in beta- and gamma-proteobacteria, regulated by specific sigma-factor σ^H and repressor HrcA (Permina and Gelfand, 2003b).

Table -2. Regulation of respiration in gamma-proteobacteria

Regulated gene	Regulator		
	FNR	ArcA	NarPQ/LX
fnr	E P V	- P V	- P -
arcAB	- E -	- E -	- - -
narL/narP	E P V	- - V	- - V
Escherichia coli (nuo)	FNR	ArcA	NarL
Yersinia pestis	FNR	ArcA	—
Yersinia entercolitica	FNR	—	—
Pasteurella multocida	FNR	ArcA	NarP
Actinobacillus actinomycetemcomitans	—	ArcA	NarP
Haemophilus influenzae	FNR	ArcA	—
Haemophilus ducreyi	FNR	ArcA	NarP
Vibrio vulnificus	—	ArcA	—
Vibrio parahaemolyticus	—	ArcA	—
Vibrio cholerae	FNR	ArcA	—
Vibrio fischeri	—	ArcA	—
Yersinia pestis	FNR	—	—
Yersinia entercolitica	FNR	ArcA	—
Pasteurella multocida	FNR	ArcA	—
Actinobacillus actinomycetemcomitans	FNR	—	NarP
Haemophilus influenzae	FNR	—	NarP
Haemophilus ducreyi	FNR	ArcA	NarP
Vibrio vulnificus	—	—	NarP
Vibrio parahaemolyticus	—	—	NarP
Vibrio cholerae	—	—	NarP
Vibrio fischeri	—	ArcA	NarP

(Top) Regulatory cascades. Notation: E, Enterobacteriaceae; P, Pasteurellaceae; and V, Vibrionaceae. (Middle) Respiratory chain operons *nuo* (*E. coli*) and *nqr* (other genomes). (Bottom) Molibdate cofactor biosynthesis operon *moa*.

Changes in regulatory systems. Even more radical path of the regulon evolution is a complete change in a regulatory system. For instance, zinc repressors in most genomes are homologous proteins ZUR, whereas they are absent in streptococci, and zinc repressor is the AdcR protein from a different family (Panina et al., 2003a).

One of the most remarkable examples of this kind is regulation of the methionine biosynthesis in firmicutes (Rodionov et al., 2004; Figure 3). The ancestral system, S-box riboswitch (Grundy and Henkin, 1998), exists not only in firmicutes, but also in some other taxa, in particular,

Actinobacteriaceae, Thermotogales, and some proteobacteria (*Xanthomonas* and *Geobacter*). S-boxes bind S-adenosyl-methionine, and the resulting change in the arrangement of heiices regulates premature termination of transcription. This system was retained in bacilli and clostridia and lost in the common ancestor of streptococci and lactobacilli: none of the extant genomes from these taxa contains S-boxes. The regulatory role in lactobacilli was assumed by a different RNA-based system, T-boxes, that normally regulate aminoacyl-tRNA-synthetase genes (Henkin, 1994). In streptococci, the methionine pathway is regulated by the transcription factor MtaR.

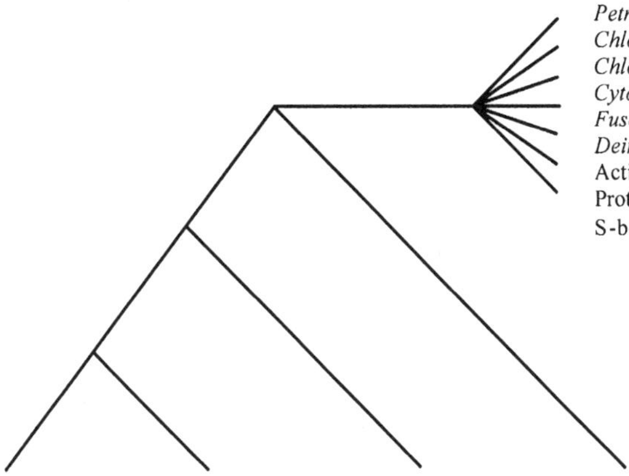

Figure -3. Regulation of methionine biosynthesis in firmicutes.

A similar situation occurs in the aromatic amino acid biosynthesis system, regulated by T-boxes, RNA-binding protein TRAP, and two unknown transcription factors whose signals have been identified by computational analysis (Terai et al., 2001; Panina et al., 2003b).

Co-evolution of regulators and signals. Analysis of protein–DNA interactions 'from the DNA point of view' demonstrated that the consensus positions, that is, the positions that clearly prefer one nucleotide, form significantly more contacts with transcription factors than non-consensus positions (Mirny and Gelfand, 2002). On the other hand, specificity-determining positions in transcription factor families cluster in three regions: the ligand-binding pocket, the subunit contact region, and DNA-binding helices (Kalinina et al., 2004).

Above, we have mentioned the ribose repressor RbsR, whose signal has changed in several gamma-proteobacterial families. Analysis of the LacI family of transcription factors, to which RbsR belongs, demonstrates that it is a common situation: factors with the same specificity form different branches, whereas signals are similar within branches (Gelfand and Laikova, 2003).

Sometimes, one can observe simultaneous changes in factors and signals. Transcription factors from the FNR/CRP family have similar sequences that allow for a reliable alignment. There are two groups of positions in the protein–DNA contact zone that demonstrate universal correlations (Table 3). In one such group, arginine in the protein yields TG in the binding signal, whereas in the second group, glutamate and one more arginine situated at the same side of the alpha-helix recognize the GA dinucleotide.

Table -3. Correlation between amino acid sequences of transcription factors from the FNR/CRP family and their signals

Genome	Factor	Fragment of protein alignment	Binding signal
DD	CooA	altteqlslhmgat**R**QtvsTllnnlvr	n**TG**TCGGCnnGCCGA**CA**n
DV	CooA	eltmeqlaglvgtt**R**QtasTllndmir	
EC	CRP	kitrqeigqivgcs**RE**tvg**R**ilkmled	T**TG**T**GA**nnnnnn**TC**A**CA**A
YP	CRP	kxtrqeigqivgcs**RE**tvg**R**ilkmled	
VC	CRP	kitrqeigqivgcs**RE**tvg**R**ilkmlee	
DD	HcpR	dvsksllagvlgta**RE**tls**R**alaklve	T**TG**T**gA**nnnnnn**Tc**A**CA**A
DV	HcpR	dvtkgllagllgta**RE**tls**R**clsrmve	
EC	FNR	tmtrgdignylgltV**E**tis**R**llgrfqk	nnTT**GA**TnnnnA**TC**AAnn
YP	FNR	tmtrgdignylgltV**E**tis**R**llgrfqk	
VC	FNR	tmtrgdignylgltV**E**tis**R**llgrfqk	

Correlated positions are shown by single- and double- underlined symbols. Genome notation: DD and DV: epsilon-proteobacteria *Desulfovibrio desulfuricans* and *Desulfovibrio vulgaris*, respectively; EC, YP, and VC: gamma-proteobacteria *E. coli*, *Yersinia pestis*, and *Vibrio cholerae*, respectively.

Another process forming the transcription signals is changes in the spacer length between halves of palindromic signals bound by dimeric factors. In particular, binding signals of the biotin repressor BirA in gram-positive bacteria and archaea, and in proteobacteria are similar and differ mainly by the size of the non-conserved spacer between the complementary half-sites (Figure 4*a*; Rodionov et al., 2002b). Sometimes, these two processes occur simultaneously.

Thus, zinc repressors from the ZUR family have similar signals in different taxa. However, in alpha-proteobacteria, one can observe point differences from the common superconsensus, whereas in gamma-proteobacteria, the spacer length is different (Figure 4*b*; Panina et al., 2003a).

a BirA

```
wwTGTtAAC   15-16   GTTaACAww    (gram-positive bacteria and archaea)
 ////////              \\\\\\\\
tTGTaAACC   15-16   GGTTtACAa    (gram-negative bacteria)
```

b ZUR

```
GaaATGTtA-----TAACATttC          (common superconsensus)
GAAATGTTAtantaTAACATTTC          (gamma-proteobacteria)
GAtATGTTA     TAACATaTC          (Rhodobacter spp.)
GtAATGTAA     TAACATTaC          (other alpha-proteobacteria)
```

Figure -4. Evolution of binding signals. (*a*) BirA signals in bacteria and archaea; (*b*) ZUR signals in proteobacteria. Lower-case letters: weakly conserved positions (BirA) and deviations from the common superconsensus (ZUR).

2. CONCLUSIONS

Analysis of regulatory systems and their evolution is currently at the level where the protein comparison and evolution was about twenty years ago. We know a number of interesting examples and see the direction of further studies. However, we are very far from a comprehensive, or even a draft theory describing the evolution of regulation.

ACKNOWLEDGMENTS

This study was supported by grants from HHMI (No. 55000309, M. Gelfand), RFBR (No. 04-04-49361, D. Rodionov), and the Russian Academy of Sciences (programs 'Molecular and Cellular Biology' and 'Origin and Evolution of the Biosphere').

COMPARATIVE WHOLE GENOME SEQUENCE ANALYSIS OF CORYNEBACTERIA

Y. Nishio[1], Y. Usuda[1], T. Gojobori[2], K. Ikeo[2*]
[1] *Institute of Life Sciences, Ajinomoto Co., Inc., Kawasaki, Japan;* [2] *Center for Information Biology and DNA Data Bank of Japan, National Institute of Genetics, Mishima, Japan, e-mail: kikeo@genes.nig.ac.jp*
[*] *Corresponding author*

Abstract: Complete genome sequences are available for three corynebacterial species: *Corynebacterium glutamicum*, which is widely used for industrial amino acid production by fermentation; *Corynebacterium efficiens*, which produces glutamic acid at a higher temperature than *C. glutamicum*; and *Corynebacterium diphtheriae*, which is a well-known pathogenic bacterium. Comparative genomic studies highlighted the evolutionary mechanisms underlying various aspects of the functional differentiation of these species, such as the unique metabolic features and thermostability of *C. efficiens*. The GC content of the *C. efficiens* genome amounted to 63.1 %, which was approximately 10 % higher than those of *C. glutamicum* and *C. diphtheriae* were. This difference was reflected in codon usage and nucleotide substitutions. Analyzing orthologous gene pairs with 60–95 % amino acid sequence identity between *C. efficiens* and *C. glutamicum* revealed a significant bias in amino acid substitutions. In particular, accumulations of three asymmetrical amino acid substitutions (lysine to arginine, serine to alanine, and serine to threonine) were associated with the thermostability and increased GC content of *C. efficiens*. A phylogenetic tree constructed using *Mycobacterium* and *Streptomyces* as outgroups indicated that the common ancestor of the corynebacteria was likely to have possessed most of the gene sets necessary for amino acid production. *C. diphtheriae* appeared to have lost the genes responsible for amino acid production. Glutamate overproduction in *C. glutamicum* was induced by a shortage of biotin, and this bacterium showed an incomplete biotin biosynthesis pathway. By contrast, *C. diphtheriae* might have acquired the complete biotin biosynthesis pathway by horizontal gene transfer. This process could have affected metabolic regulation in the corynebacteria following the loss of the glutamate overproduction mechanism in *C. diphtheriae*. Our findings suggest that dynamic genome evolution has been a motivational force for functional differentiation among the corynebacteria.

Key words: comparative genomics; *Corynebacterium*; evolution; fermentation; gene loss; glutamic acid production; horizontal gene transfer; protein thermostability

1. INTRODUCTION

Five primary elements of taste have been described: sweet, sour, salty, bitter, and umami. The last of these, umami, was originally discovered in glutamic acid as the source of the flavor of kelp, which is a type of seaweed. Glutamate is an amino acid that occurs naturally in food and is used throughout the world as a seasoning product. Glutamate is produced by a fermentation process using the Gram-positive bacterium *Corynebacterium glutamicum* for more than 50 years. The production of glutamate is increasing and now, the global yield exceeds one million tons per year (Kimura, 2003). Improving the production yield of glutamate is important not only economically but also in terms of the environment, as global warming, which might be caused by increased carbon dioxide levels, has become a serious problem. One approach to reducing the level of industrial carbon dioxide production is to improve the efficiency of fermentation. During the general fermentation process, the temperature of the fermenter is increased by heat that is generated by the growth of the bacteria (Figure 1). A chilling unit is therefore used to keep the fermenter at the optimal temperature. Electronic power, which is produced at power plants that generate carbon dioxide, is required to drive the chilling unit. The fermentation process for glutamate production using *C. glutamicum* is usually carried out at around 30 °C. Hence, if this fermentation process could be carried out at a higher temperature, it might be possible to reduce the associated electric power consumption and carbon dioxide generation (Adachi et al., 2003).

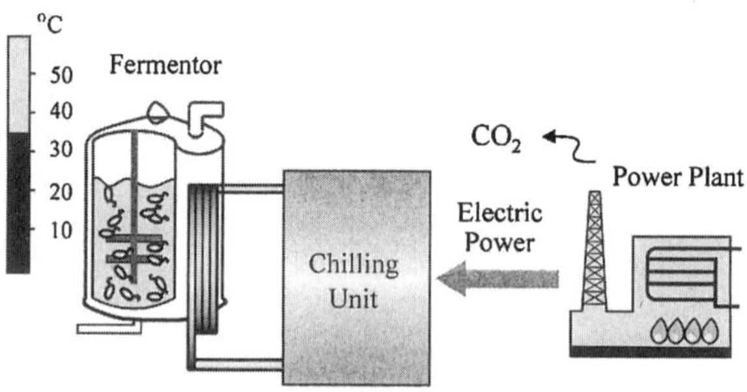

Figure -1. Heat generation in fermentation process.

Corynebacterium efficiens was originally isolated and identified as *Corynebacterium thermoaminogenes* by Yamada and Seto (1987). It was subsequently reclassified as a new species, *C. efficiens*, the nearest relatives of which are the glutamic acid–producing species *C. glutamicum* and *Corynebacterium callunae* (Fudou et al., 2002). *C. efficiens*, unlike *C. glutamicum*, can grow and produce glutamic acid at temperatures above 40 °C (Fudou et al., 2002). This feature of *C. efficiens* could help to decrease carbon dioxide production by reducing the energy needed to drive the chilling units during the fermentation process.

Differences in the growth temperature, protein stability, and GC content of *C. efficiens* and *C. glutamicum* can be investigated through comparative genomics using the complete genome sequences of these bacteria. Furthermore, as these two species are phylogenetically closely related, more than 1,000 orthologous genes with 60–95 % amino acid sequence identity can be compared individually. This is advantageous for a comparative genomic study—previous genome-wide comparisons between thermophilic archaea and mesophilic bacteria have been hindered by the fact that the amino acid residues did not correspond on a one-to-one basis.

This study focused on thermostability and, particularly, the various physiological characteristics that can be understood using a comparative approach. We aimed to elucidate the mechanism underlying the thermal stability of *C. efficiens* using a genome-wide comparison of amino acid substitutions. Our ultimate goal was to identify a general method for protein thermostabilization.

It is also important in applied biotechnology studies to understand the relevant metabolic pathways and their evolutionary history. The glutamic acid–producing species of corynebacteria are known to overproduce glutamic acid under a variety of conditions, such as biotin limitation (Kimura, 2003), although the mechanism of this phenomenon remains unclear.

Another member of this genus, *Corynebacterium diphtheriae*, is a well-known pathogen that does not produce glutamic acid. It is therefore of great interest to investigate the evolutionary processes that are related to the glutamic acid overproduction mechanisms in *C. glutamicum* and *C. efficiens* through considering the genome evolution of high GC Gram-positive bacteria.

Here, we discuss the evolutionary mechanisms involved in the differentiation of metabolic pathways and their regulation based on a comparative genome analysis of high GC Gram-positive bacteria, including *Mycobacterium* and *Streptomyces*.

2. METHODS AND ALGORITHMS

The complete genome sequences of *C. efficiens* (Nishio et al., 2003), *C. glutamicum* (Ikeda and Nakagawa, 2003; Kalinowski et al., 2003), *C. diphtheriae* (Cerdeno-Tarraga et al., 2003), *Mycobacterium tuberculosis* (Cole et al., 1998), *Mycobacterium leprae* (Cole et al., 2001), and *Streptomyces coelicolor* A3(2) (Bentley et al., 2002) were obtained from DDBJ/EMBL/GenBank (accession numbers BA000035, BA000036, BX248353, AL123456, AL450380, and AL645882, respectively). Orthologous genes were defined as the best pair of homologues identified in comparisons between two organisms (Tatusov et al., 1997). The BLAST (Altschul et al., 1997) and FASTA (Pearson, 2000) programs were used for database searches and ClustalW (Thompson et al., 1997), for multiple alignments. Phylogenetic trees were constructed using the neighbor-joining method with the p-distance or Kimura's distance (Saitou and Nei, 1987).

3. RESULTS AND DISCUSSION

3.1 Thermostabilization in *C. efficiens*

The features of the three corynebacterial species are summarized in Table 1. The genomes of these bacteria consist of a singular circular chromosome and several plasmids. In order to gain an overview of the corynebacterial genome structure, we compared the GC content profiles of the three species. *C. glutamicum* had a GC content between 50 and 60 % in most regions of the chromosome, with an average GC content of 53.8 %. By contrast, the average GC content of *C. efficiens* was 63.1 %, which was higher than that of *C. glutamicum*, over the entire chromosome. *C. diphtheriae* was used as an outgroup of *C. efficiens* and *C. glutamicum*. The window-analysis profile of the GC content of *C. diphtheriae* was more similar to that of *C. glutamicum* than to that of *C. efficiens*. This suggests that the ancestral genome structure of the corynebacteria might be closer to that of *C. glutamicum* than to that of *C. efficiens*. In addition, this implies that the thermostability of *C. efficiens* might have been acquired after divergence from the common ancestor of *C. glutamicum* and *C. efficiens*. In order to estimate the mutation that was responsible for the protein thermostability of *C. efficiens*, we analyzed asymmetrical amino acid substitutions between *C. efficiens* and *C. glutamicum*. Orthologous open reading frames (ORFs) with amino acid sequence identities of less than 60 % were omitted from the analysis, because of the large distance sequences (p-distance value = 0.4) and the need to take account of backward and parallel mutations (Nei and Sudhir, 2000). Thus, a total of 1,619 orthologous pairs of genes with identities ranging from 60 to 95 % (p-distance value = 0.2) were used to examine position-specific mutations. The most frequently observed asymmetrical amino acid substitution between *C. glutamicum* and *C. efficiens* was from Lys in *C. glutamicum* to Arg in *C. efficiens*. The next most common substitutions were Ser→Ala, Ser→Thr, and Ile→Val. Amino acid substitutions between Leu, Ile, Val, and Met are relatively common in nature. Thus, because the Ile→Val substitution is observed frequently in situations that are independent of thermostabilization, the asymmetrical substitutions (Lys→Arg, Ser→Thr, and Ser→Ala) were assumed to be specific amino acid substitution patterns between *C. efficiens* and *C. glutamicum*. Accordingly, these were judged to be the best candidates for protein thermostabilization. Many previous studies suggested that the Lys→Arg substitution might affect thermal stability (Vieille and Zeikus, 2001). Thus, if the evolutionary development of the thermal stability of proteins is responsible for the overall thermostability of *C. efficiens*, then these amino acid substitutions can be viewed as adaptive mutations that lead to overall thermostability.

Table -1. Summary of characteristics of the three corynebacterial species

	C. efficiens	*C. glutamicum*	*C. diphtheriae*
Upper temperature limit for growth (°C)	45	40	–
Glutamate production at 32 °C (%)*	80	100	–
Glutamate production at 3 °C (%)	78	40	–
Genome size (bp)	3,147.090	3,309.401	2,488.635
GC content (%)	63.1	53.8	53.5
Number of predicted genes	2.942	3.099	2.320

* Glutamate production in typical experiments using the biotin-limitation method expressed as a percentage of the production by *C. glutamicum* at 32 °C.

In a separate study, the thermal stability of 13 paired enzymes from the glutamic acid and lysine biosynthetic pathways of the two species were compared on the basis of the enzymatic activities remaining after heat treatment of the crude extracts. For each of the 13 sequences, the numbers of each of the three types of substitution within the amino acid sequence were assigned points depending on their mutational directions. The number of calculated points was then compared with the experimental results of the thermal stability test for each enzyme (Table 2). We assumed that if a point was positive, the protein in *C. efficiens* was more thermostable than that in *C. glutamicum*. If a point was negative, the protein in *C. glutamicum* was more thermostable than that in *C. efficiens*. Overall, 9 out of the 13 enzymes with measured thermostabilities agreed with the calculated points, three could not be determined, and only one did not coincide with the predicted value (Table 2). These results imply a significant correlation between the three types of amino acid substitutions and the thermal stability of the proteins. Thus, the thermostability of *C. efficiens* is proposed to be the result of the increase in GC contents after divergence from its sister species.

Table -2. Test of predicted values against actual measurements

Entry	Enzyme	Thermostable species	Point[1]	Result[2]
1	2-Oxoglutarate dehydrogenase	*C. efficiens*	0	–
2	Glutamate dehydrogenase	*C. efficiens*	1	Yes
3	Isocitrate lyase	*C. efficiens*	2	Yes
4	Phosphofructokinase	*C. efficiens*	–3	No
5	Fructose-1-phosphate kinase	*C. efficiens*	5	Yes
6	Isocitrate dehydrogenase	*C. efficiens*	4	Yes
7	Aconitase	*C. efficiens*	0	–
8	Phosphoenolpyruvate carboxylase	*C. efficiens*	10	Yes
9	Citrate synthase	*C. efficiens*	2	Yes
10	Aspartate kinase	*C. glutamicum*	–1	Yes
11	Dihydrodipicolinate synthase	*C. efficiens*	0	–
12	Diaminopimelate dehydrogenase	*C. glutamicum*	–2	Yes
13	Diaminopimelate decarboxylase	*C. efficiens*	2	Yes

[1] Defined as the difference between the sum of the three types of substitution from *C. glutamicum* to *C. efficiens* (Lys→Arg, Ser→Ala, and Ser→Thr) and the sum of the reverse substitutions (Point = {number of (Lys→Arg + Ser→Ala + Ser→Thr)} – {number of (Arg→Lys + Ala→Ser + Thr→Ser)}). [2] The results were classified as follows: 'Yes' indicated that the enzyme from *C. efficiens* was more thermostable and the point was positive, or that the enzyme from *C. glutamicum* was more thermostable and the point was negative; '–' indicated that the point was zero; and 'No' indicated all other cases.

3.2 Differentiation of metabolic pathways within the corynebacteria: genome evolution, gene gain, and gene loss

The *C. diphtheriae* genome comprises 2,488.635 base pairs (bp) and is smaller than the genomes of the other two corynebacterial species (Cerdeno-Tarraga et al., 2003). The evolutionary origin of this small genome could have involved either massive gene loss in *C. diphtheriae* or massive gene acquisition by the other two species.

In order to clarify the evolutionary event responsible for this difference, we identified the common orthologous genes among all five species of high GC Gram-positive bacteria (*C. efficiens*, *C. glutamicum*, *C. diphtheriae*, *M. tuberculosis*, and *S. coelicolor*) using the reciprocal best-hit method with BLAST (Mineta et al., 2003). We also identified the orthologous genes that were common among any four of these species when one corynebacterial species was excluded. *M. tuberculosis* and *S. coelicolor* were used as outgroups. A total of 748 orthologous genes were identified among all the five bacteria: this value was increased to 768 by the exclusion of *C. glutamicum*; to 773 by the exclusion of *C. efficiens*; and to 831 by the exclusion of *C. diphtheriae*. These results suggest that *C. diphtheriae* lost many of the orthologues that were present in the other four bacteria after it diverged from the common ancestor of the corynebacteria, and this event was the main factor for the loss of amino acid biosynthesis pathway (Nishio et al., 2004).

Considering the evolutionary changes among these three corynebacterial genomes, a dynamic change in the GC content was observed in *C. efficiens*, massive gene loss was detected in *C. diphtheriae,* and gene acquisition by horizontal gene transfer was found in *C. glutamicum* (Ikeda and Nakagawa, 2003; Kalinowski et al., 2003). The order of orthologous genes in the corynebacteria is well conserved, which was attributed to the lack of a RecBCD pathway in the homologous recombination system (Nakamura et al., 2003).

Thus, the genome rearrangement in corynebacteria might not contribute to the genome evolution and functional differentiation. A detailed comparison of the metabolic pathways and gene content of the corynebacteria indicated that the common ancestor had the ability to overproduce amino acids, including glutamate, which must therefore have been lost in *C. diphtheriae*. Differences between the metabolic pathways of the corynebacteria could result from genome evolution and particularly from gene loss and horizontal gene transfer (Nishio et al., 2004). These processes might affect not only the metabolic pathway organization, but also metabolic regulation in the corynebacteria.

One biologically important characteristic of *C. glutamicum* is the biotin requirement for growth, which is closely related to glutamate overproduction (Kimura, 2003). This biotin requirement was also observed in *C. efficiens*. Both these bacteria lack the complete biotin biosynthesis pathway from pimelate to biotin (Figure 2).

Pimelate —*bioW*→ Piomelonyl-CoA —*bioF*→ 8-Amino-7-Oxononanonate —*bioA*→ 7,8-Diamino-nonanonate —*bioD*→ Dethiobiotin —*bioB*→ Biotin

Figure -2. Biotin biosynthesis pathway.

By contrast, *C. diphtheriae* has the complete biotin biosynthesis pathway. In addition, DIP1381, encoding 6-carboxyhexanoate-CoA ligase (BioW), which is the first enzyme in biotin biosynthesis, might have been acquired by horizontal gene transfer in this species (Table 3). This is suggested by the fact that none of the bacteria that are closely related to *C. diphtheriae* possesses orthologues of DIP1381. BirA is a bifunctional protein that exhibits biotin ligase activity and acts also as the DNA-binding transcriptional repressor of the biotin operon, which is conserved in many organisms. The regulatory sequence of BirA might be conserved among many bacteria (Rodionov et al., 2002). However, the corynebacteria have lost the DNA-binding region in the orthologous *birA* gene.

Table -3. Biotin biosynthesis genes in high GC Gram-positive bacteria

	bioW	*bioF*	*bioA*	*bioD*	*bioB*
C. glutamicum			Cgl2604	Cgl2605	Cgl0072
C. efficiens			CE1421	CE1420	CE0089
C. diphtheriae	DIP1381	DIP1382	DIP1191	DIP1189 DIP1192	DIP0105 DIP1124
M. tuberculosis		Rv0032 Rv1569	Rv1568	Rv1570	Rv1589
M. leprae		ML1217	ML1216	ML1218	ML1120
S. coelicolor		SCO1243	SCO1245	SCO1246	SCO1124

Glutamate overproduction in *C. glutamicum* is induced by a shortage of biotin (Kimura, 2003). However, the regulatory sequences that are associated with the biotin biosynthesis–related genes and glutamate production remain to be identified. Comparing the regulatory regions of glutamate overproduction–related genes between glutamic acid producing and non-producing species might help to elucidate the regulatory mechanism of glutamate production. In *C. glutamicum,* a lack of biotin attenuated the 2-oxoglutarate dehydrogenase complex (ODHC) activity and the initiation of glutamate production simultaneously (Kawahara et al., 1997; Shimizu et al., 2003). By contrast, enhanced glutamate dehydrogenase (GDH) activity might not contribute to glutamate production (Shimizu et al., 2003) and showed no response to biotin limitation (Kawahara et al., 1997). The *odhA* gene, which encodes the OdhA subunit of the ODHC, has a lineage-specific structure in the corynebacteria and mycobacteria (Usuda et al., 1996), while the structure of the *gdh* gene is common among a wide range of bacteria (Bormann et al., 1992). Here, we focused on the conservation of the regulatory regions of these genes among the corynebacteria. The regulatory regions of *odhA* gene were more strongly conserved between *C. efficiens* and *C. glutamicum* than between *C. diphtheriae* and either *C. glutamicum* or *C. efficiens* (Figure 3). The accumulation of mutations in *C. diphtheriae* might explain this pattern of conservation. By contrast, the regulatory regions of *gdh* gene were equally conserved among all three species (Figure 3). These results suggest that the decreased ODHC activity induced by biotin limitation might be regulated at the gene expression level. Moreover, the loss of the glutamate overproduction ability in *C. diphtheriae* might have originated with the acquisition of the complete biotin biosynthesis pathway through horizontal gene transfer. In this case, it is assumed that the important parts of the regulatory regions were conserved despite the differences in the genome GC content, which was 10 % higher in *C. efficiens* than in either *C. glutamicum* or *C. diphtheriae* throughout the genome (Nishio et al., 2003).

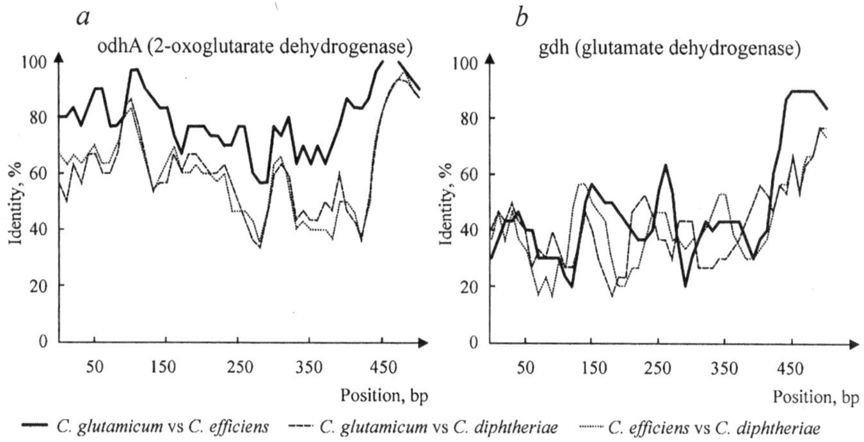

Figure -3. Window analysis of the regulatory region: (*a*) *odhA* gene and (*b*) *gdh* gene. The 500-bp sequence upstream of the start codon of each gene was analyzed according to the 30-bp window size and 10-bp step size. After alignment of the regulatory region plus the coding region, the gaps were removed from the multiple alignment and the identity was calculated.

In conclusion, the differentiation of the metabolic pathways among the corynebacteria appears to be caused by dynamic genome evolution involving not only amino acid substitutions, but also gene loss and gene gain. This comparative genomics study indicates that the dynamic genome evolution within the corynebacteria is associated with the major biological features of each species, that is, glutamate overproduction in *C. glutamicum*, thermostability in *C. efficiens*, and pathogenesis in *C. diphtheriae*. Further comparative studies, particularly of gene expression and metabolic regulation, will help to realize an ecological fermentation process.

ACKNOWLEDGMENTS

Tables 1 and table 2 were modified with permission from Cold Spring Harbor Laboratory Press© 2003 (Nishio et al., 2003).

NEW LTR RETROTRANSPOSABLE ELEMENTS FROM EUKARYOTIC GENOMES

O. Novikova[1*], M. Fursov[2], E. Beresikov[3], A. Blinov[1]
[1] *Institute of Cytology and Genetics, Siberian Branch of the Russian Academy of Sciences, prosp. Lavrentieva 10, Novosibirsk, 630090, Russia, e-mail: novikova@bionet.nsc.ru;*
[2] *Novosibirsk Center of Information Technologies 'UniPro', prosp. Lavrentieva, Novosibirsk, 630090, Russia;* [3] *Hubrecht Laboratory, Netherlands Institute for Developmental Biology, st. Uppsalalaan Utrecht, The Netherlands*
[*] *Corresponding author*

Abstract: LTR retrotransposons have been identified in the genomes of many eukaryotic organisms. Several approaches can be used in order to identify new LTR retrotransposons from a wide range of organisms. One of them is a systematic search of genomic sequences generated from whole genome sequencing projects. Here, we performed a computer analysis of 14 genomes with the aim of extracting LTR retrotransposons using a software based on the hidden Markov models. Altogether, 24 new LTR retrotransposable elements from 4 subclasses Ty3/gypsy, DIRS, Bel, and Ty1/copia were identified in 9 different genomes. A new family of retroelements was obtained after the analysis of the genomes of ascomycetes and basidiomycetes fungi. Considerable differences in the amount and diversity of retroelements were detected in the distantly related nematodes *B. malayi* and *C. briggsae*. Finally, our analysis did not reveal the presence of LTR retroelements in the genomes of parasitic protozoan.

Key words: mobile elements; LTR retrotransposons; distribution; evolution; computer analysis

1. INTRODUCTION

Retrotransposons comprise a significant fraction of many eukaryotic genomes. They constitute mobile genetic elements, which propagate themselves by reverse transcription of an RNA intermediate. Two major classes of retrotransposons that differ structurally and mechanistically have been identified: (1) LTR retrotransposons, which present long terminal repeats (LTRs) and have a transposition mechanism similar to that of retroviruses, and

(2) non-long terminal repeat (non-LTR) retrotransposable elements, which do not present terminal repeats and employ a simpler target-primed reverse transcription (TPRT) mechanism for retrotransposition (Luan et al., 1993; Malik and Eickbush, 2001).

LTR retrotransposons can be grouped into four subclasses: Ty3/gypsy, Ty1/copia, Bel, and DIRS (Havecker et al., 2004). Each subclass presents a distinctive structural organization of open reading frames (ORFs) and enzymatic modules. The majority of LTR retrotransposons share structural features and contain long flanking terminal repeats, which are present in the same orientation, and an internal region, which generally contains two open reading frames (ORFs). One of the ORFs encodes the Gag protein, which is a constituent of cytoplasmic particles within which the reverse transcription takes place. The second ORF, *pol* ORF, encodes a protein with different enzymatic activities, including reverse transcriptase (RT), ribonuclease H (RNase H), integrase (Int), and protease (Pro) activities, which in most elements cleaves the Pol polyprotein.

An initial phylogenetic analysis of retrotransposons based on reverse transcriptase sequences revealed that LTR retrotransposons are more recent than non-LTR retrotransposons (Xiong and Eickbush, 1990). Subsequent phylogenetic analysis based on the RT, RNase H, and Int domains indicated that the Ty3/gypsy LTR retrotransposons could be clearly subdivided into several clades (Malik and Eickbush, 1999).

At the moment, it is clear that the genomes of most organisms contain LTR retrotransposons that originated from multiple distinct lineages. The sequencing of eukaryotic genomes has further revealed the diversity among LTR retrotransposons. In the present study, we developed a novel approach for retrieval of LTR retrotransposons and were able to isolate new LTR retrotransposons from various available genomes. In our analyses, we have selected nine genomes representative of the main eukaryotic groups, such as fungi, chordates, nematodes, and insects. In these organisms, the LTR retrotransposons have not been extensively investigated or not studied at all. As a result, we were able to identify 24 new LTR retrotransposons in the analyzed genomes.

2. METHODS AND ALGORITHMS

The HMMER 2.1.1 (Eddy, 1998; http://hmmer.wustl.edu/) software was employed to identify sequences of LTR retrotransposable elements. The multiple sequence alignment of reverse transcriptase sequences present in the LTR retrotransposons was performed with CLUSTAL W software (Thompson et al., 1994). Genome sequences were obtained from the available databases. A six-frame translation of all the sequences was generated. The results of the HMMER searches, with scores above zero, were analyzed using the specially designed scripts that group the identified sequences into families and classify them based on the similarity to known retrotransposons and the number of stop codons (Berezikov et al., 2000). A given sequence was classified as a putatively new LTR retroelement, if there were no known LTR retrotransposons from the same organism in the family into which the sequence was assigned after the analysis.

Phylogenetic trees were generated by neighbor-joining method using the MEGA2 software package (Kumar et al., 2001). Statistical support for the trees was evaluated by bootstrapping in (100 replications; Felsenstein, 1985). The sequences of known LTR retrotransposons and retroviruses were extracted from the GenBank database, namely, *Maggy* (L35053), *grh* (M77661), *REAL* (AB025309), *Pyggy* (AF533703), *MGLR-3* (AF314096), *skippy* (L34658), *Cgret* (AF264028), *CfT-1* (AF051915), *sushi-ichi* (AF030881), *Tf2* (L10324), *Ty3* (L28920), *Skipper* (AF049230), *gypsy* (X72390), *mag* (X17219), *CER-1* (U15406), *mdg3* (X95908), *Ylt1* (AJ310725.2), *MarY1* (AB028236), *osvaldo* (AJ133521), *Tetraodon* (AF442732), *TED* (M32662), *ZAM* (AJ000387), *Saci-2* (BK004069), *Saci-3* (BK004070), *moose* (AF060859), *Pao* (AB042119), *SURL* (M75723), *celegmag1* (Z38112), *copiamel* (M11240), Ty1 (S40908), *FFV* (feline foamy virus; NC_001871), *HIV1* (human immunodeficiency virus 1; NC_001802), *HIV2* (human immunodeficiency virus 2; NC_001722), and *ASV* (avian sarcoma virus; NC_001618) and from the database of the Genetic Information Research Institute (http://www.girinst.org/), namely, *Romin1* (ROMIN1_I), *del* (DEL_LH), *Peabody* (PEABODY_I), *mdg1* (MDG1_I), *Zebra* (DIRS1_DR), *Cigr-1* (Cigr-1-I), *Bel* (BEL_I), *ninja* (NINJA_I), *DIRS-1* (DIRS1), *Tas* (TAS_I), *Pat* (PAT), and Cer13 (CER13-I_CE).

3. RESULTS AND DISCUSSION

Our aim was to identify new LTR retrotransposons in the nine chosen genomes. A list of the analyzed genomes and the results of our bioinformatics analysis are summarized in Table 1. Our study resulted in the identification of 24 new LTR elements.

New LTR retrotransposable elements in fungi. In our analysis, we included representatives of the two major groups of fungi, Ascomycota (*Aspergillus fumigatus*, *Aspergillus nidulans,* and *Neurospora crassa*) and Basidiomycota (*Phanerochaete chrysosporium*). The initial analysis resulted in 109 putative LTR retrotransposon sequences in *A. fumigatus*; 25 in *A. nidulans*; 57 in *N. crassa;* and 71 in *P. chrysosporium.*

Table -1. The list of species and groups of elements investigated in the present study

Kingdom/ Phylum	*Species*	Ty1/copia	Bel	DIRS	Ty3/ gypsy	New elements
Fungi/ Basidiomycota	*Phanerochaete chrysosporium*	+(1)	.	.	+(3)	4
Fungi/ Ascomycota	*Aspergillus fumigatus*	.	.	.	+(2)	2
	Aspergillus nidulans	.	.	.	+(2)	2
	Neurospora crassa	.	.	.	+(1)	1
Animalia/ Nematoda	*Caenorhabditis briggsae*	.	+(1)	+(1)	+(2)	4
	Brugia malayi	.	+(1)	.	.	1
Animalia/ Arthropoda	*Aedes aegypti*	+	+(1)	.	+(5)	6
Animalia/ Chordata	*Ciona intestinalis*	.	+(1)	.	+(2)	3
	Danio rerio	+	+	+	+(1)	1

The sign '+' indicates the presence of a given group of elements in the investigated species. The numbers in brackets correspond to the number of newly identified elements in each group. The total number of newly identified elements for each species is given in the last column.

A comparative analysis revealed that the sequences from *A. fumigatus* could be initially subdivided into four distinct groups according to their similarity to each other; sequences from *A. nidulans,* into two groups; and sequences from *P. chrysosporium,* into four groups. Sequences from each set of putative fungal LTR retroelements were selected for further analysis. The majority of sequences from *N. crassa* proved to be extensively degenerated. Only one sequence from *N. crassa* showed a high similarity to newly identified elements in *A. nidulans* and *A. fumigatus* (see below).

Further analysis revealed that two groups of elements from *A. fumigatus* comprise copies of the previously described *Afut1* and *Afut2* retroelements (Neuveglise et al., 1996; Paris and Latge, 2001). The other two groups contain representatives of the new families of LTR elements identified in *A. fumigatus*. The putative retroelements obtained from the genome of *A. nidulans* form two new families. They did not show any significant similarity to any previously known LTR retroelements of *A. nidulans*. The retrotransposons from the *P. chrysosporium* genome were not studied previously, and therefore, the four identified groups represent new putative retrotransposable elements. The identification of four new groups of LTR retroelements in basidiomycetes fungi is probably related to the fact that they have not been so widely investigated as ascomycetes retroelements. Several Ty3/gypsy and Ty1/copia elements were identified in basidiomycetes; however, their diversity is less known when compared to previous work performed with ascomycetes (Goodwin and Poulter, 2001; Diez et al., 2003).

After the initial analysis, we selected one representative from each of the newly identified groups based on its relative integrity and defined the subclass to which each representative belongs. In ascomycetes, all the newly identified elements belong to the Ty3/gypsy subclass. In *P. chrysosporium*, the only basidiomycete investigated, three of the four newly identified elements also belong to the Ty3/gypsy subclass, whereas the remaining element belongs to the Ty1/copia subclass.

Phylogenetic analysis was also performed for both the Ty3/gypsy and Ty1/copia elements. The phylogenetic tree of the Ty3/gypsy elements, which was built based on the sequences of the reverse transcriptase domains, indicates that the three newly identified elements from the investigated ascomycetes (*asperfum16*, *anidulans2*, and *neuro1*) and the one newly identified element from *P. chrysosporium* (*whiterot12*) form a common branch in the phylogenetic tree (Figure 1).

In addition, two families of Ty3/gypsy elements were found in the genomes of *A. fumigatus* (*asperfum1*) and *A. nidulans* (*anidulans1*). The remaining three newly identified elements of *P. chrysosporium* belong to the Ty3/gypsy subclass (*whiterot1* and *whiterot7*, Figure 1) and Ty1/copia subclass (*whiterot15*, data not shown). In summary, our analysis resulted in the identification of nine new elements in the four analyzed fungal genomes, which belong to the Ty3/gypsy and Ty1/copia subclasses of LTR retroelements.

New LTR retrotransposable elements in nematodes. LTR retrotransposons have been previously characterized in detail in the nematode *Caenorhabditis elegans* (Bowen and McDonald, 1999). However, the related nematode *C. briggsae* and another distantly related nematode, *Brugia malayi*, have not been investigated for the presence of retroelements. In our initial analysis, we identified 124 putative LTR retroelements in the *C. briggsae* genome and 11 sequences in the B. malayi genome.

Further analysis revealed that all the sequences from *B. malayi* constituted a single group. We selected a single representative (brugia1) from this group for our phylogenetic analysis. The evolutionary tree, built based on the sequences of the known Ty3/gypsy, DIRS, and Bel retrotransposons domains, places the brugia1 element in the Bel subclass of LTR retrotransposons, closer to the related TAS element from *Ascaris lumbricoides* (Felder et al., 1994; Figure 2).

The putative LTR retroelements from *C. briggsae* comprised four groups according to their similarity to each other and to the previously identified elements from *C. elegans*.

Elements belonging to the Ty1/copia subclass were not identified in the genome of *C. briggsae.* One group of newly identified elements was classified in the Bel subclass (selected individual element—*cbrigg154*) and is similar to the *Cer11* and *Cer13* elements from *C. elegans* (Bowen and McDonald, 1999).

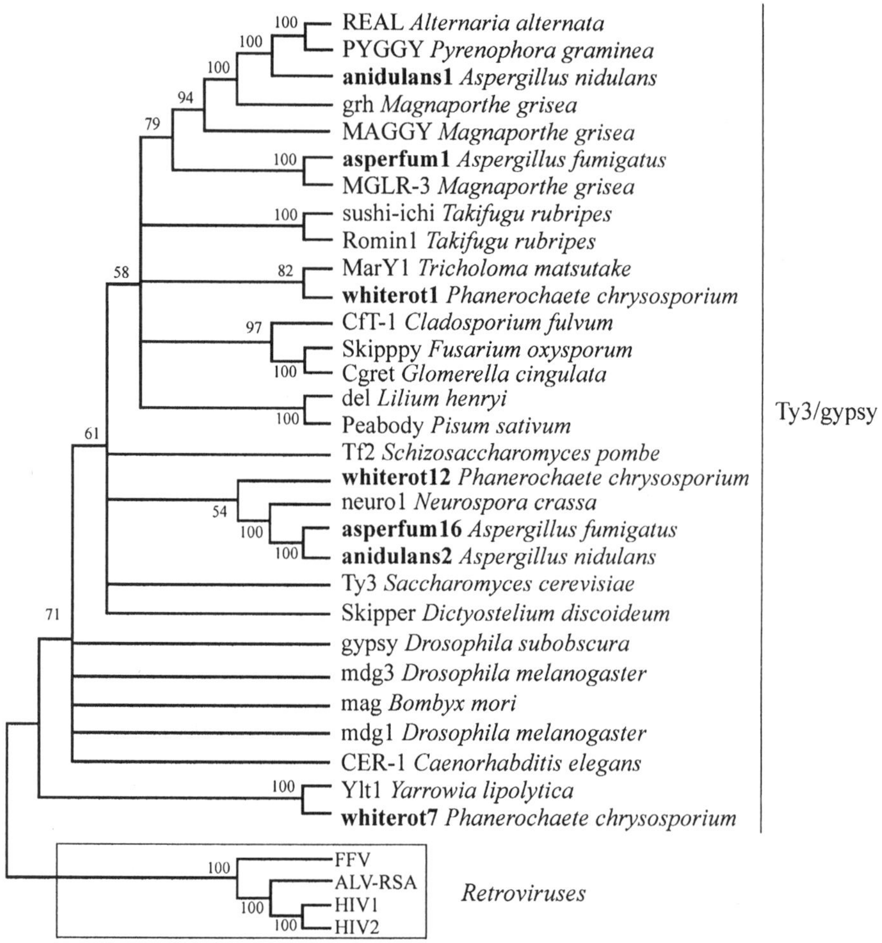

Figure -1. Phylogenetic tree based on the RT domain of the known Ty3/gypsy retrotransposons showing the position of the newly identified elements of *Aspergillus fumigatus, A. nidulans, Neurospora crassa,* and *Phanerochaete chrysosporium* (bold type). The tree was constructed by the neighbor-joining method using the MEGA2 software package (Kumar et al., 2001). Percentages of bootstrap support, from 100 replications, are indicated for branches receiving > 50 % support. Retroviral RT domain sequences were employed for rooting the tree.

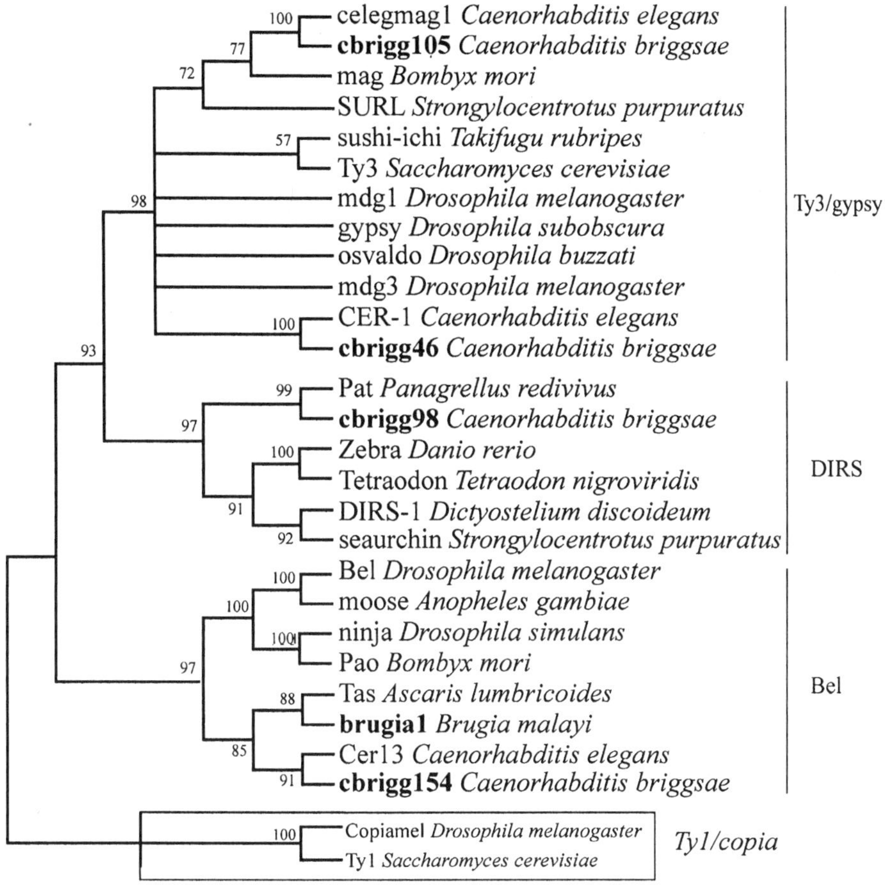

Figure -2. Phylogenetic tree based on the RT domain of the known Ty3/gypsy, DIRS, and Bel retrotransposons and the newly identified elements of *Caenorhabditis briggsae* and *Brugia malayi*. The tree was built by the neighbor-joining method using the MEGA2 software package (Kumar et al., 2001). Percentages of bootstrap support, from 100 replications, are indicated for branches receiving > 50 % support. RT domain sequences of Ty1/copia elements were employed for rooting the tree.

One of the groups (selected individual element—*cbrigg98*) showed a significant similarity to *Pat*, an element from the nematode *Panagrellus redivivus* (de Chastonay et al., 1992). This is an unexpected result, since *Pat*-like elements have not been described in *C. elegans*. The *Pat* element belongs to the DIRS subclass of LTR retroelements. We preformed a BLAST search using the *cbrigg98* sequence as a query against the *C. elegans* genome, which did not result in identification of any sequences with a significant similarity. The remaining newly identified sequences belong to the Ty3/gypsy subclass. Two previously identified elements from *C. elegans*, *CER-1*, and *celegmag1* were classified in the Ty3/gypsy subclass (Britten, 1995; Bowen and McDonald, 1999). Our analysis resulted in the identification of groups of *C. briggsae* sequences represented by *cbrigg46* and *cbrigg105*, which showed a significant similarity to the previously described *CER-1*-like sequence and the *celegmag1* sequence from *C. elegans*, respectively.

In summary, our analysis with nematodes resulted in identification of one LTR retroelement in the *B. malayi* genome and four LTR retroelements in the genome of *C. briggsae*. Furthermore, our analysis has shown that three subclasses of LTR retroelements are present in nematodes, namely, Bel, DIRS, and Ty3/gypsy, different to what has been previously described in *C. elegans*, which does not present LTR retroelements belonging to the DIRS subclass.

LTR retrotransposable elements in the other selected genomes. The analysis of the *Aedes aegypti* genome revealed 166 sequences similar to LTR retroelements. Previous studies described only one retroelement in *A. aegypti*, the *Mosqcopia*, which belongs to the Ty1/copia subclass (Tu and Orphanidis, 2001). Our investigation confirmed that the *Mosqcopia*-like elements are the only retroelements belonging to the Ty1/copia subclass in the *A. aegypti* genome. The remaining sequences could be subdivided into six distinct groups (Figure 3).

Five of the six remaining *A. aegypti* groups belong to the previously characterized clades of LTR retrotransposons from insects. The first group, represented by *aegypti907*, is a *moose*-like element, which stood out as a separate group. The *Moose* element belongs to the Bel subclass and was originally described in *Anopheles gambiae*, another mosquito model (Biessmann et al., 1999). The five remaining *A. aegypti* groups belong to the Ty3/gypsy subclass. Four of these groups displayed high similarities to elements from the following clades: Mdg3 (individual element *aegypti67*), Mag (*aegypti187*), Gypsy (*aegypti93*), and Mdg1 (*aegypti76*), which was described earlier by Malik and Eickbush (1999). Unexpectedly, the remaining sixth group formed a cluster with the recently described elements from *Schistosoma mansoni*, *Boudicca*, and *Saci-3* (individual element *aegypti97*; Copeland et al., 2003; DeMarco et al., 2004).

Finally, our analysis also included the genomes from two chordates. These distantly related organisms, for which the genome sequences are available, constituted the sea squirt *Ciona intestinalis* and the zebrafish *Danio rerio*. A previous systematic search of 1 Mb of genomic sequence from the sea squirt revealed the presence of *Cigr-1*, a Ty3/gypsy retrotransposon (Simmen and Bird, 2000). In our investigation, we identified not only *Cigr-1*-like LTR retroelements (*ciona 16*), but also elements belonging to the Bel subclass (*ciona83*) and two new families belonging to the Ty3/gypsy subclass (*ciona3* and *ciona22*).

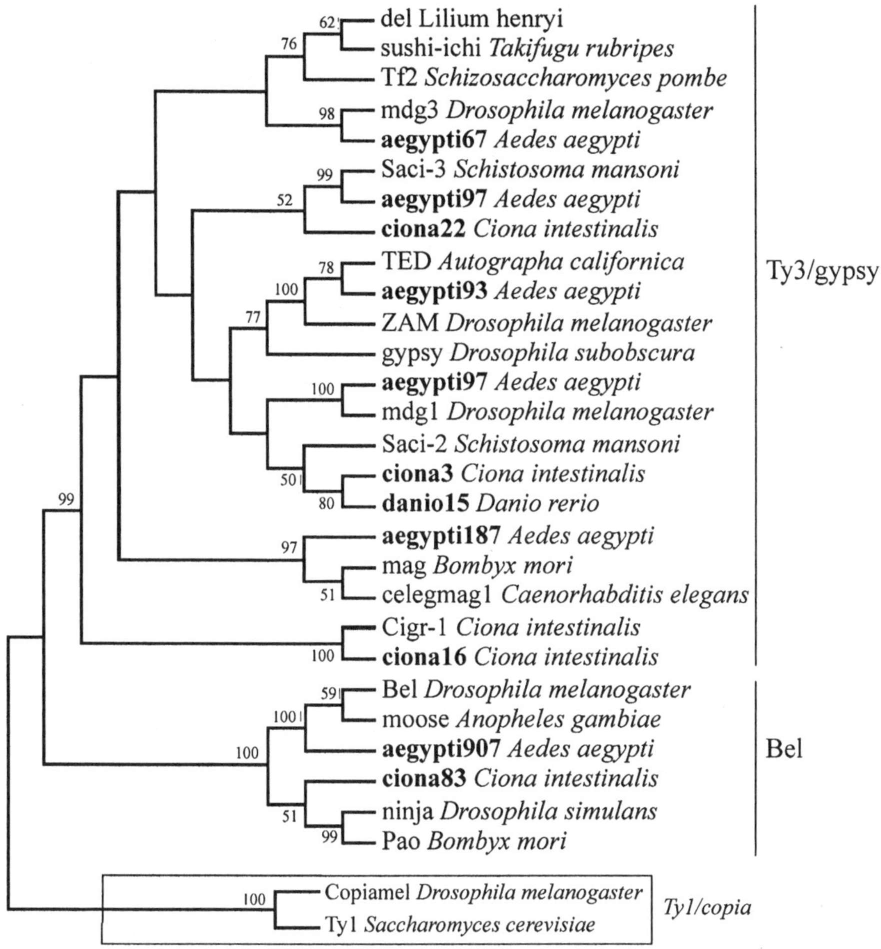

Figure -3. Phylogenetic tree based on the RT domain of the known Ty3/gypsy and Bel retroelements and the newly identified elements from Ciona intestinalis, *Danio rerio*, and *Aedes aegypti*. The tree was built by the neighbor-joining method using the MEGA2 software package (Kumar et al., 2001). Percentages of bootstrap support, from 100 replications, are indicated for branches receiving > 50 % support. Sequences of RT domain of Ty1/copia elements were employed for rooting the tree.

Interestingly, the new Ty3/gypsy families from *C. intestinalis*, *ciona3* and *ciona22*, formed clusters in the phylogenetic tree with the blood flake *S. mansoni Saci-2* and *Saci-3* elements, respectively (DeMarco et al., 2004). Additionally, a group of elements that belonged to the same cluster, which contains the *Saci-2* element, was obtained from *D. rerio* (*danio15*). Apparently, several distinct lineages of LTR retrotransposons can be distinguished in genomes of vertebrates (Miller et al., 1999; Volff et al., 2001).

4. CONCLUSIONS

In summary, we identified 24 new elements belonging to the 4 subclasses of LTR retrotransposons in the genomes of 9 distinct eukaryotic organisms. Our findings suggest the presence of multiple lineages of Ty3/gypsy retrotransposons in fungal genomes, especially, in basidiomycetes. In addition, a new family of Ty3/gypsy retroelements, which is significantly different from other fungal Ty3/gypsy elements (data not shown), was identified in both ascomycetes and basidiomycetes fungi. We detected three distinct elements in the sea squirt *C. intestinalis* genome in addition to the recently found *Cigr-1* LTR retroelement (Simmen and Bird, 2000). Furthermore, the new LTR retroelement identified in the *D. rerio* genome has formed a common cluster with one element from *C. intestinalis* and *Saci-2* from *S. mansoni*.

The genomes of chordates and insects contain distinct families of retroelements, which belong to the Ty3/gypsy subclass. However, a broader diversity of Ty3/gypsy families is found in insects. Further studies should provide more information about the diversity of LTR retrotransposons in chordate genomes.

The two nematodes that were examined showed different distributions of LTR retrotransposons. Only one family of LTR retroelements, belonging to the Bel subclass, was found in the genome of *B. malayi*, whereas four different families were found in the *C. briggsae* genome, which belong to the Bel, DIRS, and Ty3/gypsy subclasses. The *B. malayi* and *C. briggsae* have approximately the same genome sizes (~ 100-Mb total genome sequence).

At present, it is not possible to explain the observed differences in both the diversity and amount of the LTR retrotransposons. Phylogenetic analysis indicates that *B. malayi* and *C. briggsae* are distantly related, with divergence estimates ranging between 300–400 Myr (Blaxter et al., 1998). On the other hand, these two nematode species occupy extremely different ecological niches: *C. briggsae* is a free-living nematode, while *B. malayi* is a human parasitic nematode, the causative agents of lymphatic filarisis.

Finally, we have identified six clusters of elements in the genome of the mosquito *A. aegypti*. With the exception of one interesting finding, five of the clusters correspond to the previously known insect clades of LTR retrotransposons. This newly identified element formed a common cluster with the elements present in the non-insects metazoans *S. mansoni* (*Saci-3*) and *C. intestinalis* (*ciona22*).

ACKNOWLEDGMENTS

We thank Nadia Monesi for comments on the manuscript.

A GENOME-WIDE IDENTIFICATION OF MITOCHONDRIAL DNA TOPOISOMERASE I IN *ARABIDOPSIS*

A. Katyshev[1], I.B. Rogozin[2], Yu. Konstantinov[1*]
[1] *Siberian Institute of Plant Physiology and Biochemistry, Siberian Branch of the Russian Academy of Sciences, Irkutsk, Russia, e-mail: yukon@sifibr.irk.ru;* [2] *Institute of Cytology and Genetics, Siberian Branch of the Russian Academy of Sciences, prosp. Lavrentieva 10, Novosibirsk, 630090, Russia*
[*] *Corresponding author*

Abstract: Topoisomerases are conserved enzymes that play an important role in multiple cellular processes, such as DNA recombination, DNA replication, and cell cycle checkpoint control (Pommier, 1998). It is likely that at least three different plant topoisomerases function within nucleus, in mitochondria, and in chloroplasts. This hypothesis was partially supported by biochemical experiments (Daniell et al., 1995; Balestrazzi et al., 2000). Moreover, the enzymes that are of different genetic origin and structure—prokaryotic topo IA and eukaryotic topo IB type topoisomerases—may function in chloroplasts or mitochondria. Thus, there may be more than one gene encoding chloroplast and mitochondrial topo I in plant nuclear genomes. A genome-wide analysis of the *Arabidopsis thaliana* nuclear genome suggested that there was only one gene encoding non-nuclear topo I. Phylogenetic analysis of prokaryotic topo I orthologs indicated that the candidate mitochondrial topo I might have been acquired from alpha-proteobacteria, which are believed to be ancestors of eukaryotic mitochondria. Interestingly, we identified two paralogous genes for mitochondrial topo I in the rice nuclear genome. Some explanations of this fact are provided, and unique features of the gene product are discussed.

Key words: DNA topoisomerase I; mitochondria; chloroplasts

1. INTRODUCTION

Members of the DNA topoisomerase I superfamily are well characterized in prokaryotes and animals. In plant nuclei, such mechanisms are well known, and they are similar to those in animals, but it is not the case of mitochondria and chloroplasts. Moreover, the coexistence of three genomes may affect the functioning of topo I enzymes. Previously, we showed that maize nuclear and mitochondrial topo I DNA-binding and relaxation activities were regulated in different ways (Konstantinov et al., 2003). This result suggested different structures of these enzymes. To verify this hypothesis, we carried out database searches of topo I homologs in the *Arabidopsis thaliana* nuclear genome. Three candidate genes were identified. Two of them are highly similar to each other and encode a nuclear topo I. The third gene encodes a candidate mitochondrial/chloroplast topo I. It is likely that this gene encodes mitochondrial enzyme; however, we cannot reject the hypothesis that this gene encodes a chloroplast topo I. The possible theoretical explanations of the existence of only one gene for organellar topo I are that (a) the enzyme contains a dual targeting sequence allowing it to be transported into both compartments; (b) the topo I functions in chloroplasts are completely substituted by topo III functions; or (c) chloroplast topo I sequence does not have a detectable similarity to the known topo I. The most plausible are the first two considerations; the data proving their validity will be discussed. The candidate mitochondrial/chloroplast topo I is a prokaryotic type IA topoisomerase, which is also known as a *Bacillus subtilis*–like DNA topoisomerase I or eukaryotic topoisomerase IIIα (KOG1956, http://ncbi.nlm.nih.gov/COG/new/).

2. METHODS AND ALGORITHMS

Protein sequences of *Arabidopsis thaliana* DNA topoisomerase I and DNA topoisomerase I–like enzymes were obtained from the TAIR (The Arabidopsis Information Resource) peptide dataset (http://www.arabidopsis.org/index.jsp). Sequences of topo I and III from other organisms used in alignment were obtained from the GenBank database (http://www.ncbi.nlm.nih.gov/entrez/query.fcgi?CMD=Search&DB=protein) and the COG/KOG database (http://ncbi.nlm.nih.gov/COG/new/). COGs/KOGs (Clusters of Orthologous Groups of prokaryotic and eukaryotic proteins) were constructed from the results of all-against-all BLAST comparison of proteins encoded in complete genomes by detecting consistent sets of genome-specific best hits (Tatusov et al., 1997; Koonin et al., 2004). Alignments of topo I proteins were constructed using a ClustalW algorithm-based program (Thompson et al., 1994) from the Vector NTI5 package (Bethesda Inc., USA). Neighbor-joining trees were constructed using the MEGA2 program (http://www.megasoftware.net; Kumar et al., 2001). Homologous sequences were searched for in

GenBank database by BLAST service from http://www.ncbi.nlm.nih.gov/blast/ (Schaffer et al., 2001).

To predict subcellular localization of the proteins of interest, we used Internet resources available at the http://www.expasy.org/ molecular biology tools server: (a) the MITOPROT program (Claros and Vincens, 1996) at http://www.mips.biochem.mpg.de/cgi-bin/proj/medgen/mitofilter; (b) the Predotar program at http://www.inra.fr/predotar/; (c) the TargetP V1.0 program (Emanuelsson et al., 2000) at http://www.cbs.dtu.dk/services/TargetP/; and (d) the PSORT program (Nakai and Kanehisa, 1991) at http://psort.nibb.ac.jp/form.html. Data on domain organization of proteins and their motif structure were obtained by InterProScan sequence search package (http://www.ebi.ac.uk/InterProScan/) and SMART research tool (http://smart.embl-heidelberg.de/).

3. RESULTS AND DISCUSSION

The *Arabidopsis* genome contains two highly homologous genes encoding nuclear topo I, located near each other on chromosome 5 (GenBank accession Nos. NP_200342 and P30181). The significance of topo I enzymes in cell genetic processes is well known, and the existence of duplicated genes for the nuclear topo I in *Arabidopsis* supports importance of this enzyme. Hence, it is reasonable to expect additional genes encoding chloroplast and mitochondrial enzymes. Interestingly, BLASTP searches suggested only one candidate gene for mitochondrial/chloroplast enzyme in the *Arabidopsis* genome (GenBank accession No. NP_194849). Alignments of these three proteins with amino acid sequences of topo I from other organisms indicated that the candidate gene for mitochondrial/chloroplast enzyme is of prokaryotic origin whereas the nuclear enzymes are eukaryotic type IB topoisomerases. This observation suggested the organellar localization of the candidate gene for mitochondrial/chloroplast topo I, which is also known as the *Bacillus subtilis*–like topoisomerase I (*Bsu*-like topo I). The term *Bacillus subtilis*–like topoisomerase I was proposed by Brandt et al. (unpublished). To determine localization of *Bsu*-like topo I in cells, we used programs for prediction of nuclear/mitochondrial/chloroplast targeting signals in protein molecules. Results of signal searches are shown in the Table 1.

These results strongly suggest mitochondrial localization of the enzyme. However, the PSORT program suggested that the *Bsu*-like topo I is a dual targeted enzyme that functions also in chloroplasts. There are some proteins of dual targeting in *Arabidopsis* (Chow et al., 1997). However, we cannot exclude that a topo I enzyme was substituted by a topo III enzyme in chloroplasts (Pommier, 1998).

Table -1. Probabilities of mitochondrial and chloroplast localization of *Bacillus subtilis*–like topoisomerase I predicted by different programs

Program	Probability of mitochondrial localization	Probability of chloroplast localization
MITOPROT	0.9996	–
Predotar	0.822	0.001
TargetP V.1.0	0.759	–
PSORT	0.861	0.870

Phylogenetic analysis of prokaryotic topo I orthologs (COG0550, http://www.ncbi.nlm.nih.gov/COG/new/release/cow.cgi?cog=COG0550) suggested that the candidate mitochondrial topoisomerase I (mt-topo I) might have been acquired from alpha-proteobacteria (Figure 1), which are believed to be ancestors of eukaryotic mitochondria (Gray, 1992; Kurland and Andersson, 2000). This result is an additional indication that the *Bsu*-like *Arabidopsis* topo I may still function in mitochondria. Interestingly, all other known eukaryotic topo I genes (including *Arabidopsis* nuclear topo I genes) were shown to be of archaeal origin (Tse-Dinh, 1998; Grabowski and Kelman, 2003; Koonin et al., 2004) (Figure 1).

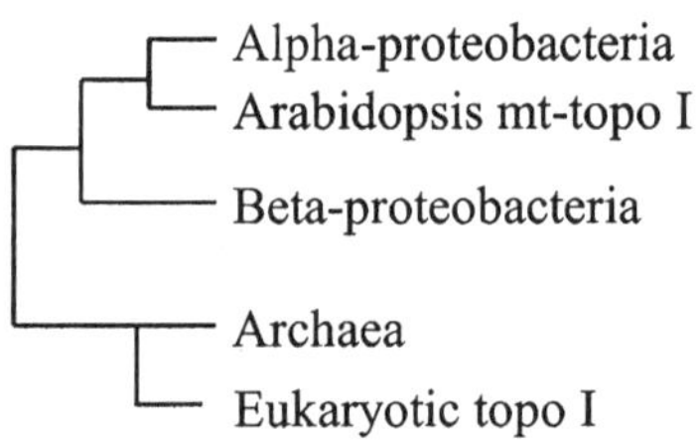

Figure -1. Schematic representation of a phylogenetic tree for topoisomerase I homologs.

The primary structure of the *Bsu*-like topo I from *Arabidopsis* contains several domains characteristic of prokaryotic topo I (SM00436, SM00437, and SM00493). The enzyme contains a conservative prokaryotic topo I functional motif (ProSite accession No. PS00396) adjacent to the active Tyr residue (Figure 2) in the core domain indicating that this is a functional DNA topo I enzyme. To identify orthologs of *Arabidopsis Bsu*-like topo I genes in plant genomes, we searched through gene database and found two other genes encoding bacterial type topo I in rice genomic sequences. The first one is located on rice chromosome 6. For this gene, there are full-length cDNA (GenBank acc. No. AK121044) and genomic DNA (PAC clone acc. No. AP005822). Another gene was identified in the genomic sequence of PAC clone corresponding to a fragment of rice chromosome 2 (GenBank acc. No. AP005067). Both genes share the same intron–exon organization but slightly differ in both nucleotide and amino acid sequences. Proteins encoded by rice genes and the *Arabidopsis Bsu*-like topo I contain conserved functional domains (Figure 2) and strongly differ in their *N*-terminus. The data obtained showed clearly that the two rice genes encoded mitochondrial topoisomerases and did not contain chloroplast targeting sequences.

```
D.pse.   LYS---------KGFIS Y PRTETNQFSK-FEEL-APLVQLQA--------GHSDWGAF 387
S.cer.   LYQ---------KGFIS Y PRTETDTFPH-AMDL-KSLVEKQAQLDQLAAGGRTAWASY 394
A.th.nc  LYQ---------AGFIS Y PRTETDSFSS-RTDL-RAMVEEQTR--------HPAWGSY 372
E.coli   LYE---------AGYIT Y MRTDSTNLSQDAVNMVRGYISDN----------------F 343
H.inf.   LYE---------AGYIT Y MRTDSTNLSQDALNMARSYIENH----------------F 349
B.sub.   LYEGIDLGREGTVGLIT Y MRTDSTRISNTAVDEAAAFIDQT----------------Y 322
E.rum.   LYEGVDIG-GEIVGLIT Y MRTDGVYISDEAVEHIRSVISSM----------------F 329
A.th.mt  LYEGVQLSDGKSAGLIT Y MRTDGLHIADEAIKDIQSLVAER----------------Y 801
O.sat.1  LYEGINLSSEEATGLIT Y IRTDGFHISDGAAEDILSLVKQR----------------Y 628
O.sat.2  LYEGITLSSEDATGLIT Y IRTDGFHISDVAAEDILSLVKQR----------------Y 616
```

Figure -2. Alignment of pro- and eukaryotic topo I protein sequence fragments containing a conservative functional motif. Species abbreviations: B.sub., *Bacillus subtilis*; D.pse., *Drosophila pseudoobscura*; S.cer.; *Saccharomyces cerevisiae*; E.coli, *Escherichia coli*; H.inf., *Haemophilus influenzae*; E.rum., *Ehrlichia ruminantium*; A.th.nc (nuclear topo I), *Arabidopsis thaliana*; A.th.mt (mt-topo I), *Arabidopsis thaliana*; O.sat.1 (mt-topo I, chromosome 6), *Oryza sativa*; and O.sat.2 (mt-topo I, chromosome 2), *Oryza sativa*. The active Tyr residue is underlined.

Plant mitochondrial/chloroplast enzymes involved in organellar genetic processes are still incompletely characterized. These enzymes could be of phage, bacterial, or nuclear origin. There is strong evidence that some proteins, e.g. phage-type RNA polymerases, are of dual targeting to both mitochondria and chloroplasts (Hedtke et al., 2000; Binder and Brennike, 2002). As for *Bsu*-like topo I gene product in *Arabidopsis thaliana*, we suggest the dual targeting of this enzyme to both types of organelles. The identification of two genes encoding mitochondrial topo I in rice genome is an additional evidence of the functional significance of topo I superfamily proteins in functioning of plant mitochondria. Further theoretical and experimental studies are needed to identify chloroplast topo I enzymes. In conclusion, the genome-wide search results allowed us to identify the gene encoding the candidate mitochondrial DNA topoisomerases I in *Arabidopsis thaliana* and rice. This gene may have been transferred from alpha-proteobacteria whereas all other eukaryotic *Bsu*-like topo I were shown to be of archaeal origin. A chloroplast topo I gene is still not found. The experimental data should be obtained to verify hypothesis on the dual targeting of *Bsu*-like topo I.

ACKNOWLEDGMENTS

The work was supported by RFBR Grant No. 01-04-48162.

CHANGE IN CpG CONTEXT IS A LEADING CAUSE OF CORRELATION BETWEEN THE RATES OF NON-SYNONYMOUS AND SYNONYMOUS SUBSTITUTIONS IN RODENTS

G. Bazykin[1*], A. Ogurtsov[2], A. Kondrashov[2]

[1] Department of Ecology and Evolutionary Biology, Princeton University, Princeton, NJ 08540, USA, e-mail: gbazykin@princeton.edu; [2] National Center for Biotechnology Information, NIH, Bethesda, Maryland 20894, USA

[*] Corresponding author

Abstract: Correlation between the rates of synonymous (silent) and non-synonymous (amino acid-changing) nucleotide substitutions in genes is a widespread and yet unexplained genome-level phenomenon, which is in disagreement with the neutral theory of molecular evolution (Kimura, 1983). Comparison of 7732 orthologous genes of mouse and rat confirms the previously observed correlation between the rates of substitutions in non-synonymous and synonymous nucleotide sites. In rodents, this correlation is primarily caused by tandem substitutions and, in particular, by CpG mutation bias leading to doublet nucleotide substitutions. The nature of correlation between the rates of synonymous and non-synonymous substitutions in seven pairs of prokaryotic genomes is unclear.

Key words: evolution; point substitution rate; mutation bias; CpG deamination; dinucleotides

1. INTRODUCTION

Synonymous (silent) nucleotide sites are often assumed to evolve 'neutrally' and, therefore, are frequently used as a reference point of neutral substitution rate. This assumption, however, conflicts with the well-described phenomenon of variation in rates of synonymous substitutions across the genome and, in particular, of correlation between rates of non-synonymous and synonymous substitutions. Selection for translation efficiency (Chamary and Hurst, 2004) or RNA structure (Smith and Hurst, 1999) acting on silent sites were suggested as possible explanation as well as the methodological biases in distance estimation (Bielawski et al., 2000).

It has been claimed that the correlation of the rates of synonymous and non-synonymous substitutions is dependent upon the particular method used for estimation of substitution rates (Bielawski et al., 2000). Therefore, to reveal the leading cause of this correlation, it is preferable to use closely related species. At low evolutionary distances, substitutional saturation is negligible, and different methods of estimation of divergence converge.

2. METHODS

Mouse and rat coding sequences were obtained from version 30 of the mouse genome (Mouse Genome Sequencing Consortium, 2002) and version 2 of the rat genome (Rat Genome Project Sequencing Consortium, 2004) from NCBI. Orthologs were identified according to the two-directional best-hit approach using protein BLAST (Altschul et al., 1997). Alignments of the amino acid sequences for each pair of the orthologs was made using ClustalW (Thompson et al., 1994) and reverse transcribed to get the nucleotide alignments.

Rates of nucleotide substitutions in different groups of sites were obtained using a PERL script available from the authors. All suitable pairs of bacterial genomes were obtained from the NCBI Entrez database and processed analogously. Genes with doublets removed are those in which the adjacent nucleotide sites were excluded from analysis if both carried substitutions.

A substitution at site 1 of the doublet was assumed to change the CpG context of the following site 2 when one of the species carried 'C' at site 1 and the other species carried some other nucleotide. A substitution at site 2 of the doublet was assumed to change the CpG context of the preceding site 1 when one of the species carried 'G' at site 2 and the other species carried some other nucleotide.

Outliers can have a profound effect on the value of correlation coefficient. In order to ensure that only high-quality (unambiguous) alignments are included in the analysis, we excluded all genes with divergences in non-degenerate sites exceeding 1.5 average amino-acid divergences between the corresponding species, and divergences in 4-fold-degenerate sites exceeding 10 average amino-acid divergences (therefore, the abrupt left and top boundaries of region with data points at Figure 1*a*). This approach is conservative in regard to determination of correlation.

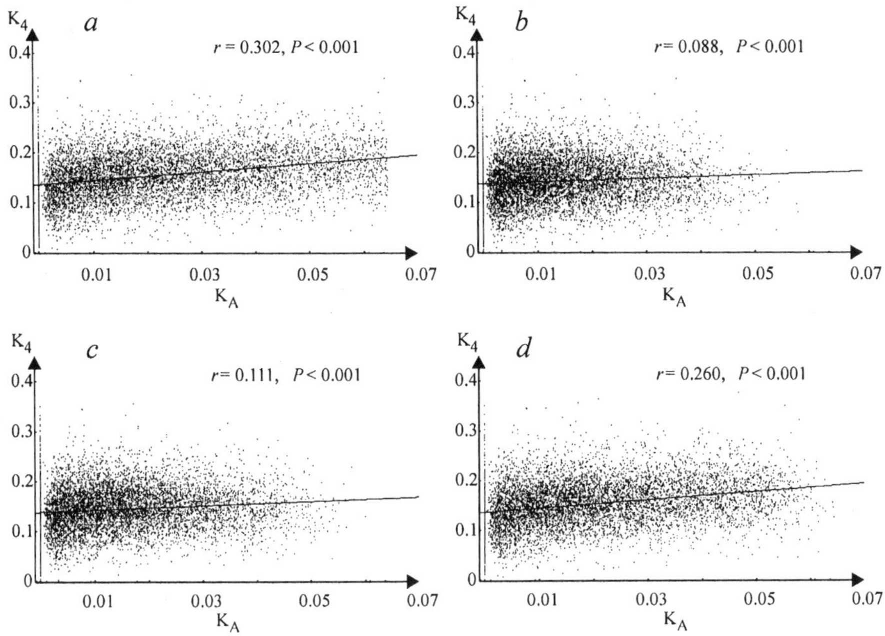

Figure -1. The relationship between per gene divergences in non-degenerate (K_A) and fourfold degenerate (K_4) nucleotide sites between mouse and rat (*a*) with all sites included into analysis, (*b*) doublets removed, (*c*) doublets with change in CpG context removed, (*d*) and doublets without change in CpG context removed.

3. RESULTS AND DISCUSSION

Our data confirms the previously observed significant correlation between per gene substitutions rates in non-synonymous (non-degenerate, K_A) and synonymous (fourfold degenerate, K_4) nucleotide sites (Figure 1*a*). This correlation, however, is primarily caused by doublet substitutions occurring in adjacent nucleotides. When sites with double substitutions were excluded from analysis, the magnitude of correlation was greatly reduced (Figure 1*b*).

Correlation between substitutions in adjacent sites can arise, if one mutational event simultaneously affects two successive nucleotides. However, such double substitutions are extremely rare (Kondrashov, 2003), and the observed effect has to be caused by separate point mutation events. Such correlation can also be due to selection on silent substitutions that restore the codon bias following an amino acid change (Lipman and Wilbur, 1984).

The nature of correlation is revealed by consideration of the sites of adjacent substitutions in which one of the substitutions can affect the CpG context of the neighboring nucleotide site. Removal of the subset of such sites is sufficient to achieve a strong reduction in correlation (Figure 1*c*). Conversely, only a minor reduction in the correlation coefficient is achieved by removal of the sites of neighboring substitutions in which both substitutions leave the CpG context of the other one invariant (Figure 1*d*).

The simplest explanation for the correlation between K_A and K_4 that is consistent with these findings is interdependence of mutational events in adjacent nucleotides due to CpG deamination. CpG dinucleotide is hypermutable in vertebrates. If the first substitution (regardless of whether it occurs in a non-synonymous or synonymous site) creates the CpG dinucleotide, the second substitution at the adjacent nucleotide site is facilitated. This is expected to result in the observed pattern of substitutions coupling.

Table -1. Correlation coefficients between divergences in non-degenerate and fourfold degenerate nucleotide sites in eight pairs of genomes

	No. of genes	Fraction of amino acid differences[1], %	All sites	Doublets removed[2]	Doublets with change in CpG context removed[2]	Doublets without change in CpG context removed[2]
Muridae	7732	4.3	0.3	0.09	0.11	0.26
Bacillus	1915	3.5	0.44	0.34	0.36	0.42
Bordetella	2696	0.4	0.12	0.10	0.11	0.12
E. coli	3122	1.2	0.27	0.21	0.23	0.25
Salmonella	2531	0.8	0.21	0.17	0.17	0.20
Staphylococcus	1591	0.5	0.23	0.20	0.21	0.21
Streptococcus	1065	0.7	0.28	0.20	0.22	0.26
Vibrio	579	0.7	0.29	0.25	0.26	0.28

The following pairs of genomes were analysed: Muridae (*Rattus norvegicus* and *Mus musculus*); *Bacillus* (*B. cereus* ATCC 14579 and *B. anthracis* strain Ames); *Bordetella* (*B. parapertussis* and *B. bronchiseptica* RB50); *Escherichia* (*E. coli* O157:H7 and *E. coli* K12); *Salmonella* (*S. typhimurium* LT2 and *S. enterica enterica* serovar Typhi Ty2); *Staphylococcus* (*S. aureus aureus* Mu50 and *S. aureus aureus* MW2); *Streptococcus* (*S. pyogenes* M1 GAS and *S. pyogenes* MGAS315); and *Vibrio*: (*V. vulnificus* YJ016 and *V. vulnificus* CMCP6).

[1] Fraction of mismatches in alignments of orthologous proteins between genomes. [2] See Methods for details. All correlations were significant at $P < 0.05$.

This explanation is further supported by the analysis of seven pairs of closely related bacterial genomes. All the pairs of bacterial species indicated significant correlation between K_A and K_4 of various magnitudes. However, removal of doublets and, in particular, of doublets involving the change in CpG context did not lead to a profound decrease in the correlation comparable with that observed in rodents. Therefore, some other factor has to be responsible for correlation between K_A and K_4 in prokaryotes.

An obvious next step would be to reveal the order of substitutions—whether the change in the context in non-synonymous site facilitates the synonymous substation or *vice versa.* This can be achieved, if a third orthologous gene from an outgroup species (e.g., human) is employed.

ACKNOWLEDGMENTS

G.B. gratefully acknowledges support from the Burroughs Wellcome Fund Training Program in Biological Dynamics (award No. 1001781) and the Pew Charitable Trusts Program in Biocomplexity (award No. 2000-002558).

UNIVERSAL SEVEN–CLUSTER STRUCTURE OF GENOME FRAGMENT DISTRIBUTION: BASIC SYMMETRY IN TRIPLET FREQUENCIES

A. Gorban[1], A. Zinovyev[2], T. Popova[3*]

[1] Centre for Mathematical Modeling, University of Leicester, Leicester, UK; [2] Institutes des Hautes Etudes Scientifiques, Bures-sur-Yvette, France; [3] Institute of Computational Modeling, Krasnoyarsk, Russia, e-mail: tanya@icm.krasn.ru

[*] *Corresponding author*

Abstract: We found a universal seven-cluster structure in bacterial genomic sequences and explained its properties. Based on the analysis of 143 completely sequenced bacterial genomes available in GenBank in August 2004, we show that there are four 'pure' types of the seven-cluster structure observed. The type of cluster structure depends on GC content and reflects basic symmetry in triplet frequencies. Animated 3D-scatters of bacterial genomes seven-cluster structure are available on our web site: http://www.ihes.fr/~zinovyev/7clusters.

Key words: triplet frequencies; genome fragments; codons; mean-field approximation; symmetry; visualization

1. INTRODUCTION

Coding information is the main source of statistical inhomogeneity in bacterial genomes. There exist well-known compositional differences between codon positions in coding regions, which we observed using pure data exploration strategy and determined as universal seven-cluster structure. We considered 64D vectors of non-overlapping triplet frequencies in sliding window within the direct strand of bacterial DNA sequence (see details in section 2.1). Visualization of 64D vectors data set in the subspace of the first three principal components shows a clear cluster structure presented in Figure 1 by examples of two genomes and in the form of the general pattern.

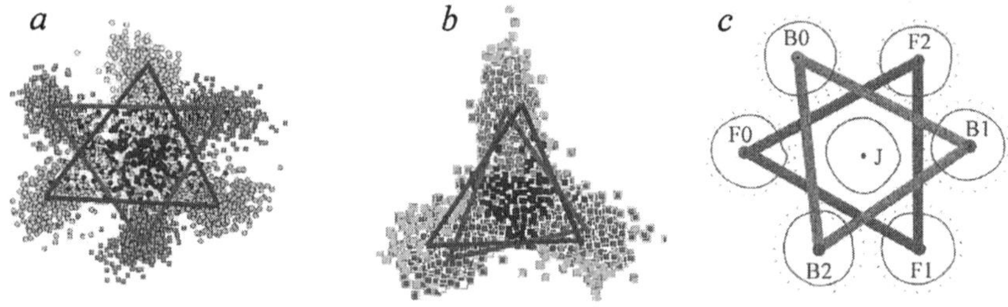

Pseudomonas aeruginosa *Buchnera aphidicola* Structure pattern

Figure -1. Universal seven-cluster structure: (*a* and *b*) visualization of genome fragment distribution and (*c*) pattern of the cluster structure.

Biologically, there are seven significantly different positions of a sliding window according to coding information: three possible reading frames of coding regions in two complementary strands plus non-coding regions. The obtained cluster structure corresponds to biologically relevant one with a higher than 90 % accuracy at the nucleotide level (Gorban et al., 2003).

This cluster structure is universal in the sense that it is observed in any bacterial genome and with any type of statistic, which takes into account three possible reading frames. The structure is basic in the sense that it is revealed in the analysis in the first place, reflecting the principal source of sequence non-randomness. The structure is well represented by a 3D-plot, while initially we have 64D vectors of frequencies. It has a symmetric and appealing flower-like pattern, hinting that there should be a symmetry in our statistics (triplet frequencies) governing the pattern formation.

The seven-cluster structure was implicitly used since long time ago in gene recognition problem (Borodovsky et al., 1993; Salzberg et al., 1998). Specific clustering of relatively short genome fragments is used in entropic or Hidden Markov Modeling (HMM) statistical approaches (Audic et al., 1998; Baldi, 2000; Bernaola-Galvan et al., 2000; Nicolas et al., 2002), which are effective due to non-randomness in DNA sequence being reflected by the seven-cluster structure. However, the structure itself was described explicitly and visualized for several genomes only recently (Zinovyev et al., 2003). We refer to the structure itself because of simplicity and formality of the presented approach: it is based on the 64 frequencies of non-overlapping triplets in sliding window and data exploration strategy regardless of any model of genome organization.

Several particular cases of flower-like pattern were observed in the 9D space of Z-coordinates (Ou et al., 2003). However, the structure was reported to pertain to GC-rich genomes only, while Zinovyev (2002) and Gorban et al. (2003) demonstrated that AT-rich genome of *H. pylori* had a flower-like cluster structure. This fact shows this simple and basic structure to be far from being completely understood and described.

In this paper, we show that the seven-cluster structure is determined by a single parameter: the genomic GC content. Based on the analysis of 143 completely sequenced bacterial genomes, available in GenBank in August 2004, we describe four 'pure' types of the structure and basic symmetries in triplet frequencies that they reflect.

2. SEVEN-CLUSTER STRUCTURE FOR COMPACT GENOMES

2.1 Algorithm of data table construction

To visualize the seven-cluster structure for some bacterial genome, a data set is prepared as follows.

1. The complete genome sequence and its annotation are extracted from GenBank. Let N be the length of a given sequence. One defines a step size p (~ 10–100 bp) and a fragment size W (odd number ~ 300–400 bp).
2. For $i = 1\ldots [N/(W + 1 + p)]$, a fragment of the length $W + 1$ centered at position $S_i = ip + W/2$ is clipped from the DNA sequence.
3. According to genome annotation, each clipped fragment is labeled by one of the F0, F1, F2, B0, B1, B2, and J labels with the letter F for S_i being inside the forward strand CDS, the letter B for S_i being inside the complementary strand CDS, and J for S_i of inter-CDS regions. Here, indices 0, 1, or 2 are equal to shift modulo 3 of the first base pair in the clipped fragment relative to the first base pair of the start codon in the corresponding CDS frame.
4. Frequencies of non-overlapping triplets are counted within each clipped fragment forming the table of 64D vectors of frequencies, which are characterized by window position S_i and annotation label.
5. Standard principal component analysis (PCA) is performed, and the first three principal components are calculated. Each vector is projected into the 3D basis of principal components and visualized (Figure 1).

2.2 Overall properties of seven-cluster structure

Typical 3D plots of data distribution obtained for bacterial genomes are shown in Figure 1. Distribution has well-detectable seven-cluster structure: each type of annotated points constitutes their own cluster, and clusters are separated from each other with visible gaps. Automatic clustering by the method of k-means with Euclidean distance attributes a data point to its annotated cluster with a more than 90 % accuracy (see Gorban et al., 2003 for more details), which corresponds to efficiency of automatic gene identification methods for bacterial genomes (Mathe et al., 2002). The seven-cluster structure reflects well-known differences in three letter word distributions between seven types of fragments. However, we show that these seven types of fragments are extracted primarily without any preliminary knowledge of genome organization, as a main source of sequence heterogeneity. Further, we consider some features of triplet frequencies in bacterial genomes that govern the cluster structure formation and its particular shape.

2.3 Phase triangles

According to annotation labels in the data table, we calculate an arithmetic mean vector of frequencies for each type of genome fragments, denoted here as f, $f^{(1)}$, and $f^{(2)}$ for coding regions of the forward strand (labels F0, F1, and F2) and as $\hat{f}$, $\hat{f}^{(1)}$, and $\hat{f}^{(2)}$ for coding regions of the complementary strand (labels B0, B1, and B2). Referring to the seven-cluster structure, these six vectors should be the centers of corresponding clusters. These centers constitute two *phase triangles* in 64D space of frequencies (Figure 1): forward strand triangle (f, $f^{(1)}$, $f^{(2)}$) and complementary strand triangle ($\hat{f}$, $\hat{f}^{(1)}$, $\hat{f}^{(2)}$).

Having codon frequencies $f = (f_{\mathrm{AAA}}, f_{\mathrm{AAC}}, \ldots, f_{\mathrm{TTG}}, f_{\mathrm{TTT}})$, one can easily calculate estimations $P^{(1)}f$ and $P^{(2)}f$ of the shifted distributions $f^{(1)}$ and $f^{(2)}$ under the assumption that no correlation exists in codon order:

$$P^{(1)}f_{ijk} \equiv \sum_{lmn} f_{lij} f_{kmn}\,,\; P^{(2)}f_{ijk} \equiv \sum_{lmn} f_{lmi} f_{jkn}\,,\; i, j, k, l, m, n \in \{\mathrm{A,T,G,C}\}. \quad (1)$$

An estimation of complementary strand phase triangle vertex $\hat{f}$ on the basis of f (denoted here as $C^R f$) consists in rearrangement of f coordinates according to reverse reading and complementary translation of the corresponding codons: $C^R f_{ijk} = f_{\hat{k}\hat{j}\hat{i}}$, $i, j, k \in \{\mathrm{A,T,G,C}\}$, where $\hat{i}$ is complementary to ith nucleotide. Estimations of shifted distributions $\hat{f}^{(1)}$ and $\hat{f}^{(2)}$ are calculated as $C^R P^{(1)} f$ and $C^R P^{(2)} f$.

Five calculated distributions $P^{(1)}f$, $P^{(2)}f$, $C^R f$, $C^R P^{(1)} f$, and $C^R P^{(2)} f$ appeared to be very close to the centers of corresponding clusters obtained according to genome annotation (Gorban et al., 2003; see also section 2.5). It means that (1) the *codon frequencies* determine seven-cluster structure and (2) between-codon correlations are, in average, much less than within-codon.

2.4 Mean-field approximation of codon frequencies

In order to reveal the properties of codon frequencies that guarantee the clusters appearance, we are to consider a *mean-field* approximation of f. Mean-field approximation, *mf*, assumes 64 codon frequencies to be modeled by 12 position-specific nucleotide frequencies as follows:

$$(mf)_{ijk} = p_i^1 p_j^2 p_k^3, \quad i, j, k \in \{A, T, G, C\}, \tag{2}$$

where

$$p_i^1 = \sum_{jk} f_{ijk}, \; p_j^2 = \sum_{ik} f_{ijk}, \; p_k^3 = \sum_{ij} f_{ijk}, \; i, j, k \in \{A,C,G,T\}.$$

Mean-field approximation models codon frequencies under the hypothesis of independent position-specific nucleotide generation in codon. This approximation is widely used in literature (Bernaola-Galvan et al., 2000).

Two another vertexes of the mean-field phase triangle, $P^{(1)}mf$ and $P^{(2)}mf$, can be easily calculated under the same hypothesis:

$$P^{(1)}(mf)_{ijk} = p_i^2 p_j^3 p_k^1, \; P^{(2)}(mf)_{ijk} = p_i^3 p_j^1 p_k^2, \quad i, j, k \in \{A,T,G,C\}. \tag{3}$$

There exists exactly triangle (mf, $P^{(1)}mf$, $P^{(2)}mf$) in 64D space of triplet frequencies because no more than three different triplet distributions can be produced under the accepted model.

Assuming coding regions in the complementary strand to have the same position-specific frequencies, one can easily get complementary strand phase triangle of mean-field approximation: ($C^R mf$, $C^R P^{(1)} mf$, $C^R P^{(2)} mf$).

Thus, the differences in nucleotide frequencies dependent on their position in codon provide existence of six possible 64D vectors of triplet frequencies. They are mf, $P^{(1)}mf$, $P^{(2)}mf$, $C^R mf$, $C^R P^{(1)} mf$, and $C^R P^{(2)} mf$ in mean-field approximation notation. Theoretically, some vectors can coincide, but only in such a way that makes the resulting set to consist of six (non-degenerated case), three (partially degenerated case), two, or one (completely degenerated cases) vectors.

Non-degenerated case corresponds to the presence of six 'coding' clusters, which is typical of a number of real genomes. Coincidence of phase triangles of the forward and complementary strands in any combination of their vertexes produces the partially degenerated case, which is an ordinary case too for certain bacterial genomes (see section 3.1).

The completely degenerated cases appear, if true and shifted codon distributions are identical. Referring to Eq. (3), they correspond to position-independent distribution of nucleotides in codons. Denoting this distribution as $\boldsymbol{m}$, one calculates it using four position-independent nucleotide frequencies:

$$m_{ijk} = p_i\, p_j\, p_k, \quad p_i = 1/3(p_i^1 + p_i^2 + p_i^3), \quad i, j, k \in \{\mathrm{A, T, G, C}\}. \qquad (4)$$

It corresponds to the simplest zero order model of coding regions and constitutes approximately the center of phase triangle. Completely randomized distribution $\boldsymbol{m}$ and its complementary reversion $C^R\boldsymbol{m}$ would coincide iff $p_A = p_T$ and $p_C = p_G$. Thus, the number of degenerated clusters depends on the interstrand symmetry in nucleotide frequencies. However, the cases were called 'degenerated' because of unusual for real genomes coincidence of coding and shifted distributions.

2.5 Information content in the triplet distributions

Visual illustration of the information content of some true and modeled triplet distributions is shown in Figure 2 by the examples of two bacterial genomes. Pairwise distance between the two triplet distributions $\boldsymbol{g}$ and $\boldsymbol{h}$ was calculated according to symmetrized Kullback–Leibler distance

$$D^{SYM}(\boldsymbol{g};\boldsymbol{h}) = \frac{1}{2}\left(\sum g_i \ln\frac{g_i}{h_i} + \sum h_i \ln\frac{h_i}{g_i}\right).$$

Metric multidimensional scaling (MDS) technique was applied to visualize the distributions on 2D plane on the basis of the obtained pairwise distances.

Figure 2 shows relative information content of true phase triangle distributions ($\boldsymbol{f}, \boldsymbol{f}^{(1)}$, and $\boldsymbol{f}^{(2)}$) and mean-field approximation ones ($\boldsymbol{mf}$, $P^{(1)}\boldsymbol{mf}$, and $P^{(2)}\boldsymbol{mf}$). The calculated shifted distributions $P^{(1)}\boldsymbol{f}$, $P^{(2)}\boldsymbol{f}$ and the center of true phase triangle $\boldsymbol{f}^{(av)}$ (which is arithmetic mean of $\boldsymbol{f}$, $\boldsymbol{f}^{(1)}$, and $\boldsymbol{f}^{(2)}$) are shown as well with the origin set at the $\boldsymbol{m}$ point. Information content of all shown distributions is proportional to their distance to the origin.

The maximum of information is contained in the codon distribution $\boldsymbol{f}$, which is the most distant point from the origin. Its high information content gives more contrast cluster structure and better quality of unsupervised gene recognition. Calculated shifted distributions $P^{(1)}\boldsymbol{f}$, $P^{(2)}\boldsymbol{f}$ are very close to real shifted distributions $\boldsymbol{f}^{(1)}$, $\boldsymbol{f}^{(2)}$, confirming the fact of small correlation in codon order for bacterial genomes.

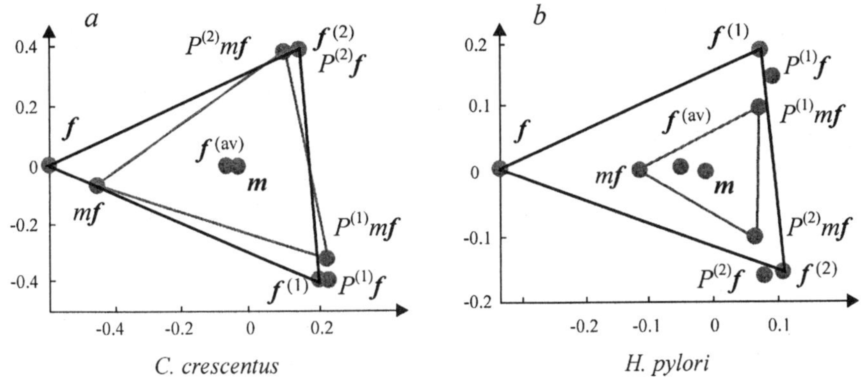

Figure -2. MDS plots representing relative information content in triplet distributions.

The difference in sizes between true phase triangle and mean-field one reflects the presence of correlation in the order of nucleotides. This difference is small for *C. crescentus* genome and considerable for *H. pylori* genome. Among all considered bacterial genomes, *H. pylori* demonstrates the largest difference between *f* and *mf*, while *C. crescentus* is in the very middle. It agrees with the fact that bacterial codon usage is reasonably well approximated by its mean-field distribution (Bernaola-Galvan et al., 2000).

3. TYPES OF SEVEN-CLUSTER STRUCTURE

The skeleton of seven-cluster structure is created by positional relationship of two phase triangles: the forward strand and complementary strand ones. The types of mutual position of cluster triangles refer to the classical problem of symmetry (or asymmetry) between the forward and backward DNA strands (Mrazek et al., 1998; Lobry and Sueoka, 2002) as well as to the pattern of symmetric properties of codon usage.

3.1 Four 'pure' types of seven-cluster structure

Among the seven-cluster structures of all considered bacterial genomes, we picked out four 'pure' types of positional relationship of two phase triangles, which are shown in Figure 3 by the examples of corresponding genomes.

The first pattern—'parallel triangles' (Figure 3*a*)—corresponds to the AT-rich genome of *Fusobacterium nucleatum* (GC content is 27 %). The phase triangles exhibit an opposite rotation of the vertex indices with the F1 vertex meeting the B1 one. This pattern is commonly observed in AT-rich genomes.

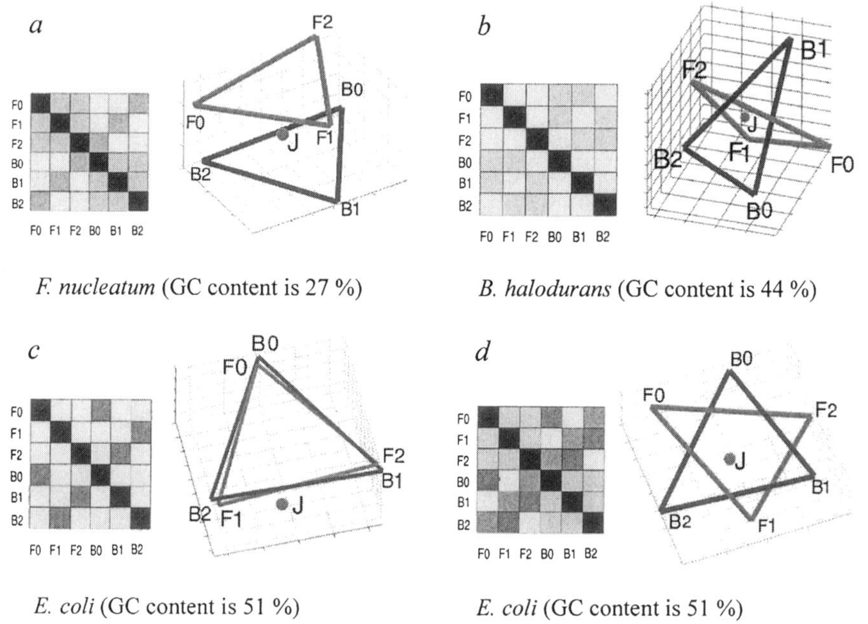

F. nucleatum (GC content is 27 %)

B. halodurans (GC content is 44 %)

E. coli (GC content is 51 %)

E. coli (GC content is 51 %)

Figure -3. Four 'pure' types of universal seven-cluster structure and corresponding matrices of pair distances between cluster centers in 64D space (the distances are shown by gray scale intensity: the darker color corresponds to the less distance): (*a*) 'parallel triangles', (*b*) 'perpendicular triangles', (*c*) coinciding triangles, and (*d*) flower-like structure.

The second pattern—'perpendicular triangles' (Figure 3*b*)—belongs to the genome of *Bacillus halodurans* (GC content is 44 %). The 'perpendicular triangles' structure is only an approximate picture; the real configuration is almost 6-dimensional due to the distance matrix symmetry: all non-diagonal elements have similar big value.

The third pattern (Figure 3*c*) represented by the *Escherichia coli* genome (GC content, 51 %) corresponds to partially degenerated case of coinciding phase triangles of the forward and complementary strand with F0–B0, F1–B2, and F2–B1 pairs of coinciding vertexes.

The fourth flower-like pattern (Figure 3*d*) being represented by the GC-rich genome of *Streptomyces coelicolor* (GC content, 72 %) is close to plane regular hexagon with non-coding cluster slightly displaced in the direction perpendicular to the hexagon plane. The displaced J cluster position is connected with the CG content of non-coding regions, which is less than that of coding regions. The same situation was observed for the third pattern of coinciding triangles. The four patterns are typical of triplet distributions of bacterial genomes observed in nature by the moment. The other ones combine features of these four 'pure' types.

3.2 Genomic GC content and type of seven-cluster structure

To represent distribution of seven-cluster structure types over bacterial genomes, we created a data table of 64D vectors of *codon* frequencies (codon usage) for all the 143 bacterial genomes and visualized it in 2D space of the first two principal components. It is a well-known fact that many properties of the codon usage are correlated with genomic GC content (Lobry, 1997; Wan et al., 2004). The first principal component explains near 60 % of the total variance in codon usage, and factor scores reflect GC content: coding, genomic, and position-specific ones are equally highly correlated with factor scores having $r > 0.95$. Figure 4 shows PCA plot of bacterial genomes distribution in the space of their codon usage. Ascribing a bacterial genome to the type of its seven-cluster structure resulted from automatic classification of distance matrices. Locations of all the mentioned genomes are highlighted by big markers and denoted by their source name abbreviation.

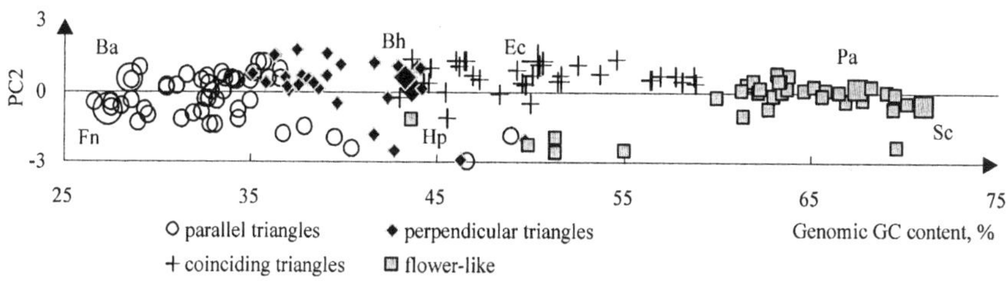

Figure -4. PCA plot of bacterial genomes distribution in the 64D space of codon frequencies with their seven-cluster structure attribution to one of the four pure types. The first principal component scores were replaced by the corresponding genomic GC content ($r > 0.95$); axes were adjusted in length to reflect approximately the explained variance ratio (60 % and 8 %).

The presented PCA plot confirms the fact that bacterial codon usage is determined essentially by the GC content and so do the type of seven-cluster structure. In 64D space of codon frequencies, bacterial codon usage located near one dimensional curve, that is almost straight line reflecting the GC content scale. In general, going along the curve, one meets at first 'parallel triangles', which transform gradually to 'perpendicular triangles'. On this way, one can meet flower-like patterns in one of the 2D projections, like that of *H. pylori* genome (Zinovyev et al., 2003). Then, the pattern goes to the coinciding triangles with genomic GC content around 50 %. Further pairs F0–B0, F1–B2, and F2–B1 diverge in the same 2D plane and after 55 % threshold in GC content, the flower-like structures are present almost exclusively.

3.3 Basic symmetry in triplet frequencies

The types of seven-cluster structure depend strongly on the genomic GC content, because bacterial codon usage is at most determined by it. The shape of mutual position of phase triangles presents a visual illustration of the peculiarities of bacterial codon usage and in particular, it reflects the symmetry (or asymmetry) between the forward and backward DNA strands and other symmetric properties of codon usage. Thus, interstrand symmetry is displayed by coincidence of the centers of phase triangles. Namely, the 'parallel triangles' type shows a strong asymmetry between DNA strands of AT-rich bacterial genomes, whereas other types reflect relative interstrand symmetry of corresponding genomes.

In terms of the mean-field approximation, the structure type reflects symmetries in the set of 12 position-specific frequencies with respect to the phase shift and complementary reverse operations. Since phase triangles exist due to the difference between position-specific frequencies p_i^1, p_i^2, p_i^3 and randomized ones p_i, $i \in \{A,T,G,C\}$, some features of cluster structure could be explained in terms of these differences.

Plane structures like hexagon and coinciding triangles are easy to observe in 3D space of the differences in coding GC content between position-specific GC frequency and the mean one: $\Delta_{GC}^k = p_C^k + p_G^k - (p_C + p_G)$, $k = 1, 2, 3$. GC-rich bacterial genomes perform the special pattern of Δ_{GC}^k: $\Delta_{GC}^1 \approx 0, \Delta_{GC}^2 < 0, \Delta_{GC}^3 > 0$, or (0 – +), if denoting them by triplet of signs. Phase shifts operation rotate the signs, while complementary reverse one only reads them from back to front (GC content is not changed under C^R transformation). The forward strand phase triangle (0 – +, – + 0, + 0 –) and the complementary strand one (+ – 0, 0 + –, – 0 +) constitute exactly hexagon in 3D space of Δ_{GC}^k, $k = 1, 2, 3$. Note that none of the vertexes coincide.

Coinciding triangles that appear in the vicinity of 50 % of genomic CG-content have a (+ – +) symmetric pattern of Δ_{GC}^k signs. It obviously gives two identical phase triangles. Moreover, the pattern of coinciding vertexes (Figure 3*c*) reflects symmetric features of position-specific nucleotide frequencies with respect to complementary reversion: $p_i^1 \approx p_{\hat{i}}^3$, $p_i^2 \approx p_{\hat{i}}^2$, $i \in \{A,T,G,C\}$.

Similar features can be observed for genomes with 'parallel triangles' structure pattern. 'Parallel' location of triangles with F1–B1 'coincidence' reflects $p_i^3 \approx p_{\hat{i}}^3$ and $p_i^1 > p_{\hat{i}}^2$ $p_i^2 > p_{\hat{i}}^1$, $i \in \{A, G\}$ properties of codon usage in the corresponding genomes.

More complicated interrelations of position-specific frequencies determine the pattern of perpendicular triangles. Complementary reversion symmetry in A and T frequencies $p_i^1 \approx p_{\hat{i}}^3$, $p_i^2 \approx p_{\hat{i}}^2$, $i \in \{A,T\}$ together with special asymmetry in G and C frequencies $\Delta_C^k = 0$, $\Delta_G^k \neq 0$, $k = 1, 2, 3$ produce distance matrix like that in Figure 3*b*.

A more detailed description of the symmetries in codon usage, which become apparent from the cluster structure, is available in (Gorban et al., 2005).

4. CONCLUSION

In this paper, we prove the universal seven-cluster structure in triplet distributions of bacterial genomes that reflects the main source of sequence heterogeneity. We showed the seven-cluster structure to be determined by a single parameter: genomic GC content. Based on the analysis of 143 completely sequenced bacterial genomes, we describe four 'pure' types of the structure and basic symmetries in triplet frequencies they reflect.

NEW METHODS TO INFER DNA FUNCTION FROM SEQUENCE INFORMATION

I. Abnizova[1*], R. te Boekhorst[2], K. Walter[1], W.R. Gilks[1]
[1] *MRC-BSU, Robinson Way, Cambridge, UK, e-mail: irina.abnizova@mrc-bsu.cam.ac.uk.,*
[2] *University of Hertfordshire, College Lane, Hatfield, UK*
[*] *Corresponding author*

Abstract: We present a new computational approach to infer DNA function from eukaryotic DNA sequence information. It is based on the fact that exons, regulatory regions, and non-coding non-regulatory DNA exhibit different statistical patterns. We suggest capturing and measuring these patterns by the following suite of statistical tools: (1) the 'fluffy-tail' test, a bootstrap procedure to recognize statistically significant abundant similar words in regulatory DNA; (2) an algorithm to assess the density of patches of low entropy as a new measure of homogeneity. This measure can be used to distinguish coding from non-coding and regulatory regions; (3) an adaptive window technique applied to rescaled range analysis and entropy measurements. This is an optimization technique to segment DNA into homogeneous parts (that are therefore likely to be coding), of which the outcomes are independent of the size of the sliding window and hence avoids averaging. The application of our methods to several annotated data sets from six eukaryotic species enables a clear separation of coding, regulatory, and non-coding non-regulatory DNA. We propose that established computational methods complemented by our new statistical tests and augmented with the novel optimization technique for sliding windows create a powerful tool for the characterization and annotation of DNA sequences. The software is available from the authors on request.

Key words: regulatory regions; coding DNA; heterogeneity; statistical methods; information entropy; long-range correlations, motif abundance

1. INTRODUCTION

Annotation of genomic DNA is one of the most important problems in bioinformatics. A number of computational algorithms using evolutionary comparisons, whole-genome data, and putative co-regulated genes have been successfully demonstrated in recent years. Unsupervised search algorithms for analyzing the structure of the genome form two broad classes: methods for characterizing the composition and methods for assessing the serial dependency of DNA sequences. Nucleotide composition is commonly investigated with tools from information theory (i.e., various ways to estimate the entropy of parts of the genome), self-organizing maps (Abe et al., 2003), complexity analysis (Li et al., 1997; Gusev et al., 1999, Wan et al., 2003; Chuzhanova et al., 2004; Orlov and Potapov, 2004), and statistical linguistics (Mantegna et al., 1994; Bolshoy, 2003).

Statistical dependencies between nucleotides/amino acids have been analyzed using mutual information functions (Azbel et al., 1995), Markov models (Krogh et al., 1994), spectra (Voss et al., 1992), latent periodicity (Korotkov et al., 1997; Chechetkin et al., 1998; Korotkov and Kudryaschov, 2001), and methods derived from random walk dynamics, such as detrended fluctuation and rescaled range analysis (Peng et al., 1994). In particular, the detection of long-range correlations (LRCs) has attracted much attention (Mantegna et al., 1994; Peng et al., 1994; Azbel et al., 1995; Li et al., 1997; Herzel et al., 1997; Guo et al., 2003) and correlations ranging from a few base pairs up to 1000 bp have been found. Results from the application of these methods indicate that non-coding and coding DNA have distinguishable statistical properties: long range correlations have been reported for non-coding but not for coding regions (Peng et al., 1994; Herzel et al., 1997; but see Voss et al., 1992). Unfortunately, when these techniques are used in the conventional way, their results will be highly dependent on the size of the sliding window.

We describe here three new methods to characterize coding, regulatory, and non-coding non-regulatory DNA regions. (1) When considering regulatory region recognition, we assume that the abundance of regulatory motifs within regulatory regions leaves a distinct 'signature' in nucleotide composition and that it is possible to capture this 'signature' statistically. More specifically, we hypothesize that it takes the form of an over-representation of 'similar words' (which are not simple repeats). This over-representation should show up as outliers in the right tail of the distribution of similar word lists of variable length. (2) We use an adaptive window technique (te Boekhorst et al., 2005) to recognize putative coding regions due to a high entropy and a low Hurst coefficient, assuming they are likely to be the most homogeneous DNA parts. (3) We suggest using the density of low-entropy patches as another complementary way to recognize homogeneous DNA parts. The performance of the methods is tested on data sets derived from six eukaryotic species. The results are consistent with those of a number of established methods (see Cox et al., 1997; Korotkov et al., 1997; Stern et al., 2001; Berman et al., 2002; Markstein et al., 2002; Nazina et al., 2003; Wan et al., 2003; Lifanov et al., 2004; Orlov and Potapov, 2004), and outperform some approaches based on the application of a fixed window size (as used, for instance, by Peng et al., 1994; Li et al., 1997) in rescaled range analysis and the estimation of informational entropy (see Results). All our three methods are unsupervised and could be used in cases where data from multiple genes is not available and as a complementary tool when such data is available.

2. METHODS AND ALGORITHMS

2.1 Adaptive window technique applied to rescaled range analysis and entropy measurements

In previous work, we developed methods for assessing the degree of local DNA homogeneity in a sequence (see for details Abnizova et al., 2003; te Boekhorst et al., 2005) using rescaled range analysis and measuring informational entropy. A genomic sequence is considered to be homogeneous if it is characterized by a low Hurst exponent (as a measure for the degree of sequential persistence, see definitions below) and a high value for entropy. We found that sequences of coding DNA have lower Hurst exponents and higher entropy compared to non-coding DNA. Below, we will briefly describe these two measures and a problem associated with the conventional way of their assessment. To circumvent this drawback, we suggest using the two statistics in combination with an optimization procedure that adapts the size of the sliding windows to the local structure of the DNA sequence under consideration.

Random walks and rescaled range analysis. One way to characterize the succession of nucleotides in a sequence is to imagine a 'walk' along the DNA string by moving up each time a pyrimidine (a T or C) occurs and by moving down whenever a purine (an A or G) is encountered. One can also encode DNA in any other way such as CG versus AT, A versus CGT, and so on. The result of the walk is a 'landscape' of which the contours depend on the way the coded elements alternate with each other. If the probability of the occurrence of a pyrimidine equals that of a purine and is independent of the position in the string, then such a 'DNA walk' would actually be a 'random walk'. In that case, increments of the walk would form a series of independent and identically distributed events with constant mean and finite variance. The sequence of increments would therefore be stationary and furthermore characterized by a flat spectrum and the absence of autocorrelations ('white noise'). The random walk itself, being an integrated white noise, is typically non-stationary. This is manifest in the 'hillyness' of the landscape, largely non-vanishing spikes in the autocorrelation function and the predominance of low frequency components in the power spectrum of a random walk.

One way to quantify such long-range correlations is by means of 'rescaled range analysis'. Rescaled range analysis can be applied to any sequence of data in time or space. Setting $x_k = +1$ for k = T, C, and $x_k = -1$ for k = A, G, the sequence $\{x_k\}$ can be characterized by

$$\langle x \rangle_n = \frac{1}{n}\sum_{i=1}^{n} x_i \, , \, X(i,n) = \sum_{m=1}^{i} [(x_m - \langle x \rangle_n)]$$

$$R(n) = \max_{i \le n} \mathrm{X}(i,n) - \min_{i \le n} X(i,n),$$

$$S(n) = \left(\frac{1}{n} \sum_{i=1}^{n} (x_i - \langle x \rangle_n)^2 \right)^{1/2} \quad \text{for any } 2 \le n \le N.$$

N is the length of the DNA sequence, n is window size. For scale-free data, $R(n)/S(n) \sim (n/2)^H$. Hence, the Hurst exponent H can be computed from the least squares fit of the regression of $\log[R(n)/S(n)]$ on $\log[n]$. The test-statistic (H) should be compared with a Hurst exponent obtained under the null hypothesis that the cumulative data (i.e., the time series after integration) are from a random walk and therefore that the original data are white noise. In that case, H would equal 0.5. It is unlikely that the increments are independently and identically distributed, if the Hurst exponent deviates significantly from $H = 0.5$. When long stretches of purine alternate with long stretches of pyrimidine, for example, the sequence shows 'persistence' and H will be larger than 0.5. Conversely, a Hurst coefficient below 0.5 indicates 'anti-persistence'.

Obviously, changes in persistence are not accounted for by the conventional application of rescaled range analysis, i.e., by using a (too) large, fixed window. We therefore suggest a window that increases until the Hurst exponent reaches a minimum. This would optimize the window for detecting the least persistent (most homogeneous) parts. Note that minimal size of the window should be not less than 250 bp.

Entropy-based DNA segmentation. The conventional procedure for measuring DNA entropy (Li et al., 1997) in a given subsequence typically consists of calculating a frequency vector of nucleotide composition for a sufficiently large but subjectively defined area, and then subjecting it to the well-known Shannon function:

$$E_{seq,M} = -\sum_{i=1}^{M} P_i \cdot \log(P_i),$$

where M is the length of the frequency vector, which in case of single nucleotides is 4 and in case of dinucleotides (the choice we adopted in our work) is 16, etc. The entropy of a frequency vector is maximal when all elements occur with equal probability, in which case $E_{seq,\ M}$ = log(*1/M*), and hence measures the 'evenness' of the composition. In contrast, low entropy indicates the 'dominance' of a few of the elements. Clearly, when increasingly larger fragments of the same window of sequence are used, the entropy of the fragments will asymptotically approach the entropy of the entire sequence. Such an overall estimate does not capture the possibly deviating entropy of small but functionally important subparts. Too large and fixed windows therefore overlook local differences in nucleotide composition.

For a more powerful method, we therefore suggest to optimize the length of the local windows. To this aim, we move a sliding window of varying length along the DNA sequence, optimizing its length at each position to find the first local maximum in entropy. If coding DNA is more homogeneous than other parts of the genome, then regions typified by high entropy are likely to be exon locations.

2.2 Density of low entropy patches as a measure of heterogeneity

DNA heterogeneity is generally believed to be caused by a gradual change (bias) in nucleotide composition in different parts of the sequence or to be brought about by short runs of repetitive (low-entropy) patches. The latter view implies that once such patches are removed, a string of homogeneous DNA remains.

To calculate the density of low entropy patches, we developed a simple entropy-based algorithm, which is able to detect these patches in a given stretch of DNA. The algorithm is based on a bootstrapping procedure: a large number of random sequences with the same base composition and length as the original sequence are generated. Next, the dinucleotide entropy is calculated in every window for both the original and the random sequences. From the randomized sequences, we set up a threshold below which the probability of finding a low-entropy patch is $p < 0.005$. For example, for a sequence of 150 bp, a fixed window size of 20 bp and with a base composition $(p_a, p_c, p_g, p_t) = (0.3, 0.2, 0.2, 0.3)$, the significant threshold in dinucleotide entropy will be 1.4. The density of low-entropy patches is computed as the number of 20-bp patches with entropy under the pre-calculated threshold divided by the sequence length.

Thus, we used the density of these low entropy patches as a measure of local heterogeneity of DNA.

2.3 Measure of similar word abundance: 'fluffy-tail' test

It is known (Berman et al., 2002; Markstein et al., 2002; Papatsenko et al., 2002) that multiple binding motifs and multiple binding sites for the same motifs are often present within regulatory regions. To quantify this motif abundance, we developed a method based on an exhaustive counting of similar words (Abnizova et al., 2005).

We call two words of the same length, m, k-**similar**, if they differ with less or equal k mismatches. We conducted analyses for $m = 3, 5, 7, 9, 12$ with corresponding $k = 0, 1, 2, 3, 4$.

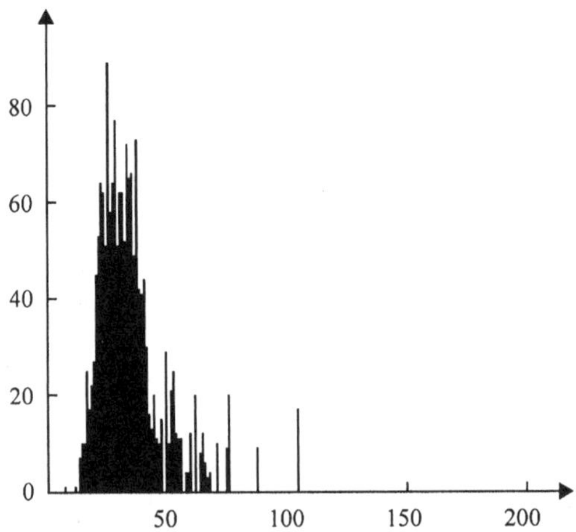

Figure -1. Histogram of similar word list length for the *abdominant Anterior* regulatory region, *Drosophila melanogaster*, 1745 bp length; $m = 5$, $k = 1$. The horizontal axis shows the similar word list length; the vertical axis is the number of lists.

To construct a distribution of similar words, we start with the first m-word in the sequence, and search for all words in the sequence k-similar to this word, building up a list of words similar to this first word. We then shift the position by one base pair, and repeat the same procedure for second word, etc. In the end, for each *m*-word in the sequence, we have a list of *k*-similar words. The number of words in the list is its length. We compute the number of lists having different lengths, and plot them as a histogram, as in Figures 1 and 2 (the data used for these figures are from Papatsenko et al., 2002). The presence of over-represented words (motifs) shows up as outliers in the right tail of the distribution of word list length.

From this plot, it can be seen that most lists contain 10 to 40 words, but there are outliers: some very large lists form a long, 'fluffy' tail. We call a list having the largest size a *maximal similar word list* (MSWL). If the original sequence is characterized by the presence of an unusually high number of over-represented words, we expect it to contain more long lists in comparison with a random sequence.

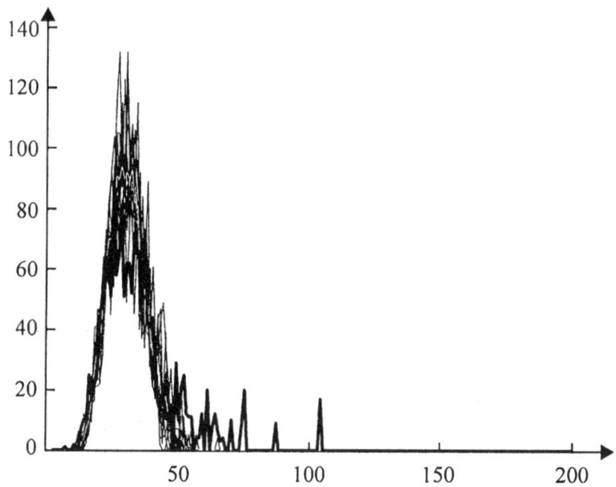

Figure -2. Example of a fluffy sequence: The solid curve represents the distribution of word list size for the regulatory *abdominant Anterior* gene of *Drosophila melanogaster* (1245 bp). The series of dotted curves are the histograms of 10 different randomly shuffled sequences. All dotted tails are shorter than the solid one, demonstrating the significance of the solid tail. The horizontal axis shows the similar word list length; the vertical axis is the number of lists.

To assess the statistical significance of the MSWL length, we compare the distribution of similar words for a null random background. The null background is generated by shuffling the sequence, retaining its single nucleotide composition and length. The significance is measured with fluffiness coefficient F_r:

$$F_r = (L_{\text{max, original}} - \bar{L}_r) / \sigma_r,$$

where $L_{\max,\,\text{original}}$ is the number of words in the MSWL of the original sequence, $\overline{L}_r$ and σ_r are the mean and standard deviations of the MSWL size in each of shuffled sequences. The coefficient F_r measures the number of standard deviations from the mean of MSWL size for original compared with randomized sequences. If $F_r > 2$, we call the sequence 'fluffy' and the MSWL in this sequence is significant.

Figure 2 shows an example of 'fluffy' sequence: an original similar words distribution (solid line) for the *abdominant Anterior* regulatory region contains much longer tails in contrast with randomized similar words distribution (dotted lines).

3. RESULTS

3.1 Application of the adaptive window technique in rescaled range analysis and entropy

In previous analyses (te Boekhorst and Abnizova, in prep.), we found a Hurst exponent value of about 0.5 for exons based on Pu–Py coding. However, using CG–AT coding, the Hurst coefficient was below 0.45. This anti-persistence might be caused by the preponderance of C/G nucleotides at the third codon position and the resulting latent periodicity within coding regions.

For the application of the adaptive window technique, we collected 25 annotated sequences from 5 species: human, mouse, fugu, fruit fly, and yeast; each sequence contained one or more exons with annotated start positions (data from the Ensembl Genome Browser http://www.ensembl.org/ and the DoTs database http://www.allgenes.org/).

For the rescaled range analysis, we used both a CG versus AT (CG–AT) and a purine–pyrimidine (Pu–Py) binary coding for each annotated DNA stretch. We segmented each stretch into heterogeneous and homogeneous parts (defined as having a Hurst exponent $H \leq 0.5$ for Pu–Py).

In 80 % of the sequences considered in this paper, annotated exons coincided with homogeneous parts identified by the adaptive window algorithm. Figure 3 shows an example of the segmentation applied to a sequence containing two internal exons (positions 700–1700 bp and 1950–2700 bp, relative coordinates) of the CG0123 gene of *Drosophila melanogaster*.

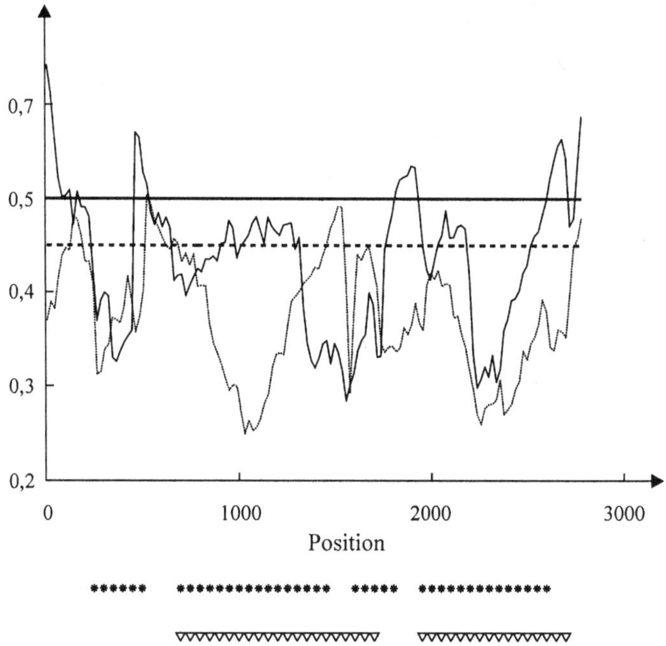

Figure -3. (Upper panel) The values of Hurst exponent (vertical axis) measured for the gene CG0123 of *Drosophila melanogaster*: solid line, based on Pu–Py coding; dashed line, based on CG–AT coding. The lengths of windows are not shown. (Lower panel) Dotted lines show predictions that are the areas where both solid (Hurst Pu–Py coding) and dashed (Hurst CG–AT coding) curves are lower than 0.5 and 0.45 correspondingly. Triangled lines show the actual exons locations.

We also ran the adaptive window algorithm for entropy measurements on the same data (Figure 4). The positions with highest entropy values (solid curve) correspond to actual start exon positions (dotted lines). The lengths of optimal windows are not shown on the plots, but they approximately correspond to actual lengths of the exons. Note that if we would combine the results for both Hurst and Entropy methods (as in a Bayesian framework), more robust predictions could be obtained, which might reduce the number of false positives (as, for instance, found in the rescaled range analysis illustrated in Figure 3).

3.2 Density estimation of low-entropy patches

Results of density estimation of low-entropy patches are summarized in Table 1. In Table 1, **'diverged'** denotes non-coding non-conserved DNA. These sequences were picked at random from the Fugu whole genome shotgun assembly v. 3.0 (August 26, 2002); **'CNE'** stands for conserved non-coding element and are considered as putative regulatory regions.

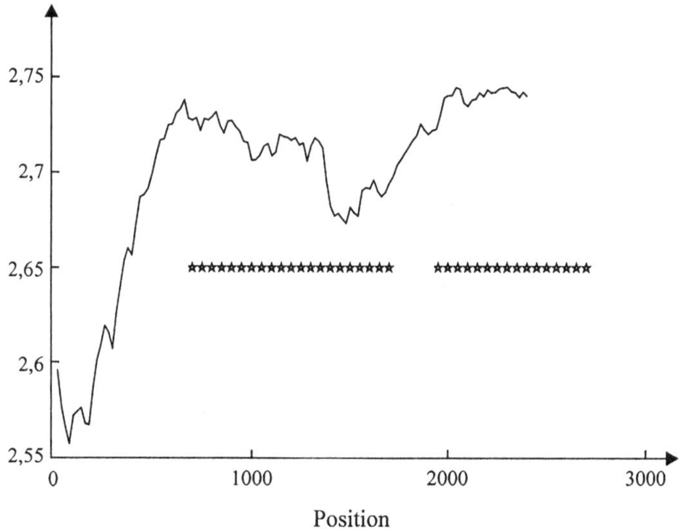

Figure -4. The solid curve is the dinucleotide entropy value measured along the annotated sequence, same data as in Figure 3. Dotted lines show actual exons positions. Entropy values are the highest at the starting positions of the two exons.

CNE's elements were collected by Woolfe et al. (2005) using MegaBLAST multiple alignment between sequences from fugu, mouse, rat, and human. Exons were randomly picked from Scaffolds 1 and 2 of the fugu whole genome shotgun assembly v. 3.0 (August 26, 2002. www.ensembl.org). Since CNEs are short, we had to concatenate them into longer fragments; the lengths of the fragments were selected in a random way in order to be compared with other two data sets.

The data in this table show that DNA sequences from fugu can be separated into functional parts due to the density of low-entropy patches (defined in subsection 2.2):. exons typically have minimum density, non-coding non-regulatory DNA has maximum density, and putative regulatory DNA (in this case represented by CNEs) has intermediate density.

3.3 The 'fluffy-tail' test

To investigate the performance of the 'fluffy-tail' test, we apply it to a collection of experimentally verified functional regulatory regions from the *Drosophila* genome (Nazina et al., 2003). We also compiled two negative test sets from the Ensembl Genome Browser, one consisting of randomly picked exons the other, of stretches of non-coding non-regulatory DNA (see Abnizova et al., 2005).

Table -1. Discrimination between diverged (non-conserved non-coding), conserved non-coding (CNE), and coding DNA due to density of low entropy patches

Functional type of sequence	Density of low-entropy patches	C + G	P(CG)	D(CG)	Length, bp
		Diverged DNA			
Diverged1	0.34	0.4300	0.028	0.60	5156
Diverged2	0.44	0.4100	0.018	0.41	15221
Diverged3	1.00	0.5400	0.030	0.44	755
Diverged4	0.66	0.3500	0.020	0.65	1000
Diverged5	0.56	0.3900	0.014	0.38	4819
Diverged6	0.62	0.4000	0.020	0.48	4750
Diverged7	0.37	0.4000	0.020	0.53	8501
Mean	**0.58**	**0.4200**	**0.020**	**0.46**	
		CNEs			
Cne2	0.032	0.3900	0.018	0.44	14782
Cne3	0.000	0.3800	0.012	0.34	3423
Cne5	0.018	0.3700	0.012	0.33	18876
Cne6	0.078	0.3800	0.010	0.29	14500
Cne7	0.020	0.4100	0.017	0.39	23258
Cne8	0.050	0.3900	0.013	0.34	10378
Mean	**0.070**	**0.3900**	**0.014**	**0.37**	
		Exons			
Ex1_fugu	0.00	0.5400	0.048	0.64	4295
Ex2_fugu	0.00	0.5100	0.036	0.54	22373
Ex3_fugu	0.03	0.5200	0.040	0.57	19940
Ex4_fugu	0.00	0.5500	0.058	0.73	1224
Ex5_fugu	0.00	0.5200	0.045	0.63	3489
Ex6_fugu	0.00	0.5300	0.042	0.58	2835
Ex7_fugu	0.00	0.5300	0.043	0.60	3714
Ex8_fugu	0.05	0.5200	0.035	0.51	5740
Mean	**0.007**	**0.5300**	**0.041**	**0.58**	

P(CG) is the proportion of base pairs in the sequences that are CG nucleotides; mutual dependence of CG is measured with D(CG) = P(CG)/P(C) × P(G), where P(C) is the proportion of C's in the sequence and P(G) is the proportion of G's in the sequence.

For each region, we computed the fluffiness coefficient, F_r. We found significant differences in the shape of the distributions from various DNA types. A high percentage, 85 % of the experimentally verified regulatory sequences scored as having a significant fluffy tail ($F_r > 2$), implying an over-representation of similar words, while this percentage was much lower in the negative training sets (1.6 % for exons and 16 % for non-coding non-regulatory sequences). In Figure 5, the separation between regulatory DNA and other DNA types on the basis of the value of the Fluffiness coefficient is visualized.

We also tested several verified human and yeast regulatory regions (not shown here), most of them (87 %) passed the test with $F_r > 2$. The results show that the test can distinguish regulatory DNA from other types.

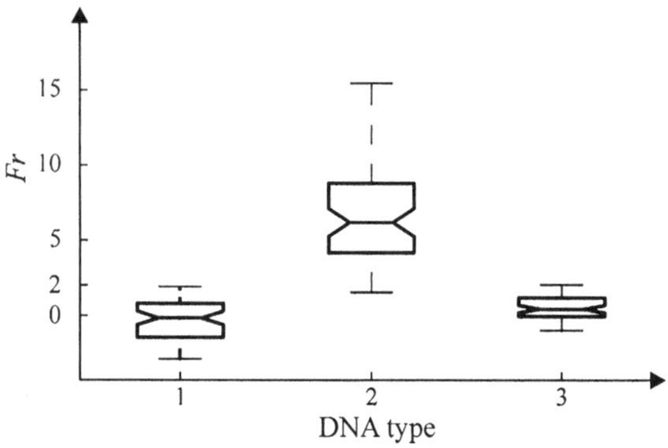

Figure -5. Separation of (2) regulatory DNA from (1) exons and (3) non-coding non-regulatory due to F. The vertical axis shows F_r coefficient.

4. DISCUSSION AND CONCLUSION

In this paper, we present a collection of independent statistical methods to characterize genomic DNA. Our collection includes two novel statistical tests—one for assessing the density of low-entropy patches, the other for measuring over-representation of similar words (the 'fluffy-tail' test)—as well as a new way to deal with established measurements: the adaptive window technique. We have applied the latter in combination with rescaled range analysis and measuring of information entropy. We have demonstrated the effectiveness of our tools on several real annotated data sets derived from eukaryotic species.

Our main finding is that the serial structure of exons is more homogeneous than that of regulatory and non-coding non-regulatory regions. This in line with a number of investigations that have shown that coding regions are characterized by high entropy, a 'random' sequence of base pairs, and are linguistically complex (Peng et al., 1994; Cox et al., 1997; Herzel et al., 1997; Stern et al., 2001; Orlov and Potapov, 2004). However, the works of Korotkov et al. (1997) and Chechetkin et al. (1998) indicate latent periodicity within coding DNA as well as our discovery that these regions are anti-persistent ($H_{cg} < 0.5$, H_{cg} is the Hurst exponent for CG–AT binary encoding here) seems to contradict the 'white noise' model for exons. Possible explanations for this discrepancy are that:

1) exon motifs actually do occur, but they are much less abundant in comparison with regulatory regions or non-coding regions containing tandem repeats. This might be due to evolutionary reasons. Arguably, DNA duplication has the strongest impact on the occurrence of motifs, and this is

assumed to be less evolutionary constrained within non-coding than within coding DNA (where mainly point mutations are accumulated);

2) the latent periodicity may be too weak to be detected without specifically designed methods. In our study, for example, anti-persistence was only found if the data were binary encoded as CG versus AT and not in case of a pyrimidine–purine coding (te Boekhorst et al., in prep.). Conceivably, had we applied the 'fluffy-tail' test to similarly coded data, exon motifs might have been discovered. We are presently investigating this hypothesis.

Our methods can be refined in several ways. First, we plan to combine their outcomes in a Bayesian way to reduce the number of false positives and to increase the likelihood of any particular DNA region to have specific function. Second, we plan to incorporate information about 2D and 3D DNA structure as well as cross-genomic comparison with already known coding and regulatory regions.

Our three methods could be used in cases where data from multiple genes are not available and as a complementary tool when such data are available.

The software written in MATLAB is available from authors on request.

ACKNOWLEDGMENT

We would like to acknowledge Yvonne Edwards, Tanya Vavouri, Adam Woolfe, Krys Kelly, Gayle McEwen, Greg Elgar, Carlo Berzuini, Yury Orlov, and two anonymous reviewers for valuable comments, discussions, and support.

REVELATION AND CLASSIFICATION OF DINUCLEOTIDE PERIODICITY OF BACTERIAL GENOMES USING THE METHOD OF INFORMATION DECOMPOSITION

A.A. Shelenkov*, M.B. Chaley, K.G. Skryabin, E.V. Korotkov
Bioengineering Center, Russian Academy of Sciences, prosp. 60-letiya Oktyabrya 7/1, Moscow, 117312, Russia, e-mail: fallandar@mail333.com
* *Corresponding author*

Abstract: We have applied information decomposition to show the presence of many sequences in prokaryotic genomes that possess latent dinucleotide periodicity but have not been found before. More than two hundred DNA sequences having such a latent periodicity were found. These sequences belonged to the 94 different microbial genomes. The classification made has shown that all of the sequences could be assigned to the 45 classes according to the type of latent dinucleotide period. A probable functional and evolutional meaning of the latent dinucleotide periodicity in different prokaryotic genomes is discussed.

Key words: latent dinucleotide periodicity; prokaryotic genomes; information decomposition; classification of periodic sequences

1. INTRODUCTION

Short tandem repeats (one–six nucleotides) or microsatellites that are long known to exist in eukaryotic genomes are highly polymorphous (Tautz, 1989; Weber, 1990). They are repeated from two to several tens times (Vogt, 1990) and are considered to have functional importance for the evolution of genetic regulation (Moxon and Wills, 1999). The study of microsatellites in prokaryotic genomes was started upon completion of the sequencing of full genomes of some microorganisms (Field and Wills, 1996, 1998; Gur-Arie et al., 2000; Mertzgar et al., 2001). The mechanism of origin of microsatellites is considered connected with the DNA strand sliding and mispairing of neighboring repeats at the time of replication (Coggins and O'Prey, 1989). The interest in studying the prokaryotic microsatellites is to apply the results of such studies to single out the markers of polymorphous loci to detect microbial organism strains and to characterize the pathogenic strains.

It has been shown earlier for the genome of Escherichia coli that dinucleotide microsatellites (with a maximal length equal to 12 nucleotides) are the most widespread after mononucleotide ones (Gur-Arie et al., 2000). Such dinucleotide microsatellites in E. coli genome are distributed evenly between the coding and non-coding regions, and this distribution is proportional to the length of these regions, while for mononucleotide microsatellites, the longer the tract is, the more microsatellites tend to appear within non-coding regions. In this work, we have applied the method of information decomposition (Korotkov et al., 2003) for searching for the latent periodicity in the prokaryotic genomes, since this method allows one to find a periodicity starting with a period length equal to two symbols. The aim of our research was to find whether the latent periodicity (i.e., not only perfect, but also a highly diverged periodicity, and dinucleotide periodicity that can be revealed only by appearance of nonrandom frequencies of individual bases in the period positions) actually exist in prokaryotic genomes. We studied the tract length of dinucleotide periodicity and in what regions (coding or non-coding) such tracts appear. We also performed the classification of latent periodicity found. The ultimate goal of these studies is to understand what functional and evolutional meaning the dinucleotide periodicity has in prokaryotic genomes and how it can be used to select new polymorphous markers.

2. METHODS AND ALGORITHMS

2.1 An information decomposition method for symbolic sequences

We have shown earlier (Korotkov et al., 2003) that the methods of finding periodicity in symbolic sequences based on a Fourier transformation and dynamic programming have a number of essential constraints that do not allow one to reveal a feebly marked periodicity in symbolic sequences.

Our previous studies have shown that the periodicity is a widespread phenomenon in various prokaryotic and eukaryotic genomes (Korotkov and Korotkova, 1995; Chaley et al., 1999, 2003). We have developed the method of information decomposition (ID method) for revealing a feebly marked or latent periodicity in symbolic sequences (Korotkov et al., 2003). This method involves four general steps:

1. Generating artificial numeric sequences possessing the periodicity with a period length of 2 to $L/2$, where L is the length of the sequence to be analyzed (a, artificial numeric sequence and b, source sequence):

```
a.    12…n12…n12…n12…k12…n…
b.    at…ggc…cta…agt…atg…t….
```

2. Calculating the mutual information between artificial and source sequences:

$$I = \sum_{1}^{n}\sum_{1}^{k} M'(i, j) \ln M'(i, j) - \sum_{1}^{n} x(i) - \sum_{1}^{k} y(j) \ln y(j) + L \ln L.$$

Here, the elements of matrix $M'(i, j)$ are the position-specific nucleotide frequencies ($a_i \in \{a,t,c,g\}$ is nucleotide type and j, position of latent period of the source sequence). The elements $x(i)$ and $y(j)$ are marginal for matrix $M'(i, j)$; n is the period length of artificial sequence; k, the alphabet size of sequence under consideration (here, $k = 4$). The matrix $M'(i, j)$ can be considered as a type of the latent period of n nucleotides.

3. Using the Monte Carlo method to estimate the statistical significance of the periods found:

$$Z(n, k) = \left\{ I(n, k) - \overline{I(n, k)} \right\} / \sqrt{D\left(I(n, k)\right)}.$$

4. Selecting the matrices of periodic regions that have their statistical significance greater than the threshold value $Z = 5.0$.

We used the ID method to search for the latent periodicity in bacterial loci taken from the GenBank. The ultimate goal is to understand the biological function of the periodic sequences found. Thus, the study of relationship between the periodic sequences found should be done.

2.2 Algorithm of the latent periodicity classification

The following algorithm was applied to classify the matrices $M'(i, j)$. Since the regions of the latent periodicity were of different length, all the compared matrices were normalized to the unity. Denoted the period length as N ($N = 2$), each matrix of the latent periodicity was represented as a vector of nucleotide frequencies distributed over $4N$ ranks. The pairwise comparison was done between the vectors as shown in Figure 1. The lower index in Figure 1 corresponds to the period position; the upper, reflects an ordinal number of compared vector. Thus, a matrix $M1$ was formed with marginal frequencies $X(i) = \Sigma_j M1(i, j)$, and $Y(j) = \Sigma_i M1(i, j)$, where $\Sigma_i X(i) = \Sigma_j Y(j) = 2$.

A_1^1	T_1^1	C_1^1	G_1^1	A_2^1	T_2^1	C_2^1	G_2^1	X(1)
A_1^2	T_1^2	C_1^2	G_1^2	A_2^2	T_2^2	C_2^2	G_2^2	X(2)
Y(1)	Y(2)	Y(3)	Y(4)	Y(5)	Y(6)	Y(7)	Y(8)	

Figure -1. A scheme of comparison between the two latent dinucleotide periodicity matrices. Both matrices are presented as $4N$-dimensional vectors.

The matrix $M2$ was constructed as expected one over a set of the random matrices having the same marginal quantities $X(i)$ and $Y(j)$ as $M1$:

$$M2(i,j) = \frac{1}{2} X(i) \times Y(j).$$

The Pearson statistics, whose value distribution follows the χ^2, allows estimating the deviation of quantities in matrix $M1$ from expected ones in $M2$ matrix:

$$U = \chi^2 = \Sigma_{i,j}\{(M1(i,j) - M2(i,j))^2\}/M2(i,j). \tag{1}$$

A number of the χ^2 freedom degrees was equal to $2 \times 4N - 1$, that is, the number of comparison ranks (the number of matrix $M1$ or $M2$ elements) minus the number of independent linkages – a single claim on constancy of marginal elements: $X(1) = X(2) = 1$.

A comparison of the original periodicity matrices was done taking into consideration all cyclic permutations of their columns, which was necessary because of an uncertainty of the period start position. These permutations were adequately reflected in original vector representations of the matrices. A possibility of classic DNA inversions was also considered. In such case, the original vector was replaced with the complementary and inverse variant. The general comparison scheme between the vectors was as follows. The first vector from a set was compared with the others, as described above, taking into consideration all the cyclic permutations and possible inversion. The least value of the χ^2 found over all the comparisons was fixed. If the value was corresponded to accidental probability of less than or equal to 5 %, then two corresponding vectors were combined via recapitulation of their elements. The elements of a new vector were calculated as weighted sums of the elements of two source vectors. The contribution of the specific vector to the sum was the greater, the higher was the number of vectors that had been already merged into it. Such a new vector was normalized again to the unity. A cyclic permutation, fixed inverse, and complementary transformation were considered in vectors combination. The process of vector comparison was continued until the χ^2 values corresponding to 5 % level were found. Thus, the classes of compared vectors (periodicity matrices) were revealed.

Let us note that critical level of the χ^2 value was estimated as the result of all $2N$ trials in searching for pairwise vector similarity. An accidental probability of similarity found in $2N$ trials, $\alpha = 1 - (1 - p)^{2N}$, should be less or equal to 5 %. From this point, a critical level of accidental probability in one trial p was calculated by using the inverse χ^2 function.

2.3 Building similarity of the dendrogram of classes

The χ^2 value was chosen as the measure of dissimilarity between the matrices in pairwise comparison, as it was described above (Methods and Algorithms, chapter 2.2), according to Eq. (1). We used the Statistica 6 package to build the tree diagram of classes similarity.

3. RESULTS AND DISCUSSION

Making a search for the latent dinucleotide periodicity in prokaryotic genomes from the GenBank-137 by using the ID method, we have found 2377 loci possessing the periodicity of such a type at the level of Z-score ≥ 4. Since our aim was to consider the most reliable cases possessing the greatest functional and evolutional significance, we selected the loci with the highest value of statistical significance of dinucleotide periodicity found (Z-score ≥ 5). Their number was as great as 455.

All the 455 loci were classified by the latent period type, as it was described above (Methods and Algorithms, chapter 2.2).

As a result, 45 classes were discriminated; each of them combined three or more loci of dinucleotide periodicity. The total number of loci belonging to the classes was 221. The dendrogram of class similarity is shown in Figure 2.

The largest class combined 19 loci of latent dinucleotide periodicity (mx10 on the dendrogram); the next two largest classes contained 11 loci each (mx32 and mx38 on the dendrogram). The latent period type of the largest class is shown in Figure 3. As we can see, cytosine and guanine are clearly predominating in both positions. Thus, the period consensus may be conventionally described as {c,g}{c,g}. The classes of latent dinucleotide periodicity shown in Figure 2 are combined according to the similarity in the period type. Let us consider, for example, two extreme left and right groups of classes. First of them combines the classes mx45, mx14, mx5, mx31, mx19, and mx4; the second, mx38, mx10, mx13, mx11, mx35, and mx1. The aggregation of classes in the first group takes place due to a significant frequency of adenine appearance in the first period position. The conventional consensus of combination is {a}{n}, where *n* is any nucleotide from the set (a,t,c,g). The aggregation process in the second group is caused by the significant values of frequencies of cytosine and guanine at the first position of the period. The combination conventional consensus in this case is {c,g}{n}.

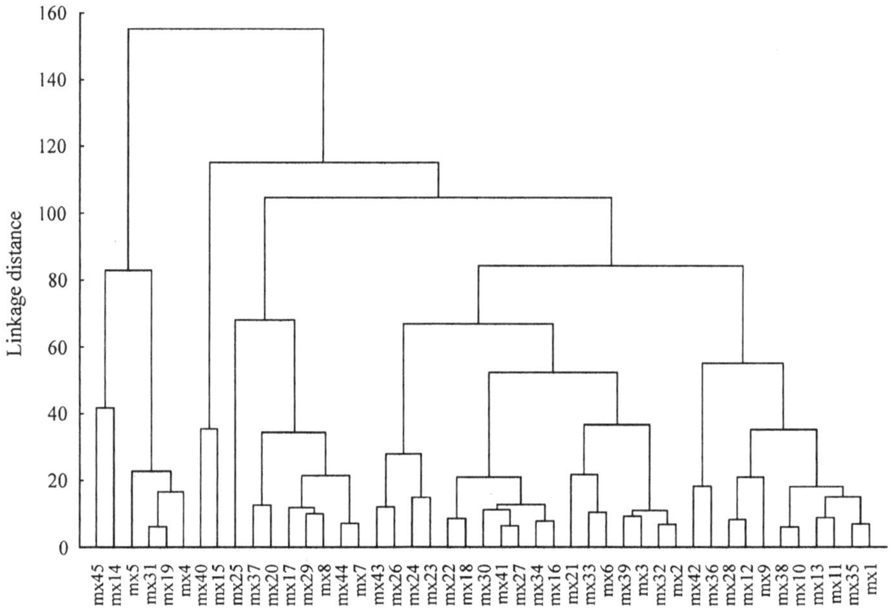

Figure -2. A tree diagram for the 45 classes of dinucleotide periodicity.

	1	2
A	0.2	0.1
T	0.1	0.2
C	0.4	0.3
G	0.3	0.4

Figure -3. A type of the latent period of the biggest class of dinucleotide periodicity (mx10 in Figure 2). This class is a product of merging of 19 periodicity matrices. The decimal numbers show the frequencies of the corresponding nucleotides at period positions.

The loci belonging to 45 classes were found in 94 different prokaryotic genomes relating to different taxonomic categories. The numbers of organisms of certain categories, in genomes of which the latent dinucleotide periodicity loci were found, are shown in Table 1.

Table -1. The numbers of prokaryotic organisms of different categories, in genomes of which the latent dinucleotide periodicity was revealed

Category	Subcategory	Number of representatives
Archaea		
	Crenarchaeota	3
	Euryarchaeota	7
Bacteria		
	Actinobacteria	6
	Bacteroid	1
	Cyanobacteria	2
	Green sulfur bacteria	1
	Firmicutes	
	Firmicutes Bacillales	8
	Firmicutes Lactobacillales	5
	Firmicutes Clostridia	2
	Planctomyces	1
	Spirochaetes	2
	Proteobacteria	
	Alphaproteobacteria	10
	Betaproteobacteria	6
	Deltaproteobacteria	1
	Epsilonproteobacteria	7
	Gammaproteobacteria	32

From Table 1, we can see that the latent dinucleotide periodicity of prokaryotic genomes is not the unique property of genome of some organism or group of organisms. It is rather the common phenomenon in the prokaryotic genomes. Considering the organism composition of the classes, we found that the largest classes (combining 19, 11, and 8 loci each) were heterogeneous, while the classes of smaller size usually combined different strains of a single organism, or the organisms of the same genus, or simply the loci of a single organism.

The analysis of the genome regions where the classified dinucleotide periodicity loci were found showed that 184 of the 221 loci (i.e., 83 %) were located in the coding regions. It is not surprising, if we take into account the fact that coding regions constitute about 80 % of the prokaryotic genomes. The perfect short dinucleotide repeats with a length not more than 12 nucleotides were found in both coding and non-coding regions of *E. coli* genome with a total number as great as 8000 tracts and were described earlier (Gur-Arie et al., 2000). However, the real tract length of dinucleotide periodicity that we found varies in a range of 36 to 1888 nucleotides (Figure 4). The earlier studies of dinucleotide periodicity of *E. coli* genome by means of Fourier analysis of positional autocorrelation function (Hosid et al., 2004), which allows finding the periodicity on the length of several hundreds of nucleotides, revealed a tandem dinucleotide periodicity in the intergenic regions only. For the coding regions, only three-nucleotide periodicity was marked out. Thus, our studies show the presence of lengthy dinucleotide periodicity tracts in prokaryotic genomes. In Figure 4, the dinucleotide periodicity tract length distribution is shown. It is easy to see that the tracts with a length more than 100 nucleotides are the most widespread in prokaryotic genomes. The longest tracts of the latent dinucleotide periodicity were found in the genes listed in Table 2.

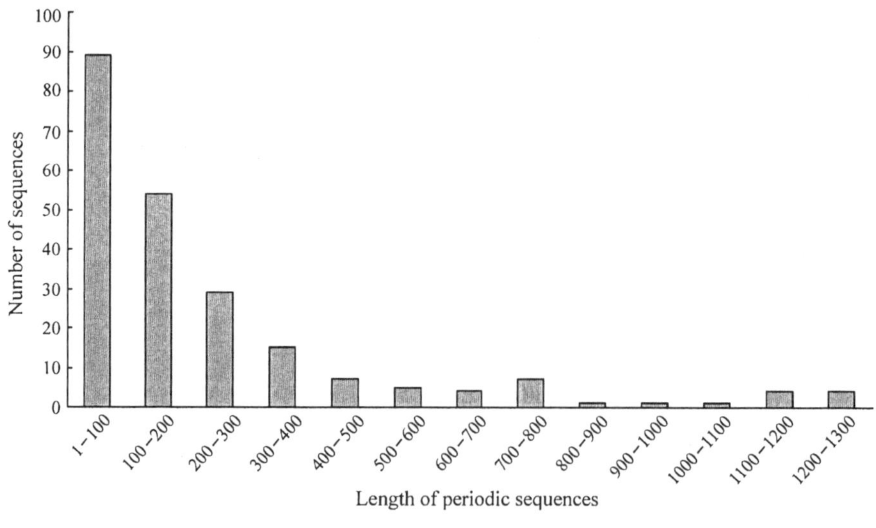

Figure -4. The length distribution of sequences possessing the latent dinucleotide periodicity found in bacterial genomes.

The presence of lengthy regions of latent dinucleotide periodicity in prokaryotic genes allows one to think that many prokaryotic genes show a high variability in those regions that have a little (if any) influence on the functionality of the protein being coded. The mechanism of microsatellites origin seems connected with DNA strand sliding and with mispairing of neighboring repeats at the time of replication (Coggins and O'Prey, 1989). Due to a short cycle of prokaryotic organism reproduction, the number of dinucleotide repeats grows rapidly, which leads to appearing of the lengthy tracts. The nucleotide mutation rate should be high in the dinucleotide periodicity tracts because of the lack of selective constraints, and thus, the dinucleotide periodicity erodes rapidly, becoming the latent periodicity. In addition, the existence of lengthy dinucleotide periodicity tracts can facilitate the improvement of genome physical properties, e.g., the raise in its flexibility, or, conversely, the rigidity for the certain DNA regions. Thus, the latent dinucleotide periodicity could be stabilized under the influence of yet other forces of natural selection.

Table -2. The latent periodicity tracts in prokaryotic genes with a length of more than 1000 nucleotides

Tract length, nucleotides	Organism	Gene	Coded protein
1024	*Escherichia coli* O157:H7 EDL933; Gammaproteobacteria	Z1379	Putative tail component encoded by cryptic prophage CP-933M; partial
1101	*Aeropyrum pernix*, Crenarchaeota	APE1445, APE1446	276 aa long hypothetical autoantigen, 182 aa long hypothetical protein
1131	*Escherichia coli* CFT073; Gammaproteobacteria	c1264	Hypothetical protein
1137	*Streptomyces avermitilis* MA-4680; Actinobacteria	cydCD	Putative ABC transporter ATP-binding protein
1176	*Listeria innocua*, Firmicutes	lin1691	CAC96922.1, similar to ABC transporter (ATP-binding protein)
1266	*Bordetella parapertussis*, Betaproteobacteria	BPP0551	CaiB/BaiF family protein
1431	*Escherichia coli* O157:H7 EDL933; Gammaproteobacteria	Z3095	Putative transposase encoded within prophage CP-933U
1632	*Shigella flexneri* 2a str. 2457T; Gammaproteobacteria	S2687	Hypothetical protein
1888	*Helicobacter pylori* 26695; Epsilonproteobacteria	HP0145	Cytochrome c oxidase, monoheme subunit, membrane-bound (fixO)

4. CONCLUSION

The searching for the latent dinucleotide periodicity in prokaryotic regions for the first time has revealed the presence of lengthy (with a length more than 36 nucleotides) tracts of such a type of periodicity in coding regions. From the 455 sequences found, 221 were systematized in 45 classes regarding the repeat type. In such a way, the dinucleotide periodicity classes for the genomes of 94 prokaryotic organisms belonging to different taxonomic categories (archeobacteria, actinobacteria, cyanobacteria, proteobacteria, etc.) have been singled out. The conducted research shows the commonality of the dinucleotide periodicity phenomenon in the genes of different bacteria and the possibility of its application to the development of species-specific and strain-specific prokaryotic markers.

ALGORITHMS TO RECONSTRUCT EVOLUTIONARY EVENTS AT MOLECULAR LEVEL AND INFER SPECIES PHYLOGENY

V. Lyubetsky, K. Gorbunov, L. Rusin, V. V'yugin*
Institute for Information Transmission Problems, Russian Academy of Sciences, Bolshoy Karetnyi per. 19, Moscow, Russia, e-mail: vyugin@iitp.ru
* *Corresponding author*

Abstract: Mathematical methods and models for comparative analysis of large sets of protein phylogenies are described. The processes modeled are gene duplication, loss, gain, and horizontal transfer. Initially, a species tree is constructed as a consensus of the corresponding gene trees using probabilistic distribution on source data. Algorithms are further implemented to identify vertices accounting for topological disparities between the gene and species trees, with possibility to infer underlying evolutionary events. The analysis is illustrated on case studies of a prokaryotic protein family and a set of protein phylogenies deduced from families from the COGs database (NCBI). The potential of the described methods to infer phylogeny and gene evolution events is discussed.

Key words: evolution; phylogenetic tree; consensus tree; gene duplication; gene loss; horizontal gene transfer; mathematic models of evolution; stochastic optimization

1. INTRODUCTION

Methods and algorithms described here are aimed at implementing two tasks: reconstruction of prokaryotic species trees and analyzing hypotheses about gene evolution. The main emphasis is placed on original algorithms and their performance, although, due to space limits, only general descriptions are provided along with the necessary references.

Events in gene evolution are usually viewed as gene divergence during species differentiation, gene duplication, gene gain, loss, and horizontal gene transfer (HGT). Molecular data is protein sequences grouped according to their amino acid and functional similarity into clusters of orthologous groups of proteins (COGs; Tatusov et al., 2001).

The general approach to reconstruct gene evolution events has long been defined (Goodman et al., 1979; Eulenstein et al., 1998). A protein gene family is selected, usually from among COGs, with subsequent assembling of multiple sequence alignment and reconstruction of the gene tree G (also referred to as a protein tree or COG tree). Further analyzed are topological similarity and disparity between the gene trees from the set $\{G_i\}$ in order to reconstruct the species tree and infer gene evolution events, respectively. Topological differences are reconciled to produce the species tree S. Alternatively, when inferring gene evolution events, considerable topological differences between a particular gene tree G (often pertaining to the family $\{G_i\}$) and the species tree S are the basis of the analysis.

Mathematic models of gene evolution are formulated to accommodate the observed differences, and optimization of model parameters is used as a tool to reconstruct evolutionary history of a microbial gene family. The evolutionary model is defined as a procedure of comparing the gene and the species trees, while its parameters are defined as sets of tree vertices with assigned evolutionary events. An optimized model has parameters corresponding to the extremes of the relevant evolutionary characteristics.

2. METHODS AND ALGORITHMS

2.1 Reconstructing the gene tree

For a given protein family (usually, for a COG), a multiple sequence alignment is assembled (routinely we use the program PROBCONS v. 1.09). Sequences with a low level of the overall detectable homology with respect to the other family members are identified using the CORE index (Notremade et al., 2003), which scores each residue for the amount of positional consistency it contains with respect to the other residues occurring in the same column (index computed with the program T-COFFEE v. 2.11). Sequences with the CORE value below the recommended threshold are removed from the alignment.

At the next step, a list of reliable phylogenetic clades is defined. For this purpose, a standard bootstrap analysis is applied. Sufficiently large numbers of bootstrapped replicates are generated for primary data using the program *seqboot* from the PHYLIP v. 3.63 package and further used to estimate the ML distance matrices under selected evolutionary model using the program PUZZLEBOOT. Neighbor joining is used to construct the trees that are further reconciled to produce a 70 % consensus (facilitated by the programs *neighbor* and *consense*, respectively, from PHYLIP v. 3.63). The groups retained in the consensus comprise the list of *reliable clades*. An ML model to be used whenever else needed is selected from more that 50 empirical models of protein evolution on the basis of significant improvement in the data likelihood according to the likelihood ratio test (Akaike, 1974; Goldman, 1993) and the Bayesian (Schwarz, 1978) information criteria. Model selection is implemented with the program ModelGenerator.

High evolutionary rates often lead to mutational saturation and loss of phylogenetic signal in highly variable regions of the protein molecule. We introduce several functions of conditional entropy in order to range the columns of the *initial alignment* according to the amount of consistency they possess with respect to the list of reliable clades and subsequently screen out for the non-informative ones.

In order to detect the amount of columns needed to be removed from the initial alignment to achieve maximum performance of phylogenetic inference, we implement a criterion based on two statistics. After eliminating a subsequent portion of highest entropy positions, we compute for the resulting alignment (1) the percentage of unresolved quartets of taxa and (2) g_1-statistic. Procedures of estimating the statistics were modified as follows.

(1) Maximum-likelihood mapping. The quartet analysis was conducted so that the phylogenetic signal related to robust clades does not contribute to the percentage of unresolved quartets. Namely, sequences corresponding to the taxa in a reliable clade from the list were substituted with an ancestral sequence reconstructed with ML at the root of the clade, thus defining a *reduced* alignment. Maximum likelihood mapping (Strimmer and Haeseler, 1997) was performed with the program TreePuzzle v. 5.02 and ancestral sequence reconstruction, with the PAML v. 3.14 package.

(2) g_1-statistic. Under the maximum parsimony, the tree length is defined as a minimum number of the changes required to explain its topology. If aligned data are phylogenetically structured, the percentage of shorter trees among a large set of the randomly generated ones will skew the tree length distribution to the left (Hillis and von Huelsenbeck, 1992). The distribution skewness is measured with the g_1 statistic. To preclude the phylogenetic signal related to well-resolved groups from contributing to the distribution skewness, we constrained analysis by generating random topologies in the areas remaining unresolved in the 70 % consensus.

Alignment columns are removed until both statistics reach extreme values. In the cases when the statistics diverge in detecting the optimal alignment, phylogenies are estimated with both alignments and further reconciled in a strict consensus. In the resulting tree, the evolutionary distances are computed as branch lengths with ML according to the selected evolutionary model.

The entire procedure is iterated until an optimal alignment is found. Phylogenetic trees inferred with the described approach always possess a higher likelihood with respect to the primary data than do the trees estimated with the initial alignment and often do not constitute a confidence set with them.

2.2 Constructing the bacterial species tree

A species tree is produced by reconciling a set of the gene trees $\{G_i\}$. It is defined as such S from the space of all *suitable* species trees that maximizes a certain parameter, e.g., the similarity between S and all G_i. There exist several natural definitions of this 'similarity' (examples are provided below), while *a priory* suitability requirements to be imposed on species trees are not biologically straightforward.

Mapping of and the cost for dissimilarity of trees were introduced by Goodman et al. (1979), Guigo et al. (1996), and Page and Charlstone (1997). This cost definition was modified by V'yugin and Lyubetsky (2002) by substituting the number of edges with the sum of the corresponding edge lengths, introducing edge length normalization, parameter γ, and probabilistic distribution over the primary sequence data (details are discussed below). The interpretation of the edge length in the tree G depends on the tree inference method. It can either be an estimate of the edge robustness or evolutionary distance between vertices.

Let α denote the conventional mapping of the gene tree G into the species tree S and let $c(G, S)$ denote the cost of such mapping, a measure of dissimilarity between α and identical mapping of trees, i.e., the extent to which the tree G is not identical to the tree S. Remember that a duplication event can be thought of as a pair (g, s), where g is a vertex in the gene tree G and s is a vertex in the species tree S satisfying the condition $\alpha(g) = \alpha(g')$ for one or both immediate descendants g' of the gene g (one to the left is designated as cg; one to the right, as Cg). The vertex s of the species tree S is g intermediate, if it is situated exactly between the vertices $\alpha(g)$ and $\alpha(pg)$, where pg stands for the last common ancestor of the vertex g. Let us denote $M(G, S)$ as a set of all g intermediate vertices for all gs from G. A member of the set $M(G, S)$ is also called a gap. The gap corresponds to the edge (g, pg) with the length $l_{(g,\ pg)}$ in the gene tree G. Guigo et al. (1996) proved a theorem stating that the total number of gene losses equals to the total number of one-side duplications and gaps.

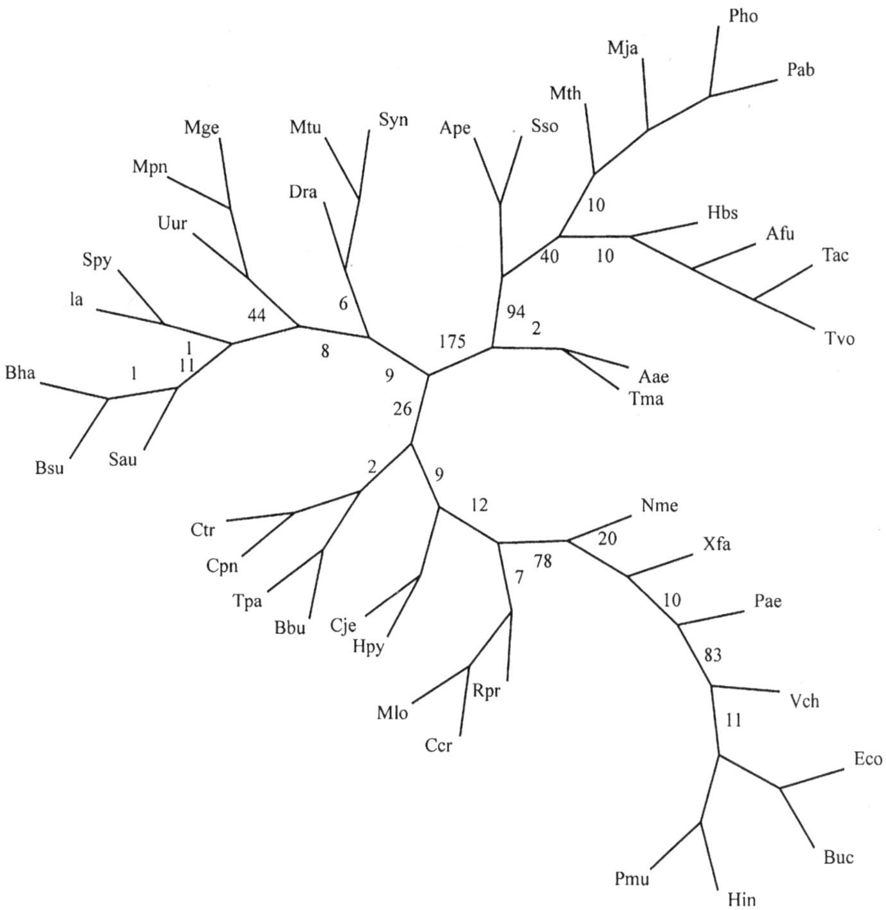

Figure -1. Evolutionary tree of 40 microorganisms from the following groups: Archaea—(Afu) *Archaeoglobus fulgidus*, (Hbs) *Halobacterium* sp. NRC-1, (Mja) *Methanococcus jannaschii*, (Mth) *Methanobacterium thermoautotrophicum*, (Tac) *Thermoplasma acidophilum*, (Tvo) *Thermoplasma volcanium*, (Pho) *Pyrococcus horikoshii*, (Pab) *Pyrococcus abyssi*, (Ape) *Aeropyrum pernix*, and (Sso) *Sulfolobus solfataricus*; Gram-positive bacteria—(Spy) *Streptococcus pyogenes*, (Bsu) *Bacillus subtilis*, (Bha) *Bacillus halodurans*, (Lla) *Lactococcus lastis*, (Sau) *Staphylococcus aureus*, (Uur) *Ureaplasma urealyticum*, (Mpn) *Mycoplasma pneumoniae*, and (Mge) *Mycoplasma genitalium*; Alpha-proteobacteria—(Mlo) *Mesorhizobium loti*, (Ccr) *Caulobacter crescentus*, and (Rpr) *Rickettsia prowazekii*; Beta-proteobacteria—(Nme) *Neisseria meningitidis* MC58; Gamma-proteobacteria—(Eco) *Escherichia coli* K12, (Buc) *Buchnera* sp. APS, (Pae) *Pseudomonas aeruginosa*, (Vch) *Vibrio cholerae*, (Hin) *Haemophilus influenzae*, (Pmu) *Pasteurella multocida*, and (Xfa) *Xylella fastidiosa*; Epsilon-proteobacteria—(Hpy) *Helicobacter pylori* and (Cje) *Campylobacter jejuni*; Chlamydia—(Ctr) *Chlamydia trachomatis* and (Cpn) *Chlamydia pneumoniae*; Spirochetes—(Tpa) *Treponema pallidum* and (Bbu) *Borrelia burgdorferi*; and DMS—(Dra) *Deinococcus radiodurans*, (Mtu) *Mycobacterium tuberculosis*, (Syn) *Synechocystis*, (Aae) *Aquifex aeolicus*, and (Tma) *Thermotoga maritima*. Vertices are assigned the total number of duplications for 132 protein families. The list of the families is taken from Wolf et al. (2001).

Remember that the duplication (g, s) is one-sided if either of the conditions $\alpha(g) = \alpha(cg)$ or $\alpha(g) = \alpha(Cg)$ is true. A one-side duplication (g, s) corresponds to the edge (g, cg) or (g, Cg) with the length $l_{(g,\ cg)}$ in the gene tree G. A set of all one-side duplications is designated as $O(G, S)$.

The duplication (g, s) is considered to have occurred in the vertex s. The number of such pairs under fixed s defines the number of duplications in the vertex. The total number of duplications in the genome assigned to the vertex s is the sum of all one-side duplications in the vertex over all gene families from a fixed set of families (Figure 1). The statement 'in the genome' implies that the set is assembled to be maximally representative. For an individual protein family, the total number of duplication in descendants of the vertex s is estimated as a sum of one-side duplications in all vertices of the clade contained in s. A more sophisticated procedure is used to infer the number of gene losses in the vertex. Remember that a gene loss in the vertex s corresponds to the pair (g, s), where g contains a duplication, s descends from $\alpha(g)$, and the clade s does not contain either of genes from the clade g', g' being an immediate descendant of g, while the clade ps does contain genes from both clades g' (Eulenstein et al., 1998). This definition is sometimes made more complex with additional conditions imposed on the pair (g, s). The number of losses in s is defined as the number of all such pairs (g, s) under fixed s. Other types of evolutionary events are treated analogously. The total estimates are considered as important characteristics of vertices of the species tree, protein families (genomes), and phylogenetic clades. HGT is considered as a special case of gene gain when its origin can be traced.

The cost of mapping of the gene tree G into the species tree S is defined as

$$c(G, S) = |O(G, S)| + \gamma \cdot |M(G, S)|,$$

where $|\{\cdot\}|$ stands for the cardinality $\{\cdot\}$, i.e., the number of set members. Otherwise, it can be given by two sums:

$$c(G,S) = \sum_{g \in O(G,S)} l_{(g,cg)} + \gamma \cdot \sum_{g \in M(G,S)} l_{(g,pg)} \cdot$$

By minimizing the value of $c(G, S)$ under $\gamma = 1$, the total number of gene losses is minimized. If $\gamma < 1$, the cost favors duplications over gaps. In some cases, only the number of duplications is minimized (Page and Charlstone, 1997).

Thus, the species tree S is produced by minimizing the value

$$c = c(S) = c(G_1, S) + c(G_2, S) + \ldots + c(G_n, S),$$

where all gene trees G_i are already obtained, and the unknown species tree S is being produced under certain *a priory* imposed conditions. This value will also be referred to as a cost of mapping of the gene tree set $\{G_i\}$. From the mathematical standpoint of computational complexity theory, finding the minimum of $c(S)$ is a highly nontrivial task. Importantly, more robust edges of the trees G_i have more impact on minimization of the function $c(S)$. The edge lengths (robustness or divergence times) of trees G_i can be induced on the resulting species tree S.

V'yugin et al. (2003) proposed a partial solution of the known issue with reconstructing species trees caused by long branch attraction artifact. Namely, the minimization of function $c(S)$ is preceded by normalization of the edge lengths in gene trees from $\{G_i\}$. Edge lengths are re-estimated using the formula

$$l(g) = (l(g) - l_{cp})(1 + m)^{-(l(g)/l_{cp})} + l_{cp},$$

where l_{cp} stands for the mean edge length of gene trees. The normalization procedure reduces the impact of extreme edge lengths in G_i on the resulting tree S.

Let us now concern the selection of value for parameter γ, which determines the ratio between the numbers of duplications and losses. Many of the loss events, especially in vertices close to the root of the species tree S, may represent false predictions incurred from incorrect topologies of the source trees. Apart from that, mapping α does not accurately account for the gene gain events (particularly, HGT). A putative gene gain event can be alternatively explained by the topological disparities between the gene and species trees caused by a small number of gene duplications compared to a magnitude-larger number of gene losses. Therefore, a species tree constructed with optimization of an α-based model may be improved by assigning more weight to duplications, which are predicted more accurately. In our experiments, we generally assumed $\gamma = 0.1$.

Our algorithm constructs an optimal tree S as a *specific* local minimum. Since the algorithm produces a local minimum depending on the initial species tree S_0, we developed an *ad hoc* approach to construct the initial S_0. Namely, a probability distribution in the set of all initial species trees is built; it is defined automatically by the family $\{G_i\}$ of gene trees as follows. For any species a and b, the distribution $p(b|a)$ is defined as a probability for both b and a to form an elementary tree (i.e., to be located at a distance less than or equal to some fixed r, for example, $r = 2$). Let N_a be the number of the gene trees containing species a, and let $N_{a,\ b}$ be the number of trees containing a and b located at a distance r. Then, $p(b|a) = N_{a,\ b}/N_a$, and

$1 - \sum_b p(b|a)$ is the probability of the event that there is no occurrence of $b \neq a$ in any elementary tree containing a (i.e., there is no species b located at a distance r from a). We considered small species trees defined with the distances r = 2, 3, 4, etc., although larger distances require larger sets of primary data. The random binary tree S_0 is generated with the distribution and is taken as an initial tree in the search algorithm. The final output is a consensus tree computed on a subset of the resulting trees with a sufficiently small value of the function c. Edges of this consensus tree are assigned values of support of the corresponding clusters.

2.3 Identification of vertices introducing incongruence between gene and species trees

Optimizing of parameters of the above-described models requires identification of the tree vertices informative with respect to the inferring events of gene evolution. A substrate for this type of analysis is a topological incongruence between some gene trees G_i and the consensus species tree S, which can be accounted for by actual events in gene evolutionary history or artifacts in reconstruction of the source trees, the latter representing a problem of its own.

Three algorithms for detecting sets of incongruent vertices are described. The first algorithm is based on identification of the subset G' of terminal vertices (leaves) in the gene tree G representing the gene gain events. The evolutionary model is two mappings α with different domains: initially, α is defined on the gene tree G and, subsequently, on its subset of the leaves $G \backslash G'$ obtained by excluding the leaves with putative HGTs. The subset $G \backslash G'$ is transformed into a binary tree using a standard procedure. Liberally speaking, $G \backslash G'$ can be considered as a subtree of G. The set G' is a parameter of the model, and its selection (optimization of parameter G') is carried out by maximizing a set-dependent value. Other evolutionary events are defined via mapping α as described above.

The first algorithm consists of the two segments.

The first segment. The terminal gene g in gene tree G is considered as putative HGT, if all genes $g_1, g_2, \ldots, g_n$ in its proximity except for g itself are mapped with α onto the species $s_1, s_2, \ldots, s_n$ distant from the species $s = \alpha(g)$. It is also prerequisite that the set of species $\{s_1, s_2, \ldots, s_n\}$ is located compact enough in the species tree S, i.e., its ancestor s_0 is close enough to the leaves and the distance between s_0 and s is considerably large in S. The genes $g_1, g_2, \ldots, g_n$ are defined as a set of terminal vertices without g separated in G with a distance less than r, where the distance is the length of the path from g to g_i; it either takes into account the edge lengths or does not if those are unit lengths. The gene set $\{g_1, g_2, \ldots, g_n\}$ is also called the punctured neighborhood of the gene g with a radius r. Usually, under unit lengths, we assumed $r = 4$. Let us provide some more details.

If two terminal genes g and g_1 are located at a small distance in G but species $\alpha(g)$ and $\alpha(g_1)$ in mapping α are at a great distance in S, it may suggest an abnormal position of one of the genes. Hence, the distance $r(g, g_i)$ in the gene tree and the distance $r(s, s_i)$ in the species tree are calculated, where $i = 1, \ldots, n$, and thus the average values are

$$r(g) = (1/n) \sum_i r(g, g_i) \text{ and } r(s) = (1/n) \sum_i r(s, s_i).$$

The value of $R_g = r(s)/r(g)$ determines the extent to which the size of the species set $\{s_1, s_2, \ldots, s_n\}$ is larger than that of the gene set $\{g_1, g_2, \ldots, g_n\}$. Large values of R_g can be interpreted as suggesting abnormality in location of the gene g in the species tree. Conventional p-values are calculated for the statistic $p(.)$ using the formula

$$p(g) = \left|\left\{g' \middle| R_{g'} \geq R_g\right\}\right| / m,$$

where m is the number of all terminal vertices. The computer program selects all genes g with $p(g) \leq p_0$, where p_0 is a threshold. Such genes are considered as abnormally positioned.

The algorithm also selects all cases when the species $s_1, s_2, \ldots, s_n$ are part of a taxonomic group that does not contain species s and its ancestor s_0 is sufficiently separated from s in S. This suggests that this group is a putative origin of a horizontally transferred gene g.

The second segment. Suppose that each abnormally located gene generates a series of invalid duplications and losses under mapping α, which are required to explain incongruence between the gene G and species S trees in the model. Therefore, temporarily omitting the transferred gene g from G and re-estimating α after the deletion entails an essential reduction in the cost $c(G, S)$ of mapping G into S. Therefore, we calculate $c(G, S)$ and subsequently remove each gene g from G to obtain the reduced gene tree G_g and compute the cost c_g of mapping of the new gene tree G_g into the same species tree S. The relative change in the mapping cost is $F_g = (c_g - c)/c$. As above, we use p-values for the statistic F_g for all genes g from the given COG G. Similarly, the computer program selects all the genes g for which $p(g) \leq p_0$.

The mean and standard deviation of the statistic F_g can be used, if the empirical distribution of the statistic F_g is normal. Interestingly, our studies reveal a considerably high support for the hypothesis of normality of the empirical distribution of F_g and log-normality of R_g for most COGs.

The genes selected at the second segment of the algorithm are interpreted as gained (not only due to HGT, as its origin is not always determined). The genes selected in both segments are considered as gained during an evolutionary event, probably, a HGT.

The first algorithm is designed to detect the recent HGTs, when the recipient and donor species did not diverge greatly in evolution. Gorbunov and Lyubetsky (2005) proposed two novel algorithms as generalization of the first algorithm to be able to detect deeper ancestral HGTs. In this sense, *ancestral* genes are those existing in an internal vertex of the phylogenetic tree. To stress this discrimination, *extant* genes are sometimes referred to as those existing in terminal vertices.

The second algorithm implements a juxtaposition of the gene tree G with the species tree S using graph β instead of mapping α in the first algorithm. Let us define some terminology.

Each *vertex* g in a tree corresponds to the *set* K of all leaves contained in the vertex g, in which sense the vertex g and *clade* K are mutually deterministic. The graph β contains all clades in G and all clades in S as vertices, with each clade K in G connected via one edge with each clade K' in S, edge K, K'; the graph contains no other edges. Let us define the *components* of the edge K, K' as two sets $M = K\backslash K'$ and $M' = K'\backslash K$.

For each edge K, K', we calculate the ratio of the cardinality of component M to the cardinality of the components containing clade K, the $\frac{|M|}{|K|}$ value, and, analogously, the $\frac{|M'|}{|K'|}$ value for the clade K'. Let us remember that the cardinality $|M|$ of set M is defined as the number of its members. The probability of the component M on edge K, K', we define as $1-\frac{|M|}{|K|}$, and, analogously, the probability of the component M' as $1-\frac{|M'|}{|K'|}$. Each edge K, K' in the graph β is assigned the two probabilities, which we define as *probabilities* of the edge K, K'. The edge K, K' can be viewed as an analogue to the pair < g, $\alpha(g)$> in the first algorithm.

Let us define *the workmate* M^* of the set M of leaves (terminal genes) as a complement of M to the set of all leaves in a certain subtree of G. Two alternatives of defining the subtree are considered: the subtree is rooted in the last common ancestor of all members of the set M; otherwise, it is rooted in the node parental to this ancestor. Let us call these the *first* and *second* workmates. The Sets M and M^* usually are not clades.

The algorithm described tests the possibility of HGT between the ancestor of set M and the ancestor of its workmate M^* in the species tree S. The algorithm is as follows. A list of all edges K, K' in the graph β is defined, for which at least one of the probabilities is above a certain threshold. For each nonempty component M with such probability, both workmates M^* are analyzed. The pair $< M, M^*>$ is called a *candidate pair*, if three simple conditions are satisfied:

(1) *Similarity of the candidate pair* $< M, M^*>$, measured as a mean distance between the elements of the two sets in the gene tree G (if edge lengths are present) or as a percent identity in pairwise alignments of the corresponding sequences, is under a certain threshold.

(2) *Compactness* of the set M in species tree S, defined as the ratio of the cardinality of M to the cardinality of the leaf set in a subtree of S rooted in the last common ancestor of all leaves from M as well as the analogous compactness of the set M^*, are above certain threshold.

(3) *The distance* between the last common ancestor of the set M and the analogous ancestor of M^* in the species tree S exceeds a certain considerable threshold. (If the ancestors are close in the tree S, conditions (1) and (2) may be true simply due to relatedness of M and M^*). This requirement is supplementary to the requirement that the compactness of the union of all species from M and M^* is below a certain threshold. Low values of this compactness, to the contrary, suggest a HGT event.

The more edges are in the graph β that imply the pair $<M, M^*>$ with higher probability, the higher weight is given by the algorithm to the pair as a candidate HGT between ancestors of M and M^*.

Performance of the second algorithm can be assessed on a case example of two trees, species tree $(((a,(e, b)),(3,(4, 5))),((1, 2),(c, d)))$ and gene tree $((((a, b),(c, d)),e),((3,(4, 5)),(1, 2)))$, with $M = \{a, b\}$ and its second workmate $M^* = \{c, d\}$.

For reasons of conciseness, *the third algorithm* will be described for the case when a gene copy persists in the source lineage after HGT. The algorithm is not sensitive to this constraint. It is based on analysis of fuzzy gene sets from a fixed COG.

The fuzzy gene set R is defined by a credibility function, which estimates the 'credibility of membership' in R of each gene from a fixed COG. Let K be a clade in the species tree and P be the set of all genes from the COG belonging to K. The fuzzy set R is given by P, i.e., given is a string of numbers, credibilities p_g, for all genes g from the COG. In the simplest case, p_g is proportional to similarity of the gene g to its closest match g_1 from P. The similarity can be estimated from COGs multiple alignment, from a path in the COG tree, or, in absence of the two former, simply from a percent identity of pairwise alignments of the corresponding sequences (Gorbunovand Lyubetsky, 2005).

Instead of similarity, one may calculate 'informativity about the gene g contained in g_1' or 'informativity about the gene g contained in set P'. To do so, we applied the Lempel–Ziv algorithm originally modified to use the entry g_1 sequence or entry set P. For the basics, one may consult Otu et al. (2003).

Hence, for an arbitrary pair of clades K and K' in the species tree, one can calculate a pair of the corresponding fuzzy sets R and R'. Let the *quality* $Q(K, K')$ of the clade pair be the ratio of the cardinality of 'fuzzy intersection' of R and R' to the cardinality of 'fuzzy union' of R and R', i.e., by definition, $Q(K,K')=\frac{\sum_g \min(p_g,q_g)}{\sum_g \max(p_g,q_g)}$. Let the *kernel* M of two primary clades K and K' be a set of genes g from the COG, for which $\min(p_g, q_g)$ is above a certain threshold. The genes from M may be interpreted as descendants of a horizontally transferred gene. The algorithm searches for HGTs as pairs of disjoint clades in the species tree S, with their kernel M containing two gene sets M_1 and M_2, both having sufficient compactness in S and their union closely coinciding with M, and not having high compactness in S (relevant thresholds implied).

Performance of the algorithm can be illustrated on the same case study as provided above after defining reasonable distances between the genes with respect to the gene tree and assuming a simple transformation of the distance x from gene g to its closest match g_1 from P into credibility p_g as $p_g = 32 \cdot (4-x)$.

3. RESULTS AND DISCUSSION

3.1 Reconstruction of bacterial species phylogeny

Consider a typical output of the algorithm described in section 2.1. It was run to infer the phylogeny of 40 microorganisms with 132 protein families. A detailed description of the primary data is provided in V'yugin et al. (2003; Figure 1). The search algorithm runs on a set of 5000 generated initial species trees S_0 as described above. The minimum value of c was 42 648. Robustness of the algorithm can be judged from the observation that two groups of resulting species trees selected by the algorithm, a set of 48 trees with $42\,648 < c < 42\,991$ and a set of 182 trees with $42\,648 < c < 44\,861$, have identical consensus topologies. The same holds true for a number of subsequently constructed species trees. The incongruence between the species tree thus obtained and the best species tree published in Wolf et al. (2001, approach (v)) is negligible and occurs only with respect to the relative position of groups of epsilon-proteobacteria, Aae, and Tma.

3.2 Deciding between two alternative hypotheses

Consider a typical case when decision is to be made in favor of either a small number of HGTs or a considerable number of gene losses. Application of the first algorithm to COG0272 (NAD-dependent DNA ligase) returns the following result: the initial mapping α onto the species tree detects 5 duplications (with 4 existing in the vertex of species tree) and 17 gaps, thus giving 22 gene losses in total. It identifies the gene *yicF* from *E. coli* as largely accounting for the incongruence between the protein and species trees.

After omitting gene *yicF* from the gene tree, the number of duplications reduces to two and the number of gaps, to five, giving a total of seven gene losses.

Thus, assuming HGT with *yicF* decreases the number of losses by 15 (Figure 2). The first algorithm concludes with a high confidence that the gene *yicF* was horizontally transferred from some spirochaete bacteria.

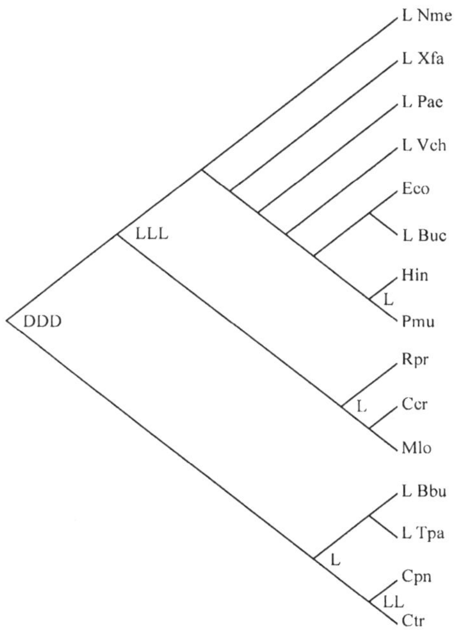

Figure -2. A part of the evolutionary history of COG0272 (NAD-dependent DNA ligase). The hypothesis about the absence of HGT events for gene *yicF* requires assuming 3 additional duplications and 15 additional losses of this and other genes. Duplications are marked with D; losses, with L.

3.3 Reconstruction of ancestral events in gene evolution

We conducted mass analyses of COGs using the tree algorithms (for more detail, refer to V'yugin et al., 2003; Lyubetsky et al., 2003a, b). Consider a typical result obtained for the above-mentioned 132 protein families. Initially, we tested all genes from each COG and selected 365 of those that contribute the most to the incongruence between the gene and species trees. Subsequently, all the 365 genes selected were omitted from their gene trees, and the mapping α of each of the gene trees into the species tree was re-estimated. For both cases, we counted the numbers of gene duplications and losses for each COG. In the first case, called *non-GAIN scenario*, the algorithms detect 1558 gene duplications and 9009 gene losses. The second case is called *GAIN scenario* and produces 1392 gene duplications, 7400 gene losses, and 365 GAIN events. The hypothesis about single GAIN event reduces the number of losses by an average of 4.4 (the difference between 9009 losses in non-GAIN and 7400 losses in GAIN scenario divided by 365 gains). The distribution of total estimated duplications under the GAIN scenario across prokaryotic families is as follows: Archaea, 154 (94 in the root); gram-positive bacteria, 65 (8 in the root); alpha-proteobacteria, 7 (all in the root); beta-proteobacteria, 0; gamma-proteobacteria, 124 (20 in the root); and chlamydias and spirochetes, 2 (both in the root; Figure 1).

Large total numbers of gene duplications (comparable to the number of protein families) assigned to a vertex of the gene tree might suggest *whole genome duplications*. Such are the group of 92 duplications in the root of Archaea and the group of 83 duplications in the root of (((Pmu,Hin),(Eco,Buc)),Vch).

Table -1. Selected number of reconstructed evolutionary events

	non-GAIN scenario			GAIN scenario		
1	2	3	4	5	6	7
COG	Dupl	Loss	Gain	Dupl	Loss	Gain
COG0012	12	63	0	9	48	1
COG0102	13	71	0	11	53	1
COG0143	16	102	0	14	80	2
COG0198	18	99	0	13	67	1
COG0215	16	71	0	11	46	2
COG0272	8	51	0	5	28	1
COG0290	9	41	0	8	28	2
COG0343	14	70	0	13	65	1
COG0544	4	30	0	5	25	1
COG0571	9	59	0	4	22	3
COG0653	8	39	0	7	29	2
COG1160	5	27	0	4	23	1

The computer programs also output mappings of each COG tree into the species tree for purposes of evolutionary history reconstruction under both scenarios for each of the 132 protein families (Lyubetsky et al., 2003b). Selected numbers of evolutionary events thus reconstructed are given in Table 1; columns 2–4 contain inferences under non-GAIN scenario and columns 5–7, those under GAIN scenario.

To continue, let us provide some details on the event reconstructions for selected COGs.

COG0012 (predicted GTPase). *Buchnera aphidicola*, a member of the gamma-proteobacteria group, occurs in the species tree in the same cluster with *E. coli*, but its gene *bu191* is found close to chlamydial genes in the gene tree. We suggest that this group is the source of HGT. Also suggested is that the gene *sll0245* is horizontally transferred to the genome of *Synechocystis* sp. from spirochetes.

COG0215 (aminoacyl-tRNA synthetases and alternative system for amino acid activation). It is suggested that the gene *vng1095G* from *Halobacterium* sp. (halophilic archaebacteria originating from eubacteria) is horizontally transferred from the genome of an organism similar to *Deinococcus radiodurans*. It is likely that the gene *xf0995* from the organism *Xylella fastidiosa*, which occurs in the same cluster, is transferred from some alpha-proteobacteria similar to *Caulobacter crescentus.*

COG0143 (methionyl-tRNA synthetase). The *mlr5926* gene from *Mesorhizobium loti* (alpha-proteobacteria) is a putative HGT from some archaebacteria. Moreover, this event entailed subsequent divergence of paralogous genes in this genome.

COG0102 (ribosomal protein, large subunit) provides an example of a ribosomal gene HGT. The *dr0174* gene from *Deinococcus radiodurans* (L13 protein) is likely transferred from a genome of some gamma-proteobacteria.

COG0198 (ribosomal protein, large subunit). The *bb0489* gene from *Borrelia burgdorferi* (Spirochaeta; encodes L13 protein) is transferred from some gamma- or beta-proteobacteria.

COG0272 (basal replication machinery). The *yicF* gene from *E. coli* (NAD-dependent DNA ligase) is horizontally transferred from spirochaete bacteria. In addition, the *E. coli* genome contains gene *lig*, bearing the same function as *yicF*.

COG0343. The *af1485* gene from *Archaeoglobus fulgidus* (queuine/archaeosine-tRNA ribosyltransferase) is likely to be transferred from eubacteria.

The second and third algorithms converge in inferring the same ancestral HGTs. Thus, for COG0180 (tryptophanyl-tRNA synthetase), the algorithms predicted putative HGTs between the ancestors of groups {Bha, Bsu, Sau} and {Vch, Eco, Buc, Hin, Pmu}. The predictions corresponded to 6 edges in graph β, high densities and 6 pairs of clades in species tree producing the same kernel M = {Bha, Bsu, Sau, Vch, Eco, Buc, Hin, Pmu, Hpy, Mtu} (refer to descriptions of the second and third algorithms).

Putative HGTs can be alternatively identified with non-phylogenetic approaches based on comparative analyzes of codon usage, frequencies of genomic features, and other contextual characteristics (Garcia-Vallve et al., 2003).

PART 2

COMPUTATIONAL STRUCTURAL AND FUNCTIONAL PROTEOMICS

MINING FROM COMPLETE PROTEOMES TO IDENTIFY ADHESINS AND ADHESIN-LIKE PROTEINS: A RAPID AID TO EXPERIMENTAL RESEARCHERS

S. Ramachandran*, P. Jain, K. Kumar, G. Sachdeva
G.N. Ramachandran Knowledge Center for Genome Informatics, Institute of Genomics and Integrative Biology, Mall Road, Delhi, 110 007, India, e-mail: ramu@igib.res.in
* *Corresponding author*

Abstract: Adhesins are microbial surface proteins that mediate the adherence of microbial pathogens to host cell surfaces. This interaction is often the first step in the establishment of a disease. Identification of novel adhesins and their characterization are important for studying host–pathogen interactions and for testing new vaccine formulations prepared from adhesins. Currently, experimental methods are used for detecting and characterizing adhesins, which is a time-consuming task and demands large resources. The availability of a software program specifically focused to identifying adhesins from the predicted proteomes of microbial pathogens can aid experimenters in simplifying the complexity of this problem. We have employed artificial neural networks to develop an algorithm SPAAN, which predicts the probability of a protein being an adhesin P_{ad} based on 105 compositional properties computed from its sequence. SPAAN had optimal sensitivity of 89 % and specificity of 100 % on a defined test data set and could identify 97.4 % of the known adhesins at a high P_{ad} value from a wide range of bacteria. Data mining using SPAAN not only identified the known adhesins, but also guided in improvement of annotation of several proteins as adhesins. Several novel adhesins were identified in many pathogenic organisms causing diseases in humans and plants. These results offer new leads for rapid experimental testing.

Availability: SPAAN is freely available from ftp://203.195.151.45 or ftp://203.90.127.75

Contact: ramu@igib.res.in

Key words: virulence factors; adhesins; vaccine; neural networks

1. INTRODUCTION

Microbial pathogens encode several proteins known as adhesins located on their surfaces that mediate the adherence of these pathogens to host cell surface receptors, membranes, or extracellular matrix for successful colonization (Boyle and Finlay, 2003). Investigations in this primary event of host–pathogen interaction over the past decades have revealed a wide array of adhesins in a variety of pathogenic microbes (Finlay and Falkow, 1997). New approaches to vaccine or drug target development focus on targeting adhesins to abrogate the colonization process (Wizemann et al., 1999; Ofek et al., 2003). However, the specific roles of many adhesins in several pathogens remain to be elucidated.

Generally, adhesins are of two types, namely, fimbrial and nonfimbrial. One of the best-understood mechanisms of bacterial adherence is attachment mediated by pili or fimbriae. Several adhesins of this type are well studied. Examples include FimH and PapG adhesins of *Escherichia coli* (Hahn et al., 2002); the type IV pili adhesins in *Pseudomonas aeruginosa*, *Neisseria*, *Moraxella*, enteropathogenic *Escherichia coli,* and *Vibrio cholerae* (Strom and Lory, 1993); and many others (for further details, see http://www.igib.res.in/data/seepath/spaan_data.html).

The currently approved vaccine for whooping cough, caused by *B. pertussis*, contains a preparation of the adhesins filamentous hemagglutinin and pertactin proteins (Halperin et al., 2003). Immunization with the adhesin FimH is being evaluated for protective immunity against pathogenic *E. coli* (Langermann et al., 2000). The pneumococcal surface adhesin PsaA is being investigated as a potential vaccine candidate against pneumococcal disease (Rapola et al., 2003). Likewise, immunization with outer membrane vesicle preparations including BabA adhesin of *H. pylori*, the causative agent of gastric ulcer, show promise for developing a vaccine against *H. pylori* (Prinz et al., 2003). A synthetic peptide anti-adhesin vaccine is being evaluated for protection against *Pseudomonas aeruginosa* infections (Cachia and Hodges, 2003).

Clearly, identification of adhesins and adhesin-like proteins through data mining with the aid of specialized software programs is likely to complement researchers investigating the mechanisms of host–pathogen interactions. The usual step in computational analysis of predicted proteomes is the use of BLAST family of programs (Altschul et al., 1990). However, this procedure suffers from limitations when the homologues are not experimentally characterized or when the sequence divergence is high. An alternative successful methodology is to use sequence composition properties combined with the power of the Artificial Neural Networks (ANNs). Given a data set of positives and negatives, ANNs are able to extract patterns from these relatively simple numerical data, which can in turn be used to classify a sequence of unknown function into either the positive group or the negative group with little ambiguity. This approach is inherently non-homologous, which could, in principle, overcome the sequence diversity between species that widely differ in their relative phylogenetic positions.

Although many examples are available in the literature that document the successful application of this approach, only a few are close to this work: algorithms for predicting secretory proteins in bacteria and apicoplast targeted proteins in *Plasmodium falciparum* (Schneider, 1999; Zuegge et al., 2001). We describe an algorithm SPAAN (Software for Prediction of Adhesins and Adhesin-like proteins using Neural Networks) for prediction of adhesin and adhesin-like proteins and its application for a wide range of pathogens. Data mining of several proteomes revealed that SPAAN identified the well-known adhesins from a wide range of pathogens causing diverse diseases and offered a list of proteins with high probability of being adhesins. This list enabled us to improve the annotation of many proteins as adhesins by re-verification using BLAST and CDD searches. Furthermore, several predictions were supported by another computer program called BETAWRAP, which predicts the beta helix motifs found to be associated with many virulence factors and toxins. Finally, a few proteins had no complementary supporting evidence suggesting them as novel adhesin-like proteins.

2. METHODS AND ALGORITHMS

Detailed description of the methodology is provided in Sachdeva et al. (2005) and the definitions are provided in Brendel et al. (1992).

The five attributes

The five attributes used were (1) Amino Acid frequencies; (2) Multiplet frequencies; (3) Dipeptide frequencies of the dipeptides NG, RE, TN, NT, GT, TT, DE, ER, RR, RK, RI, AT, TS, IV, SG, GS, TG, GN, VI, and HR; (4) Charge composition and their distribution in terms of statistical moments; and (5) Hydrophobic composition and their distribution in terms of statistical moments. A sum total of 105 compositional properties were used to predict the adhesin-like characteristics of a given protein sequence.

Positive and Negative data sets

Adhesins (Positive dataset). Protein sequences were retrieved from http://www.ncbi.nlm.nih.gov using the keyword 'adhesin'. This primary retrieval was subjected to manual curation to remove unrelated entries to produce the adhesin database that contained well-annotated proteins, many of which have been experimentally verified.

Non-adhesins (Negative data set). We collected sequences of enzymes and other proteins that function within the cell, and therefore they are unlikely to be present on the surface. Because of large size of this data set and the requirement of both positive and negative datasets to be of equal or nearly equal size, we selected sequences from three organisms *Escherichia coli*, *Methanococcus jannaschii* and *Saccharomyces cerevisiae*, representing the three primary kingdoms of life: bacteria, archaea, and eukarya. This selection offers a diverse set for obtaining a broad range of limits for the detection of non-adhesins.

Eliminating redundant entries. We used CLUSTALW (Thompson, 1994) to remove redundant entries by retaining one sequence among pairs with CLUSTALW score of 100. Partial sequence entries were also removed.

Neural network

The feed forward error back propagation neural network algorithm was used. The program was downloaded from the web site (http://www.cs.colostate.edu/~anderson). This was a kind gift from Charles W. Anderson, Department of Computer Science, Colorado State University, Fort Collins, CO 80523, anderson@cs.colostate.edu

Algorithm

Neural network architecture and the P_{ad} value. The 'fully connected' neural network used here has a multi-layer feed forward topology. It consists of an input layer, a hidden layer, and an output layer, as shown in Figure 1.

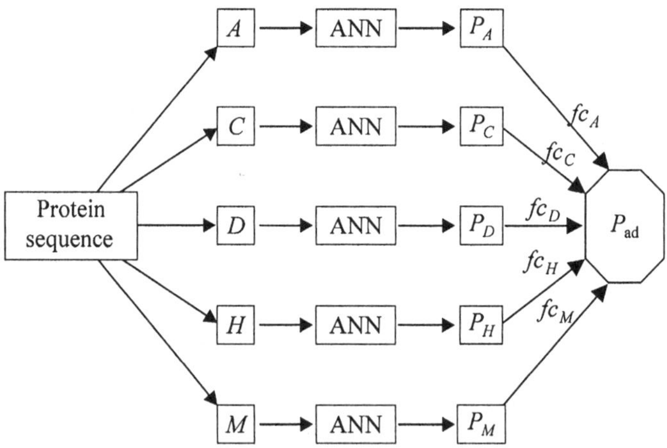

Figure -1. Architecture of SPAAN. A given protein sequence is first processed through the five modules *A*, *C*, *D*, *H*, and *M* to quantify the five types of compositional attributes. *A*, amino acid frequencies; *C*, charge composition; *D*, dipeptide frequencies; *H*, hydrophobic composition; and *M*, multiplets frequencies. The directions of arrows show data flow.

The weight of connection between them is denoted by w_{ij}. The state I_i of each neuron in the input layer is assigned directly from the input data, whereas the states of hidden layer neurons are computed from the states of input layer neurons using the sigmoid function,

$h_j = 1 / (1 + \exp(-(w_{j0} + \sum w_{ij} I_i))$, where, w_{j0} is the bias weight.

The back propagation algorithm was used to minimize the differences between the computed output and the target value. The target value for adhesins was set as '1' and for non-adhesins as '0'. Ten thousand cycles (epochs) of training iterations were performed, the best epoch with minimum error on validate set was identified, and the corresponding weight matrix was used for the prediction.

A network was trained optimally for each attribute. Thus, five networks were prepared. The schematic diagram displayed in Figure 1 shows the procedure adopted. The number of neurons in the input layer was equal to the number of input data points for each attribute. The optimal number of neurons in the hidden layer was determined through experimentation for minimizing the error at the best epoch for each network individually. An upper limit for the total number of weight connections was set to half of the total number of input vectors to avoid overfitting, as suggested previously (Andrea and Kalayeh, 1991). During predictions, the network was fed with new data from the sequences that were not part of training set. Each network assigned a probability value of being an adhesin to a given sequence. The final probability of a given protein sequence being an adhesin (the P_{ad} value) is given by

$$P_{ad} = \frac{(P_A * fc_A + P_C * fc_C + P_D * fc_D + P_H * fc_H + P_M * fc_M)}{(fc_A + fc_C + fc_D + fc_H + fc_M)},$$

here, fc_i is fraction of correlation of ith module of the trained neural network, where $i = A$ (Amino acid frequencies), C (Charge composition), D (Dipeptide frequencies), H (Hydrophobic composition), or M (Multiplet frequencies). The fraction of correlation fc_i represents the fraction of total entries that were correctly predicted ($P_{i,\text{adhesin}} > 0.5$ and $P_{i,\text{ non-adhesin}} < 0.5$) by the trained network on the validate set (C.W. Anderson, http://www.cs.colostate.edu/~anderson); $fc_A = 0.84$, $fc_C = 0.71$, $fc_D = 0.84$, $fc_H = 0.79$, and $fc_M = 0.83$.

Performance assessment

We set the True Positives (*TP*) as 'adhesins' and the True Negatives (*TN*) as 'non-adhesins'. For a given threshold of P_{ad} value, *TP* are known adhesins with P_{ad} greater than the threshold and *TN* are known non-adhesins with P_{ad} lower than the threshold. False Negatives (*FN*) are those cases wherein a known adhesin had P_{ad} value lower than the threshold and False Positives (FP) are non-adhesins with a P_{ad} value higher than the threshold.

The sensitivity, Sn, is given by $\left(\frac{TP}{TP+FN}\right)$ and specificity, Sp, is given by $\left(\frac{TP}{TP+FP}\right)$. The Matthew's correlation (Matthews, 1975) is defined as

$$Mcc = \frac{(TP*TN)-(FP*FN)}{\sqrt{(TN+FN)(TN+FP)(TP+FN)(TP+FP)}}.$$

System Requirements

Computer programs to compute individual compositional attributes were written in C and executed on a PC with operating system Red Hat Linux v. 7.3 or 8.0. SPAAN accepts input sequence files in FASTA format. Multiple sequences can be presented in one file. Protein sequences with ambiguous amino acids (other than 20 amino acids) and/or of a length less than 50 amino acids are filtered out. Amino acids must use the single letter code according to IUPAC-IUB nomenclature system.

3. RESULTS AND DISCUSSION

3.1 Performance

SPAAN could identify 89 % of the known adhesins at 100 % specificity when examined at probability of being an adhesin (P_{ad}) ≥ 0.51 using a clearly defined test data set. At this threshold value of P_{ad}, the Matthew's correlation coefficient was observed to be highest (0.94). We observed that the combination of five modules provided the best results. Assessment of performance in individual modules showed that they performed poorly when compared to the combination. Our experience is in agreement with others who have used similar sequence composition–based approaches. SPAAN is also able to detect 97 % of the known adhesins with high P_{ad} value (Sachdeva et al., 2005).

3.2 Application of SPAAN to whole genomes

An example of application of SPAAN to the pathogen *Escherichia coli* O157:H7 is displayed in Table 1. Details are available at http://www.igib.res.in/data/seepath/spaan_data.html. We used stringent criterion of P_{ad} > 0.7 to minimize the detection of false positives and further restricted our analysis to a maximum of top scoring 50 proteins to identify top scoring adhesins and adhesin-like proteins with high confidence. Several of the predicted adhesins are supported by complementary evidence by the most commonly used computer programs such as Conserved Domain Database search (RPS-BLASTP), BLASTP, and beta helix predictor beta-wrap (Altschul et al., 1990; Bradley et al., 2001; Marchler-Bauer et al., 2002). A third to three fourths of these top scoring proteins also contain beta helix motif. The beta helix motif has been found to be associated with several adhesins, toxins, virulence factors, and surface proteins (Bradley et al., 2001). SPAAN guided the improved annotation of a number of adhesins by suggesting re-examination of these proteins using the most commonly used softwares listed above.

Table -1. Data mining of predicted proteomes of two important pathogens

Species	Disease caused	Total No. of proteins analyzed[1]	No. of these supported by complementary evidence CDD/BLASTP[2]/PubMed	No. of these supported by complementary evidence BetaWrap	No. of adhesin like proteins	No. of false positives
Escherichia coli O157:H7	Diarrhea	50	37[3]	33	12	1
Helicobacter pylori	Peptic ulcers	50	25	36	24	1

[1] After selecting proteins with $P_{ad} > 0.7$, only the top scoring proteins (upper limit set to 50) were analyzed further for complementary evidence using CDD search, BLASTP, and beta-wrap for adhesin characteristics to minimize detection false positives; [2] only the top scoring similar sequences with $e < 0.001$ were considered for assessing sequence based relationships. The low complexity filter was 'off'. The stringent criterion of selecting sequences with $e < 0.001$ eliminated the possibility of identifying unrelated sequences as similar; [3] includes fimbrial adhesins (nine proteins), AidA-I, gamma intimin, hemagglutinin, translocated intimin receptor, putative tail fiber protein, and putative major tail protein.

The top scoring proteins in *E. coli* O157:H7 include nine fimbrial adhesins, AidA-I, gamma intimin, hemagglutinin, and translocated intimin receptor. We classified proteins with high P_{ad} value identified by SPAAN as 'adhesin-like' for which either limited or no complementary evidence exists. These could serve as new leads for experimental testing.

The lone false positive is a protein that displays high similarity to a protein involved in polysaccharide metabolism in *Salmonella typhimurium*. At this time, there is no evidence for its location on the surface of the bacterium. While it is apparent that increasing the stringency further by choosing even higher P_{ad} values could eliminate false positives, this exercise would reduce sensitivity. A judicious approach for experimental characterization could be developed by considering the total number of proteins to be analyzed, prioritizing proteins with other complementary evidence while keeping the number of false positives to as low as possible. The success of SPAAN is another example in the group of computational approaches that use compositional properties for addressing biologically interesting issues, such as prediction of protein secondary structure (Rost and Sander, 1993) and identification of secretory proteins in bacteria (Schneider, 1999) and apicoplast targeted proteins in the malarial parasite *P. falciparum* (Zuegge et al., 2001).

ACKNOWLEDGMENTS

SR thanks The Council of Scientific and Industrial Research, Govt. of India for a grant under the New Millennium Indian Technology Leadership Initiative (NMITLI) program. We thank Prof. G. Padmanaban, Prof. Samir K. Brahmachari, and Dr. Ramesh Sonti for useful discussions and Charles W. Anderson for the base Neural Network program.

CENTRAL MOMENTS BASED STATISTICAL ANALYSIS FOR THE DETERMINATION OF FUNCTIONAL SITES IN PROTEINS WITH THEMATICS

L.F. Murga[1], J. Ko[2], Y. Wei[1], M.J. Ondrechen[1*]

[1] *Department of Chemistry and Chemical Biology, Northeastern University, Boston, MA 02115, USA, e-mail: +mjo@neu.edu;* [2] *Department of Chemistry, Indiana University at Pennsylvania, 975 Oakland Avenue, Indiana, PA 15705, USA*

[*] *Corresponding author*

Abstract: One of the big challenges in genomics is to obtain information related to protein function. Computational methods that allow fast and accurate determination are needed. We present here the application of a statistical analysis coupled with THEMATICS that allows the fast identification of functional sites in proteins. The method is amenable for automation and thus for the purpose of high throughput analysis. We show the detailed analysis of two proteins and a summary of the application of the method to a set of 15 proteins with diverse functions.

Key words: active site; functional genomics; titration curves; THEMATICS

1. INTRODUCTION

The advent of the 'genomic revolution' has made available thousands of new protein structures in the last few years (Westbrook et al., 2003). Currently, the PDB databank (Berman et al., 2000) lists of the order of 10^3 protein structures annotated as 'hypothetical' or 'unknown function'; this number is steadily increasing as structural genomics projects proceed and better methodologies are developed. Thus, it is necessary to have fast and reliable computational methods that can produce functional information for these proteins.

We have recently reported on THEMATICS (Ondrechen et al., 2001; 2003; Shehadi et al., 2002; Murga et al., 2004; Ringe et al., 2004)—Theoretical Microscopic Titration Curves—as a technique for the determination of functional information in proteins starting from the three-dimensional structure only. THEMATICS is based on well established Poisson–Boltzmann methodologies (Bashford and Karplus, 1991; Bashford and Gerwert, 1992; Gilson, 1993; Antosiewicz et al., 1994) for the determination of the electrical potential function of proteins followed by a hybrid Monte Carlo procedure for the determination of the protonation state as a function of the pH for the ionizable residues (Arg, Lys, Tyr, Cys, Glu, Asp, and His). These ionizable residues behave as acids/bases, and most of them can be described well by the Henderson–Hasselbach (HH) equation rewritten here in a slightly different way to express the dependence of the residue mean net charge on the pH as

$$C(x) = \pm\,(10^{\pm x} + 1)^{-1}. \tag{1}$$

In Eq. (1), the positive sign applies for residues that form a cation upon protonation (Arg, Lys, and His) and the negative sign applies to residues that form an anion upon deprotonation (Tyr, Cys, Glu, and Asp). The variable x is defined as $x = pH - pKa$.

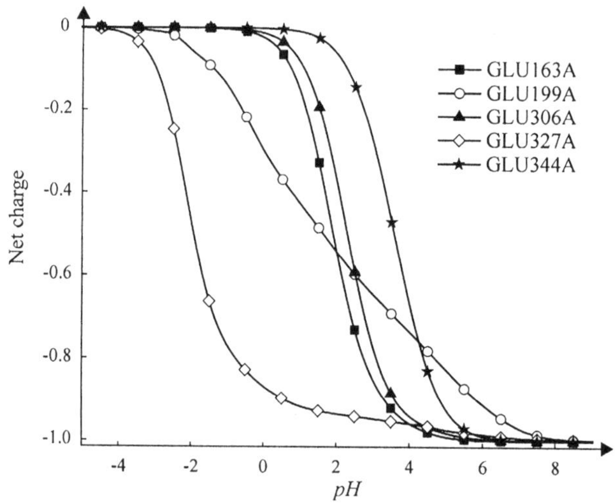

Figure -1. Predicted titration curves for selected glutamate residues of acetylcholinesterase (PDB ID 1EA5) from *Torpedo californica*.

Figure 1 shows selected predicted titration curves computed for glutamate residues of acetylcholinesterase (ACE) from *Torpedo californica*. Most of the curves exhibit the typical sigmoidal shape predicted by Eq. (1). However, the known catalytic residues Glu 199 and Glu 327 show a non-sigmoidal or perturbed shape. These deviations from typical HH behavior arise from interactions between different ionization events for residues that are both close in pKa and close in space similar to the ones that occur in small polyprotic acids (Onufriev et al., 2001). We have argued (Ondrechen et al., 2001; Ondrechen, 2002; Shehadi et al., 2002; Ringe et al., 2004) that this perturbed titration behavior can provide an advantage in catalysis and/or reversible recognition because reversible protonation occurs over an extended pH range. In previous papers (Ondrechen et al., 2001; Ondrechen, 2002; Shehadi et al., 2002; 2004; Murga et al., 2003; 2004; Ringe et al., 2004), we have shown that clusters of residues with non-HH curves tend to occur with such regularity and with sufficient exclusivity in the functional sites of proteins that they are reliable markers for the identification of such sites.

After a THEMATICS calculation has been performed, the critical step is the determination and selection of the residues with perturbed behavior (also called THEMATICS positives). Originally (Ondrechen et al., 2001), this was done visually by plotting and comparing the titration curves for all the ionizable residues of a given protein. This procedure was inconvenient albeit successful. First, it was slow and represented a barrier for high-throughput application, as it required human intervention. Second and perhaps more important, it introduced an element of subjectivity, as different users could report slightly different lists of perturbed residues. This is particularly true in borderline cases where perturbations are subtle and difficult to spot by visual inspection.

In this chapter, we present a new approach for the determination of perturbed behavior based on simple statistical criteria. The new methodology is fast and amenable to automation. It also provides the basis for an objective, rigorous criterion for the selection of THEMATICS positive residues.

2. METHODS AND ALGORITHMS

2.1 Statistical Selection Based on Central Moments

A common feature that appears in perturbed titration curves is an extended flat region. Glu 199 and Glu 327 from ACE are examples (Figure 1). The shallow slopes enable these residues to act as both acids and bases over an extended *pH* range. Thus, the titration curves can be characterized by their first derivative curves. For HH curves, this derivative is negative everywhere; therefore, we evaluate the negative of the first derivative. Figure 2 shows these $-dC/d(pH)$ functions for the same residues as Figure 1.

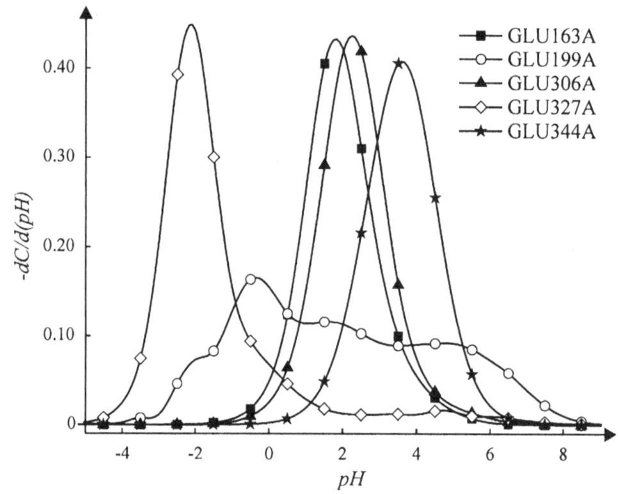

Figure -2. First derivative function $-dC/d(pH)$ for the same ACE residues shown in Figure 1.

For ordinary residues, the first derivative functions $-dC/d(pH)$ resemble Gaussian distribution functions. These curves are peaked and symmetrical with a maximum located at the *pKa*. In contrast, perturbed residues exhibit asymmetry, broadening, or multiple peaks in their derivative curves. Examples of this behavior are shown by residues Glu 199 and Glu 327 in Figure 2. Note that the curves are normalized, since the area under the curve is always unity.

One way to describe the characteristics of these titration curves is with moments, used in the characterization of Gaussian-like distribution functions. The *n*th central moment of the first derivative function is defined by

$$\mu_n = \int (pH - M_1)^n\ [-dC/d(pH)]\ \mathrm{d}(pH), \tag{2}$$

where M_1 is the first raw moment defined by the expression of the *n*th raw moment as

$$M_n = \int (pH)^n\ [-dC/d(pH)]\ \mathrm{d}(pH). \tag{3}$$

The integrals in Eqs. (2), (3) are over all space ($-\infty$ to $+\infty$). For a residue with titration curve described by the HH equation, the first derivative function f is given by

$$\ln 10/[10^x + 10^{-x} + 2], \tag{4}$$

where $x = pH - pKa$. The curves defined by Eq. (4) are sharply peaked functions with a maximum at $x = 0$ ($pH = pKa$). For residues obeying Eq. (4), the first raw moment is equal to the pKa. Thus, the corresponding odd moments are equal to zero. Eqs. (2)–(4) give the values of 0.62 and 1.62, respectively, for the second and fourth moment of an HH residue. Residues that exhibit perturbed behavior have values of their moments that deviate from the HH nominal values.

It has been observed (Ko et al., 2005) that the third and fourth central moments are the best selectors of abnormal behavior, because their average values for active site residues differ the most from the average values for all residues. In Table 1, we show the values of the third and fourth central moments for the residues whose curves are shown in Figures 1 and 2.

Table -1. Value of the third and fourth central moments for the selected residues displayed in Figures 1 and 2

Residue	$\|\mu3\|$	$\mu4$
Glu 163	0.52	5.36
Glu 199	6.13	126.75
Glu 306	0.59	5.93
Glu 327	20.8	185.61
Glu 344	0.04	3.99

The third central moment is related to the asymmetry (skewness) of the function *f*. Residues like Glu 344 in Figure 2 with a highly symmetrical curve will have a small third central moment, while a residue like Glu 327, whose curve possesses a tail on the high *pH* side, will have a large third central moment due to the asymmetry. The fourth central moment is related to the kurtosis. A distribution with a high kurtosis is one with higher population in the tails than in the center of the distribution near the mean. Glu 199 is an example of a curve with high kurtosis. Note that a high third moment does not necessarily imply a high fourth moment and vice versa. Therefore, consideration of both metrics gives better selection of the active residues.

We define the *Z* score of the *n*th central moment as

$$Z_n = (\mu_n - \langle\mu_n\rangle)/\sigma_n \,(n \text{ even}) \tag{5}$$

$$Z_n = (|\mu_n| - \langle|\mu_n|\rangle)/\sigma_n \,(n \text{ odd}) \tag{6}$$

In Eqs. (5) and (6), σ_n is the standard deviation of the values of the *n*th central moment for the set of all ionizable residues of a given protein. Note that for the odd moments, we use the absolute values, as they can be either positive or negative. Z_n thus represents how far the nth central moment of a particular residue deviates from the mean value in units of the standard deviation. In a study (Ko et al., 2005) of 44 well characterized proteins, we found that the criterion $Z_4 > 1$ is able to select most of the ionizable residues that belong to the active site. Those not chosen by this criterion were found often to have a $Z_3 > 1$. Thus, our composite criterion for the selection of perturbed residues that are functionally important is $Z_3 > 1$ or $Z_4 > 1$. For example, in the case of deoxyribonuclease from *Bos taurus* (PDB code 1DNK), the criterion $Z_4 > 1$ is able to choose the known catalytic residues (Bartlett et al., 2002) H134, D212, and H252. However, E39 and E78, also known to be important in catalysis (Weston et al., 1992), are left out by $Z_4 > 1$ but are selected by $Z_3 > 1$.

2.2 Computational Procedure

The protein structures for the calculations were obtained from the PDB (http://www.rcsb.org/) and hydrogens were added using the program TINKER (Ren and Ponder, 2003) using the force field OPLS-UA (Jorgensen et al., 1983; Jorgensen and Tirado-Rives, 1988).

Titration curves were calculated using the program UHBD (Madura et al., 1995) followed by the program HYBRID (Gilson, 1993). Substrates, cofactors, water molecules, and other entities, designed as HETATM in the PDB files, were excluded in the calculations. The values for the ionic strength and temperature were 150 mM and 293K, respectively. The dielectric constant was set at 20 for the protein and at 80 for the solvent. Justification for these values has been presented elsewhere (Antosiewicz et al., 1996a; 1996b; Murga et al., 2004).

The first derivative functions and the values of the third and fourth central moments were calculated using standard formulae (Press et al., 1992). Values of Z_3 and Z_4 were calculated using Eqs. (5) and (6). For each analyzed protein, those residues with $Z_3 > 1$ or $Z_4 > 1$ were designated as THEMATICS positive. For the purpose of an UHBD calculation, the charge on an ionizable residue is assumed to reside on a particular atom of the ionizable group, designated as the 'charge center'; thus, for the case of Asp or Glu residues, the charge is assumed to be associated to the central C atom of the carboxyl group. Using the coordinates of these charge centers, clusters of ionizable residues were defined such that the distance of the charge center of a particular residue must be within a 9 Å of at least one other charge center from another residue in the cluster. These clusters are the THEMATICS positive clusters. Clusters with two or more members are considered predictive.

3. RESULTS AND DISCUSSION

We present here the detailed analysis of the application of the $Z_3 > 1$ or $Z_4 > 1$ rule to two proteins. We then present a summary of the results for the application of the rule to a set of 15 proteins spanning different functionalities.

3.1 Human Leukotriene A4 Hydrolase

Leukotriene A4 hydrolase (LTA4H) is a bifunctional zinc metalloenzyme. Its best characterized biological function is the catalysis of the last step of the formation of leukotriene B_4 (LTB_4) by hydrolysis of an unstable epoxide derivative of arachidonic acid (Rudberg et al., 2002). LTB_4 has been shown (Haeggstrom, 2000) to play an important role in inflammation, immune response, and platelet aggregation among others.

In addition to the epoxide hydrolysis activity, LTA4H also has an anion-dependent aminopeptidase activity (Thunnissen et al., 2001). The biological role of this activity has not been completely established; it is suspected that it is probably related to the processing of peptides involved in inflammation processes (Thunnissen et al., 2001).

We performed the calculations on the biologically active monomer structure from *Homo sapiens* complexed with the inhibitor bestatin and other ligands (PDB code 1HS6) determined at a 1.95 Å resolution. As is standard in THEMATICS calculations, all ligands were removed from the input PDB file prior to the calculation. The monomer consists of 611 residues.

Application of the $Z_3 > 1$ or $Z_4 > 1$ rule followed by the clustering algorithm with a 9 Å cutoff produced the clusters [**E271**, **H295**, **E296**, **H299**, **E318**, **D375**, **Y378**, **Y383**, **K565**] [C135, C140, D148, C199]. Residues with known catalytic activity are indicated in boldface. THEMATICS correctly identifies most of the residues involved in both the epoxide hydrolysis and aminopeptidase functions (Haeggstrom, 2000; Thunnissen et al., 2001). The selection criterion fails to recognize R563, involved in carboxylate recognition, and Y267, involved in the hydrolysis mechanism. However, R563 is about 4 Å from K565 and Y267 is at 3.8 Å from D375.

Thus, although not every functionally important ionizable residue is selected by the Z_3 or Z_4 rule, the ones that are not selected are still within the small local region specified by the predicted cluster.

The second cluster identified by THEMATICS is not reported by any of the references as functionally important. This cluster is about 12 Å away from the first one, and thus, it is likely not involved in any of the known catalytic activities of LTA4H. Surface solvent accessibility calculations show that none of the residues in this cluster are accessible, and thus, it may be eliminated as a potential active site. This cluster appears to arise from the proximity of the buried cysteines to each other.

3.2 *L*-Arabinose Binding Protein

L-Arabinose binding protein (LABP) is a member of a group of proteins found in the periplasm of Gram-negative bacteria. LABP mediates the high affinity uptake of *L*-arabinose in *Escherichia coli* (Newcomer et al., 1981). Calculations were performed on the structure (PDB code 1ABE) from *E. coli* determined at 1.7 Å resolution. This crystal structure contains an *L*-arabinose molecule in it, so the interactions of the ligand with the protein are well characterized.

Application of the $Z_3 > 1$ or $Z_4 > 1$ criterion followed by clustering with a cutoff of 9 Å gives the single cluster [**E14**, D206, D235, E20, D89, **D90**, H259]. Residues E14 and D90 are within hydrogen-bond distance of the *L*-arabinose molecule. Application of the statistical criterion does not select R151 and K10, which are within hydrogen-bond distance of the substrate and presumably are important for binding. The other residues, while not directly bonded to *L*-arabinose, completely surround the molecule and may be involved in binding.

3.3 Summary for fifteen proteins

Table 2 presents the results of the application of the $Z_3 > 1$ or $Z_4 > 1$ criterion for 15 proteins. The examples chosen include at least one enzyme from each of the EC classes 1 to 6, representing a wide spectrum of chemical functions. Table 2 also contains a few examples of proteins with binding, transport, or regulation roles that are not known to possess catalytic activity.

THEMATICS combined with our simple selection rules does a remarkable job at finding the active site region for all the cases presented. In a larger study (Ko et al., 2005), we found an overall success rate of 91 % in the identification of the active site zone. Note that the protein region where THEMATICS predicts the location of the active site is very specific and localized. This is in contrast to simple cleft searching, which in general produces much larger regions for the location of potential active sites. For example, in the case of citrate synthase, the results of Table 2 show that the larger cluster contains eight members, two of which are known catalytic residues; cleft search using the program CASTp (Liang et al., 1998) indicates that the largest pocket (to which 5 of the 8 residues of the largest cluster belong) is composed of 77 residues, 22 of which are ionizable. Thus, the active site is located in the largest pocket as Laskowski et al. (1996)

reports to be true for 83 % of monomeric enzymes. However, THEMATICS focuses the active region to a small volume in or around the 8 residues of the predicted cluster, while cleft search indicates a much larger volume contained within the 77 residue set.

Table -2. Results for 15 proteins using the $Z_3 > 1$ or $Z_4 > 1$ criterion. Residues belonging to the same cluster are shown together inside square brackets. Residues in bold are confirmed to be functionally important for binding and/or catalysis. Clusters containing two or more residues are considered predictive

PDB ID	Protein Name (classification)[1]	THEMATICS Positive clusters[2] (reference[3])
1A99	Putrescine receptor (Potf) (binding protein)	[E66,E184,**E185,D247,D278,Y314**] [H123] (Vassylyev et al., 1998)
1ABE	*L*-Arabinose binding protein (binding protein)	[**E14**,D206,D235,E20,**D89,D90**,H259] (Newcomer et al., 1981)
2AID	HIV protease (3.4.23.16)	[**D25a,D25b**,D29a,D29b] (Prabu-Jeyabalan et al., 2000)
1DNK	Deoxyribonuclease I (DNase I) (3.1.21.1)	[**E39,E78,H134**,D168,**D212,H252**] (CatRes) and (Weston et al., 1992)
1AL6	Citrate synthase (2.3.3.1)	[Y231, H235, D237, H238, **H274**, R329, **D375**, R401], [Y318, Y330], [Y190, Y219], [E86], [D174] (CatRes)
1BYK	Tetrahalose repressor (gene regulation)	[D73,**R71,E77**,E99,**Y157**,D159,D241,**Y284**] [D94a,H110b] [H274] (Hars et al., 1998)
1PSO	Pepsin	[**D32,D215**,D303,R307] (Fujinaga *et al.,* 1995)
1HS6	Leukotriene A4 hydrolase (3.3.2.6)	[**E271,H295,E296,H299,E318,D375,Y378,Y383, K565**] [C135,C140,D148,C199] (Thunnissen et al., 2001)
1AOP	Sulfite reductase hemoprotein (1.8.1.2)	[R117,**K215,K217**,K306,K308,Y309,R342,C434, R475,**C483**,R485] (CatRes)
2PLC	Phosphatidylinositol diacylglycerol lyase (4.6.1.13)	[**H45, D46**, D82, H236], [Y71, K115], [D127] (Moser et al., 1997)
1CTT	Cytidine deaminase (Cda) (3.5.4.5)	[**H102,E104,C129**,H131,**C132**][E138,H203] [Y252] (CatRes)
1TLY	Nucleoside-specific channel-forming protein Tsx (membrane protein)	[**E38,Y53**,Y124,Y144,K166,**K168**,Y265] [K44,Y103] [H273] (Ye and Berg, 2004)
1B73	Glutamate racemase (5.1.1.3)	[**D7a**,Y39a,**C70a**,Y123a,C139a,E147a,E148a, **C178a**, **H180a,D7b**,Y39b,**C70b**,Y123b,C139b, **E147b**, E148b, **C178b,H180b**] [K185a,K189a, E152b] [K185b,K189b] (CatRes)
1GIM	Adenylosuccinate synthase (6.3.4.4)	[**D13**,K16,**H41**,E221, K267,Y269,R305, K331], [H110], [C328], [C344] (CatRes)
1ADL	Adipocyte lipid-binding protein (lipid binding protein)	[Y19,R78,**R106,C117,R126,Y128**] (LaLonde et al., 1994)

[1] For enzymes, the EC number is specified. For other proteins, the classification has been taken from the PDB Database; [2] for multimeric proteins, identical clusters are implicit for each chain. Clusters formed by residues from different chains are explicitly indicated using a, b, etc.; [3] the given reference contains information with respect to the functional site. Information about the active site for proteins for which CatRes is given as a reference has been obtained from CATRES database (Bartlett et al., 2002).

Our studies show that THEMATICS predicts 1.8 clusters per protein subunit. It is unclear at this point whether some of the clusters predicted by our method are simply false positives or are actual sites yet to be characterized. Clusters containing the known functionally important residues usually include some other residues for which there is no information about their role. Again, we do not know if these residues are simply false positives or if their function is not yet known. We suspect that many of these cases include residues necessary for function, perhaps in supporting roles. Mutagenesis experiments are in progress to study this.

Applications of THEMATICS to this day have been more focused on enzymes. In this study, we have included a few binding proteins with no known catalytic function. For the binding proteins shown in Table 2, THEMATICS does very well in finding the functionally important regions. However, the sample is too small to draw firm conclusions about the overall performance of the method for non-catalytic binding proteins. A much wider study of these systems is currently in progress.

ACKNOWLEDGMENTS

JK thanks the Department of Chemistry and Chemical Biology and the Institute for Complex Scientific Software at Northeastern University for their hospitality during her sabbatical leave. This work was supported by the U.S. National Science Foundation under grant No. MCB-0135303 and by the Institute for Complex Scientific Software at Northeastern University.

AMINO ACIDS SURFACE PATTERNS IN PROTEIN DOMAIN FUNCTIONALITY ANALYSIS

I. Merelli[1], L. Pattini[2], S. Cerutti[2], L. Milanesi[1*]

[1] *Istituto di Tecnologie Biomediche, Consiglio Nazionale delle Ricerche, via Fratelli Cervi, 93, Segrate (MI), Italy, e-mail: luciano.milanesi@itb.cnr.it ;* [2] *Politecnico di Milano, piazza Leonardo da Vinci, Milano, Italy*

[*] *Corresponding author*

Abstract: Surface characterization of peptides may provide useful information about functionality and potential interactions with other molecules. A description of a protein site through a surface that models the shape conferred by the exposed residues is an effective tool for the analysis and modeling of proteins that may highlight similarities and relationships not detectable through comparisons at the levels of primary, secondary, and tertiary structure. This study concerns the development of a tool that extracts the residues that concur to the shape modeling of the surface of a protein or a portion of it. This task is accomplished without taking into account the order of amino acids in the primary structure but only according to the selection of a portion of the protein indicated through geometric parameters or an explicit list of amino acids belonging to the site of interest. Using this software, it is possible to identify some surface patterns that could be analyzed both in comparison with the functional domains of specific protein families or for study of particular interaction sites in macromolecular complexes.

Key words: analysis; modeling; pattern; protein representation; surface; structure; sequence; amino acids; surface patterns

1. INTRODUCTION

Much work has been carried out in the analysis of proteins at the levels of primary, secondary, and tertiary structure. Comparisons between different proteins performed on the basis of this kind of information have demonstrated structural similarities that lead to important functional and evolutionary relationships. However, protein interactions and functionalities are strictly correlated to the surface shape. Mechanisms, such as enzyme catalysis, interaction with a ligand, recognition of signals by specific binding sites, and docking usually depend on protein surface. Thus, detection of surface similarities can reveal relationships between proteins to which sequence alignments and fold comparisons are blind (Via et al., 2000) (Pickering et al., 2001). Moreover, a flexible surface description may allow the analysis of macromolecular complex interfaces to study protein–protein interactions. Shape similarities and complementarities may elicit surface patterns that are not recognizable in trivial residue motifs. Indeed, it is often the case that shape similarity is more conserved through evolution than sequence, as for the active sites of trypsin-like and subtilisin-like serine protease families, whose similarity occurs without common features in the protein folding (Fischer et al., 1994). Again, from an evolutionary point of view, surface comparison may highlight the convergence of different sequences in a unique pattern of exposed residues that specifies the function but is not detectable from a conventional alignment, because they are interspersed along the primary structure. Similarly, specificity can be improved through a surface characterization of proteins.

This study concerns the implementation of a software for the reconstruction of protein surfaces through definition of a three-dimensional model that allows the description of the protein volumes, starting from the knowledge of the atomic coordinates stored in the Protein Data Bank. From this model, the software is able to define the protein surface through the Marching Cubes algorithm. Analyzing the resulting support mesh in relation to the position of different atoms, it is possible to identify which amino acids are exposed on the protein surface. This is very important to establish a conclusive relationship between the exposed amino acids and the protein functionality. Departing from the position of the amino acids, instead, it is possible to define their position on the protein surface and to analyze the piece of mesh in which they are exposed, to establish complementarities between different macromolecules.

The data show that when analyzing a specific part of the protein surface, it is possible to find the amino acids that have a key role in the biological process. Inside the protein sequence, in fact, the exposed amino acids are limited in number because the big inside structures are often buried, while the functional sites are exposed on the surface. However, focusing the analysis on the areas of surface that are involved in enzymatic reactions by restricting the analysis only to functional domains provides a more precise and conclusive identification of the key amino acids. Another case of the study consists in marking the amino acids of the protein surface in the primary sequences during a multi-alignments process to allow the cross-correlation of functional domains in relationship to their presence on the protein surface.

2. METHODS AND ALGORITHMS

The 3D atomic coordinates of a protein, as retrieved from the Protein Data Bank, are used to generate a so-called space-filling model of it. This approach leads to an implicit modelization of the surface that is more suitable for this kind of analysis than the parametric one. Each atom is modeled as a volumetric item where a function is defined, which assumes negative values within van der Waals radius and positive values outwards, with a sign change just in correspondence with the van der Waals surface. For example, in the volume that contains an oxygen atom, the sign inversion occurs at 0.73 Å from the atom center.

These items are positioned in a uniform space grid coherently with the PDB coordinates and are summed point to point. The sum allows the function values where the spheres are contiguous or overlapped to be smoothed, avoiding introduction of a high frequency noise. The resolution grid can be varied according to the desired level of detail (Vakser et al., 1999). The 3D function defined on the grid is further low-pass filtered with a filter box with 5 as size of the convolution kernel. This operation prevents the description of non-representative details.

The extraction of the surface is accomplished through computation of an isosurface in the 3D grid function, which separates grid points with the value below a defined threshold from those above. The linear interpolation is performed on the basis of the Marching Cubes principle, which provides triangulation of the surface. Note that to investigate the interaction of macromolecules, it is important to examine their surfaces considering the solvent molecules that surround the Van der Waals surface, in particular, the protein modifying their accessibility. This is the Lee and Richards model (Lee and Richards, 1971), which is intuitively generated by rolling a probe sphere with a given radius (usually, 1.4 Å for water as the solvent, since hydrogens are neglected) around the Van der Waals surface of the molecule: the trace of the center of the probe sphere produces the surface.

A smoothed version of the Lee and Richards surface is calculated with the same algorithm used for the Van der Waals surface just modifying the sign inversion point in the volumetric item through a specific threshold that now occurs at the Van der Waals radius augmented by the solvent radius. As an example, see the rendered surfaces for $FADH_2$ protein in Figure 1.

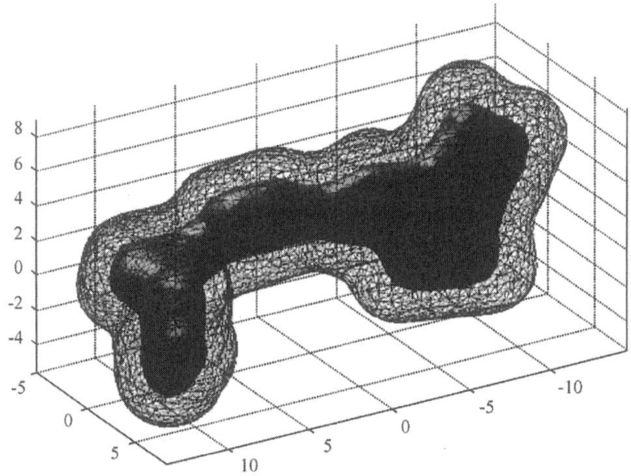

Figure -1. Van der Waals surface and solvent accessible surface for $FADH_2$.

From a particular selected portion of a surface, the subtended residues may be individuated. They are geometrically derived from the mesh that models the protein surfaces at the interface. Then, they are reported on the primary structure to enrich the information carried by the amino acid sequence. This allows amino acid sequences to be compared on the basis of spatial relationships.

There can be the case that two protein sequences are very different and a mere alignment does not show any significant similarity. However, on the respective surfaces, they display resembling patterns, and the key functional residues may be individuated. In this way, the internal structures, negligible if attention is on the interaction phenomena, are not considered and only the amino acids that concur to the composition of the envelope are retained.

A specific surface pattern can be selected by giving the list of the amino acids to be modeled or a portion of space described by geometric parameters as input to the tool, but it is even possible to make a graphical selection of the area of interest. Once a piece of surface is identified, it is possible to isolate the correspondent polygonal mesh through selection of the vertexes close to the elements chosen.

This functionality is very important, because it allows identification of different meshes in relation to specific amino acids in order to elaborate and compare them. For example, Figure 2 shows the insulin protein with highlighted active site that binds receptors on its surface and details of the mesh of the only site.

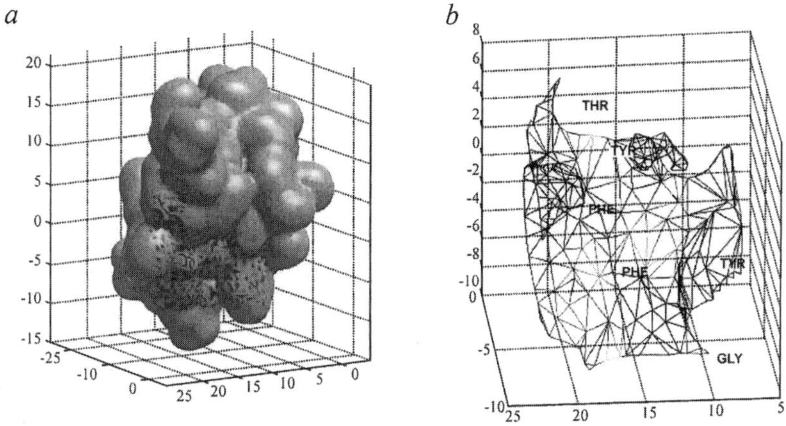

Figure -2. Insulin protein: the active site that binds the receptors (*a*) and the mesh of the only site (*b*).

3. RESULTS AND DISCUSSION

The method is implemented in Matlab, which provides an ideal framework for both numeric analysis and graphics. The software, in fact, is implemented through a series of powerful calculation libraries, which allow a 3D interpolation of the data and, therefore, extraction of the polygonal mesh with precision and efficiency. The mesh obtained could be straightforwardly visualized, and many different options for the rendering are available as well as coloring of the surface according additional information, such as the electrostatic potential. The mesh is exportable as obj format from Matlab to other softwares, such as AutoCad, 3D Studio, and VRML. After surface extraction, it is possible to isolate specific regions of it that correspond to catalytic sites, characteristic domains, or, in general, zones of interest.

3.1 Analysis of surface patterns in protein complexes

In this first analysis, the surfaces of known protein complexes were examined. In this way, an important test of the predictive potential of the software is implemented to analyze the possibilities in relation to identification of superficial patterns and their key role inside the protein domains sequence (Ma et al., 2001). The portions of surfaces that constitute the interfaces of a few protein complexes were considered. The attention is on the exposed residues that characterize the surface regions involved in the interactions.

Chain H
Bos Taurus

I V E G Q D A E V G L S P W Q V M
L F **R** K S P Q **E** L L C G A S L I S
D R W V L T A A **H** C L **L Y P W D K**
N F T V D D L L V R I G K H S R T
R Y E R K V E K I S M L D K I Y I
H P R Y **N W K E N L D** R D I A L L
K L K R P I E L S D Y I H P V C L
P D K Q T A K L L H A G F K G R V
T G W G N R R **E** T W S V A E V Q P
S V L Q V V N L P L V E **R** P V C **K**
A S **T R I R I** T D N M F C A G Y K
P G E K R G D A **C E** G D S G G P F
V M K S P Y N N R W Y Q M G I V S
W G **E G** C **D R** D G **K Y** G **F** Y T H V
F R L K K W I Q K V I D R

Chain H
Homo Sapiens

I V E G S D A E I G M **S** P W Q V M
L **F** R **K S P Q E L L** C G A S L I S
D R W V L T A A **H** C L L **Y** P **W** D **K**
N F T E N D L L V **R** I G K H S **R T**
R Y E R N I E K **I** S **M** L E K I Y I
H P R Y N W R E N L D R D I A L M
K L K K P V A F S D Y I H P V C L
P D R E T A S L L Q A G Y K **G** R V
T G W G N L K E T K G Q P S V L Q
V V N L P I V E R P V C K D S T R
I R I T D N M F C A G Y K P D E G
K R G D A C **E** G D S G G P F V M K
S P F N N R W Y Q M G I V S **W** G E
G C D R D G K Y G F Y T H V F R L
K K W I Q K V I D Q F

Chain H
Bos Taurus

I V E G Q D A **E V** G **L S** P **W** Q V M
L F R K S P Q E L L C G A S L I S
D R W V L T A A H C L L Y P W D K
N F T V D D L L V R I G K **H** S R T
R Y E R K V E K I S M L D K I Y I
H P R Y N W K E N L D R D I A L L
K L K R P I E L **S D Y** I **H** P V C **L**
P D K Q T A **K** L L H A G **F K** G **R** V
T G W G N R R E T W S V A E V Q P
S V L Q V V N L P L V E R P V C K
A S T R I R I T D N M F C A G Y K
P G E K R G D A C E G D S G G P F
V M **K** S **P Y N N R W** Y Q M G I V S
W G E G C D R D G K Y G F Y T H V
F R L K K W I Q K V I D R

Chain H
Homo Sapiens

I V E G S D A **E** I G **M S** P **W** Q V M
L F R K S P Q E L L C G A S L I S
D R W V L T A A H C L L Y P W D K
N F T E N D L L V R I G K H S R T
R Y E R N I E K I S M L E K I Y I
H P R Y N W R E N L D R D I A L M
K L K K P V A F S D Y I **H** P V C L
P D R **E** T A **S** L L Q A G **Y K** G **R** V
T G W G N L K E T K G Q P S V L Q
V V **N** L P I V E R P V C K D S T R
I R I T D N M F C A G Y K P D E G
K R G D A C E G D S G G P F V M **K**
S **P F N N R W** Y Q M G I V S W G E
G C D R D G K Y G F Y T H V F R L
K K W I Q K V I D Q F

Figure -3. Primary structure of thrombin where surface interacting amino acids are marked.

The first complex that was considered is the thrombin complex of *Homo sapiens* (PDB ID: 1a3e) and *Bos taurus* (PDB ID: 1id5). This complex is constituted by a heavy chain (H), a light chain (L), and an inhibitor (I).

Chain A
Bos Taurus

V L S A A D K G N V K A A W G K V
G G H A A E Y G A E A L E **R** M F L
S F P T T K T Y F P H F D L S H G
S A Q V K G H G A K V A A A L T K
A V E H L D D L P G A L S E L S D
L H A H K L R V D P V N F **K** L L S
H S L L **V** T L A **S H** L P S D **F** T **P**
A V **H** A S L **D** K F L A N V S T V L
T S K Y R

Chain A
Homo Sapiens

V L S P A D K T N V K A A W G K V
G A H A G **E** Y G A **E** A L E **R** M F L
S F P T T K T Y F P H F D L S H G
S A Q V K G H G K K V A D A L T N
A V A H V D D M P N A L S A L S D
L H A H K L R V D P V N F **K** L L S
H C L **L V** T L A **A** H L **P** A E **F** T P
A V **H** A S L **D** K F L A S V **S** T V L
T S K Y R

Chain B
Bos Taurus

M L T A E E K A A V T A F W G K V
K V D **E** V G G **E** A L G **R** L L V V Y
P W T Q R F F E S F G D L S T A D
A V **M** N N P K V **K** A H G K K V L D
S F S N G M K H L D D L K G T F A
A L S E L H C D K L H V D F E N F
K L **L** G **N V** L **V** V V L A **R** N F G **K**
E F T P V L Q A D F Q K V V A G V
A N A L A H R Y H

Chain B
Homo Sapiens

V H L T P E E K S A V T A L W G K
V N V D E V G G E A L G **R** L L V V
Y P W T Q R F F E S F G D L S T P
D A V M G N P K V K A H G K K V L
G A F S D G L A H L D N L K G T F
A T L S E L H C D K L H V D P **E** N
F **R** L L G **N V** L V **C** V L A **H H** F G
K E F T P P V Q A A Y **Q** K V V A G
V A N A L A H K Y H

Figure -4. Primary structure of hemoglobin where surface interacting amino acids are marked.

Using the method proposed, the trace of the interaction with the light chain can be detected on the heavy chain of both organisms segregating only the residues exposed at this interface. There is a remarkable similarity in the obtained patterns of these two species. The same procedure was followed to extract the amino acid of H involved in the interaction with I. Since I is different for the two complexes, its trace on the H primary sequences does not show any common feature (Figure 3).

Another example of interspecies comparison between the surface determinants is shown for hemoglobin, again a complex, constituted of two chains (A and B), involving human (PDB ID: 1bbb) and bovine (PDB ID: 1hda) proteins. The trace of interaction of the chain B with chain A and the trace of chain A on chain B, respectively, are described: while a strong correspondence may be found between markers on A chains, it is clear that B chains are very dissimilar (Figure 4).

3.2 Analysis of surface patterns in protein sequence multi-alignments

Another type of analysis can be conducted identifying the protein surface amino acids in a group of proteins that check their functional similarity. Making a multi-alignment of the primary sequences, it is possible to check the correspondence of the amino acids exposed in order to obtain more complete information. This test was conducted on a set of protein kinase structures to verify how the information on the functional domain completes the data regarding the surface patterns.

From the data downloaded from PDB, a volumetric reconstruction of the proteins can be performed followed by extraction of the relative surface and identification of its key amino acids. To prove the strength of the analysis conducted, it is important to notice that the software mainly identifies polar amino acids expected to be exposed on the surface. Furthermore, the second glycine of the kinase domain is frequently exposed on the protein surface, perhaps because it is the docking point for the molecules involved in the reaction (Figure 5).

However, it is not easy to identify correlations among the amino acids exposed on the protein surface and those of the catalytic domains because of the lack of protein structure entries in PDB.

In fact, although numerous 3D structures are known in detail, many of them contain incompletely modeled proteins. This is a fundamental piece of information for investigating the surface patterns, because only knowing the whole 3D structures allows defining the relative position of the amino acids. To get conclusive results, it will be necessary to implement searches on vast datasets creating special databases of entries, as has already been done for some particular classes of proteins.

```
1CM8_p38g             YRDLQPV------AVCSAVDGRTGAKVAIKKLYRPFQSELFAKRAYRELRLLKHMRH-EN
1R39_p38a             YQNLSPVGSGAYGSVCAAFDTKTGLRVAVKKLSRPFQSIIHAKRTYRELRLLKHMKH-EN
1B38_cdk2             FQKVEKIGEGTYGVVYKARNKLTGEVVALKKIVP--------STAIREISLLKELNH-PN
1F3M_pak1             YTRFEKIGQGASGTVYTAMDVATGQEVAIRQMNLQQQPK--KELIINEILVMRENKN-PN
1H8F_gsk3b            YTDTKVIGNGSFGVVYQAKLCDSGELVAIKKVLQG-----KAFKN-RELQIMRKLDH-CN
1MQ4_AurA             FEIGRPLGKGKFGNVYLAREKQSKFILALKVLFKA-----QLEKAGVEHQLRREVEI-QS
1FMK_src              LRLEVKLGQGCFGEVWMGTWN--GTTRVAIKTLKPGTMSP--EAFLQEAQVMKKLRH-EK
1MQB_EphA2            VTRQKVIGAGEFGEVYKGMLKTKKEVPVAIKTLKAGYTEKQRVDFLGEAGIMGQFSH-HN
1MP8_fak              IELGRCIGEGQFGDVHQGIYMSPPALAVAIKTCKNCTSDSVREKFLQEALTMRQFDH-PH
1IA8_chk1             WDLVQTLGEGAYGEVQLAVNRVTEEAVAVKIVDMKR----CPENIKKEICINKMLNH-EN
1JWH_ck2a1            YQLVRKLGRGKYSEVFEAINITNNEKVVVKILKPVK-----KKKIKREIKILENLRGGPN
1KWP_mapkapk2         KVTSQVLGLGINGKVLQIFNKRTQEKFALKMLQDCP-------KARREVELHWRASQCPH
                            :       *            .                   *

1CM8_p38g             VIGLLDVFTPDETLDDFTDFYLVMPFMG-TDLGKLMK--HEKLG---EDRIQFLVYQMLK
1R39_p38a             VIGLLDVFTPARSLEEFNDVYLVTHLMG-ADLNNIVK--CQKLT---DDHVQFLIYQILR
1B38_cdk2             IVKLLDVIHTEN------KLYLVFEFLH-QDLKKFMD--ASALTGIPLPLIKSYLFQLLQ
1F3M_pak1             IVNYLDSYLVGD------ELWVVMEYLAGGSLTDVVT--ETCMD---EGQIAAVCRECLQ
1H8F_gsk3b            IVRLRYFFYSSGEKKDEVYLNLVLDYVP-ETVYRVARHYSRAKQTLPVIYVKLYMYQLFR
1MQ4_AurA             HLRHPNILRLYGYFHDATRVYLILEYAPLGTVYRELQKLSKFDEQR----TATYITELAN
1FMK_src              LVQLYAVVSE-------EPIYIVTEYMSKGSLLDFLK--GETGKYLRLPQLVDMAAQIAS
1MQB_EphA2            IIRLEGVISKY------KPMMIITEYMENGALDKFLR--EKDGEFSVL-QLVGMLRGIAA
1MP8_fak              IVKLIGVITE-------NPVWIIMELCTLGELRSFLQ--VR-KYSLDLASLILYAYQLST
1IA8_chk1             VVKFYGHRREGN------IQYLFLEYCSGGELFDRIEP----DIGMPEPDAQRFFHQLMA
1JWH_ck2a1            IITLADIVKDPVS----RTPALVFEHVNNTDFKQLYQT-------LTDYDIRFYMYEILK
1KWP_mapkapk2         IVRIVDVYENLYAG--RKCLLIVMECLDGGELFSRIQ--DRGDQAFTEREASEIMKSIGE
                       :                   :.        .

1CM8_p38g             GLRYIHAAGIIHRDLKPGNLAVN---EDCELKILDFGLARQADSEMG-------VVTRWY
1R39_p38a             GLKYIHSADIIHRDLKPSNLAVN---EDCELKILDFGLT---DDEMTGY-----VATRWY
1B38_cdk2             GLAFCHSHRVLHRDLKPQNLLIN---TEGAIKLADFGLARAFGVPVRTYTH--EVVTLWY
1F3M_pak1             ALEFLHSNQVIHRDIKSDNILLG---MDGSVKLTDFGFCAQITTMVG---------TPYW
1H8F_gsk3b            SLAYIHSFGICHRDIKPQNLLLDP--DTAVLKLCDFGSAKQLVRGEPNVS---YICSRYY
1MQ4_AurA             ALSYCHSKRVIHRDIKPENLLLG---SAGELKIADFG----WSVHAPSSR---TLCGTLD
1FMK_src              GMAYVERMNYVHRDLRAANILVG---ENLVCKVADFGLAR---------------FPIKW
1MQB_EphA2            GMKYLANMNYVHRDLAARNILVN---SNLVCKVSDFGLK----------------IPIRW
1MP8_fak              ALAYLESKRFVHRDIAARNVLVS---SNDCVKLGD--------------------LPIKW
1IA8_chk1             GVVYLHGIGITHRDIKPENLLLD---ERDNLKISDFGLATVFRYNNRERLLNKMCGTLPY
1JWH_ck2a1            ALDYCHSMGIMHRDVKPHNVMIDH--EHRKLRLIDWGLAEFYHPGQEYNVR---VASRYF
1KWP_mapkapk2         AIQYLHSINIAHRDVKPENLLYTSKRPNAILKLTDFG----------------------
                      .: :       ***: . *:           :: *

1CM8_p38g             RAPEVILNWMRYTQTVDIWSVGCIMAEMIT-GKTLFKGSDHLDQLKEIMKVTGTPPAEFV
1R39_p38a             RAPEIMLNWMHYNQTVDIWSVGCIMAELLT-GRTLFPGTDHIDQLKLILRLVGTPGAELL
1B38_cdk2             RAPEILLGCKYYSTAVDIWSLGCIFAEMVT-RRALFPGDSEIDQLFRIFRTLGTPDEVVW
1F3M_pak1             MAPEVVT-RKAYGPKVDIWSLGIMAIEMIE-GEPPYLNENPLRALYLIATNG--------
1H8F_gsk3b            RAPELIFGATDYTSSIDVWSAGCVLAELLL-GQPIFPGDSGVDQLVEIIKVLGTPTREQI
1MQ4_AurA             YLPPEMIEGRMHDEKVDLWSLGVLCYEFLV-GKPPFEAN---------------TYQETY
1FMK_src              TAPEAAL-YGRFTIKSDVWSFGILLTELTTKGRVPYPGMVNREVLDQVERGY--------
1MQB_EphA2            TAPEAIS-YRKFTSASDVWSFGIVMWEVMTYGERPYWELSNHEVMKAINDGF--------
1MP8_fak              MAPESIN-FRRFTSASDVWMFGVCMWEILMHGVKPFQGVKNNDVIGRIENGE--------
1IA8_chk1             VAPELLKRREFHAEPVDVWSCGIVLTAMLAGELPWDQPSDSCQEYSDWKEKK--------
1JWH_ck2a1            KGPELLVDYQMYDYSLDMWSLGCMLASMIFRKEPFFHGHDNYDQLVRIAKVLGTEDLYDY
1KWP_mapkapk2         FAKETTSGPEKYDKSCDMWSLGVIMYILLC-GYPPFYSNHGLAISPGMKTRIR-------
                                 .    *:*  *     .

1CM8_p38g             QRLQSDEAKNYMKGLP------ELEKKDFASILTNASPLAVNLLEKMLVLDAEQRVTAGE
1R39_p38a             KKISSESARNYIQSLT------QMPKMNFANVFIGANPLAVDLLEKMLVLDSDKRITAAQ
1B38_cdk2             PGVTS--MPDYKPSFP------KWARQDFSKVVPPLDEDGRSLLSQMLHYDPNKRISAKA
1F3M_pak1             -----------TPELQ------NPEK---------LSAIFRDFLNRCLDMDVEKRGSAKE
1H8F_gsk3b            REMNPNYTEFAFPQIK------AHPWTKVFR--PRTPPEAIALCSRLLEYTPTARLTPLE
1MQ4_AurA             KRISR--VEFTFPDFV------TEG--------------ARDLISRLLKHNPSQRPMLRE
1FMK_src              ----------RMPCPP------ECP------------ESLHDLMCQCWRKEPEERPTFEY
1MQB_EphA2            ----------RLPTPM------DCP------------SAIYQLMMQCWQQERARRPKFAD
1MP8_fak              ----------RLPMPP------NCP------------PTLYSLMTKCWAYDPSRRPRFTE
1IA8_chk1             ---------TYLNPWK------KIDS------------APLALLHKILVENPSARITIPD
1JWH_ck2a1            IDKYNIELDPRFNDILGRHSRKRWERFVHSENQHLVSPEALDFLDKLLRYDHQSRLTARE
1KWP_mapkapk2         --------MYEFPNPE----------------WSEVSEEVKMLIRNLLKTEPTQRMTITE
                                                          :  .        *
```

Figure -5. Multi-alignments of protein kinase sequences containing superficial amino acids.

4. CONCLUSION

Small functional changes may be better explained by differences in the surface shape than by differences in the overall tertiary structure. Moreover, surface comparison makes negligible the trivial matching of voluminous inner structures that are less informative with respect to interactions. It is useful to segregate amino acids from a functional point of view: identification of the locus of residues that are exposed in a region of interest may constitute additional information for the amino acid sequence. In this way, protein comparisons can be performed on the basis of information carried in both the primary structure and shape models, leading to definition of superficial patterns.

The results presented are an example of what kind of information can be retrieved through this approach. The developed procedure allows for describing portions of surfaces and identifying putative functional residues that concur to the shape complementarities of interacting peptides. It is often the case that functionality is encoded in few residues, and this approach can be useful to visualize on the primary structure the particular amino acids involved in the interaction studied. These surface patterns thus identified may lead to identification of new proteins involved in a certain process and may help to describe better the mechanisms that govern the interaction phenomena.

This type of analysis has an important impact on proteomics, because it represents the conjunction between a sequence-based analysis and structural analysis. The software implemented provides a direct approach to sequence analysis from a structural morphology point of view. Joining the classical analyses of the primary sequence for the protein domain definition and identification of secondary structures with this novel approach, distribution notable results were obtained. Merging the information provided by all these approaches can indeed establish important relationships between the primary sequences and the protein functionality.

ACKNOWLEDGMENTS

This work was supported by the MIUR-FIRB projects 'Functional Genomics', 'Bioinformatics for Genome and Proteome', 'Enabling Platforms for High-Performance Computational Grids Oriented to Scalable Virtual Organizations', and MIUR (Laboratory of Interdisciplinary Technologies-LITBIO – RBLA0332RH) and NATO (grant No. LST.CLG.979816).

COMPUTER SIMULATIONS OF ANIONIC UNSATURATED LIPID BILAYER—A SUITABLE MODEL TO STUDY MEMBRANE INTERACTIONS WITH A CELL-PENETRATING PEPTIDE

A.A. Polyansky[1, 2*], P.E. Volynsky[2], A.S. Arseniev[2], R.G. Efremov[2]
[1] Department of Bioengineering, Biological Faculty, Lomonosov Moscow State University, Moscow, 119992, Russia, e-mail: newant@nmr.ru; [2] Shemyakin and Ovchinnikov Institute of Bioorganic Chemistry, Russian Academy of Sciences, Moscow, 117997, Russia
[*] Corresponding author

Abstract: Unsaturated phosphatidylserine (PS) bilayers are widely used in experimental studies of peptides and proteins in charged membranes. Atomic-scale details of peptide-membrane interactions may be assessed via molecular dynamics (MD) simulations. Unfortunately, a wide application of computational techniques to such systems is limited because of serious technical problems related to correct treatment of long-range electrostatic effects. Here we present a new model of full-atom hydrated PS bilayer. It consists of 128 molecules of 1,2-dioleoyl-sn-glycero-3-phosphoserine (DOPS), 128 Na^+ counterions, and explicit waters. The system was subjected to 15-ns MD simulations with different algorithms of electrostatics treatment (cutoff function and particle mesh Ewald summation (PME)). As a result, an optimal PME-based MD-protocol was elaborated. It provides a good agreement between the macroscopic averages calculated for the equilibrium part of the MD trajectory and those available from experiments. The model of the DOPS bilayer was used to study interactions with penetratin (pAntp). pAntp is a 16-residue peptide that is capable of passing through cell membranes and negatively charged phospholipid vesicles without their leakage. Unfortunately, the mechanism of penetration is still insufficiently understood. During the simulations, a free adsorption of pAntp on the water–lipid interface was observed. The membrane binding of pAntp was accompanied by distortion of its initial α-helical conformation. The critical roles of individual amino acid residues and surface charge of the DOPS bilayer were delineated. It was shown that the peptide-induced perturbation of the membrane had a local character.

Key words: molecular dynamics; DOPS; lipid–water interface; cell-penetrating peptides; peptide–membrane interactions

1. INTRODUCTION

Protein–membrane and peptide–membrane interactions are important for many processes in the living cell. That is why they attract growing interest in the field of structural biology and bioengineering. The novel and interesting aspect of such interactions is their internalization by peptide-based cellular transporters. Penetratin (pAntp) is the first member of the rapidly expanding family of cell penetrating peptides (CPP) originating from either natural or synthetic source. This is a short peptide (16 residues) isolated from Antennapedia homeodomain protein of *D. melanogaster* (Derossi et al., 1998). Experiments demonstrate that pAntp is capable of passing through cell membranes and model negatively charged phospholipid vesicles without their leakage. Moreover, pAntp acts as a nonspecific cell transporter of covalently bound charged molecules (proteins, peptides, oligonucleotides, and drugs). However, the mechanism of penetration is yet to be understood. Two possible models of such process (endocytosis and receptor-independent mechanism) were proposed (Letoha et al., 2003). Therefore, clarification of this issue represents an intriguing challenge. Unfortunately, experimental techniques (NMR, ESR, and CD) often give only an overall picture of behavior of a model peptide–membrane system, while molecular details of peptide-membrane interactions in many cases are missing. Important insights into the problem may be gained *via* computer experiments. Among such techniques, molecular dynamics (MD) of peptides in explicit lipid bilayers system is the most powerful approach. Numerous studies of saturated and unsaturated zwitterionic bilayers composed of lipids with phosphatidylcholine (PC) and phosphatidylethanolamine (PE) headgroups appeared during the last decade (Forrest and Sansom, 2000; Efremov et al., 2004). However, just few MD simulations of charged bilayers are successful (Lopez Cascales et al., 1996; Pandit and Berkowitz, 2002; Mukhopadhay et al., 2004). Possible reasons for this are the serious technical problems, which often occur during the simulations. They are mainly related to improper description of long-range electrostatic interactions (Anezo et al., 2003; Patra et al., 2003; Rog et al., 2003). The most common algorithms used for electrostatics treatment are cutoff functions and the particle mesh Ewald (PME) summations (Essmann et al., 1995). Previously, it was shown that cutoff-based MD simulations of explicit bilayers led to computational artifacts, such as decrease in area per lipid, asymmetry of dipole potential, anomalous ordering of acyl chains, and polar headgroups (Anezo et al., 2003; Patra et al., 2003). Such conclusions seem to contradict a number of recent studies revealing feasibility of cutoff-based schemes for MD-simulations of zwitterionic bilayers. Moreover, Rog et al. (2003) compared the results obtained using both protocols for treatment of electrostatics in MD calculations of zwitterionic (POPC) and negatively charged (POPG/POPE) bilayers. Consequently, several artifacts related to ordering of acyl chains, mobility of charged lipids, and values of area per lipid were observed; however, only when the cutoff algorithm was applied to charged bilayers. Therefore, a special care is necessary in MD studies of such systems.

Among the objectives of this work are elaboration of a model bilayer composed of unsaturated negatively charged lipids, optimization of computational protocol for its MD study, and analysis of possible artifacts caused by improper treatment of electrostatic interactions. With this aim in view, a bilayer containing 128 lipids of 1,2-dioleoyl-sn-glycero-3-phosphoserine (DOPS) was built and subjected to long-term MD simulations. Nowadays, DOPS bilayers are commonly used in experimental studies of interactions of peptides, proteins, and drugs with model membranes. Physicochemical properties of 'pure' DOPS bilayers were also studied in experiments (Petrache et al., 2004). The MD data obtained in this work reveal several artifacts caused by application of cutoff function in MD simulations of the DOPS bilayer. They are related to uncorrected treatment of interactions between Na^+ counterions and charged lipid headgroups. The model of the DOPS bilayer that agreed well with the experimental data was used to study penetratin–membrane interactions. The starting conformation of pAntp was the conformation solved by NMR spectroscopy in TFE–water mixture (Czajlik et al., 2002). Analysis of 18-ns MD trajectory detected no essential alterations in the structural and dynamical properties of the DOPS bilayer. Nonetheless, a specific local membrane response induced by the peptide was detected. During the simulations, a spontaneous adsorption of pAntp on the water–lipid interface was observed. It was shown that the membrane binding was accompanied by distortion of the initial α-helical conformation of the peptide. The critical role of individual amino acid residues and the membrane surface charge was delineated.

2. METHODS AND ALGORITHMS

Construction of bilayer systems. MD-simulation protocol. The spatial structure of DOPC molecule was obtained via X-ray and neutron scattering experiments[1]. It was further used as a scaffold to build a 3D model of DOPS molecule: its PC headgroup was changed to PS with the help of molecular editor. Topology and GROMOS87 force field (van Gunsteren et al., 1987) parameters for DOPS were obtained basing on the available characteristics for the POPC molecule[2]. Partial atomic charges of PS were computed with a help of semi-empirical quantum chemistry methods (ZINDO) and optimized for the

[1] *http://blanco.biomol.uci.edu/Bilayer_Struc_frames.html*
[2] *http://moose.bio.ucalgary.ca/Downloads/*

united atom GROMOS87 force field. Based on this structure, the model bilayer of 128 DOPS lipids (64 in each monolayer) was constructed and placed into rectangular box (64 × 64 × 80 Å^3, the last value is the size along the normal to the membrane plane) with the same number of Na^+ counterions and 5624 waters. In the bilayer system, the water/lipid ratio was about 40. This corresponds to a fully hydrated state of bilayer: the experimentally measured values are about 30 (Petrache et al., 2004). Before MD simulations, the system was subjected to energy minimization (1000 conjugate gradients steps). Water molecules were equilibrated via 20 ps MD runs with fixed lipid atoms. At this stage, 18-Å cutoff for electrostatics was employed. Each system was subsequently heated to 300 K during 60 ps. MD simulations were carried out in the NPT ensemble—at T = 300 K and isotropic pressure of 1 bar. Van-der-Waals interactions were truncated using the twin range (12/20 Å) cutoff. Electrostatic interactions were treated in two different ways: using the same cutoff scheme and the PME algorithm (1.2 Å Fourier spacing). The equilibrated structure of DOPS bilayer (after 15-ns MD run) was employed for MD simulations with a presence of pAntp. Starting conformation of pAntp was obtained from NMR-data in TFE–water mixture (Czajlik et al., 2002). The DOPS + pAntp system was studied via free MD simulations (without any additional potentials and restraints) using the PME-based protocol and the same computational setup. In addition, 8-ns PME-based MD simulations of pAntp were carried out in water to obtain the reference system. The calculations were performed with the help of the GROMACS v. 3.14 software (Berendsen et al., 1995). Parameters of the resulting MD trajectories are summarized in Table 1.

Table -1. Parameters of generated MD trajectories

System	Electrostatics treatment	Length, ns	Abbreviation
128 DOPS/ 5624 waters/ 128 Na^+	PME	15	PS-PME
128 DOPS/ 5624 waters/ 128 Na^+	cutoff	16	PS-Cut
1 pAntp/ waters / 7 Cl-	PME	8.3	Wt-pAntp
128 DOPS/ 6704 waters/ 128 Na^+/1 pAntp	PME	18.5	PS-pAntp

Analysis of MD-trajectories. The computed MD trajectories were analyzed with a help of original software and modified utilities supplied with the GROMACS package. To compare the simulation results with the experimental data, several important macroscopic averages were estimated for the equilibrium parts (last 5 ns) of all MD trajectories. These are the area per lipid molecule (A_L), the order parameter of acyl chains (S_{cd}), the distance between the planes determined by phosphorus atoms of lipids in different monolayers ($D_{p\text{-}p}$), and the function of electron density distribution along the normal to bilayer plane (ρ(z)). (Hereinafter, the axis Z is perpendicular to the membrane plane.) In addition, 3D radial distribution functions (RDF, g(r)) were calculated for 'pure' bilayer. Energies of electrostatic and van-der-Waals interactions of lipid headgroups with waters, Na^+ counterions, and pAntp were computed using the g-nbi program, specially written for this purpose. The depth of insertion of a given pAntp residue into the bilayer was estimated as average distance (D_i) between this residue and the plane determined by phosphorus atoms of lipids in 'upper' (closest to peptide) monolayer. Landscape of the membrane surface was constructed basing on deviations in Z-positions of phosphorus atoms of lipids from their mean value. Secondary structure of the peptide was assigned using the do_dssp program from the GROMACS package.

3. RESULTS AND DISCUSSION

Table -2. Equilibrium macroscopic averages for DOPS bilayer: MD simulations and experimental data

Parameter[1]	PS-Cut	PS-PME	PS-pAntp	Experiment[2]
A_L, Å^2	63.33 ± 0.07	63.31 ± 0.08	63.03 ± 0.06	64
$\langle D_{p\text{-}p}\rangle$, Å	37.7 ± 0.6	39.0 ± 0.2	39.4 ± 0.2	39
D_{HH}, Å	44	39	4	39
$\langle S_{cd}\rangle$	0.10 ± 0.03	0.14 ± 0.05	0.14 ± 0.05	0.15[2]

[1] Description of parameters: A_L is the average area per lipid molecule; $\langle D_{p\text{-}p}\rangle$, average bilayer thickness (distance between the planes determined by phosphorus atoms of lipids in different monolayers); D_{HH}, distance between the peaks of electron density for phosphorus atoms; and $\langle S_{cd}\rangle$, average value of order parameter for acyl chains. [2] Data taken from Petrache et al. (2004).

3.1 'Pure' bilayer simulations

The calculated equilibrium macroscopic averages of structural parameters for DOPS bilayers along with the experimental data are listed in Table 2. These parameters were estimated for the trajectories obtained using two different algorithms of electrostatics treatment.

Structural properties of the DOPS bilayer. Area per lipid molecule (A_L) is one of the most important structural parameters of lipid bilayers. Comparison of calculated equilibrium average values of A_L with the experimental values is often used to assess the validity of MD protocol and the force field parameters employed. The values of A_L were estimated from the analysis of two types of trajectories (PS-Cut and PS-PME, Table 1). The average values of A_L are equal to 63.3 Å^2. The experimental values for them are greater by about 1 % (Table 2). However, different algorithms of electrostatics treatment do not influence the time-dependent behavior of A_L (Figure 1*a*). On the equilibrium parts of different types of trajectories (PS-cut and PS-PME), the magnitude of A_L fluctuations is negligible (less than 0.1 Å^2). The average D_{p-p} value ($<D_{p-p}>$) was selected as a parameter that describes the bilayer thickness. The values of D_{p-p} for the PS-Cut trajectory demonstrate a wider distribution than those in the PS-PME simulation (Figure 1*b*).

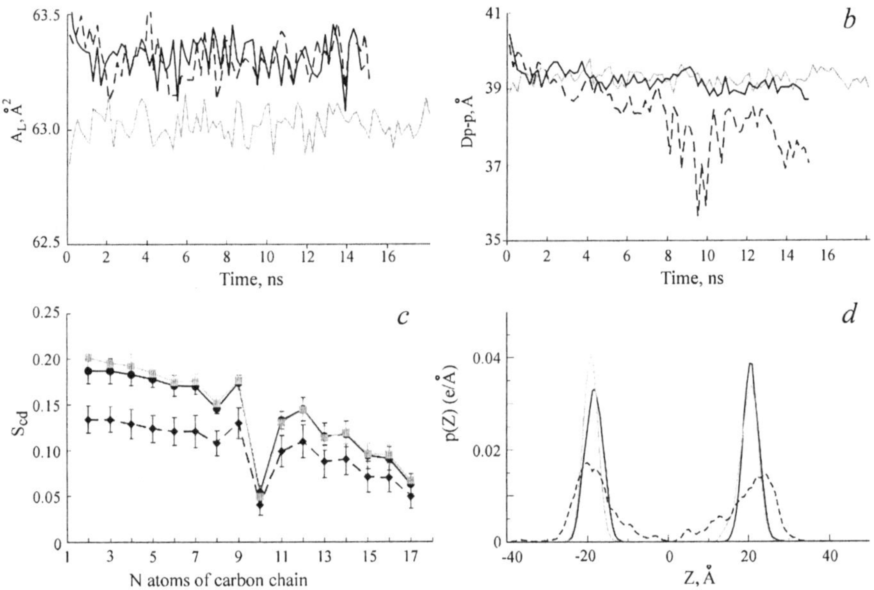

Figure -1. Structural parameters of DOPS bilayer obtained via different MD simulations: PS-Cut (dashed black lines), PS-PME (solid black lines), and PS-pAntp (solid gray lines). (*a*) Area per lipid molecule; (*b*) bilayer thickness (average distance between the planes determined by phosphorus atoms of lipids in different monolayers); X axis, the simulation time; (*c*) profiles of average order parameter for acyl chains of DOPS (standard deviations are shown with vertical bars); (*d*) electron density profiles for phosphorus atoms of bilayer. Z is the distance along the normal to the membrane plane. The bilayer center corresponds to Z = 0 Å. Electron density is measured in the number of electrons per cubic angstrom.

This demonstrates a less stable behavior of the bilayer in such case. Moreover, the $<D_{p-p}>$ value for the PS-PME trajectory is very close to the experimental one—the distance between the peaks of electron density for phosphorus atoms from different monolayers is 39.0 ± 0.2 Å. The average order parameters of acyl chains (S_{cd}) for various carbon atoms are shown in Figure 1*c*. In general, the curve shapes are similar to those reported previously (Chiu et al., 2001) for the oleoyl acyl chains. A typical feature of the S_{cd} curves of unsaturated lipids is the presence of a highly disordered region, which corresponds to unsaturated carbon atoms. Analysis of MD data obtained for PS-Cut and PS-PME trajectories shows that the average values of S_{cd} ($<S_{cd}>$) and the shapes of S_{cd} curves strongly depend on the scheme employed to treat electrostatics (Figure 1*c*, Table 2). Thus, acyl chains are more disordered in the PS-Cut trajectory as compared to PS-PME (Figure 1*c*). Moreover, in the latter case, the $<S_{cd}>$ value is closer to experimental data than in the PS-Cut trajectory (Table 2). The electron density profiles obtained for phosphorus atoms of the two systems (trajectories PS-PME and PS-Cut) are shown in Figure 1*d*. Note that the electron density distribution is sensitive to the scheme of electrostatics treatment. Thus, the profiles are more diffused in the PS-Cut trajectory than in the PS-PME one. In the last case, phosphorus atoms demonstrate compact spatial distributions along the normal to the bilayer plane: this is reflected in prominent peaks on the profiles. Moreover, strong overlap of the peaks corresponding to counterions demonstrates that Na^+ ions are preferentially localized near the phosphate groups of lipids (data not shown). Furthermore, the distance between the peaks obtained in the PS-PME trajectory is very close to the experimental value (Table 2).

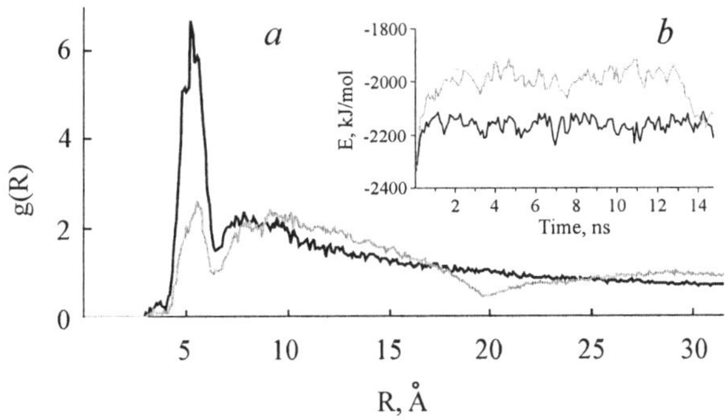

Figure -2. Interactions of PS headgroups with Na^+ counterions: results of MD simulations of DOPS bilayer with different schemes of electrostatics treatment—cutoff function (gray) and PME algorithm (black). (*a*) 3D radial distribution function of phosphorus atoms relative to Na^+ ions; (*insert*) Time-dependence of the energy of electrostatic interactions between PS groups and Na^+ counterions.

Interactions of Na^+ counterions with polar headgroups of DOPS. Time dependence of the average energy of electrostatic interactions of lipid polar headgroups ($<E_{electr}>$) of DOPS molecules with water and Na^+ is shown in Figure 2 (*insert*). The contribution of DOPS–water interactions to $<E_{electr}>$ does not depend on the scheme of electrostatics treatment (data not shown). On the contrary, lipid headgroups interact with Na^+ more efficiently in MD simulations with the PME algorithm than with the cutoff. This is also reflected in the computed 3D radial distribution function (RDF) profiles for phosphorus and Na^+ atoms. Thus, analysis of the PS-PME trajectory permits delineation of a prominent peak on the Na^+-P RDF profile (Figure 2*a*). It demonstrates that Na^+ counterions are preferentially localized near the phosphate groups of lipids and have favorable interactions with them. For the PS-Cut trajectories, the Na^+-P RDF profile is diffused, and no prominent peaks can be delineated. The same RDF profiles for combinations of carboxyl C, N, and P atoms of lipid headgroups with each other and with water molecules are similar in both types of MD trajectories. Moreover, the shapes of Na^+-carboxyl C and Na^+-N RDFs are insensitive to the current algorithm of electrostatics treatment.

Therefore, analysis of the aforementioned artifacts permits a conclusion that the interaction of Na^+ with phosphate groups of PS is critical for the bilayer organization. It is significant that similar effects of counterions on the structure of another charged bilayer (DPPS) were recorded in previous studies by Pandit et al. (2002).

3.2 Binding of pAntp to bilayer

Peptide interactions with bilayer. The model of DOPS bilayer that agrees well with the experimental data was selected to study its interactions with pAntp. After first 0.5 ns of MD simulations, the peptide adsorbed on the water–lipid interface. This process was driven by strong electrostatic interactions between positively charged residues of pAntp and negatively charged DOPS headgroups. During the rest of the simulation time, the peptide was localized to the bilayer surface. The presence of bilayer influences evolution of the peptide's secondary structure. As is seen in Figure 3*b*, during the last 5 ns of the PS-pAntp trajectory, the peptide's residues spent ~ 80 % of time in unordered conformation (coil, β-bend, and β-turn), although some residues in the region 8–14 might adopt a β-sheet conformation.

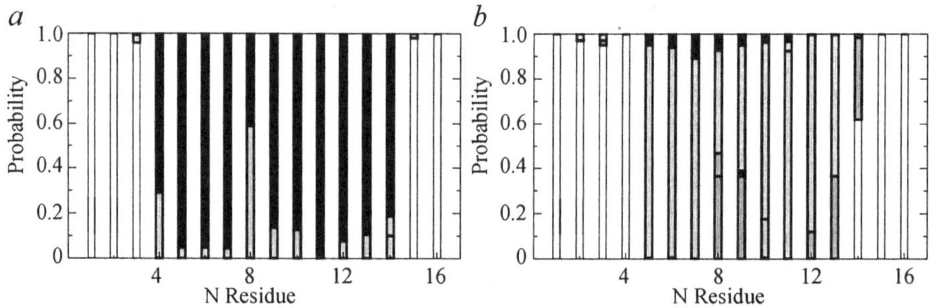

Figure -3. Secondary structure of pAntp. The probability of peptide's residues to form the elements of secondary structure calculated over the last 5 ns of MD simulations: black, α-helix; light gray, β-turn and β-bend; dark gray, β-sheet; and white, random coil. pAntp (*a*) in water and (*b*) in the presence of DOPS bilayer.

On the other hand, simulations in water show that during the last 5 ns of Wt-pAntp trajectory, the region 4–14 is mainly presented by an α-helical conformation (Figure 3*a*). Thus, interaction of pAntp with DOPS bilayer leads to destabilization of its initial α-helical structure and to increase in β-sheet content. These results agree well with the following experimental observation: increase in the surface charge density of POPC/POPG vesicles promotes β-structure of pAntp (Magzoub et al., 2002). Interestingly, in some studies, the membrane perturbation correlated with formation of a β-sheet conformation upon binding of pAntp (Magzoub et al., 2003).

In addition to the time evolution of pAntp secondary structure, the degree of flexibility of different peptide's parts on the water–membrane surface represents an important issue. The values of root-mean-square fluctuations (RMSF) of coordinates of the residues near their average equilibrium positions, calculated over the last 5 ns of MD, show that the peptide in the presence of bilayer has a more rigid structure than in water (Figure 4*a*).

Thus, binding of the peptide on the bilayer interface has a stabilization effect. Interestingly, even for largely disordered peptide structure, the RMSF values are relatively small (< 1.0 Å). This indicates that stabilization of a peptide on the membrane surface may be reached not only due to secondary structure formation, but also can be induced by strong interactions of completely or partly disordered peptide with the water–lipid interface. For example, the termini of pAntp display strong van-der-Waals and electrostatic interactions with the bilayer interface (Figure 4*c*, *d*). This correlates well with a deep insertion of terminal residues into bilayer (Figure 4*b*).

Thus, the most buried pAntp residues are Arg1, Met12, Lys13, Trp14, and Lys15-16. They have the smallest D_i values. On the other hand, the D_i profile strongly correlates with the energy chart of peptide bilayer van-der-Waals contacts (Figure 4*b*, *c*). To summarize, we can conclude that interactions of the terminal parts of pAntp (especially, Arg1, Trp14, Lys13, and Lys15) with the DOPS interface are important for peptide binding. These regions demonstrate the lowest energies of electrostatic and van-der-Waals interactions with lipids, the most prominent decrease in RMSF values compared to those in water, and the largest insertion depth compared with the rest part of the peptide.

Moreover, experimental data demonstrate that substitutions of Arg1, Trp14, Lys13, and Lys15 with Ala lead to essential decrease in the percentage of cell associations and cell uptake of pAntp (Drin et al., 2001). It is significant that almost all of these basic residues (except for Trp14) have strong electrostatic interactions with negatively charged lipid headgroups. Therefore, the surface charge of DOPS bilayer is critical for such interactions. This corresponds to the experimentally observed preference of pAntp to associate with anionic model membranes (Christiaens et al., 2002).

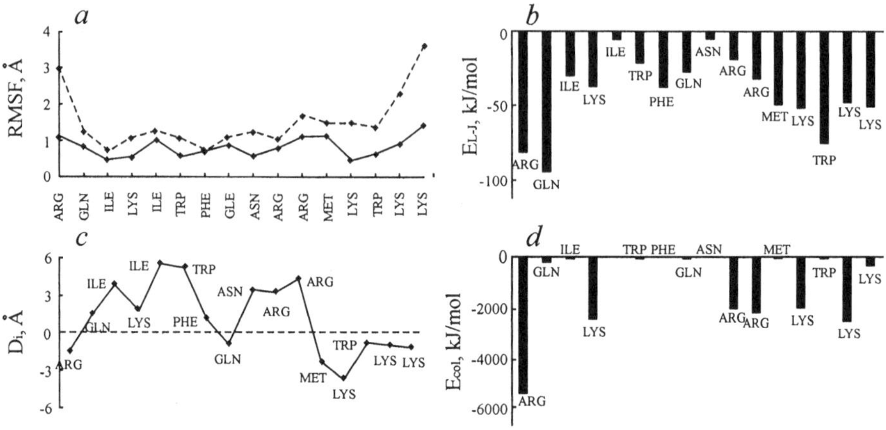

Figure -4. Interactions pAntp with DOPS bilayer. (*a*) RMSF of peptide's residues calculated at the last 5 ns of MD trajectories: Wt-pAntp, dashed line and PS-pAntp, solid line. (*b*) Insertion depth (D_i) of pAntp residues into the bilayer. The plane determined by phosphorus atoms of lipids is indicated by dashed line. Average energies of (*c*) van-der-Waals and (*d*) electrostatic interactions of peptide's residues with lipid headgroups.

Membrane response to peptide binding. Analysis of the accumulated 18 ns MD trajectory reveals no essential alterations in structural properties of the overall DOPS bilayer (Table 2, Figure 1). Thus, only small (by 0.3 $Å^2$) decrease in the average values of A_L was observed. A possible reason for this is that the interactions of pAntp with DOPS bilayer lead to local changes in the lipid organization.

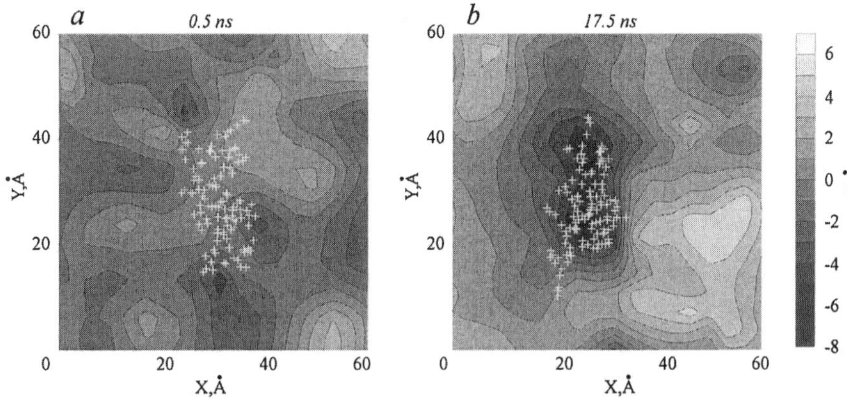

Figure -5. Landscape of the membrane surface after (*a*) 0.5 ns and (*b*) 17.5 ns of MD simulations. Positions of the peptide atoms are shown with white crosses. Deviations of Z-positions of lipid phosphorus atoms from their mean value are colored according to the scale given to the right.

To describe these effects, time evolution of the surface landscape of the 'upper' (closest to peptide) monolayer was studied. As is seen in Figure 5, in the starting conformation of the peptide–bilayer complex (after first 0.5 ns of MD), the landscape is smooth and does not show any prominent grooves and/or ridges. In contrast, after 17.5 ns MD, the surface becomes rough—a significant groove and ridge appear. They are located in the vicinity of the peptide's binding site.

This indicates that pAntp induces local perturbation of the membrane interface, although for the overall bilayer, the average value of its thickness and pattern of phosphorus distribution remain virtually unchanged (Figure 1*b*, *d*; Table 2). Perturbation of the membrane surface is accompanied by disordering of lipids. Such effects also have a local character: decrease in the S_{CD} values is detected only for 10–20 neighboring lipids in the peptide 'shadow', while the order parameters for the entire bilayer are quite similar to those obtained for pure DOPS bilayer (PS-PME trajectory; Figure 1*c*, Table 2).

4. CONCLUSIONS

MD simulations with different algorithms of electrostatics treatment performed for the bilayer composed of DOPS lipids permit delineation of the detailed microscopic picture of its organization. It was shown that behavior of the charged bilayer depends strongly on the way of electrostatics treatment. Unlike the cutoff scheme, the PME-based MD protocol provides a correct description of interactions between Na^+ counterions and charged lipid headgroups. As a result, the model of DOPS bilayer was obtained that agreed well with the experimental data. It was further employed to study behavior of pAntp in anionic unsaturated lipid membrane. During the MD simulations, the peptide spontaneously adsorbs on water–lipid interface. This leads to distortion of the initial pAntp α-helical conformation and increase in its β-sheet content. A number of residues playing an important role in membrane insertion were delineated. All of them are located at the *N*- (Arg1) and *C*- (Lys13, Trp14, and Lys15) terminal regions and have strong interactions with lipid headgroups. Due to this, the terminal parts of peptides are deeply buried inside the bilayer interface. Binding of pAntp provokes perturbations of the membrane, which have a local character. However, note that further analysis of details of the membrane response is required. Hopefully, this will permit delineation of a possible mechanism explaining the functioning of the cell-penetrating peptides.

ACKNOWLEDGMENTS

This work was supported by the Programme RAS MCB, Russian Foundation for Basic Research (grant No. 04-04-48875-a), and Ministry of Science and Technology of the Russian Federation (grants Nos. SS-1522.2003.4, 01.106.11.0002, and MK-4415.2004.4). R.G.E. and D.E.N. are grateful to the Russian Science Support Foundation for the grants awarded. Access to computational facilities of the Joint Supercomputer Center (Moscow) is gratefully acknowledged.

THE ROLE OF WATER IN HOMEODOMAIN–DNA INTERACTION

A. Karyagina[1], A. Ershova[1], S. Spirin[2], A. Alexeevski[2*]
[1] *Institute of Agricultural Biotechnology, ul. Timiryazevskaya, Moscow, 127550, Russia;*
[2] *Belozersky Institute, Moscow State University, Vorobiovy Gory, Moscow, 119992, Russia, e-mail: aba@belozersky.msu.ru*
[*] *Corresponding author*

Abstract: Interfacial water molecules in homeodomain–DNA complexes were analyzed involving all available 3D structures of homeodomains. The main results are as follows: (i) eight conserved water bridges between homeodomains and DNA phosphate groups were identified, as compared to ten 'direct' conserved hydrogen bonds with phosphates; (ii) conserved water molecules mediating contacts with DNA phosphate groups seem to be preorganized on an isolated homeodomain surface; (iii) water-mediated contacts of homeodomains with the DNA major groove could contribute to homeodomain specificity in addition to the direct hydrogen bonds and hydrophobic contacts; and (iv) almost all places for potential water bridges on the homeodomain–DNA interface are occupied by water molecules identified by X-ray crystallography.

Key words: homeodomain; 3D structure; water-mediated contact; water bridge

1. INTRODUCTION

Water molecules play an important role in the interaction between proteins and nucleic acids. In crystal structures solved to sufficiently high resolution (2.5 Å or better), the presence of water molecules on DNA–protein interface is routinely reported. The role of interfacial water molecules in DNA–protein interaction is discussed in many original papers describing structures of DNA–protein complexes (see, for example, Otwinowski et al., 1988; Wilson et al., 1995; Tucker-Kellogg et al., 1997). Several reviews (Schwabe, 1997; Janin, 1999; Jayaram and Jain, 2004) reveal common mechanisms of water behavior in the protein–DNA interaction. Here, we present some data illustrating the role of water in the interaction of homeodomains with DNA.

Homeodomains are homologous ~ 60 amino acid residue long DNA-binding protein domains from eukaryotic transcription factors. The homeodomain globule is formed by three alpha-helices. The third helix (called the recognition helix) realizes contacts with the DNA major groove via hydrogen bonds (residues 51, 54, and 55; see below for the residue numeration) or hydrophobic contacts (residues 47, 50, and 54). Preceding the first helix is an *N*-terminal arm (residues 1–8), which binds the DNA minor groove. In resolved structures, homeodomains form numerous hydrogen bonds with phosphates (residues 3, 6, 8, 13, 25, 28, 31, 43, 44, 46, 51, 53, 55, and 57; Ledneva et al., 2001).

A number of water bridges with phosphates and DNA bases were reported for homeodomain–DNA complexes. The most detailed comparative analysis of water molecules in homeodomain–DNA complexes was presented in a review of M. Billeter (1996). This analysis was based on comparison of three X-ray structures; only the water-mediated contacts between DNA and residues 50 and 51 of homeodomain were discussed. During last years, a lot of information on homeodomain–DNA structures was accumulated, which allows us to perform a more detailed analysis of interfacial water molecules in homeodomain–DNA complexes. The main results of this analysis are as follows. (i) Water molecules play an important role in binding of homeodomain with phosphates: eight conserved water bridges between homeodomains and DNA phosphate groups were identified, as compared to ten 'direct' conserved hydrogen bonds with phosphates. (ii) Conserved water molecules that mediate contacts with phosphate groups seem to be preorganized on the surface of an isolated homeodomain. (iii) Water-mediated contacts of homeodomains with the DNA major groove could contribute to homeodomain specificity in addition to direct hydrogen bonds and hydrophobic contacts. (iv) Almost all places for potential water bridges on the homeodomain–DNA interface are occupied by water molecules identified by X-ray crystallography.

2. METHODS AND ALGORITHMS

The following software was used: RasWin 2.7.2 (Sayle and Milner-White, 1995), SwissPDBviewer 3.7 (Kaplan and Littlejohn, 2001), and a program for detection of hydrogen bonds between nucleic acid and protein (http://monkey.belozersky.msu.ru/~mlt/cgi-bin/nuc_prot.pl).

The structure data used in the work are all the 20 currently available Protein Data Bank entries that have a resolution of 2.5 Å or better and contain homeodomains (9ANT, 1B72, 1B8I, 3HDD, 1DU0, 2HDD, 1JGG, 1AKH, 1LE8, 1YRN, 1K61, 1MNM, 1IG7, 1FJL, 1E3O, 1GT0, 1HF0, 1AU7, 1PUF, and 1ENH). The entry 1ENH contains an isolated homeodomain; other entries contain homeodomain–DNA complexes. These entries hold structures of 14 different homeodomains representing 9 homeodomain classes. Each entry carrying more than one homeodomain chain was transformed to files containing one double-stranded DNA (dsDNA) and one homeodomain chain. Table 1 of the Supplementary Materials are available at http://monkey.belozersky.msu.ru/hd/. lists these 32 files. All protein residues and DNA nucleotides were renumbered using the numeration of homeodomains and their recognition sites given in (Ledneva et al., 2001). The numeration of the recognition site is illustrated in Figure 1:

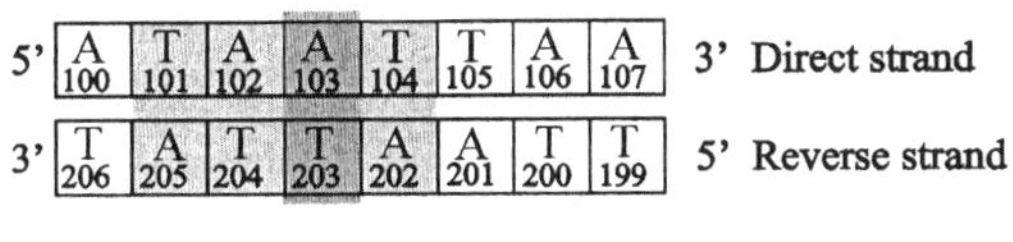

Figure -1. The numeration of the recognition site.

All nucleotides are numbered relative to the nucleotide A103, which forms a bidentate hydrogen bond with absolutely conserved Asn51 in all the known homeodomain–DNA complexes. Highlighted with gray nucleotides 101–104 of the direct strand and nucleotides 202–205 of the reverse strand correspond to the so-called *core* of the homeodomain recognition sequence. Aligned homeodomain sequences and their recognition sequences are given in Supplementary Materials.

'*Water bridges*' were identified as water molecules that form hydrogen bonds with DNA and protein simultaneously. A *hydrogen bond* was detected if the distance between the water oxygen and the donor or acceptor group of DNA or protein was in the range of 2.5 to 3.5 Å.

The following procedure was used to identify places within cavities between the homeodomain and DNA surfaces where potential DNA–protein water bridges could exist. A lattice with a node separation of 0.5 Å was put onto each structure. The nodes that are at a distance in the range of 2.0 to 3.5 Å from nitrogen or oxygen atoms of DNA and protein simultaneously were considered as potential places for water bridges.

Using SwissPDBviewer, the 32 structures of homeodomain, dsDNA, and water molecules were superimposed by optimal fitting of the backbones of the recognition helices. Water molecules from different structures were considered as a *conserved water group* if (i) they are located within a sphere with a diameter of < 1.5 Å in the superimposed structure; (ii) they mediate contacts between residues and nucleotides at the same positions, of the same type, and the same atom groups are involved in water bridge formation; and (iii) the number of water molecules in a group is at least 10.

3. RESULTS AND DISCUSSION

Structurally conserved water molecules could be considered as functionally more important (Ogata and Wodak, 2002). In the superimposed structures, several compact conserved groups of water molecules on the homeodomain–DNA interface are evident (Supplementary Materials).

3.1 Conserved water-mediated homeodomain contacts with DNA phosphate groups

Eight conserved water groups mediating contacts of homeodomain with DNA phosphate groups were found (Supplementary Materials and Figure 2). Nine backbone atoms and three atoms of side chains are involved in these interactions. Three-amino acid residues (Trp48, Asn51, and Arg53) whose side chains are involved in formation of water bridges are completely conserved among homeodomains. Only two phosphate groups in the direct DNA strand (102 and 103) and two phosphate groups in the reverse strand (199 and 200) form water-mediated contacts with homeodomains.

A surprisingly large amount of well-conserved water molecules mediating contacts between protein backbone polar atoms and DNA phosphate groups supports the idea of a significant role of water molecules in fixation of homeodomains by the DNA backbone.

3.2 Water bridges between homeodomain and DNA bases

Asn51 plays a central role in the homeodomain interaction with the DNA major groove. In all known complexes, it forms a bidentate hydrogen bond with A103. There are no other conserved homeodomain contacts with the DNA major groove. It is convenient to subdivide all water-mediated contacts with the DNA major groove into two groups: (i) water molecules included into a network around Asn51 residue and (ii) all the rest water bridges (Supplementary Materials).

The most conserved water bridges are grouped around the Asn51 side chain, forming a complicated network of hydrogen bonds. For example, there are two conserved water bridges between Oδ1 of Asn51 and DNA bases of the complementary nucleotide pair T104–A202 (Figure 3*a*).

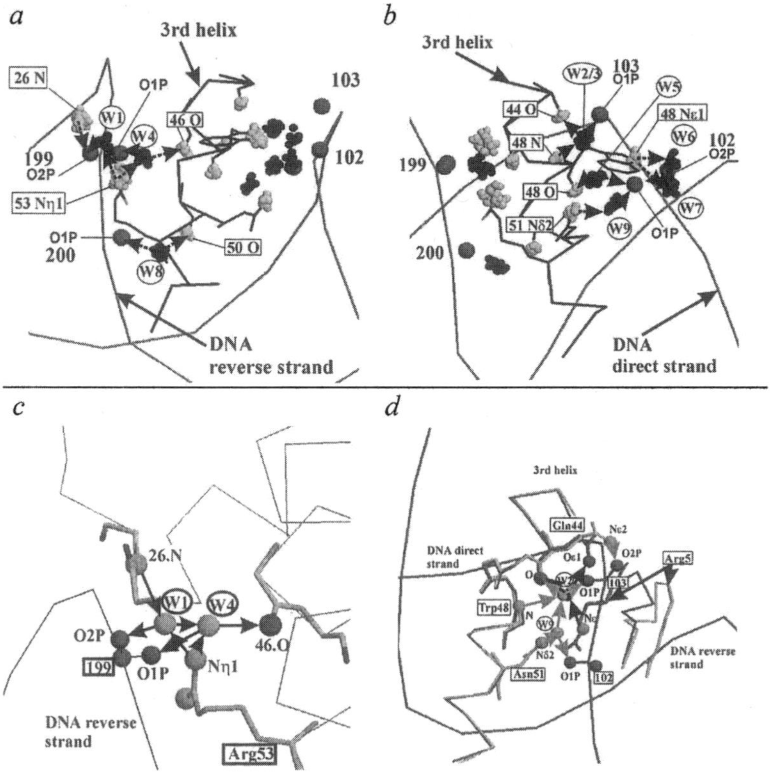

Figure -2. Conserved water-mediated contacts between homeodomain and phosphate DNA groups. (*a*) all conserved water-mediated contacts with phosphates of the reverse DNA strand; (*b*) all conserved water-mediated contacts with the direct DNA strand. In (*a*) and (*b*), the recognition helix backbone (black) and DNA (grey) of 9ANT, chain A, are shown. The side chains involved in the interaction are shown in black. Big dark gray balls represent oxygen of DNA phosphate groups. Small balls represent atoms of 32 superimposed homeodomain–DNA complexes: black balls correspond to water oxygen; light gray balls, to protein atoms. Arrows connecting the atom groups indicate hydrogen bonds; (*c*) scheme of the hydrogen bonds network around water molecules from the groups W1 and W4. Backbone of the protein and wireframe representation of side chains and/or backbone of the interacting residues and water are from 1ENH. Backbone of DNA reverse strain with interaction phosphate group is from 3HDD, chain A. Interacting atoms are shown as balls; (*d*) scheme of hydrogen bonds network around water molecules from the groups W2 and W9. DNA, protein, and water colored in grey are from 3HDD, chain A. Protein and water colored in black are from 1ENH.

Atoms O4 of T104 and N6 of A202 are involved in these interactions. Simultaneously, Asn51.Oδ1 is hydrogen bonded with N6 atom of A103 base. Three polar atoms in locations appropriate for hydrogen bonding with Asn51.Oδ1 could be explained by switching hydrogen bonds. The water molecule mediating Asn51.Oδ1–T104.O4 contact was observed in representatives of seven homeodomain families. In Matα2 recognition site,

T104 is substituted with C104. Notably, water molecules in two Matα2 structures were observed in the same location. They mediate Asn51.Oδ1–C104.N4 contact. In DNA complexes of homeodomains from Antp, Prd, Eve, and En classes, the water molecule that mediates the contact between Oδ1 of Asn51 and O4 of T104 simultaneously forms a hydrogen bond with Nε2 of Gln50 (Supplementary Materials, and Figure 3*c*). A more complicated network was observed in Prd homeodomain, where this water is additionally hydrogen bonded with the water molecule mediating the contact between Nε2 of Gln50 and O6 of G201 (Figure 3*c*).

In En and TALE homeodomain classes, the above-mentioned conserved water bridges are accomplished with additional conserved water molecules, which are also connected with each other (Figure 3*b*). The complete network of six water molecules connecting the third helix of Engrailed homeodomain with the DNA major groove is shown in Figure 3*h*. Three of these water molecules mediate contacts of Asn51 and Gln50 with T104, A201, and A202.

A possible role of water bridges in DNA binding specificity is demonstrated in Figure 3*d–f*. Each shown water molecule is hydrogen bonded with four atoms from protein and DNA. As a water oxygen could be double proton donor and twice proton acceptor, such contacts might be considered as specific contacts.

Variable water bridges connect non-conserved homeodomain residues with DNA bases in the major groove (Supplementary Materials). Except for few examples, these water bridges are not conserved even within a single homeodomain class. The example of interfamily conservation is water-mediated contact between Arg46, T200, and G105 in complexes of Pit-1 and Oct-1 homeodomains from POU homeodomain class: one of Nη atoms of Arg46 via one water molecule interacts with both O4 of T200 and O6 of G105. In Oct-1 homeodomain-DNA complex, this water molecule is bound with another water molecule mediating contact between Ser50, G105, and C201.

Three more water molecules in 1E3O_C complex of Oct-1 homeodomain is involved in the network of hydrogen bonds connecting Arg58, Gln54, C102, G204, T203, and A202 (Figure 3*g*).

Thus, the data presented show a complicated network of water-mediated contacts of homeodomain with the DNA major groove. The observed conservation of the network correlates with types of residues and nucleotides in contact zone and could partially explain the specificity of homeodomain–DNA interaction.

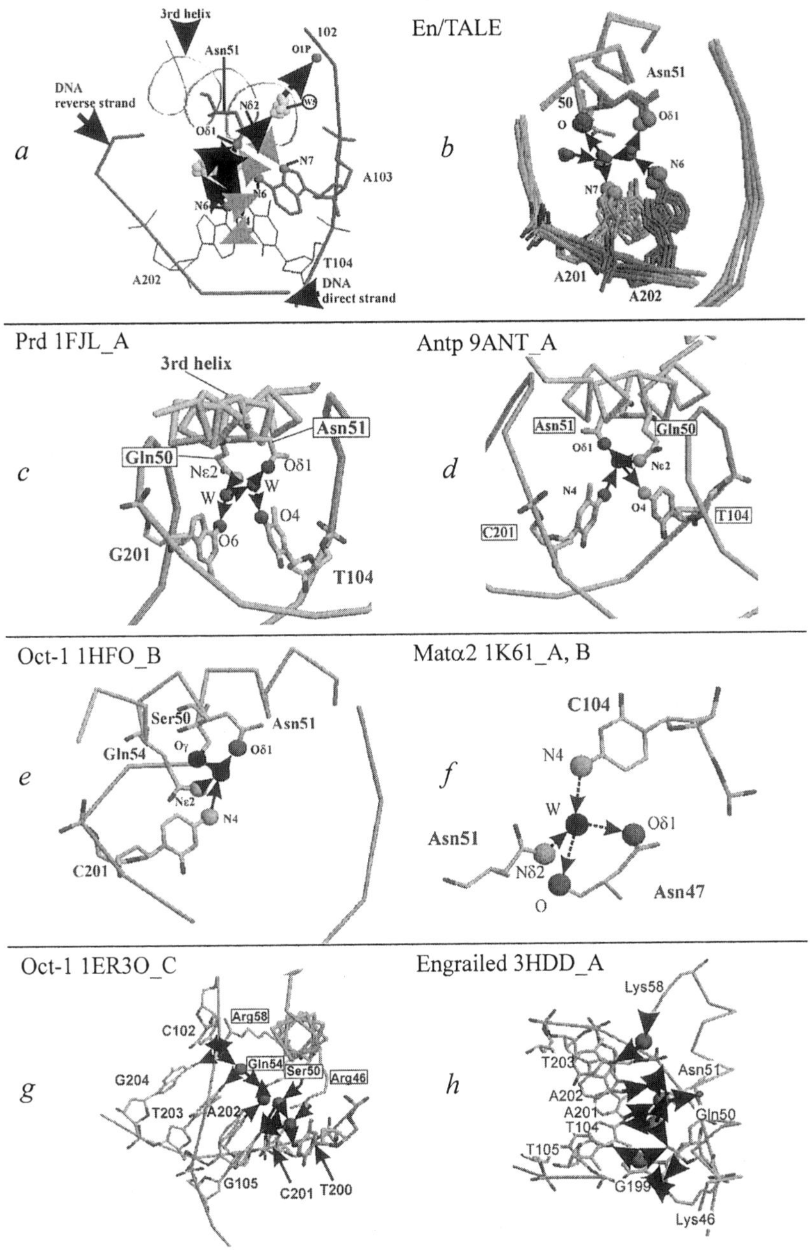

Figure -3. (*a–f*) contacts of Asn51 with DNA. Water-mediated hydrogen bonds are shown as black dotted arrows directed from acceptor to donor; direct hydrogen bonds, shown as gray arrows. (*a*) spatially coinciding water molecules that mediate contacts of Asn51 with DNA in several structures. Shown are the 3rd helix; side chain of Asn51; two DNA fragments; and A103, T104, and A202 DNA bases of Antp homeodomain (9Ant_A structure). Interacting atoms of DNA and protein are shown as dark gray balls. Groups of light gray balls represent water molecules mediating the same contacts in superimposed structures of the homeodomain–DNA complexes; *(b–f)* hydrogen bonds of Asn51 network in different homeodomains. Black balls correspond to oxygen atoms; gray balls, to nitrogen; *(g–h)* all water-mediated contacts of amino acid residues from the recognition helix of the homeodomains Oct-1 (*g*) and Engrailed (*f*) with DNA bases in major groove.

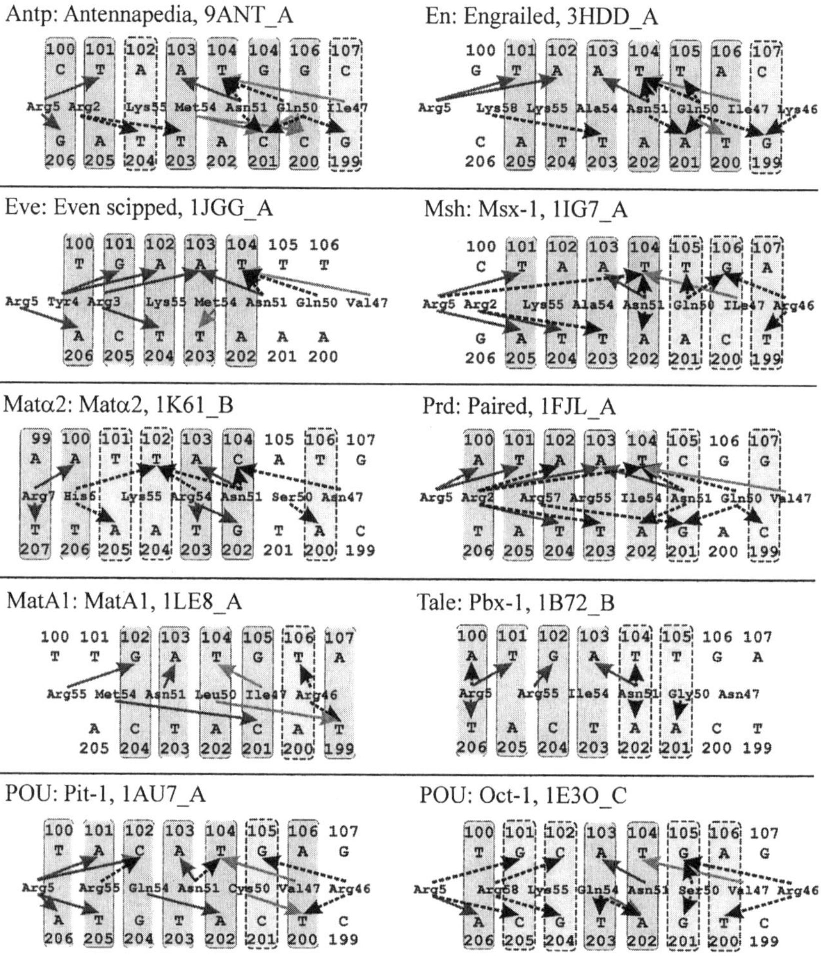

Figure -4. Schemes of contacts with DNA bases in the major and minor grooves in 10 homeodomains from different homeodomain classes. Residues 2–7 provide contacts with DNA bases in the minor groove; residues 46–58, in the major groove. Gray and light gray arrows represent direct hydrogen bonds and hydrophobic contacts, respectively; dotted arrows, water-mediated contacts. Nucleotide pairs involved in the direct and water-mediated contacts with homeodomains are highlighted with dark and light gray, respectively.

In all 32 structures of homeodomain–DNA complexes containing water, there are only 16 water molecules that mediate contacts with DNA bases in the minor groove (the data not shown).

The schemes of direct and water-mediated contacts of homeodomains with DNA bases are shown in Figure 4 for 10 structures from different homeodomain classes. All these schemes are proved to be different. Water-mediated interactions of homeodomains with the DNA major groove are less conserved as compared with the contacts with phosphate groups. This fact can be explained by the variability of nucleotide sequences recognized by homeodomains. Indeed, a few structurally conserved water molecules on the homeodomain–DNA interface are located mainly around the unique absolutely conserved interacting pair Asn51–A103.

The specificity of homeodomain-DNA binding cannot be explained by protein–DNA hydrogen bonds and hydrophobic interactions mainly in the DNA major groove (Jayaram and Jain, 2004). There are at least two additional factors of homeodomain specificity. The first is an 'indirect' recognition of sequence-specific DNA conformation via precise contacts of homeodomain with DNA phosphate groups (Gruschus et al., 1997). Our data demonstrate an essential role of water-mediated contacts in phosphate binding; thus, they could contribute to indirect recognition. The second additional factor is water-mediated contacts with DNA bases in the major and minor grooves. The schemes of contacts, shown in Figure 4, demonstrate an essential enlargement of the quantity of nucleotide pairs involved in the interaction due to inclusion of water-mediated contacts into consideration. At least some of these contacts may be considered as specific. They represent the enlarged hydrogen bonds in the cases when side chains of certain residues are not sufficiently long to reach a DNA base. The examples of such contacts are shown in Figure 3*b–f*. In many such cases, water molecules could distinguish between proton donors and acceptors at DNA and thus provide the specificity of DNA recognition.

3.3 Identification of potential water bridges in homeodomain–DNA complexes

The identification procedure of potential water bridges is described in Materials and Algorithms. No potential water bridges were identified that were common for all 32 complexes. The places of potential water bridges common for most (20 to 30) complexes spatially coincide with eight conserved water molecules mediating contacts between homeodomains and DNA phosphate groups (Figure 2*a*, *b*).

The picture of potential water bridges on the major groove interface is unique for each homeodomain–DNA complex (data not shown). To find whether any additional water molecules can be placed on the homeodomain–DNA interface except for water molecules identified by X-ray crystallography, the following procedure was performed. First, the places of potential water bridges on the major groove interface were identified for each homeodomain–DNA complex. Then, these places were restricted by exclusion the water molecules located closer than 3.5 Å to the real water molecules in a given structure. The results of such analysis for Engrailed homeodomain structures 3HDD_A and 1DU0_A are shown in Figure 5 (Supplementary Materials). Although there are six interfacial water molecules in 3HDD_A structure and only three water molecules in 1DU0_A structure, no places for potential water bridges in both structures are observed except for two or three places at the edge of the homeodomain–DNA interface.

These data demonstrate that, in crystal structures of complexes, practically all places for potential water bridges on the homeodomain–DNA interface are occupied by water molecules. This suggests that these water molecules in solution are relatively stable.

3.4 Identification of 'preorganized' water on homeodomain surface

Only one X-ray resolved water-containing structure of free homeodomain is available (Engrailed homeodomain, 1ENH). Comparison of this structure with structures of the same homeodomain in complexes with DNA allowed several 'preorganized' water molecules to be described (Fraenkel et al., 1998).

To compare preorganized water molecules with conserved water groups, we superimposed the structures of free Engrailed homeodomain (1ENH) with three Engrailed–DNA complexes (2HDD_A, 3HDD_A, and 1DU0_A). The 11 of overall 33 water molecules from 1ENH spatially coincide (within 1-Å limits) with at least 1 water molecule from the structures with DNA. Seven of these molecules are located on the DNA–protein interface; six of them are included in one of the conserved water groups W1, W2/3, W4, W5, W7, or W8. All these groups except for W1 contain water molecules from all the four analyzed Engrailed structures. The group W1 contains water molecules from three structures. Two examples of preorganized water molecules are illustrated in Figure 2*c*, *d*.

Thus, six of the eight conserved water molecules mediating homeodomain contacts with DNA phosphate groups are 'preorganized' on the engrailed homeodomain surface.

The observed high coincidence of 'preorganized' and conserved interfacial water molecules supports the idea of a functional role of water molecules in homeodomain–DNA interaction. The most probable role of 'preorganized' water molecules on the homeodomain surface is to improve fitting of the homeodomain and DNA surfaces and thus simplify the binding process (Jayaram and Jain, 2004). This is in accordance with the concept of conserved hydration sites in protein–DNA complexes (Shakked et al., 1994) and the analysis performed by Woda et al. (1998). In the latter work, it was shown that the protein atoms forming hydrogen bonds to DNA in protein–DNA complexes are located close to the hydration site predicted for free DNA.

ACKNOWLEDGMENTS

The work was supported by Russian Foundation for Basic Research (grants Nos. 03-04-48476 and 03-07-90157) and Ludwig Institute for Cancer Research (CRDF GAP grant No. RB0-12771).

MOLECULAR MODELING OF HUMAN MT_1 AND MT_2 MELATONIN RECEPTORS

A. Chugunov[1, 2*], P. Chavatte[3], R. Efremov[2]

[1] Shemyakin–Ovchinnikov Institute of Bioorganic Chemistry, Russian Academy of Sciences, ul. Miklukho-Maklaya 16/10, GSP Moscow, 117997, Russia; [2] Department of Bioengineering, Biological Faculty, Lomonosov Moscow State University, Vorobiovy Gory, 119899, Moscow, e-mail: volster@nmr.ru; [3] Faculte des Sciences Pharmaceutiques et Biologiques, BP 83, 59006 Lille Cedex, France

[*] Corresponding author

Abstract: Malfunction of G-protein coupled receptors (GPCR) provokes large amount of diseases. Their adequate treatment requires rational design of new high affinity and high selectivity drugs targeting these receptors. Molecular modeling represents a powerful tool to solve the problem. In this work, we present three-dimensional models of human MT_1 and MT_2 melatonin receptors, members of the GPCR family. The models were based on the X-ray structure of bovine rhodopsin. The modeling approach employs an original computational procedure for optimization of receptor–ligand structures. It includes rotation of one of transmembrane α-helices around its axis with simultaneous assessment of the quality of complexes according to a number of developed criteria. The optimal geometry of the receptor–ligand binding was selected based on analysis of complementarity of hydrophobic/hydrophilic properties between the ligand and its protein environment in the binding site. The resulting 'optimized' models were applied to inspect the details of protein–ligand interactions for melatonin and a number of its analogs with known affinities for MT_1 and MT_2 receptors. The results permit rationalization of experimental data on affinities and selectivities of studied compounds towards both receptor subtypes. Prospects of the constructed models in drug design are discussed.

Key words: GPCR; homology modeling; hydrophobic interactions; receptor–ligand binding

1. INTRODUCTION

Melatonin is an indole-derived neurohormone produced by pineal gland according to circadian rhythm. It is believed to be one of the most widespread hormones in the organism; it parties involved in almost all vital activities—sleep, functioning of the cardiovascular and immune systems, etc. It is used for synchronization of diurnal rhythms and sleep normalization when shifting time zones. The main disadvantage of melatonin as a drug is its low half-life time in a blood vessel (15–20 min) and poor oral bioavailability. The features of melatonin were exhaustively reviewed in (Mor et al., 1999).

According to different pharmacological profiles, melatonin receptors are divided into two classes: MT_1 and MT_2. They belong to the A-family (rhodopsin family) of GPCRs and represent integral membrane proteins that share common fold in their membrane-bound domain—a bundle of seven transmembrane (TM) helices. Properties of ligands acting on melatonin receptors have been widely studied, although the problem of design of new potent and selective compounds—potential future drugs—is yet to be solved.

Mutagenesis studies have revealed residues critical for binding of melatonin by its receptors. These are a histidine residue in TM-helix 5 (TM5)—His5.46$^{(195MT1,\ 208MT2)}$, forming a hydrogen bond with the 5-metoxy group of the ligand, and two serines in TM3—Ser3.35$^{(110MT1,\ 123MT2)}$ and Ser3.39$^{(114MT1,\ 127MT2)}$, interacting with NH- and CO- groups of the acetamide moiety of an agonist (Conway et al., 1997, 2001). Residues are numbered according to the nomenclature proposed in (Ballesteros and Weinstein, 1995).

Understanding of intimate molecular mechanisms of receptor–ligand interactions requires knowledge of three-dimensional (3D) structure of a receptor, especially in the vicinity of the binding site. Unfortunately, up to now, the spatial structure of only one member of the GPCR family, namely, the bovine visual rhodopsin (Rh), is known with atomic resolution (Palczewski et al., 2000). Therefore, development of theoretical models for GPCRs seems to be indispensable. Such models may be very helpful in rationalization of experimental data as well as in search for and design of new potent ligands. That is why during last decade, *in silico* technologies became popular for the structure-based drug design. Most of these methods employ homology modeling for construction of 3D models of the receptors. In particular, computational techniques have already been used to elaborate atomic-scale models of melatonin receptors. This was done using bacteriorhodopsin (Sugden et al., 1995; Grol and Jasen, 1996; Navajas et al., 1996) or Rh X-ray structures (Ivanov et al., 2004) as a template.

In this work, we present the 3D models of human MT_1 and MT_2 melatonin receptors built by homology with Rh. To satisfy the geometric constraints on ligand binding imposed by mutagenesis data, the models were optimized using a specially elaborated computational procedure. It includes 'manual' rotation of helix TM3 around its axis and multistep energy minimization accompanied by assessment of the quality of complexes according to a number of developed criteria. The resulting 'optimized' models were applied to inspect the details of protein–ligand interactions for melatonin and a number of its analogs with known affinities for MT_1 and MT_2 receptors. The results permit rationalization of experimental data on affinity and selectivity of studied compounds towards both receptor subtypes.

2. METHODS AND ALGORITHMS

Building and optimization of starting models. Sequence alignment for Rh (Swiss-Prot code *opsd_bovin*), MT_1 (code *ml1a_human*), and MT_2 (code *ml1b_human*) was generated using the GCG software. The resulting alignment (Figure 1) was slightly corrected manually to avoid gaps in the TM regions. It served as a basis for building the model of TM parts of the receptors MT_1 and MT_2 using homology modeling. This was done with the MODELLER v. 6 (Marti-Renom et al., 2000). 3D structure of Rh (PDB entry 1l9h; Teller et al., 2001) was used as a template for modeling. For each receptor, five slightly different models were generated; and the 'best' models were chosen for further inquiry based on the analysis of the resulting spatial violations. The models were initially optimized as described elsewhere (Chugunov et al., in press). After minimization, the extramembrane loops were removed, and only TM regions (Figure 1) were future used. Secondary structure of the models was calculated with the DSSP program (Kabsch and Sander, 1983). The quality of the models was estimated using the Profiles_3D (Bowie et al., 1991) and ENVIRON (Koehl and Delarue, 1994) software.

Hydrophobicity and variability analysis. Variability and hydrophobicity moments for TM α-helical regions were calculated by the methods proposed in (Du and Alcorta, 1994) and (Donelly et al., 1993), respectively. Hydrophobicities of residues were taken from (Rees et al., 1989). Spatial hydrophobic and hydrophilic properties of TM segments were estimated using the molecular hydrophobicity potential (MHP) approach, as described elsewhere (Furet et al., 1988). The values of atomic hydrophobicity constants were taken from (Viswanadhan et al., 1989). MHP distributions on surfaces of α-helical segments were visualized by one-dimensional MHP plots in polar coordinates (Efremov and Vergoten, 1995).

Building and quality estimation of models with rotated TM3. In the starting models (generated by MODELLER), the TM helix 3 was rotated by a given angle around its α-helical axis in a clockwise direction as seen from extracellular side of the membrane. The resulting conformation was subjected to two-stage energy minimization of melatonin–receptor complexes according to the protocol described elsewhere (Chugunov et al., in press). Note that during the energy minimization, a cavity that permits accommodation of the melatonin molecule was formed, although absent in the starting models. In total, 12 models with the starting rotation angles (θ) of TM3 from 0° to 60° were generated for each receptor. The quality of complexes was assessed using H-bond geometry violation penalties (bee below) and GOLDSCORE values obtained from series of docking runs (for details, see Chugunov et al., in press).

Penalty on H-bond geometry violation. In the resulting models of complexes, distances (r) and angles (α) corresponding to the H-bonds that presumably mediated receptor–ligand interactions (Table 1) were estimated. Deviations of the H-bond geometry from the 'ideal' one (r_0 = 2.1 Å; α_0 = 150°, where r_0 is distance between H atom and acceptor atom, and α_0 is donor–H atom–acceptor atom angle), were penalized as follows. If $r \geq r_0$, the penalty was equal to $r - r_0$. If $\alpha \leq \alpha_0$, the penalty was equal to $\alpha_0 - \alpha$. The total distance and angle penalties for three putative H-bonds were used to characterize the total violation of H-bonding geometry for each receptor model.

Complementarity of hydrophobic properties. Complementarity of hydrophobic/hydrophilic properties of a ligand and its binding site was estimated based on the MHP approach (see above). MHP created in the surface points of the ligand by its own atoms (MHP_{ligand}) as well as by protein atoms of the binding site (MHP_{site}) were calculated for each complex. Every surface point was attributed either to polar, neutral, or non-polar classes. The fraction of surface points that simultaneously belong to the same class based on their MHP_{ligand} and MHP_{site} values was defined as complementarity.

3. RESULTS AND DISCUSSION

3.1 Building and verification of the starting models

The main objective of this paper is elaboration of 3D models for MT_1 and MT_2 and their application to study receptor–ligand interactions, so we assume that the influence of extramembrane regions may be omitted (with regard to membrane-buried location of melatonin binding site). The only available high-resolution spatial structure of Rh (Palczewski et al., 2000) was chosen as a template to build the models. Predicted TM regions of MT_1 and MT_2 align well with the TM segments of Rh observed in the X-ray structure (Figure 1).

```
TM1: opsd : 35   WQFSMLAAYMFLLIMLGFPINFLTLYVTVQ     64
     MT1  : 25   WLASALACVLIFTIVVDILGNLLVILSVYR     54
     MT2  : 38   WVAPALSAVLIVTTAVDVVGNLLVILSVLR     67

TM2: opsd : 71   PLNYILLNLAVADLFMVFGGFTTTLYTSLH     100
     MT1  : 61   AGNIFVVSLAVADLVVAIYPYPLVLMSIFN     90
     MT2  : 74   AGNLFLVSLALADLVVAFYPYPLILVAIFY     103

TM3: opsd : 107  PTGCNLEGFFATLGGEIALWSLVVLAIERYVVV  139
     MT1  : 97   YLHCQVSGFLMGLSVIGSIFNITGIAINRYCYI  129
     MT2  : 110  EEHCKASAFVMGLSVIGSVFNITAIAINRYCYI  142

TM4: opsd : 151  NHAIMGVAFTWVMALACAAP-PLVG          174
     MT1  : 142  KNSLCYVLLIWLLTLAAVLPNLRAG          166
     MT2  : 155  WHTPLHICLIWLLTVVALLPNFFVG          179

TM5: opsd : 199  NESFVIYMFVVHFIIPLIVIFFCYGQLVFT    229
     MT1  : 184  SSAYTIAVVVFHFLVPMIIVIFCYLRIWIL    213
     MT2  : 197  STQYTAAVVVIHFLLPIAVVSFCYLRIWVL    226

TM6: opsd : 247  EKEVTRMVIIMVIAFLICWLPYAGVAFYIFT    277
     MT1  : 234  FRNFVTMFVVFVL-FAICWAPLNFIGLAVAS    263
     MT2  : 247  LRSFLTMFVVFVI-FAICWAPLNCIGLAVAI    276

TM7: opsd : 286  IFMTIPAFFAKTSAVYNPVIYIMMN          310
     MT1  : 275  WLFVASYYMAYFNSCLNAIIYGLLN          299
     MT2  : 288  GLFVTSYLLAYFNSCLNAIVYGLLN          312
```

Figure -1. Sequence alignment for TM segments of visual rhodopsin (opsd) and melatonin receptors MT1 and MT2. For each TM helix, homology level exceeds 40 %. The most conservative residues in the whole GPCR family are marked with bold. The residues critical for melatonin binding are underlined.

The constructed models represent just the melatonin receptor sequences mounted on the Rh backbone and subjected to energy relaxation. They take into account neither the biological specificity of the modeled objects nor the experimental data on receptor–ligand interactions and, therefore, such models may be very 'crude'. To attain predictive power in subsequent drug design tasks, the following information is used in this original procedure: (i) point mutagenesis data on involvement of particular MT residues in melatonin binding; (ii) spatial distribution of hydrophobic/hydrophilic properties on the surfaces of TM helices; (iii) variability properties of TM helical segments; (iv) packing quality of TM helices; and (v) quality of melatonin fitting into the binding site. Application of all the aforementioned criteria is described below.

3.2 Analysis of starting models and their 'manual' optimization

According to mutagenesis data, the ligand-binding site in both receptors is located in the vicinity of residues Ser$^{3.35}$, Ser$^{3.39}$, and His$^{5.46}$ (Conway et al.,

1997; 2001). It was established that these residues form H-bonds with the ligand molecule. Therefore, distances between the corresponding atoms of the receptor and the ligand (Figure 2*b*) should not exceed 2.5÷2.8 Å. Examination of the binding site in the starting models shows that for melatonin docked into the binding site in orientation similar to that in Figure 2*b*, these distances are equal to ~3÷4.5 Å even after energy minimization of the complex with imposed distance constraints on the H-bonds. Thus, the spatial arrangement of the crucial residues is unsuitable for simultaneous forming of all three H-bonds mentioned above regardless of melatonin conformation. As illustrated in Figure 2*a*, one possibility to satisfy these H-bonds and, therefore, to enhance the ligand binding, is to rotate the whole α-helix TM3 around its axis in a clockwise direction as seen from the extracellular side of the membrane (hereinafter, the term 'rotation' is used only in this sense). As a result, both serines, being originally located outside the binding site, become rather more accessible to interactions with the ligand. This conformational change allows for satisfying experimentally derived geometrical constraints on H-bonds in the active site and preserves well both the 3D structure and overall topology of the entire TM α-helical complex—only small changes in orientation of side chains of residues in TM3 and neighboring segments may occur.

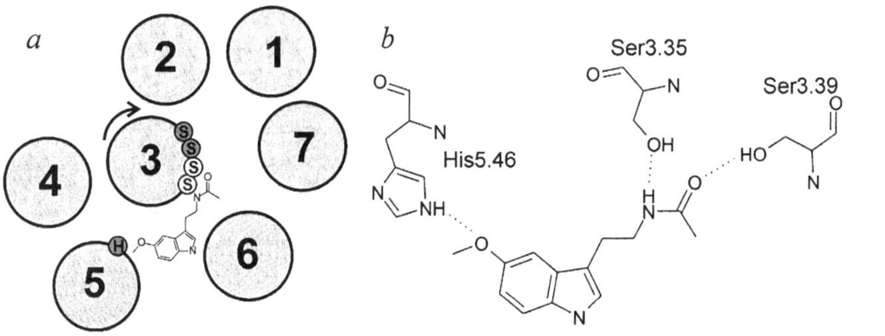

Figure -2. Melatonin interactions in the binding sites of melatonin receptors. (*a*) schematic illustration of TM domain of melatonin receptors as viewed from the extracellular side. TM helices are shown with numbered circles. Direction of the proposed rotation of helix TM3 is indicated with arrow. Residues Ser3.35, Ser3.39, and His5.46 are marked. Positions of these residues in the starting and 'optimized' models of the receptor are shown with gray and white circles, respectively; (*b*) schematic drawing of functionally important H-bonds in the active site of the receptor (Conway et al., 1997; 2001).

Hydrophobic properties of TM domain. Figure 3*a* illustrates hydrophobic properties of TM helices in the model of MT_1 (MT_2 has very similar properties; data not shown). Diagrams in polar coordinates in the center of each helix (schematically shown as circles) show angular distribution of the total molecular hydrophobicity potential (MHP) on the helix surface. The larger is the radius-vector, the higher is the hydrophobicity degree of a given side of α-helix. It is seen that the hydrophobic properties of helices are not homogeneous: their different sides possess different hydrophobicities. Moreover, the helices in some cases are strongly amphiphilic while in others, more complex ('butterfly-like') pattern exist The main tendency is that the most hydrophobic sides of TM segments are oriented either towards the lipid environment or to the proximal helix interfaces (except TM3).

Comparison of the hydrophobic organization of TM domain of melatonin receptors (starting models) with that of the calculated for Rh X-ray structure of (Figure 3*b*) shows that, in overall, they are quite similar. It is important that the hydrophobic properties of the helix TM3 in MTs agrees better with those of Rh if the segment TM3 of melatonin receptor is rotated as described above. This gives an additional argument in favor of the proposed conformational change.

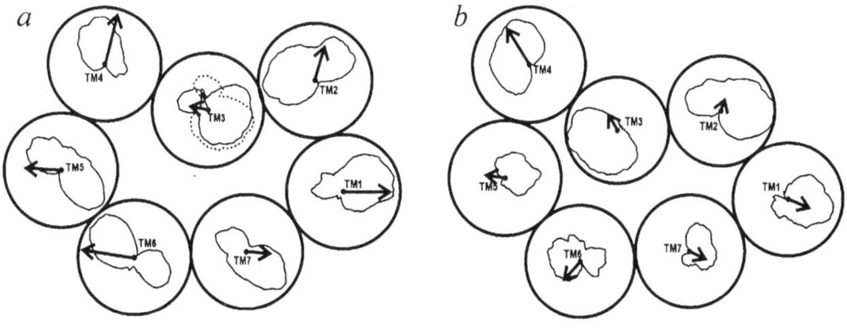

Figure -3. Hydrophobic and variable properties of TM domains of (*a*) MT1 receptor model and (*b*) bovine rhodopsin. Circles show TM α-helices as if seen from extracellular part of the membrane. Graphs in polar coordinates show total molecular hydrophobic potential (MHP) corresponding to a given angle (greater function value means greater hydrophobicity). Arrows pointing outward of helix centers indicate vectors of variability (in other words, they show most variable side of α-helix with respect to amino acid substitution). Broken lines at TM3 of MT1 model show hydrophobic and variable properties of the optimized model of the receptor.

Variability properties of TM domain. An important observation about TM domains of proteins is that, replacements of amino acid residues in a family of homologous proteins most frequently occur on the outer surface. Quantitatively, such distributions are often characterized with a help of variability moment vectors (Du and Alcorta, 1994). For α-helical segments, the vectors point out the most variable side of the helix, and their module indicates the variability degree: the larger is the most variable. Such vectors calculated for TM helices of MT_1 and Rh are shown in Figure 3. It is seen that, in contrast to hydrophobicity, each TM helix faces lipids with its most variable side, thus confirming the aforementioned empirical rule. Hence, according to this criterion, the overall packing of TM helices in MT_1 and MT_2 seems to be correct. Moreover, the rotation of helix TM3 results in a better agreement between the orientation of its variability moment vector and that in Rh.

To summarize, we may conclude that the procedure proposed to optimize the starting models of melatonin receptors can provide a better agreement with the experimental data on ligand–receptor binding as well as with general principles of hydrophobic organization of membrane proteins and their variability properties. Computational protocol employed to select the optimal value of the rotation angle for helix TM3 is described below.

3.3 Finding of the optimal values of the rotation angle for helices TM3 in MT_1 and MT_2

For each receptor, twelve models of complexes with different values of angle θ were generated. The 'best' models were chosen based on the following criteria: (i) the minimal violation of the geometrical constraints imposed on three H-bonds between melatonin and the receptor; (ii) the highest value of the GoldScore function for melatonin docking into the active site; and (iii) the 'best quality' of helix packing according to the so-called 3D-1D scoring function.

Violation of H-bonding geometry. In each model of MT_1 and MT_2 complexes, the geometry of the aforementioned H-bonds (after energy minimization with melatonin) was compared with an 'ideal' one (see Methods). It was shown that, according to this criterion, the 'best' models had indices 30 and 35 for MT_1 receptor and 0 and 10, for MT_2 receptor.

The GoldScore penalty. This criterion was used to assess the quality of melatonin docking into each receptor model. As docking simulations were carried out with the GOLD software, the GoldScore function supplied with this program was used for comparative analysis of the resulting complexes melatonin–receptor. According to this criterion, the 'best' models of MT_1 have indices 0, 10, 30, 35, 40, and 45, while the 'best' models of MT_2 have indices 0, 10, 20, and 45.

Helix packing quality. Quantitative estimations according to this criterion are based on the formalism of the so-called '3D-1D function', proposed earlier in (Bowie et al., 1991). The idea lies in calculation of propensities (3D-1D score) for amino acid residues to occur in a particular protein environment in a set of high resolution 3D structures (reference set). These statistically derived scores characterize the quality of environment for a given protein segment: the higher is the score, the better is the agreement with the environmental characteristics observed in the reference set. In this study, we applied this score to characterize the quality of helix packing in MT models with different values of angle θ. Note that the 3D-1D method was originally developed for globular but not for membrane proteins. The objective here is to estimate the differences in 3D-1D score induced by rotation of the helix TM3 for a set of MT models. 'Best quality' models for MT_1 have indices 10, 35, 40, and 55; for MT_2, 10 and 20.

Considering all three criteria, the models of MT_1 and MT_2 with indices 35 and 10, respectively, were chosen as optimal for melatonin binding. Hereinafter, only these 'optimized' models will be used. Table 1 shows the geometry parameters of H-bonds between the functionally important residues of MTs and the docked melatonin in the receptor models before and after optimization. It is seen that simultaneous formation of all three H-bonds, difficult in the starting models, becomes possible in the optimized ones.

The resulting models of both receptors were future employed to delineate differences in their active sites. Hopefully, this might explain available experimental data on the selectivity of some melatonin analogs to one of the receptors subtypes.

Table -1. Geometry of H-bonds between the functionally important residues of MTs and the docked melatonin in the receptor models before and after optimization

H-bond with	MT_1				MT_2			
	Starting model		Optimal (rot35)		Starting model		Optimal (rot10)	
	Dist.	Ang.	Dist.	Ang.	Dist.	Ang.	Dist.	Ang.
Ser3.35	2.82	109	2.14	150	3.29	136	2.02	167
Ser3.39	3.15	152	2.21	165	2.99	145	2.26	143
His5.46	3.15	155	2.06	163	2.73	140	2.31	146

3.4 Differences in the melatonin binding sites of MT_1 and MT_2 models

Apart from the overall packing of helices in models of MT_1 and MT_2 receptors (which is quite similar), it is especially interesting to compare their active sites as well as the details of melatonin binding. Binding sites contain mainly non-polar residues (including many aromatic), except for residues $Ser^{3.35}$, $Ser^{3.39}$, and $His^{5.46}$, critical for melatonin binding. In the vicinity of the active site, there are only two amino acid substitutions in MT_2 as compared to MT_1: an isofunctional replacement I3.40V and a more substantial, F5.45I. Moreover, the conservative residue W6.48 (TM6) in MT_2 has drastically different orientation in the site: it is located closer to the indole ring of the ligand. As discussed below, this provides an explanation of the MT_2 selectivity to one of melatonin analogs, 2-benzylmelatonin.

Are the structural features delineated in the active sites of the constructed models of MT_1 and MT_2 capable of explaining the experimentally observed differences in their interactions with various ligands? To address this question, we analyzed spatial hydrophobic/hydrophilic properties of the ligands and the binding sites. To characterize these interactions quantitatively, we proposed a new computational approach that utilizes complementarity of hydrophobic properties of a ligand and its binding site.

Figure 4*a* demonstrates the molecular surface of melatonin, colored according to the differences in hydrophobic environment of melatonin bound to MT_1 and MT_2 receptors. The surface regions are colored black if they correspond to the atoms whose environment in the MT_2 active site is more hydrophobic than that in MT_1. Otherwise, these regions are colored white. For example, the surface of melatonin near the atoms N1 and C2 of the indole ring and 5-metoxy group fall into more hydrophobic environment when complexed with MT_2 as compared to MT_1. It is also seen that, in overall, the ligand-binding site in MT_2 is more hydrophobic than in MT_1.

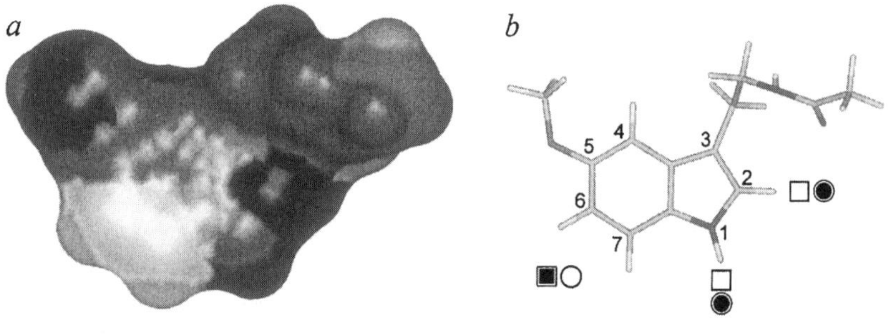

Figure -4. Difference in hydrophobic properties in the active sites of MT1 and MT2 receptors. *a* – molecular surface of melatonin colored according to differences in MHP values created by MT_1 rather than by MT_2 (where environment of melatonin in MT1 is less hydrophobic than in MT2, the color is black; *vice versa*, white); *b* – skeletal model of melatonin molecule. The symbols '○', '●', '□', and '■' near positions 1, 2, 6, and 7 of the ring indicate potential substituent groups that affect the selectivity of melatonin derivatives to MT1 (circle) and MT2 (square) receptors. Hydrophobic and polar substituents are shown with white and black color, respectively.

The found distinctions may serve to explain the experimental data on selectivity of some melatonin analogs containing hydrophobic or polar substituents introduced in the aforementioned positions. Based on the analysis of hydrophobic organization of the binding sites, we propose that introducing of a hydrophobic substituent into positions 1 or 2 of the indole ring should increase selectivity of a ligand to MT_2, while introducing of a polar group into positions 6 or 7 should raise its selectivity to MT_1. Indeed, experimental data demonstrate pronounced selectivity of 2-benzylmelatonin towards MT_2 (Table 2; Figure 5), thus confirming the theoretical prediction. Other ligands listed in the table also exhibit selectivity towards one of the receptors (except melatonin itself, which has no substantial selectivity). A detailed MHP analysis of the corresponding ligand–receptor complexes obtained via docking simulations provides a basis for rationalization and explanation of the available experimental data on selectivity of these compounds (Chugunov et al., in press).

Table -2. Experimental data on binding of some melatonin analogs to the human melatonin receptors MT1 and MT2 (For structures see Fig. 5)

Cmpd*	R_1	R_2	pK_i^{MT1}	pK_i^{MT2}	Ref.
MLT	H	H	9.63	9.43	Mor et al., 2001
2-Bz-MLT	Bz	H	7.5	9.6	Dubocovich et al., 1997
2-Br-MLT	Br	H	10.54	9.94	Mor et al., 2001
N-Bz-MLT	H	Bz	6.85	8.19	Rivara et al., 2003

* MLT, melatonin; bz, benzyl.

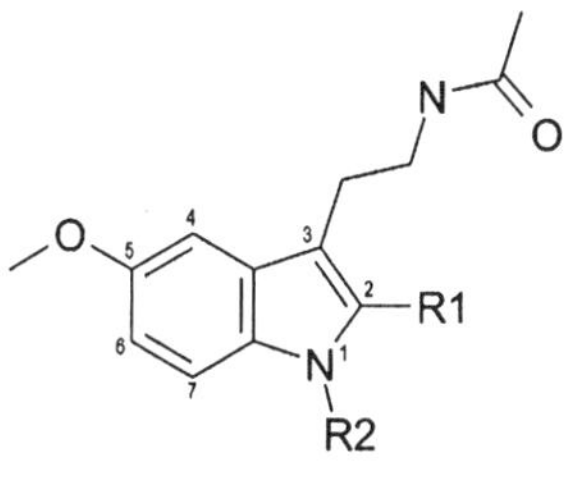

Figure -5. Melatonin analogs structure (see Table 2).

CONCLUSIONS

In this study, spatial models of TM domains of two human melatonin receptors (MT_1 and MT_2) were elaborated. The models were subjected to an original computational procedure to optimize their interactions with the natural agonist, melatonin. The method includes rotation of one of TM α-helices around its axis with simultaneous assessment of the quality of complexes according to a number of developed criteria. The constructed 'optimized' models were applied to delineate the details of protein–ligand interactions for melatonin and a number of its analogs with known affinities and selectivities to MT_1 and MT_2 receptors. It was shown that the found minor differences in the structure and hydrophobic properties of the binding sites in these receptors permitted rationalization of experimental data on ligand–receptor interactions for both receptor subtypes. In future, these models can be used to design new high affinity and high selectivity compounds acting on different subtypes of these receptors. The proposed computational procedures can be applied in modeling of other proteins from the GPCR family.

MOLECULAR DYNAMICS OF SMALL PEPTIDES USING ERGODIC TRAJECTORIES

K.V. Shaitan*, K.B. Tereshkina
Lomonosov Moscow State University, 119992, Moscow, Russia, e-mail: shaitan@moldyn.org
* *Corresponding author*

Abstract: A comparative study of the molecular dynamics of natural amino acid residues and some of their homologs and isomers is carried out. MD protocols not interfering with a principle of equidistribution of energy on degrees of freedom are used. Poincare cross-sections are considered. Dynamic properties of conformational degrees of freedom in series amino acid residues are classified.

Key words: molecular dynamics; thermostats, peptides; ergodicity; attractors

1. INTRODUCTION

Functional activity and shaping of spatial structure of natural polypeptides are intimately connected to singularities of their dynamic behavior. Now not quite clear there is a problem what and in what standard the individual properties of natural amino acid residues are important for shaping unique protein frames and how critical for protein folding can be a substitution of natural amino acid residues with their analogs, homologs, or isomers. This problem, fundamental from biophysics standpoint, can have some practical applications in building of essentially new biologically active structures.

For comprehension of possible singularities of these frames, it is important to know, basically, the conformational and dynamic properties of elementary units—amino acid residues and their modifications, in particular, in requirements of the lack of perturbing influence of the neighbors. The monopeptides, consisting of an amino acid residue bound with acetyl from the N-terminus and with N-methyl amine from the C-terminus are convenient in this respect. Relatively small amount of atoms makes it possible to study in more detail various variants of dynamic behavior of a system due to variation of external parameters. Such structures can be investigated by methods of molecular dynamics in close detail as well as separately in the presence of a solvent. Hydrophobicity of the medium is known to be important for shaping the spatial structure of a polypeptide; therefore, information on the influence of medium on dynamic conformations of amino acid residues without perturbing the neighbors can appear useful. Routinely, Ramachandran and potential energy maps use for performance of conformational possibilities amino acid residues (Jakubke and Jeschkeit, 1985; Ovchinnikov, 1987; Finkelshtein and Ptitsyn, 2002).

In this case, it is necessary to fix the remaining dihedral angles. Another disadvantage of such approach is connected with elimination of the contribution of entropic factor to stabilization of conformations. In this work, the approach to definition of probabilities of conformation occupancy based on the information received from long and statistically reliable trajectories of the molecular dynamics is used. Probabilities of conformational occupancy are defined in the space of two and three angular variables. All the remaining variables are averaged. This approach makes it possible to study in detail and compare the conformation dynamics of monopeptides and some of their analogs in the media with various hydrophobicities.

Note that for these goals, we cannot use the most popular MD protocols because of nonlinear friction thermostats (Berendsen or Nose–Hoover effects) or additional friction effect in stochastic (Langevin) approaches.

2. THERMOSTATS

It would be marked that at study of distributions of peptides on conformational states, it is important to avoid MD protocols interfering with a principle of energy equidistribution on degrees of freedom (Landau and Lifshits, 1986). MD simulation is a procedure with some tricks. If these ruses are not taken into consideration, it may lead to artifacts in MD simulations. These artifacts are not so clear for large molecules but very impressive for small ones. Obtaining of reasonable ergodic trajectories is the most important problem in MD simulations. During the calculation, the representative point has to scan all important available areas of configuration space. We especially consider artifacts because of thermostats. It may be very substantial, but it has not been taken into account until recently. The most common thermostats are based on alternating nonlinear friction (Berendsen et al., 1984; Hoover, 2001). The other type of thermostats is collisional one (Lemak and Balabaev, 1994). It uses gas of virtual particles with a certain mass and Maxwell velocity distribution at a given temperature. The average collision frequency of particles with atoms of molecules has to be indicated. Collisions are considered to be central and to happen according to the law of hard spheres. The thermalization of virtual particles is considered to happen in a moment.

Thermostats with alternating nonlinear friction. Add-on of alternating friction to equations of motion is the basic idea of the most widely used thermostats. The friction constant depends on the ratio of instantaneous temperature and the preset value of temperature T_0:

$$\gamma = \alpha\left(\frac{T}{T_0} - 1\right). \tag{1}$$

The thermostat should maintain the energy of the system over the range adequate to the given temperature T_0. However, it is shown that Berendsen thermostat (Berendsen et al., 1984) results in rather strange and physically incorrect effects. If the molecule's centre of mass is not fixed, integration process using this thermostat was remarked to result in molecular energy transfer from internal degrees of freedom to translation of molecule as a whole. Such system can be described as a flying bit of ice. If the molecule centre of mass were fixed, then molecule would rotate as a whole. Its internal energy at the same time would transfer to rotation. If rotational degrees of freedom were also fixed, then energy distribution would not be in equilibrium (Shaitan et al., 1997a). Dynamic attractor takes place in the systems under this thermostat. Specific attractive lifetime slowly (linearly) increases simultaneously with the system size. Thus, the above-mentioned thermostat leads to incorrect results even for large proteins, if the trajectory length is more than 10 ns (Golo and Shaitan, 2002). Using Nose–Hoover thermostat (Hoover, 2001), similar problems arise as well (Golo et al., 2004).

We use collision dynamics (CD) method (Lemak et al., 1994; Shaitan et al., 1997a). Temperature is constant due to collisions of atoms of molecules with the particles of virtual medium. In the numerical procedure the new terms appear. These terms describe accidental collisions between the atoms and the particles at the moment of time t_k. Dynamics of the particles in an explicit form is not interesting. Because of collisions there are jumps of atom velocities at random time moments t_k. New velocities are calculated as a result of a central-force collision of hard spheres, i.e., collisions between the atoms of the system and the virtual particles of mass m_0 and velocity $\vec{v}_0$. Velocities obey the Maxwell law.

Between serial collisions, the atoms of molecule move according to the Newton equations. In CD method, moments of time t_k (below 'moments of collisions'), when velocities jump, present random variables that take place with the Poisson probability. Collisions are independent. In other worlds, we obtain an expression of probability of n collision with a given atom during interval $[0, t]$:

$$g_n(t) = \frac{1}{n!}(\lambda t)^n e^{-\lambda t} \tag{2}$$

Intervals between serial collisions $\Delta t_k = t_{k+1} - t_k$ have the following distribution:

$$g_0(\Delta t) = e^{-\lambda \Delta t}, \tag{3}$$

where λ is the average value of collisions between a single atom and virtual particles per time unit and $\tau_c = 1/\lambda$ is the average interval between collisions.

The trajectory modeled by CD method in phase space is described below. During random interval Δt_k, the system moves along the trajectory. This movement can be represented by dynamics equations of motion on a constant total energy surface and a total momentum Π_k. Then, the system instantly switches to other surface Π_{k+1} and so on. The sudden change takes place only in momentum part of the phase space. Coordinates and hence, the potential energy remain constant during the sudden change of momentum.

If atoms of a molecule have identical mass m, we can get analytical results. In CD method, change in atom velocities after interval Δt is calculated in two stages. At first, velocities $\vec{v}_i'$ $(i = 1, \ldots, N)$ are calculated at the moment of time $t + \Delta t$ from the following equation:

$$m\frac{d\vec{v}_i}{dt} = -\frac{\partial U}{\partial \vec{r}_i}. \tag{4}$$

Then, the calculated velocities $\vec{v}_i'$ are changed randomly due to collisions with virtual particles.

Taking into account (2), we obtain an expression of probability that during the interval Δt atom s collides m times:

$$g_s^m(\Delta t) = \frac{(\lambda \Delta t)^m}{m!} e^{-\lambda \Delta t}. \tag{5}$$

According to the energy and momentum conservation laws, the velocity of an atom *s* after the collision is:

$$\vec{v}_s(t+\Delta t) = (1-\alpha)\vec{v}_s^{'}(t+\Delta t) + \alpha\vec{v}_0, \tag{6}$$

where $\vec{v}_s^{'}$ is the velocity of atom *s* before the collision, $\vec{v}_s$ is its velocity after the collision, and $\vec{v}_0$ are random values that are distributed by Maxwell law.

$$\alpha = \frac{2m_0}{m+m_0}, \tag{7}$$

where m_0 is the mass of virtual particle and T_0 is the thermostat temperature.

Let us give an example for coefficient of translational diffusion of some particle with mass *M* in virtual collisional medium:

$$D = \frac{\langle u^2(0)\rangle}{3\alpha\lambda} = \frac{k_b T}{\lambda M \alpha}. \tag{8}$$

The coefficient of friction in collision thermostat was shown to be a dynamics invariant at a given temperature as follows from comparison (8) with the famous Einstein formula. This coefficient is proportional to product of sum of atom reduced mass in molecule and effective frequency of collisions. Using (8), we can calibrate the collisional medium. For example, if we want to use the medium under water viscosity, we should take $m = 18$ amu and $\lambda = 55$ ps $^{-1}$ (Shaitan et al., 1997b; Shaitan and Saraikin, 2002).

3. METHODS

MD calculations of 20 natural amino acid residues and 5 forms of the modified Tyr with various positions of hydroxyl groups were carried out. All amino acid residues in order to prevent the end effects were been linked to N-methyl amine from the C-terminus and with acetyl c the N-terminus (Figure 1). Modification of side radicals of Tyr is given in Figure 2. In residues TY2 and TY3, the hydroxyl group was moved from a *para-* position in an *ortho-* (TY2) or a *meta*-position (TY3). In residue TYO, the second hydroxyl group was added, from residual TYS the methylene bunch was removed. In residue TYC, a side radical enlarged by one CH2- group.

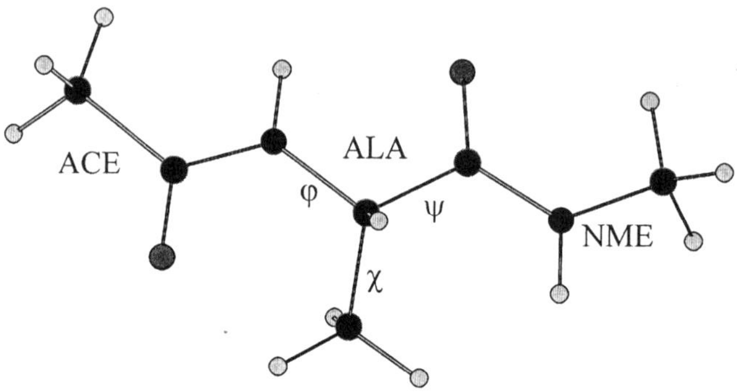

Figure -1. N-acetyl-a-alanine, the residue of alanine modified by acetyl and N-methyl amine with formation of two peptide bonds. Torsion angles j, y, and c are selected, for these angles correlation functions Poincare cross-sections were found.

The standard method of the molecular dynamics with the following parameters of the protocol was used:

1. AMBER-99 potential field (Weiner and Kollman, 1981; Weiner et al., 1984; Weiner et al., 1986; Pearlman et al., 1991; Cornell et al., 1995).
2. Length of a trajectory was taken 20 ns, temperature of a thermostat was 2000 K.
3. Berendsen and collision thermostats were used.
4. Time constant of a modification of a velocity in Berendsen thermostat $\tau = 0.5$ ps.
5. Inductivity of medium $\varepsilon = 1$.
6. Cut-off distances for electrostatic interactions $R_{el} = 20$ Å for calculations in collisional medium, $R_{el} = 10.5$ Å for calculations in solvents.
7. Cut-off distances for Van der Waals interactions $R_{VdW} = 16$ Å for calculations in collisional medium, $R_{VdW} = 8.4$ Å for calculations in solvents.
8. Mass of virtual particles $m = 18$ amu, a collision frequency of virtual particles with atoms $\nu = 55$ ps^{-1} for calculations in collisional medium, $m = 0.01$ amu and $\nu = 150$ ps^{-1} for calculations in solvents.
9. For a numerical integration algorithm, Verlet was used. The initial velocities of atoms were determined with the help of the generator of random numbers on the Maxwell distribution.
10. Integration step was taken 1 fs.
11. Entry step to trajectory file 0.1 ps.

Models of molecules were considered in a full-atomic approach. Parameters of the standard potential field AMBER-99 for the modified forms of Tyr were supplemented by experimental data (Stull et al., 1971) and also by data of quantum-chemical calculations with the help of program complex GAMESS (Schmidt et al., 1993; Granovsky, 2004). Restricted

Hartree–Fock method with a standard basis set 6–31G (2d, p) was used. Effective charges of atoms were found by means of method Mulliken.

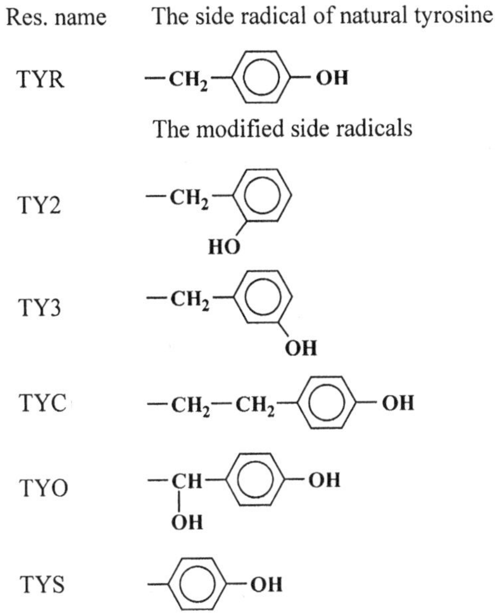

Figure -2. Side radicals of Tyr and its modifications

Three bunches of examinations were realized. The behavior of amino acid residues in an aqueous environment, in collisional medium imitating an aqueous phase (Shaitan et al., 2000; Shaitan et al., 2002), and in solution of methanol was studied. Calculations in an environment of water and methanol were carried out in a box with periodic boundary conditions of dimension 20 × 20 × 20 Å^3. The model of water TIP3P (Weiner et al., 1986) with the parameters corresponding to parameters of AMBER force field was used. In order to prevent dynamic attractor regimes during study of solutions dynamics, the collisional thermostat was used along with the Berendsen thermostat. The density of water was taken ρ (H_2O) ≈ $0.99 \cdot 10^3$ kg/m^3. The density of methanol was taken ρ (CH_3OH) = $0.7928 \cdot 10^3$ kg/m^3 (Kikoin, 1976). The temperature of 2000 K was used to accelerate the configuration space scanning procedure.

The basic contribution to a modification of a configuration is yielded with hindered rotations on torsion angles φ, ψ, and χ_1 (further, it is designated as χ).

For these angles, the one-dimensional, two-dimensional, and three-dimensional distributions of probability density and Poincare cross-sections for all combinations of these angles were calculated (Shaitan et al., 1997, 2000).

Autocorrelation functions of a special type were calculated to evaluate the individual dynamic behavior of torsion angles:

$$F_{xx} = \left\langle e^{ik(t)} e^{-ik(t+\tau)} \right\rangle. \tag{9}$$

Here, $k(t)$ is a value of torsion angle in an instant t.

Dispersion analysis was carried out for comparative analysis of dynamic behavior of amino acid residues. The Euclidean metric was used to define distinctions between maps of levels of a free energy, to detect the same type objects, and to classify conformational degrees of freedom. Metrics for a determination of distinctions between two-dimensional maps (10) and autocorrelation functions (11) was chosen as follows:

$$d_{sr} = a^2 \sqrt{\sum_i (p_{r,i}(\varphi, \psi) - p_{s,i}(\varphi, \psi))^2}$$

$$d_{sr} = a^2 \sqrt{\sum_i (p_{r,i}(j, \psi) - p_{s,i}(j, \psi))^2} \tag{10}$$

$$d_{sr} = \frac{\sqrt{\int_0^\tau \left(f_r(t) - f_s(t) \right)^2 dt}}{\max \int_0^\tau (f_i(t))\, dt}. \tag{11}$$

Here, indexes r and s correspond to two different amino acid residues, a is a parameter of a partition, p is a probability density, f is value of a real part of an autocorrelation function, and the index i designates an autocorrelation function the integral under which has maximal value on a considered field. The algorithm of a choice of minimum distances was applied for build-up of a cluster tree.

4. RESULTS

The basic variants of secondary structure on a background of the allowed and forbidden ranges for the angles φ and ψ can be submitted on the well-known Ramachandran map.

For Poincare cross-sections (or maps of free energy levels) at all monopeptides in a subspace of φ–ψ torsion angles, four types of maps can be seen. The maps 2-D Poincare cross-sections for alanine, glutamine acid, glycine, and proline are shown in Figure 3. The darkest ranges in figures correspond to the greatest population of conformations. Angles are taken from –360 up to +360 degrees to visually represent transitions between loci with a minimum free energy.

On all maps except for the proline, the most populated areas appeared to correspond to β-conformations and a right-handed α-helix. These loci are connected by collective degree of freedom, and transferring between them is realized through formation of a right-handed 27-helix. The area corresponding to the left-hand α-helix and the left-handed 310-helix is most sensitive to environment; thus, internal rotation on the angle φ is less sensitive to a choice of a solvent in comparison with the angle ψ. As a whole, presence of a solvent flattens a contour of a potential surface and increments a panel of probable conformations under the given conditions.

This effect already was considered earlier in the literature (Finkelshtein et al., 2002). In particular, the possibility of formation of hydrogen bonds inside a monopeptide molecule between NH and C=O groups and between the corresponding bonds of monopeptide and solvent molecules was considered. In collisional medium, the state in which there is a hydrogen bond inside a monopeptide appears more favorable. In solvents, more favorable appears the formation of hydrogen bindings not inside a molecule of a monopeptide but between the molecules of a solvent and a monopeptide.

The effect of influence of solvents on free energy distribution is shown using cluster analysis of two-dimensional maps of a free energy for the angles ψ and φ in collisional medium, aqueous environment, and methanol (Figure 4). The largest distinctions are observed for residues of glycine and proline. At build-up of a cluster tree, the objects distinguished from each other on the magnitude dsr ≤ 0.004 were referred to one group. In collisional medium, the following groups are detected: (1) amino acid residues having small side radicals and residues, having rings in composition of side radicals or three substituents, bound to hinged CH_2-group; (2) amino acid residues, which side radicals have positive charges; (3) amino acid residues with negatively charged side radicals (glutamic and aspartic acids); (4) glycine; and (5) proline. In methanol, use of the same criteria makes it possible to reveal four bunches of one-type Poincare cross-sections: (1) glycine; (2) proline; (3) arginine, lysine, and histidine; and (4) the remaining amino acid residues. In water, the behavior of torsion angles in monopeptides becomes more one-type. Distinctions are observed only for proline and glycine.

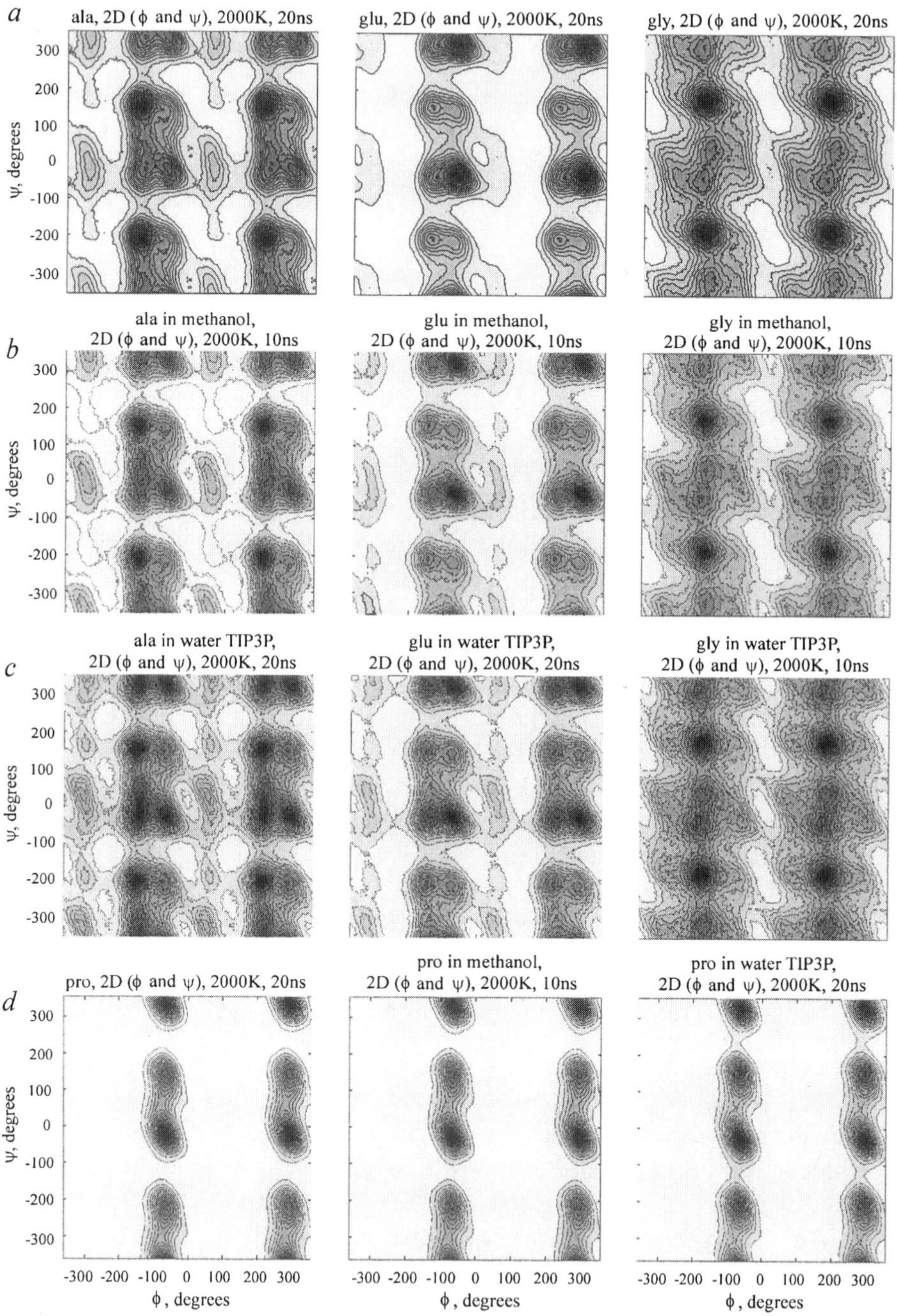

Figure -3. 2-D Poincare cross-sections for torsion angles φ (abscissa axis) and ψ (axis of ordinates) in degrees (–360, 360). The most populated areas of conformational space are shown with darker color; (*a*) peptides (Ala, Glu, and Gly) in collisional medium; (*b*) in methanol; (*c*) in water; and (*d*) proline residue in vacuum, methanol and water medium consequently.

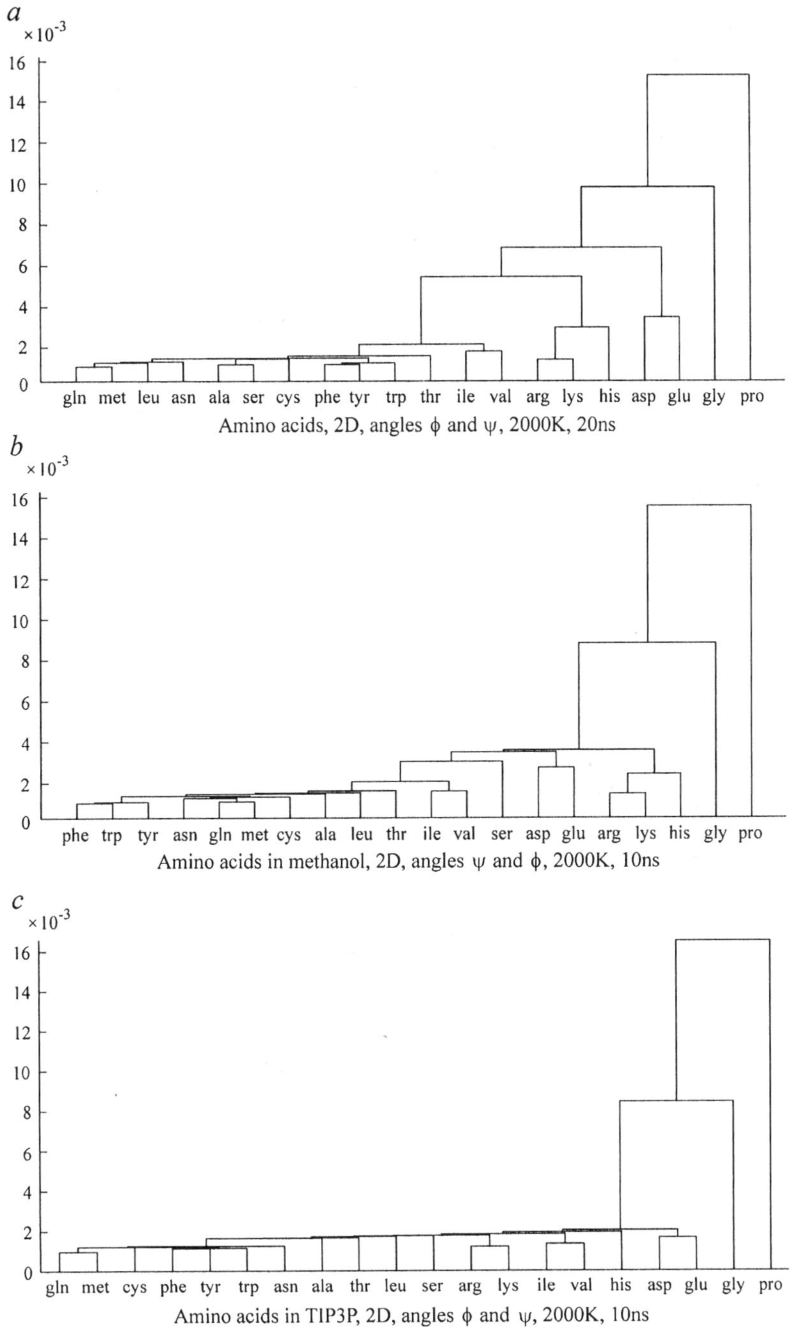

Figure -4. Cluster trees for comparison of 2-D Poincare cross-sections on the angles φ and ψ in (*a*) collisional medium, (*b*) water, and (*c*) methanol.

Similar effects can be observed having carried out dispersion analysis and on other two-dimensional maps (on angles ψ–χ and φ–χ).

The greatest distinctions in dynamic behavior are observed at monopeptides in collisional environment; the least, in aqueous. On maps of a free energy, it is clear that in going from collisional medium to aqueous phase or from collisional medium to methanol, the areas around of minimums of a free energy with coordinates $\varphi = -60 \geq°$, $\psi = 60°$, $\varphi = 60°$, $\psi = -60°$, or M and H potential wells, according to Popov's (1997) terminology, undergo the greatest modifications. These conformations fit two probable states of N-acetyl-α-amino acids with intramolecular hydrogen bonds. They differ from each other orientation of a side chain concerning the seven-membered cycle formed by a hydrogen bond between the group N-H (Nme, the third residue) and O = C (Ace, the first residual). These areas appear to be more occupied in methanol in comparison with an aqueous medium; this effect reflects major ability of water to form hydrogen bonds with a molecule of a monopeptide.

By reviewing maps of a free energy for a monopeptide in various solvents, it is evident that in a molecule of a monopeptides, there are no new spatial forms, the solvent only displaces equilibrium of conformations aside those structures that in the best way interact with molecules of a solvent.

Results on the influence of polar medium on behavior of monopeptides are in good agreement with the results obtained by Popov (1997) by analysis of conformational maps of monopeptides Gly, Ala, Pro, and Val *in vacuo* and polar medium. Really, at the presence of a solvent, population of narrow M and H potential wells decreases with increase of polarity of a solvent.

Three-dimensional Poincare cross-sections make it possible to estimate influence of a polar environment. The increase in available configuration volume is observed in going from nonpolar (collisional) to polar (methanol or water) medium. The greatest effects are in evidence for residues with the charged side radicals.

The real part of autocorrelation functions of the angles ψ, φ, and χ for twenty natural amino acid residues in collisional medium, water, and methanol are investigated as well. Dispersion analysis of autocorrelation functions reveals similar dynamic behavior for the majority of torsion angles in monopeptides. Essential distinctions are observed only for glycine and proline because of the lack of side radical in glycine and its cyclization in proline. The greatest distinctions are observed for rotations on angle χ in view of the considerable variations in structure of side radicals.

5. CONCLUSIONS

In terms of dynamic behavior in a subspace of torsion angles φ, ψ and χ_1, all monopeptides of natural amino acid residues show similar properties. The basic differences are detected for monopeptides of proline and glycine because of characteristic structure of their side radicals.

For all monopeptides, the most populated are the areas that correspond to β-conformations and a right-handed α-helix . These results are in good agreement with Popov (1997) on analysis of conformational maps of methylamides of some amino acids.

The presence of a solvent flattens a contour of a potential surface and enlarges a set of probable conformations under the given conditions. In going from non-polar (collisional) to polar (methanol and water) media, increase in the available configurational volume is observed. For residues with the charged side radicals, this effect is the most expressed.

Distinctions in dynamic behavior of monopeptides are maximal in collisional medium; in going to methanol and to aqueous medium, they accordingly decrease.

Monopeptides with small side radicals in the least degree experience modifications at a variation of a solvent.

Cyclization of a monopeptide at hydrogen bonding between atoms of oxygen of acyl residue and hydrogen of N-methyl residue has the most expression in collisional medium. This result also complies with the data of Popov (1997) at study of the monopeptides Pro, Gly, Ala, and Val in vacuum and polar mediums. We shall mark that according to Popov (1997), the solvent was simulated exclusively by increase in the medium inductivity.

In series of the modified tyrosines, the additional hydroxyl group makes conformational transitions on angle χ slower; and the presence of additional hinged group CH2, on the contrary, makes movement freer. These phenomena are most clear in collisional medium. In methanol and aqueous environment, dynamics of conformational transitions on angle χ for tyrosine without hinged methyl group in a side radical is the slowest. The requirement of ergodicity (or quasi-ergodicity for large systems) demands suitable conditions in MD calculations. The choice of these conditions substantially depends on topology of potential energy level surfaces of the system under study.

ACKNOWLEDGMENTS

This work was partially supported by RBFR (grant No. 04-04-49645), Ministry of Education and Science of the Russian Federation, and Moscow Government.

A PERIODICAL NATURE OF 94 PROTEIN FAMILIES

V.P. Turutina[1*], A.A. Laskin[1, 2], N.A. Kudryashov[2], K.G. Skryabin[1], E.V. Korotkov[1, 2]

[1] *Bioengineering Center of Russian Academy of Sciences, Prospect 60-letya Oktyabrya, 7/1, 117312, Moscow, Russia, e-mail: veratp@yandex.ru;* [2] *Moscow Physical Engineering Institute, Kashirskoe shosse 31, 115409, Moscow, Russia*

[*] *Corresponding author*

Abstract: The techniques of Information Decomposition, Cyclic Profile Alignment, and Noise Decomposition allowed us to search for latent repeats within protein families of various functions. We have found out 94 protein families with a family-specific periodicity. In each case, the periodic element was found in greater than 70 % of family members. Latent periodicity profiles with specific length and signature was obtained in each case. The possible relationship between the periodic elements thus identified and the evolutionary development of the protein families are discussed with specific reference to the possibility that there is a correlation between the periodic elements and protein function.

Key words: latent periodicity; alignment; information decomposition; noise decomposition; profile analysis; repeats

1. INTRODUCTION

Investigation of amino acid sequence periodicities may bring light to structural organization of protein sequences and protein evolution. Ohno (in 1970) offered that duplications and divergences of DNA base strands are primaries for evolution (Ohno, 1970). These evolutionary mechanisms may generate novel coding sequences via multiple duplications of relatively short DNA sequences and their subsequent divergence (Ohno and Epplen, 1983; Ohno, 1984). Therefore, if the generation of genes by multiple tandem duplication is relatively widespread, coding regions within DNA sequences should retain within their structures the traces of these tandem duplications as low-homology periodic repeats, which may carry over to amino acid sequences. This periodicity, however, would be difficult to identify due to the substantial divergence of initial repeat sequences via insertions, deletions, and base substitutions. If these primary sequences could be identified, investigation and classification of the amino acid periodicity could facilitate our understanding of structural organization and its relationship to protein evolution and structure. For example, it may be hypothesized that all gene sequences descended from a single duplicated ancestor sequence would possess a characteristic but eroded periodicity detectable at the amino acid level. Amino acid sequences with similar biological functions would be predicted to have similar periodicity patterns.

Current mathematical models are able to identify periodicity in small samples of protein sequences (Heringa and Argos, 1993; McLachlan, 1993; Makeev and Tumanyan, 1996; Heringa, 1998; Rackovsky, 1998; Benson, 1999; Andrade et al., 2000; Heger and Holm, 2000; Jackson et al., 2000; Katti et al., 2000; Neuwald and Poleksic, 2000; Landau et al., 2001; Murrey, 2004).

The level of internal homology in domains depends on the time passed since their formation. Similarity is apparent in recent repeats, but long lifetimes of proteins of fundamental types are likely to make them hidden, unseen by traditional search methods. One may question our ability to identify this hidden periodicity, which has descended from those ancient multiple tandem gene duplications. In turn, we suppose that the lack of existing data concerning periodic structure of protein families is caused by imperfection of applied periodicity search methods and their inability to find faintly marked repeats. It has previously been shown that when searching for periodicity in symbolic sequences, Fourier-based and dynamic programming–based techniques have constraints that limit their ability to identify faintly marked periodic elements (Korotkov et al., 2003).

The development method of the Information Decomposition (ID) to search for latent periodicity within symbolic sequences has been applied. This method is free from many of the shortcomings identified in Fourier-based or dynamic programming–based techniques (Korotkov et al., 2003). The ID technique is capable of identifying feebly marked periodic elements within symbolic sequences (Korotkov et al., 2003; Laskin et al., 2003) but without insertions and deletions symbols. The Noise Decomposition (ND) and Cyclic Alignment (CA) techniques were developed to deal with such insertions and deletions (Laskin et al., 2003). The ND technique is a dynamic programming–based methodology that uses a latent periodicity matrix, which is in turn based on the ID method. After completing a number of iterations, the ND technique is able to identify a family-specific latent periodicity. At present, a combination of ID and ND techniques allows the identification of weak or latent periodicity in more than 80 % of NAD-binding sites (Laskin et al., 2003).

By using the approach mentioned above, latent periodicity has been identified within 100 additional protein families (Laskin et al., 2004), providing a strong basis to suppose that protein domains could contain latently periodic elements, which may be identified by using the combination of ID and ND techniques. In this paper, we demonstrate the presence of latent periodicity in 30 protein families of various biological functions. More than 70 % of members of each of these families are shown to possess latent periodicity of common period length and signature. These results support the viewpoint that such periodicity is not just a property of individual proteins but could be specific to protein families, thus supporting the hypothesis of gene origination by multiple tandem duplications. Certain forms of amino acid latent periodicity may therefore correspond to particular biological functions of protein families.

2. METHODS AND ALGORITHMS

Latent periodicity is defined here as periodicity that is identified using ID but is not detectable at a statistically significant level using Fourier or homology search techniques. The homology search settings are often defined using a PAM or BLOSUM matrix (Henikoff and Henikoff, 1993; Holmes and Durbin, 1998), which provides weightings that are higher for similar amino acid matches and lower for dissimilar ones, or an autocorrelation function (Dodin et al., 2000). As an example, let us consider a set of sequences $S^1;S^2;...;S^N \equiv \left\{s_1^1 s_2^1 ... s_L^1; s_1^2 s_2^2 ... s_L^2; ...; s_1^N s_2^N ... s_L^N\right\}$ of equal lengths L, where s_i^j is an amino acid. To evaluate the overall similarity between these sequences, we construct their (indel-free) multiple alignment. The total weight of this multiple alignment is generally a sum of position weights:

$$W = \sum_{i=1}^{L} W_i \, . \qquad (1)$$

$$W_i = \frac{1}{2} \sum_{l,k} m(i,l)(m(i,k) - \delta_l^k) P(l,k) \, , \qquad (2)$$

where l and k are amino acid types and $m(i, l)$ is the number of amino acids of type l at position i in the multiple alignment, $P(l, k)$ is some amino acid affinity matrix, such as PAM or BLOSUM, and δ_1^k is the Kronecker function. We previously proposed another measure of similarity (Chaley et al., 1999; Korotkova et al., 1999) based on concepts of information theory and called 'information content' (Kullback, 1959):

$$W_i' = \sum_{l=1}^{20} m(i,l) \ln \frac{Km(i,l)}{x(i)y(l)}, \qquad (3)$$

where $K = NL$, $x_i = \sum_{l=1}^{20} m(i,l)$ and $y_l = \sum_{i=1}^{L} m(i,l)$. These measures are clearly different; thus, an alignment may achieve a high score using information theory measures, while achieving a low score using homology-based measures and vice versa. However, the term 'high-scoring' is of little significance, especially when comparing weights calculated with different measures.

Earlier in our works (Laskin et al., 2003), it has been shown that the measure of similarity of sequences W_i (2), used at search of homology between sequences S_i, is capable of passing the latent periodicity in length L in sequence S; and the information measure of similarity W_i' (3) allows revealing the latent periodicity at a significant level.

It is necessary to make sure that the obtained value of W is much higher than those calculated using sets of random, unrelated sequences. To ensure this, the initial sequences should be shuffled and either p-value or Z-value of the obtained alignment should be calculated. Z-value can be estimated by the Monte Carlo calculations as:

$$Z = \frac{W - E(W)}{\sqrt{D(W)}}, \qquad (4)$$

where $E(W)$ and $D(W)$ are mean and variance of W, respectively, calculated for a set of random sequences with the same length and same amino acid frequencies; low p-value or high Z-value would indicate a significant similarity between the sequences S^1, S^2, ..., S^N. One usually sets up some threshold value beyond which the similarity is not considered casual.

When sequence S consists of N studied sequences S_i (I = 1, 2, ..., N) of equal lengths, $S \equiv S^1S^2S^3 \ldots S^N$, significant similarity between S_i indicates significant periodicity in this sequence. As we said before, different similarity measures result in different weights and different significance values; in some cases, periodicity of a sequence may be apparent from an information theoretical viewpoint while omitted by homology searches. In our studies, we call this effect 'latent periodicity'. In our previous studies (Korotkov et al., 2003; Laskin et al., 2003; Laskin et al., 2004), we have shown that such latent periodicity occurs in many biologically important sequences.

Let us designate the p-value calculated using Eqs. (1), (2), and (4) as α and the p-value calculated using Eqs. (1), (3), and (4) as β. Let us consider that within that symbolic sequence S, there is a latent periodicity, if the value of probability α is greater than 0.05 (statistically insignificant value α) and the value of probability β is smaller than 0.05 (statistically significant value β). In this case, the number of homologous coincidences will be relatively small for each position in the period. This may lead to a comparatively low value of W and correspondingly high value of α. In contrast, the estimation of probability β is based on deviations of symbol frequencies at each period position from relative symbol frequencies derived from the whole sequence. These deviations may be substantial, resulting in low and significant values of β. In this sense, the definition of latent periodicity, based on the similarity measure Eq. (3) and on the probability β, reflects a more common property than the definition of tandem homological repeats, based on the similarity measure Eq. (2) and probability α.

To identify latent periodicities within protein families, we have analyzed the complete Swiss-Prot database using the Information Decomposition technique (Korotkov et al., 2003).

We identified a total of 15 000 amino acid sequences possessing periodicity with a period length being from 2 to 200 residues and Z-scores greater than 5.0. We determined periodicity matrices from the initially identified sequences; their elements represent the number of occurrences of amino acids at each position of the period in an identified latently periodic subsequence. Next, these periodicity patterns were used as profiles to provide the identification of latent amino acid periodicities with insertions and deletions. The elements of the corresponding position-weight matrix W were calculated from the periodicity matrices M using the expression (Karlin et al., 1990):

$$W_{i,j} = C \ln \frac{p_{i,j} + \varepsilon f_i}{f_i + \varepsilon}, \tag{5}$$

where $W_{i,j}$ is an element of the position-weight matrix for the symbol of type i at position j, $p_{i,j} = m(i, j)/y(j)$, and f_i is the frequency of occurrence of symbols of type i in the periodic subsequence. The small number ε was introduced in Eq. (5) to eliminate the consideration of zero values, and it was equal to 10^{-5} in our calculations. The scaling parameter C may be chosen arbitrarily, since multiplying all weights and scores by a factor changes neither the path of the alignment nor its statistical significance, provided that gap penalties are multiplied by the same factor.

We used the ND technique to obtain the position-weight matrix that would allow us to identify statistically significant latent periodicity located in most proteins of a family but not in other amino acid sequences. The essence of iterative ND is as follows. First, amino acid sequences with statistically significant periodicity ($Z > 6.0$) of a given type were selected from Swiss-Prot version 41. This database was scanned using modified profile analysis, also known as Cyclic Alignment (Chaley et al., 2003; Laskin et al., 2003), using a window of 200–300 residues.

At the second stage, the results were divided into two sets. The first set contained amino acid subsequences with a *Z*-score greater than 6.0 and with the functionality similar to that of the initial latently periodic sequences found with ID. This set was referred as the 'true alignments'. Functionalities of proteins were determined from their descriptions (Swiss-Prot DE field), keywords (KW field), and feature tables (FT field), which should be identical to the corresponding fields of initial sequence. We also formed a set of unrelated sequences called 'false alignments'. These sequences were found in the protein families with functionality different from that of the initial latently periodic sequences found with ID. The resulting set contained amino acid subsequences of 'true alignments', which were optimally aligned with matrix *W* and which had a *Z*-score greater than 6.0.

The third stage involved the modification of the initial position-weight matrix *W*. This modification pursued two aims. First, we wanted the modified profile analysis to identify as many true alignments as possible; ideally, all protein sequences being functionally identical to the initial sequence should be identified at a statistical level $Z \geq 6.0$. Second, we tried to eliminate sequences with $Z \geq 6.0$ from the false positive dataset. These aims may be termed 'sensitivity' and 'selectivity of latent periodicity identification', respectively.

To achieve these goals, Eq. (5) was modified in the following fashion:

$$\bar{W}_{i,j} = C \ln \frac{r_{i,j}}{\pi_{i,j}}, \tag{6}$$

where $r_{i,j}$ is the weighted value of $p_{i,j}$. The true positive set may contain homologous or identical sequences. This equation may cause the overrepresentation of some amino acids in some period positions. To take this effect into account, it is necessary to compare all found sequences and calculate the weights for each sequence from the true positive set. The weight of the sequence should reflect its representation in the set. Let the alignment score for sequences k and l be $S(k, l)$. These values were used to calculate $T(k)$, which represents the prevalence of k-like sequences in the true positive dataset:

$$T(k) = \sum_{l} \max(0, S(k,l)/\{\max(S(k,k), S(l,l))\}) . \tag{7}$$

Index l runs through all the true positive dataset. In the sum in Eq. (7), the term with $l = k$ always equals one (any sequence is self-similar), the terms for unrelated sequences are equal to zero, and the terms for similar sequences range from zero to one. Therefore, we get $T(k) = 1$ in the case when there are no sequences similar to k; we get $T(k) = N$, if all N sequences in the true positive set are identical; and we get a value from one to N depending on the similarity level, if sequences are similar.

The matrix M^k for each amino acid sequence k from the true positive set was calculated, and these matrices were summarized with weights equal to $1/T(k)$. We calculated the weighted periodicity matrix M for the entire true positive set as:

$$m(i, j) = \sum_{k} m^k (i, j)/T(k) , \tag{8}$$

where k runs through all sequences from the true positive set, $m^k(i, j)$ is an element of the matrix M^k, and $m(i, j)$ is an element of the matrix M. Then, we calculated the values of $r_{i,j}$ as:

$$r_{i,j} = \frac{\sum_{k} m^k (i, j)/T(k)}{\sum_{i} m(i, j)} , \tag{9}$$

where k runs through all sequences from the true positive set, $m^k(i, j)$ is an element of the M^k matrix, and $m(i, j)$ is an element of the M matrix.

Thus, we eliminated the possibility of overrepresentation of some sequences in the true positive set. We define $\pi_{i,j}$ as:

$$\pi_{i,j} = c_0 f_0 + (1 - c_0) \sum_{k} q^k_{i,j} / N_1 , \tag{10}$$

where k runs through all sequences from the false positive set. Frequencies $q^k_{i,j}$ are analogous to $m^k_{i,j}$ but defined for the kth sequence from the false positive dataset; f_i are amino acid frequencies from within Swiss-Prot. The mixing constant c_0 was experimentally chosen to provide the best selectivity of resulting position-weight matrix while keeping its sensitivity.

After calculation of the new position-weight matrix $\overline{W}_{i,j}$, we moved to the first stage, Swiss-Prot scanning with a modified profile, so that the periodicity search was iterative. The number of iterations ranged from 3 to 10. The performing of the iterations was intended to identify as many protein sequences with the functionality identical to that of the initial sequence as possible and to eliminate unrelated sequences with $Z \geq 6.0$ from the false positive set. Iterations were stopped when the true positive set stopped growing.

3. RESULTS AND DISCUSSION

We have applied the techniques of Information Decomposition, Cyclic Profile Alignment, and Noise Decomposition to search for latent repeats within protein families of various functions in Swiss-Prot release 41. Overall, 30 novel family-related types of periodicity were identified. From 70 to 100 % of the proteins within each family were found to contain a statistically significant latent period. The percentage of false alignments was generally within the range of 0–10 %, although in a few of the protein families, up to 20 % of false alignments were identified.

For approximately 10 % of the periodicity classes identified by the ID technique, the iteration procedure was unable to find that more than 70 % of the other members of the families the initial sequences belonged to possess the latent periodicity, that is, the set of true alignments was small. The same periodicity classes also contained cases for which we were unable to eliminate the prevalence of false alignments with $Z \geq 6.0$ even after performing several iterations. In our opinion, the problems experienced within this 10 % subgroup could be due to significant levels of evolutionary divergence of periodicity, possibly combined with a large number of insertions and deletions within the protein sequences analyzed.

The names of protein families, period lengths, sizes of families in Swiss-Prot, and numbers of identified periodic sequences are listed in Table 1.

All cyclic alignments and periodicity profiles for protein families may be found at the authors' website (http://bioinf.narod.ru/new1). Note that while the methodology described here has enabled the identification of these sequences, the ID technique in its present form (Korotkov et al., 2003) is unable to identify these periodicities when used in isolation due to the variety of insertions and deletions.

We believe that our approach achieves a significant level of sensitivity (not less than 70 % of proteins in a family are identified) and selectivity (not more than 10 % of results are false alignments) for searching in Swiss-Prot, which contains more than 120 000 amino acid sequences. These results would therefore appear to be of use when making functional predictions for newly identified proteins with unknown biological function.

Table -1. The list of protein families with latent periodicity identified using iterated profile analyses

	Protein family	Period length	Number of proteins in Swiss-Prot (release 41)	Number of proteins belonging to set of true alignments	Number of proteins belonging to set of false alignments
1	Homeobox protein (homeobox domain)	14	725	572	42
2	MADS box protein (domain MADS)	13	73	70	1
3	T-Box protein (T-box domain)	14	65	59	9
4	P450 protein (chain cytochrome P450, active site)	14	665	577	31
5	Pyruvate kinase (ADP binding site)	11	67	59	6
6	Protein Cpn10 (subunit hasn't been marked out)	14	133	121	8
7	Protein Cpn60 (chain haperonin CPN60, for many proteins detailed subunit hasn't been marked out)	25	245	222	0
8	Glycosyltransferase (subunit hasn't been marked out)	33	57	47	4
9	Ice nucleation protein (domain octapeptide periodicity)	8	8	8	0
10	Heat shock protein 70 family (mod_res phosphorylation)	24	310	289	13
11	Clathrin heavy chain family (domain heavy chain ARM)	3	7	7	1
12	Pheromone response proteins (chain pheromone-binding protein)	8	34	30	5
13	Adenine phosphoribosyltransferase (subunit hasn't been marked out)	13	68	66	8
14	Beta-galactosidase (chain beta-galactosidase, active site proton donor, active site nucleophile)	25	41	36	5
15	Lysozyme C (chain lysozyme C, active site)	6	64	60	8
16	Cyclin AB subfamily (subunit hasn't been marked out)	4	74	67	11
17	Heat shock protein 90 family (domain A, substrate-binding site,	14	94	84	6

	Protein family	Period length	Number of proteins in Swiss-Prot (release 41)	Number of proteins belonging to set of true alignments	Number of proteins belonging to set of false alignments
	domain B)				
18	Phosphoenolpyruvate carboxylase (active site)	50	110	108	9
19	CA_binding site EF-hand (CA-binding EF-hand site, domain ancestral calcium site)	36	545	431	26
20	Catalase (active site)	10	118	101	4
21	Chalcone synthase (subunit hasn't been marked out)	17	119	118	5
22	CF(1): ATP synthase beta chain	7	135	134	0
23	CF(1): ATP synthase alpha chain	9	92	89	4
24	Chemotaxis proteins (active site, domain response regulatory, domain extracellular, transmem, domain cytoplasmic)	7	291	198	24
25	Chalcone synthase (subunit hasn't been marked out)	17	119	117	5
26	Aspartate aminotransferase (chain aspartate aminotransferase, binding pyridoxal phosphate site)	16	54	52	6
27	Cytochrome c oxidase polypeptide I (metal copper B)	7	148	132	11
28	Cytochrome c oxidase polypeptide II (domain mitochondrial intermembrane, metal copper A, chain cytochrome C oxidase polypeptide II)	5	220	216	14
29	Cytochrome c oxidase polypeptide III (chain coproporphyrinogen III oxidase)	11	137	130	10
30	Triosephosphate isomerase(TIM) (chain triosephosphate isomerase, active site)	6	129	118	7

To all appearance, the number of latently periodic protein families is much higher than the 94 identified here. The data presented here, in conjunction with previously published work (Laskin et al., 2003; Laskin et al., 2004), provide support for the view that this may be a ubiquitous phenomenon. It is also clear from Table 1 that period lengths differ for diverse protein families. However, the same (or close) period length of two or more protein families does not imply that the types of periodicity are the same. The type of periodicity is defined by the matrix M.

In summary, we have shown that there are many biologically important protein families with a latently periodic signature within at least the majority of their members. These results support earlier hypotheses of gene evolution by multiple tandem duplications (Ohno, 1970; Ohno and Epplen, 1983; Ohno, 1984). While the presence of latent periodicity has been previously shown at the nucleotide level (Chaley et al., 1999; Korotkova et al., 1999; Korotkov et al., 2003), the methods and the data presented here may provide the means of comprehending the evolution of protein families.

Thus, the process of evolution could have a limited number of periodic elements to generate effective structures and subsequently refine protein functionality by means of tandem duplications (Elder, 2000). When subsequent evolutionary pressure was applied to the protein sequences, the required further reinforcement of functionality and stability of proteins was provided by indels and substitutions. Finally, the protein comes to its present state, and only poorly detectable traces of periodicity are left.

The obtained results show that there is a certain correspondence between some classes of amino acid periods and functions of proteins where the given periods are observed in the form of the latent periodicity. If such correspondence really exists for all protein families, it can be found by the further accumulation of data about presence of the latent periodicity at various protein families.

ACKNOWLEDGMENTS

This investigation was supported by the ISTC (grant No. 1379-2). The authors would like to thank Andrew Shelenkov for suggestions leading to improvement of the manuscript.

PREDICTION OF CONTACT NUMBERS OF AMINO ACID RESIDUES USING A NEURAL NETWORK MODEL

D.A. Afonnikov*

Institute of Cytology and Genetics, Siberian Branch of the Russian Academy of Sciences, prosp. Lavrentieva 10, Novosibirsk, 630090, Russia, e-mail: ada@bionet.nsc.ru; Novosibirsk State University, ul. Pirogova 2, Novosibirsk, 630090, Russia

* *Corresponding author*

Abstract: The number of contacts is an important characteristic of amino acid residues in proteins. This characteristic relates to the accessibility to solvent and may be used in the prediction of the protein contact matrices. Here, we propose an approach to the prediction of the number of residue contacts in proteins on the basis of protein sequence, position specific scoring matrices for homologous sequences, and neural network algorithm.

Key words: protein structure; residue contact numbers; position-specific scoring matrix; neural network

1. INTRODUCTION

The post-genomic era faces with the problem of development of algorithms and programs for the prediction of the structural and functional features of proteins. The goals of the approaches are, as much as possible, to annotate the protein moiety of newly sequenced genomes and the already known relying on their sequences (Doerks et al., 1998; Gerstein, 2000). There are several categories for protein structure prediction: (3D) atomic coordinates (Baker and Sali, 2001); (2D) contact maps (Fariselli et al., 2001); and (1D) structural profiles to which prediction of residue secondary structure is referred (Rost and Sander, 1994a; Baldi et al., 1999; Jones, 1999), solvent accessibility (Rost and Sander, 1994b), and contact number (Rodionov et al., 1981; Fariselli and Casadio, 2000). The task of contact number prediction is closely related to prediction of solvent accessibility, as the later is related to contact number (Rodionov et al., 1981; Rost and Sander, 1994b). However, Fariselli and Casadio (2000) have demonstrated that characteristics are different, and therefore, have their own specific features. The main approaches to estimation of the residue contact numbers in proteins currently use neural network algorithms (Fariselli and Casadio, 2000, Pollastri et al., 2001). This means the prediction of the contact number state depending on whether the contact number is greater or smaller than the mean value. However, there are tasks when little information can be derived from the model of the two state predictions. For example, the residue contact value can be used as an additional parameter for an efficient search for the native contact map (Fariselli et al., 2001, Kabakcioglu et al., 2002). Thus, it is obviously high time to develop the methods that would estimate a particular value of the contact number for the residue. It should be noted that, in the case of prediction of the solvent accessibility, a method for the calculation of a particular accessibility value was proposed (Ahmad et al., 2001). The current study suggests a method for the estimation of particular contact number for residues at amino acid positions. The method is similar to that of Jones (1999) and uses position-specific scoring matrices for homologous sequences (PSSM) calculated by the PSI-BLAST program (Altschul et al., 1997). Another problem dealt with is prediction of the number of the local contact residues (nearby, local contacts with the residues neighboring in primary structure). The prediction accuracy for the total and the local contact numbers are estimated.

2. METHODS AND ALGORITHMS

2.1 Contact number definition

It is proceeded on the assumption that two residues of a protein are in contact, if the distance between their CA atoms does not trespass the r_c threshold.

The total contact number $cn_t(i)$ of the ith residue is expressed as the number of residues in a protein that are in contact with it. In addition, for each residue i, the number of local (nearby) contacts is also estimated; $cn_l(i)$ is the number of residues remote in the primary structure from the ith residue by not more than w_l positions that have established contacts with it. The w_l is characterized by the threshold that separates the interactions that are local along the polypeptide chain from the distant interactions. Clearly, estimation of contact numbers of residues depends on the r_c and w_l parameters. Here, the value of the parameter $w_l = 7$ is fixed (that is about two alpha-helical turns). The r_c parameter assumes the values 6, 8, 10, and 12 Å, and the prediction algorithm was tested separately for each value.

2.2 The neural network structure

Prediction on the basis of the position-specific scoring matrix weights (PSSM) and two fed-forward neural networks were used to recognize the contact numbers (Figure 1).

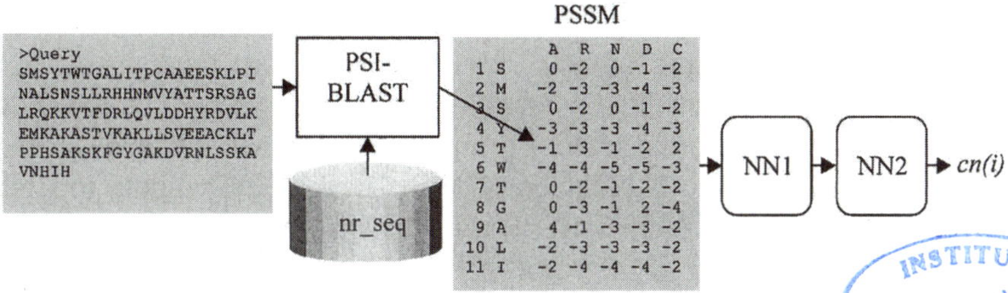

Figure -1. A scheme showing how the algorithm for contact number prediction works. The query sequence is at the input of the PSI-BLAST program for search for the homologous sequences and their multiple alignment. PSI-BLAST builds the PSSM matrix that is input into the first level neural network. The NN1 predictions are input into the second level neural network. NN2 yields the contact number estimate $cn(i)$.

The first neural network (NN1) predicts the residue contact number on the basis of the PSSM weights. The structure of the NN1 is as follows. The PSSM values are the input parameters. The PSSM is the matrix of the $L \times 20$ size, where L is the length of the query protein sequence. The PSSM matrix is built on the basis of the PSI-BLAST multiple alignment program (Altschul et al., 1997) and expresses the similarity of the amino acids of 20 different types with columns of multiple sequence alignment. The PSI-BLAST program with three iterations is utilized for the search for and alignment of homologous sequences in the *nr* databank of non-redundant protein sequences. Additionally to the PSSM values, we applied information per position, which is also output by the PSI-BLAST program.

A sliding window of the $2h + 1$ size is considered for each position. The ith residue is in the centre of the window, while the window borders include the residues numbered $i - h$ and $i + h$. The contact number is estimated for the ith residue. In the current study, h was set to five, because testing demonstrated that at $h > 5$, the prediction was not all improved.

For each position j from the sliding window, the vector numbers were assigned. These included information per jth position, 20 PSSM weights for each amino acid type at that position. The PSSM weights were set –10 for window positions outside the protein sequences in the case of i being N or C terminal residues. The complete vector of the input data for the NN1 thus composed of $(2h + 1) \times 1 = 11 \times 21 = 231$ parameters. Two internal layers with 50 and 3 neurons were implied for the NN1. A single number was at the NN1 output, the predicted value of the contact number of the ith residue.

The neural network at the second level (NN2) was built as follows. The contact numbers predicted by NN1 served as its input data; 41 values were estimated for each position, namely, the predicted contact numbers at the ith and ith ± 20 positions. The contact numbers were set –1 for window positions outside the protein sequences in the case of i being N or C terminal residues. In this way, 42 parameters were at the NN2 input; the NN2 contained one internal layer of 10 neurons. There was also one value at the NN2 output rounded up to the nearest integer value, the contact number of the ith residue that expressed the ultimate estimate of the contact number.

2.3 Training and testing of the prediction algorithm

Training and testing of the algorithm was implemented on a sample of the 234 monomeric protein chains with known spatial structures extracted from the PDB database (Berman et al., 2000) and belonging to different protein fold types according to the SCOP classification (Andreeva et al., 2004). The resolution of proteins was higher than 2 Å, and they did not contain missed CA atomic coordinates. The samples were training, control and testing, with 78 proteins in each and with approximately the same distribution along the protein length. The neural network was trained on the training sample by the backpropagation algorithm (Rumelhart et al., 1986), with 50 epochs, momentum of 0.9, and learning rate of 0.01. The training was stopped when the control sample ceased to show prediction improvement. The prediction accuracy was estimated on the testing data.

2.4 Prediction accuracy of the contact numbers

To evaluate the prediction quality of the contact number, the following parameter set was used:

- MAE is the mean absolute error given by $\text{MAE} = (1/N)\cdot\Sigma|cn_O(i) - cn(i)|$, where N is the test sample size, $cn_O(i)$ is the observed contact number for the residue *I*, and $cn(i)$ is the predicted contact number for the residue i;
- MAE_{norm} is the MAE ratio to the standard deviation s of the contact numbers in the training sample;

- Q_s is the fraction of residues in the sample for which $|cn_0(i) - cn(i)| < s$; and
- *corr* is the correlation coefficient between the predicted and observed contact numbers.

The smaller were the values for the MAE and MAE_{norm} and the greater were the Q_s, *corr* parameters, the higher was the prediction accuracy. The MAE_{norm}, Q_s, and *corr* parameters were introduced to assess the relative improvement in the qualitative prediction accuracy of the method at different values of the contact distance and the contact number types (local/total). This is because the means and variations of their distributions are different.

For comparison purposes, we also introduced the Q_{state} measure, which estimates the prediction accuracy in terms of contact number state. The state is defined depending on whether the contact number is greater or smaller than the mean value and was used as the predicted value in recent work (Pollastri et al, 2002). Q_{state} is the fraction of the states predicted correctly.

2.5 Basic algorithm

The prediction quality of neural network algorithm was compared with that of the basic algorithm (BAS) that disregards the local residue environment, being rather based on the most frequently occurring numbers of contacts for specific amino acid types (Richardson and Barlow, 1999). To illustrate, at r_c = 10 Å, $cn_t(i)$ = 22 is assigned to all alanine residues and $cn_t(i)$ = 16 for arginine and so on, whatever positions they occupy. A similar procedure was applied in the case of local contacts.

3. RESULTS AND DISCUSSION

To provide an example of how the algorithm works, we calculated the residue contact numbers of the NS5B protein of the hepatitis C virus and compared with the observed. Figure 2 shows the comparative results. As seen in Figure 2, the predicted profile tends to follow the observed contact number trend (*corr* = 0.501). However, the algorithm often gives inaccurate predictions for the extremely low and high values.

The accuracy for the contact number predictions is validated in Table 1. It follows from the data in Table 1 that (1) predictions using neural networks are consistently more accurate than those using the basic algorithm. The improvement was observed for all contact distance values, local and total contact numbers, and all qualitative evaluation. (2) In general, the prediction accuracy for the local contact number proved to be higher than for the total contact number.

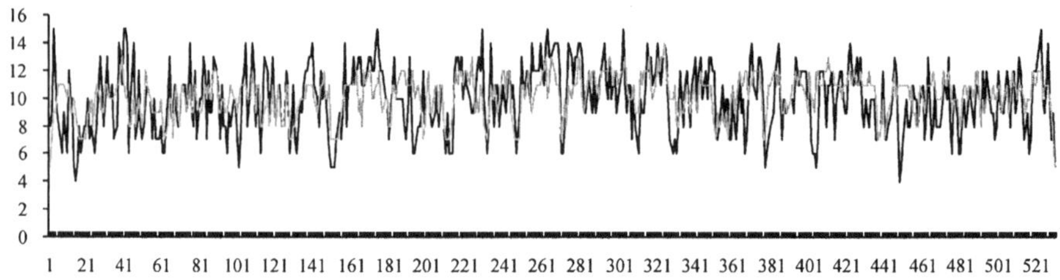

Figure -2. The observed (dotted line) and predicted (solid line) total contact number profiles for hepatitis C virus NS5B protein (PDB ID 1GX5_A). The X-axis is the residue index; the Y-axis is the contact number value. Contacts were defined at $r_c = 8$ Å.

Table -1. Comparison of the contact number prediction accuracy (columns 4–8) for different contact distances (r_c, Å), contact types (cn), and prediction algorithms (Pred)

r_c	cn	Pred	MAE	MAE_{norm}	Q_s	*corr*	Q_{state}
6	cn_l	nn1	0.966	0.554	0.769	0.663	0.775
		nn2	0.947	0.543	0.777	0.678	0.783
		BAS	1.784	1.022	0.492	0.113	0.576
	cn_t	nn1	1.088	0.659	0.723	0.511	0.718
		nn2	1.086	0.658	0.722	0.515	0.717
		BAS	1.325	0.803	0.622	0.116	0.542
8	cn_l	nn1	1.072	0.595	0.727	0.619	0.754
		nn2	1.034	0.574	0.742	0.635	0.756
		BAS	1.884	1.045	0.447	0.079	0.555
	cn_t	nn1	1.782	0.621	0.736	0.606	0.737
		nn2	1.782	0.621	0.736	0.607	0.735
		BAS	2.416	0.842	0.605	0.241	0.616
10	cn_l	nn1	1.409	0.593	0.837	0.650	0.736
		nn2	1.362	0.573	0.850	0.672	0.748
		BAS	2.167	0.911	0.639	0.088	0.544
	cn_t	nn1	3.394	0.600	0.802	0.654	0.751
		nn2	3.376	0.597	0.805	0.658	0.751
		BAS	4.582	0.810	0.677	0.286	0.641
12	cn_l	nn1	1.423	0.553	0.826	0.698	0.768
		nn2	1.345	0.523	0.853	0.730	0.788
		BAS	2.719	1.057	0.509	0.105	0.554
	cn_t	nn1	5.623	0.562	0.858	0.699	0.765
		nn2	5.578	0.558	0.862	0.705	0.764
		BAS	8.342	0.834	0.699	0.222	0.616

This was because the information about the primary protein structure within the 2h + 1 sliding window expresses more accurately the local residue interaction along the chain than interactions remote in it. The improvement is for the neural networks at both NN1 and NN2 levels.

(3) The second level neural network NN2 enables the improvement of prediction accuracy. NN2 was introduced to additionally take into consideration the interdependence of the contact number of residues neighboring along the protein chain. For example, for the surface exposed residues and, therefore, of small values of contact numbers, it may be expected that the residues nearest along the polypeptide chain would be near the protein surface and would also have small contact number value. The NN2 introduces, although slight, yet regular improvement in prediction accuracy. Its introduction is validated.

A comparison of the prediction performance for contact number by the Q_{state} with the reported elsewhere would be of interest. Relevant method for the prediction of total contact number is based on bidirectional recurrent neural network (BRNN) algorithm (Pollastri et al., 2002). The method uses a combination of six BRNN networks differing by neuron numbers. At different contact distance values, the accuracy was 0.7324 (r_c = 6 Å), 0.7095 (r_c = 8 Å), 0.7213 (r_c = 10 Å), and 0.7409 (r_c = 12 Å; Pollastri et al., 2002, Table VIII). Using the algorithm we developed, the observed values for the contact numbers were predicted and relying on it, we predicted the contact number state (1, if it was greater than the mean, 0 otherwise). Testing yielded the following: 0.717 (r_c = 6 Å), 0.735 (r_c = 8 Å), 0.751 (r_c = 10 Å), and 0.764 (r_c = 12 Å; Table 1), which outperforms Pollastri et al. (2002) estimates by 2–3 % at $r_c > 6$ Å. The difference in prediction may be partly explained by the increase in sequence amount in the databank, a recent trend. The difference may be also explained by the different sets of queried protein.

4. CONCLUSION

Here, we propose an approach to the prediction of amino acid contact numbers on the basis of protein sequence. The approach is novel in that it enables the estimation of real contact number. Prediction of the number of local residue contacts is another aspect of the work. It was demonstrated that prediction of the local contact number could be achieved with improved accuracy than the prediction of total contact numbers. It is envisaged to apply the developed algorithm to tackle the problems related to the building of protein contact maps.

ACKNOWLEDGMENTS

The author is grateful to I.V Lokhova and A.V. Morozov for technical assistance and to A.N. Fadeeva for English translation. The work is supported by the US Civilian Research and Development Foundation for the Independent States of the Former Soviet Union (CRDF) within the Basic Research and Higher Education Program (grant No. Y1-B-08-20); program of the Ministry of Education of the Russian Federation 'The Development of the Higher School Scientific Potential' (grant No. 3H-352-05); Russian Foundation for Basic Research (grants Nos. 05-04-49141-a and 05-04-49283-a); and the governmental contract with the Federal Agency for Science and Technology 'Identification of potential targets for novel medicinal drugs based on reconstructed gene networks', priority direction 'Living system' (No. 10002-251/P-25/155-270/200404-082).

FASTPROT: A COMPUTATIONAL WORKBENCH FOR RECOGNITION OF THE STRUCTURAL AND FUNCTIONAL DETERMINANTS IN PROTEIN TERTIARY STRUCTURES

V.A. Ivanisenko[1,2*], S.S. Pintus[1,2], P.S. Demenkov[2], M.A. Krestyanova[2], E.K. Litvenko[2], D.A. Grigorovich[1], V.A. Debelov[3]

[1] *Institute of Cytology and Genetics, Siberian Branch of the Russian Academy of Sciences, prosp. Lavrentieva 10, Novosibirsk, 630090, Russia, e-mail: salix@bionet.nsc.ru;* [2] *Novosibirsk State University, ul. Pirogova 2, Novosibirsk, 630090, Russia;* [3] *Institute of Calculus Mathematics and Mathematical Geophysics, Siberian Branch of the Russian Academy of Sciences, prosp. Lavrentieva 6, Novosibirsk, 630090, Russia*

[*] *Corresponding author*

Abstract: The recognition of the structural/functional determinants in proteins has broad implications for structural and functional genomics. A better understanding of these determinants, such as protein–protein, protein–DNA, and protein–RNA interaction sites, would provide insight into protein functions. A computational workbench for the recognition of functional sites in protein tertiary structure combined with molecular complex reconstruction, estimation of mutation effect on protein thermodynamic stability, as well as search for the structure–activity relationships in homologous protein set was developed. Here, we illustrate the capabilities of the workbench by providing examples of (1) search for interactions of the hepatitis C virus proteins with the human proteins, (2) analysis of the potential toxicity of molecular compounds, (3) analysis of structure–activity relationships in the disintegrin family, and (4) structural protein classification based on the structural similarity to the functional sites.

Key words: protein functional sites; site recognition; protein tertiary structure; molecular complex reconstruction; transcription factors classification; hepatitis C virus

1. INTRODUCTION

Experimental data on protein tertiary structure are growing at a rapid pace (Westbrook et al., 2003). The body of literature is extensive. With the advent of methodologies for the recognition of functional sites in primary structure (Bairoch and Bucher, 1994), tools for site recognition in tertiary structure based on structural data on it alone (Ondrechen et al., 2001; Gutteridge et al., 2003) as well as on structural similarity to related proteins of known function (Wallace et al., 1997; Jones et al., 2003) were developed. There is now a repertoire of tools for the search of functional sites using databases containing structural data on protein–ligand interactions (Hendlich, 2003).

Research increasingly focuses on proteomics in efforts to clarify how ligand–protein binding sites may be recognized and to generate their complexes. Thus, the concept of molecular docking became popular (for an overview, see Schneidman-Duhovny et al., 2004).

When doing studies in functional genomics, there looms the problem of how the molecular structure of a protein relates to its biological effects (Hughes et al., 2004). To predict functionally important residues, an approach providing the identification of conserved residues in protein groups and their related functional specificity and/or evolutionary trace is used (Livingstone et al., 1993; Mihalek et al., 2004; Kalinina et al., 2004). Quantitative structure/sequence–activity relationships analysis is also applied. For this purpose, statistical models (multiple linear regression analysis, neural networks, and projections to latent structures) are advantageous because they relate protein activity to variables that describe protein site properties, e.g., alpha-helicity, hydrophobicity/hydrophilicity, and charge, among others (Ivanisenko and Eroshkin, 1997; Sandberg et al., 1998).

We have developed the PDBSite database (http://wwwmgs.bionet.nsc.ru/mgs/gnw/pdbsite/) for the spatial structures of the protein functional sites, including the posttranslational modification and binding sites, the active enzyme centers (Ivanisenko et al., 2005a). The created PDBSiteScan program (http://wwwmgs.bionet.nsc.ru/mgs/systems/fastprot/pdbsitescan.html) provides search on the PDBSite database using pairwise protein–site structure alignment (Ivanisenko et al., 2004). Good recognition accuracy of the functional sites by screening of protein tertiary structure on the PDBSite database has been illustrated by the active enzyme centers (Ivanisenko et al., 2004).

Here, we extend and improve PDBSite by developing a PDBLigand database and a molecular complex reconstruction module to further combine them. The molecular complex reconstruction module can help to solve an important aspect of the molecular docking problem, the initial disposition of interacting molecules with respect to each other in space.

We have developed the WebProAnalyst program (http://wwwmgs.bionet. nsc.ru/mgs/programs/panalyst/) for scanning quantitative structure–activity relationships in protein families (Ivanisenko et al., 2005b). The tool allows users to search for correlations between protein activity and physicochemical characteristics (i.e., hydrophobicity or alpha-helical amphipathicity) in queried sequences.

One important requirement for protein design is its capability to predict changes in protein stability upon mutation. A problem designed to estimate the influence of mutations in proteins on their thermodynamic stability was developed. The program can be useful in search for functionally and structurally important residues in proteins and for design of protein engineering experiments. The current version of the program is not yet web-accessible.

An approach to automated structural and functional classification of proteins on the basis of their structural similarity to the functional sites from the PDBSite database is proposed. The resulting classification of the representatives of the main transcription factor families agreed well with the standard manual classification (Heinemeyer et al., 1999).

Examples are provided to illustrate the benefits of combining the programs we developed into a common workbench, namely evidence for (1) the possible role of RNA-directed RNA polymerase (NS5B) hepatitis C virus (HCV) in the regulation of host immunity; (2) the potential toxicity of molecular compounds; and (3) the structure–activity relationships in the disintegrin family.

2. METHODS AND ALGORITHMS

The workbench consists of the PDBSite and PDBLigand databases, the PDBSiteScan and WebProAnalyst programs, and the molecular complex reconstruction and protein stability prediction modules. The PDBSite database stores the data for the functional protein sites; the PDBLigand database those for the ligands of the sites.

A brief description of the PDBSite structure and the PDBSiteScan program follows (for details, see Ivanisenko et al., 2005; Ivanisenko et al., 2004). PDBSite contains more than 14 000 sites, including catalytically active centers of various enzymes; the sites of posttranslational protein modification; the sites of ion metal binding; the sites of binding organic/inorganic compounds; the sites of drug binding; and the sites of protein–protein, protein–DNA, and protein–RNA interactions. The data extracted from the PDB databank (Berman et al., 2000) on the basis of information in the SITE field of PDB indicating the amino acid residues of the functional sites; the sites of protein–protein, protein–DNA, and protein–RNA interactions were identified by analysis of the atomic coordinates in their heterocomplexes. The sites included the amino acid residues that are in contact with the ligand (protein, RNA, or DNA). A residue was accepted as contact if it had at least three atoms whose distance from any atom of the partner chain was smaller than 5 Å.

The PDBLigand database contains data on the low molecular weight ligands, proteins, DNA, and RNA that bind to the sites from PDBSite. The PDBLigand database includes the atomic coordinates of the ligands as well as their functional description extracted from the PDB databank. Every entry of the PDBLigand database contains information on a particular ligand links to an entry of the PDBSite database providing information on the binding site of the ligand.

The PDBSite database is integrated with the PDBSiteScan program for recognizing the functional sites in protein tertiary structures. PDBSiteScan provides automated search for the spatial fragments in protein tertiary structure similar in structure to the functional sites from the PDBSite database.

The molecular complex reconstruction module works as follows. The PDBSite database contains the site-templates with known atomic coordinates of their complexes with the ligands from the PDBLigand database. Draft docking is done by transfer of the ligand together with the site-template during the structural alignment of the site-template to protein. The generated draft protein–ligand complex can be accepted as a start approximation for the further docking or molecular dynamics analysis.

WebProAnalyst has been elaborated for analysis of quantitative data on protein activities (Ivanisenko et al., 2005). The program provides automated generation and verification of the hypotheses on quantitative relationships between the physicochemical characteristics in the regions of protein sequence alignments and their activities. The WebProAnalyst is multipurpose: it queries for a region whose substitutions are correlated with variations in the activities of a set of homologous proteins, the so-called activity modulating sites; it searches for the key physicochemical characteristics that affect the changes in the activities; and it enables the building of multiple linear regression models that relate these characteristics to protein activities. WebProAnalyst implements methods of multiple linear regression analysis, the sequence–activity correlation coefficient, and neural networks.

The PSP program (Protein Stability Prediction) was designed to predict whether given mutation increases or decreases the protein thermodynamic stability $\Delta\Delta G$ with respect to the native structure. Recently, a significant database ProTherm of thermodynamic data on protein stability changes upon single point mutation was generated (Bava et al., 2004). This allows developing approaches to predict free energy stability changes upon mutation starting from the protein sequence. Our task is to predict whether a given mutation increases or decreases the protein stability, without predicting the exact ΔΔG value. In this respect, the task can be a classification problem for the protein upon mutation.

Our method is of the nearest neighbor type. In the course of learning, the method divides the provided examples into subsets. The shortest open path uniting all the examples is outlined. The longest edges are removed from the obtained graph so that vertices of the same type are seen in the obtained subgraph. New elements are assigned to the subset in which the distance of the elements is shortest from the new element.

Input data are the occurrence frequencies of amino acids that are at a distance of not more than 10Å from the mutated residue in protein tertiary structure, relative solvent accessibility, temperature, and pH. Using a dataset consisting of 2001 mutations, our predictor correctly classifies > 73 % of the mutations in the database.

3. RESULTS AND DISCUSSION

3.1 Search for the potential interactions between HCV and human proteins

The RNA dependent RNA polymerase NS5B is a 65-kDa protein that resembles other viral RNA polymerases (Lohmann et al., 1997). HCV replication is thought to occur in membrane-bound replication complexes. The complexes transcribe the positive strand, and the resulting minus strand is used as a template for the synthesis of genomic RNA. Search on the PDBSite database using PDBSiteScan demonstrated that NS5B contained fragments structurally similar to the binding site to human nuclear transport factor 2 (NTF2) and human nuclear factor of activated T cells (NFAT).

NTF2, a homodimer of approximately 14-kDa subunits, stimulates efficient nuclear import of a cargo protein (Stewart, 2000). NFAT transcription factor family is involved in the expression of the cytokines IL-2, IL-3, IL-4, IL-5, granulocyte-macrophage colony-stimulating factor, and tumor necrosis factor-alpha, as well as of several cell-surface molecules, such as CD40L and FasL. NFAT proteins are also expressed in B cells, mast cells, basophils, and natural killer cells, as well as in a variety of non-immune cell types and tissues, such as skeletal muscle, neurons, heart, and adipocytes (Porter et al., 2000).

The potential complexes of NS5B with NTF2 and NFAT generated using the program designed for molecule reconstruction are shown in Figure 1. It is seen that two loops are involved in the interaction of NS5B with NTF2 (Figure 1*a*) and, hence, the contact might be close. Further calculations in terms of molecular dynamics, for example, are required to estimate the contact affinity. The second complex results from contacts between only four residues at each site (Figure 1*b*). However, NS5B can contact with the DNA bound to NFAT. It is suggested that the double contact of NS5B with NFAT and DNA can establish a stable complex. Molecular modeling is required to prove this.

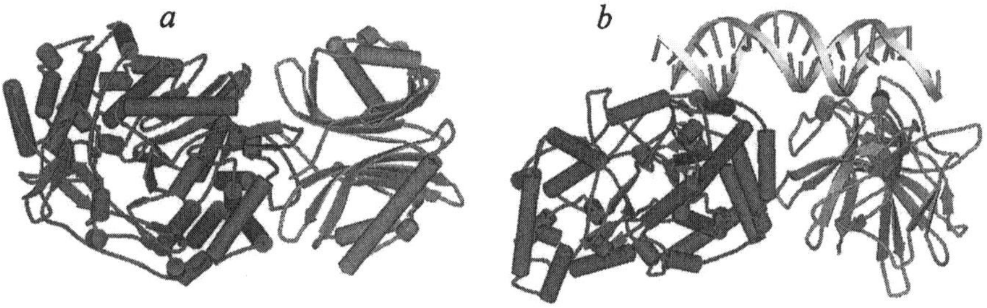

Figure -1. The potential complexes of NS5B with (*a*) NTF2 and (*b*) NFAT. The NS5B structure is dark grey, the NTF2 and NFAT structures are light grey. In the NS5B–NFAT complex, a fragment of double-stranded DNA, which interacts with NFAT and presumably with NS5B, is depicted. The atomic coordinates of NS5B, NTF2, and NFAT in the complex with DNA were extracted from PDB 1QUV, 1A2K, and 1A02, respectively.

3.2 Potential toxicity of molecular compounds

Small molecule compounds underlie the derivation of drug candidates. There is no knowledge related to mechanism of action for a large number of small molecule compounds with medicinal properties identified in the pathway-based assays or phenotypic screens. Many small molecule drugs demonstrate unexpected toxicity and side effects in humans. Understanding of proteins that bind to these drug molecules will improve our understanding of the molecular basis of efficacy or toxicity.

Leptin is a protein of great interest in medical research. We analyzed the potential capacity of biologically active molecular compounds from the PDB database to bind to leptin. Leptin is an adipocyte-derived hormone that circulates in the serum in a free and bound forms. Serum levels of leptin reflect the amount of energy stored in adipose tissue. Short-term energy disbalance as well as serum levels of several cytokines and hormones influence circulating leptin levels. Leptin acts by binding to specific receptors in the hypothalamus to alter the expression of several neuropeptides that regulate neuroendocrine function and energy intake and expenditure. Thus, leptin plays an important role in the pathogenesis of obesity and eating disorders and is thought to mediate the neuroendocrine response to food deprivation.

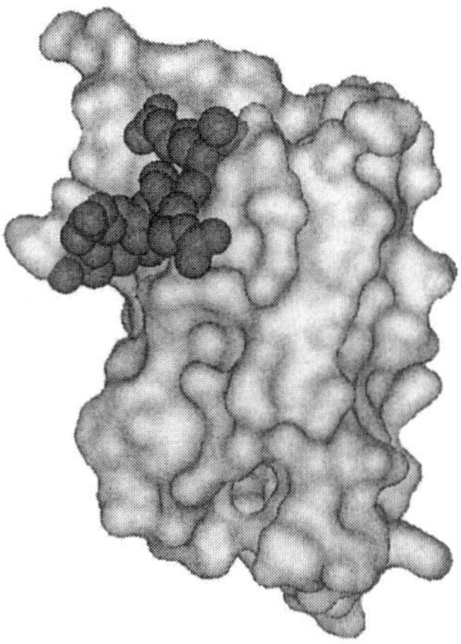

Figure -2. Potential leptin–ACE-ARG-ARG-LEU-ASN-FCL-NH peptide complex. Leptin is shown as surface molecule; the peptide is depicted using ball model.

It was found that the ACE-ARG-ARG-LEU-ASN-FCL-NH peptide, developed for the inhibition of the cyclin-dependent kinase 2/cyclin complex (Kontopidis et al., 2003) is also capable of binding to leptin (Figure 2).

Peelman et al. (2004) has demonstrated that mutation at positions 41, 115–118, 122, and 124 of leptin affect its binding to the membrane proximal cytokine receptor homology domain (CRH2). CRH2 is the domain of the leptin receptor. The binding site of ACE-ARG-ARG-LEU-ASN-FCL-NH to leptin covers these positions. It can be thus suggested that the peptide can inhibit binding of leptin to receptor. It appears that medicinal drugs with antitumor action can be derived from the peptide (Peelman et al., 2004). The drugs may have side effects because of the binding. The results are tentative and they need checking by molecular dynamics as well as experimental verification.

3.3 Structure–activity relationships of disintegrins

Disintegrins are the proteins of snake venom that inhibit the interaction of fibrinogens with blood platelet receptors (Dennis et al., 1990).

We analyzed the structure–activity relationships in these proteins. The goals were: (1) search for the region in protein sequences of the disintegrin family whose amino acid substitution are related to changes in the activity of these proteins; (2) establishment of the activity–physicochemical property relationships of amino acids of the given region; and (3) analysis of the influence of substitution in the given region on kistrin stability and prediction of substitutions increasing the activity of kistrin.

```
                                10        20        30        40
Kistrin                QCGEGLCCEQCKFSRAGKICRIPRGDMPDDRCTGQSADCPRYH
Flavoridin             QCADGLCCDQCRFKKKTGICRIARGDFPDDRCTGLSNDCPRWNDL
Applagin               QCAEGLCCDQCLFMKEGTVC-RARGDDVNDYCNGISAGCPRNPFH
Eristicophin           PCATGPCCRRCKFKRAGKVCRVARGDWNNDYCTGKSCDCPRNPWNG
Echistatin alpha       ECESGPCCRNCLFLKEGTICLRARGDDMDDYCNGLTCPCPRNPHLGPAT
Tergemenin             QCADGLCCDQCRFNKKGTVCRMARGDWNDDTCTGQSADCPRNGLYG
Triflavin              QCADGLCCDQCRFKKKRTICRIARGDFPDDRCTGQSADCPRWNGL
Bitan alpha            QCNHGECCDQCRFKKAGTVCRIARGDWNDDYCTGKSSDCPWNH
Batroxastatin          QCAEGLCCDQCRFKGAGKICRRARGDNPDDRCTGQSADCPRNRF
Elegantin              QCADGLCCDQCRFKKKRTICRRARGDNPDDRCTGQSADCPRNGLYS
Barbourin              QCADGLCCDQCRFNKKGTVCRMAKGDWNDDTCTGQSADCPRNGLYG
Trigramin beta2        QCGEGPCCDQCSFMKKGTICRRARGDDLDDYCNGRSAGCPRNPFHA
Albolabrin             QCGEGLCCDQCSFMKKGTICRRARGDDLDDYCNGISAGCPRNPLHA
Bitistatin             QCNHGECCDQCKFKKARTVCRIARGDWNDDYCTGKSSDCPWNH
Trigramin gamma        QCGEGLCCDQCSFMKKGTICRRARGDDLDDYCNGISAGCPRNPLHA
Trigramin alpha        QCGEGLCCDQCSFIEEGTVCRIARGDDLDDYCNGRSAGCPRNPFH
Trigramin beta 1       QCGEGPCCDQCSFMKKGTICRRARGDDLDDYCNGRSAGCPRNPFH
Halysin                QCAEGLCCDQCRFMKKGTVCRIARGDDMDDYCNGISAGCPRNPF
Echistatin alpha 2     QCESGPCCRNCLFLKEGTICLRARGDDMDDYCNGLTCPCPRNPHLGP
```

Figure -3. Multiple alignment of disintegrins. The conserved residues are grey. The boxes over the alignment indicate the positions at which substitutions mainly cause a decrease in kistrin stability. The boxes under the alignment indicate positions which substitutions either increase or do not affect kistrin stability.

Figure 3 shows multiple alignment of disintegrin sequences. Using the WebProAnalyst program, we established a correlation between the protein activity and two physicochemical properties in the region spanning positions 97–30 in the multiple alignment (Figure 4). These properties were charge and hydrophobicity moment. The program calculated the multiple regression equation for the established relation $Y = -2.9X_1 - 2.12X_2 + 1.4$, where Y is activity, X_1 is charge, and X_2 is alpha-helix periodicity of Eisenberg hydrophobicity. The revealed region is near the protein active centre, the RGD site. The functional importance of the revealed region has been previously demonstrated. Thus, point substitutions of amino acids in the region by alanine caused a considerable decrease in kistrin activity (Chang et al., 2001).

WebProAnalyst allows the prediction of mutant protein activities on the basis of the established structure–activity relationships. Table 1 provides examples of the predicted effect of substitutions at position 27 of the multiple alignments on the activity and thermodynamic stability of kistrin. Kistrin stability changes were calculated using the PSP program.

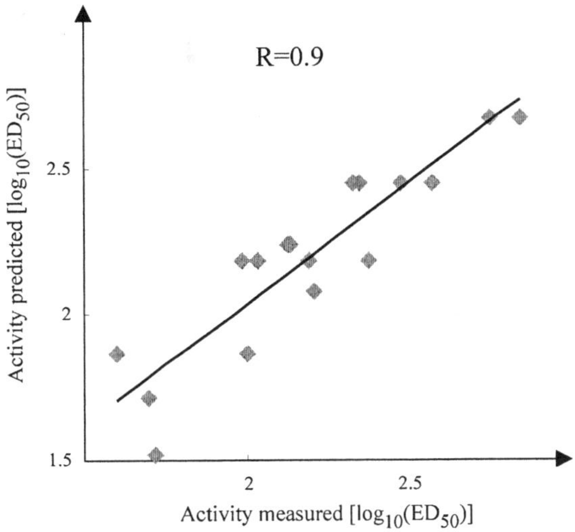

Figure -4. Correlation between the measured disintegrin activity and the activity predicted using the regression model. The correlation coefficient is 0.9 and significant at confidence level > 0.99.

Table -1. Predicted substitution effect at position 27 of the multiple alignments on kistrin activity and stability

Residue[&] (position 27)	Activity $[\log_{10}(ED_{50})]$	Thermodynamic stability[#]
*Met	2.07	–
Cys	2.18	0
Val	1.9	+
Ile	1.78	+
Asp	2.94	–

&: The positions are numbered for multiple alignments (Figure 4); * indicates wild type; #: (0) no change in thermodynamic stability, (+) stability increase, and (–) stability decrease.

As the data in Table 1 show, certain substitutions that produce an increase in kistrin activity decrease its stability. Designs of protein engineering experiments have to take into account the effects of substitution on both protein activity and stability.

3.4 Classification of transcription factors

Overall, 17 families of transcription factors were chosen for classification. Structural similarity to the functional sites from the PDBSite database was searched for every protein. The total number of functional site types we examined was 88. The maximum distance mismatch (MDM) and amino acid type match were calculated to express the similarity between a protein fragment and a site (see Ivanisenko et al., 2004). A site and a protein fragment were accepted as structurally similar if the MDM value was less then 2 Å. The fragments structurally similar to the sites were further divided into four classes: (1) completely matching the amino acids; (2) one mismatch; (3) two mismatches; and (4) three or more mismatches.

The distance between a pair of protein tertiary structures was calculated from $D_{ij} = \sqrt{\sum_{k=1}^{88}\sum_{m=1}^{4}(x_{km}^{i} - x_{km}^{j})^2}$, where i, j are the indices of protein tertiary structure; x_{km}^{i} is the variable indicating whether or not at least one fragment, structurally similar to a functional site of the kth type and assigned to the m mismatch class, is present in the ith protein structure; the values assigned to x_{km}^{i} were either 1 or 0; 1 was assigned to a particular site type if at least one fragment in the protein structure was found to be similar to at least one site from PDBSite of this type, otherwise 0 was assigned to this site type. The UPGMA method was used for clustering; and the PHYLIP package (Lim and Zhang, 1999), for constructing the hierarchical tree (Figure 5).

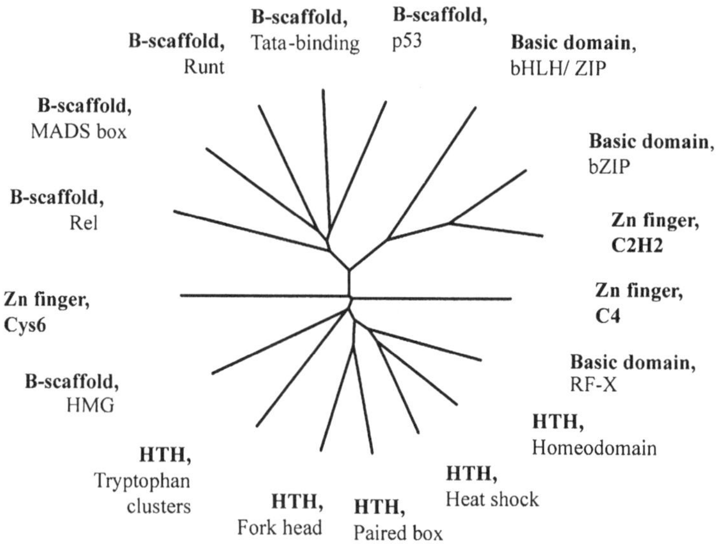

Figure -5. A hierarchical tree for a classification of the representatives of the main classes of transcription factors. The tree was built on the basis of search for the structural homology of the DNA-binding domains of these factors with the functional sites from the PDBSite database. The name of the class, the PDB ID, and a schematic representation of tertiary structure are given for every domain.

4. CONCLUSION

The developed workbench was designed for addressing problems related to the functional annotation and draft docking of proteins with low molecular weight ligands, proteins, RNA, and DNA. The FASTPROT workbench was applied to the analysis of the HCV–human proteins interactions. As a result, we identified the potential binding site NS5B of HCV to the human nuclear transport factor 2 (NTF2) and to the DNA binding domain of the human transcription factor NFAT. The results suggest that the NS5B–NTF2 interaction provides NS5B transport into the cell nucleus, where it interacts with NFAT and DNA and participates in the regulation of gene expression, thereby suppressing antiviral immunity. It should be noted that the assumption requires support by modeling of the NS5B–NTF2 and NS5B–DNA–NFAT draft complexes.

A program for the reconstruction of molecular complexes allows the assessment of the toxicity of molecular compounds. We have predicted the potential contacts of the ACE-ARG-ARG-LEU-ASN-FCL-NH peptide with leptin. Such contacts may be a cause of the side effects of drugs derived from the peptide.

Design of proteins with improved medical and biological properties requires predictions of the effects of mutations on both protein activity and structure. Integrative use of WebProAnalyst and PSP makes the dual task feasible. In this way, FASTPROT can be useful in design of protein engineering experiments.

Although the structural classification of proteins is a powerful clue to proteomics problems, there are no universal algorithms. Structure alignment methods are difficult to implement because of the vagueness of their global similarity measures. The structure alignment methods often measure similarity by the root-mean-square-deviation (RMSD) between the aligned atoms. Røgen and Fain (2003) have indicated that the RMSD of aligned atomic coordinates is a perfect measure of similarity for two shapes that are nearly identical. However, the RMSD is a poor measure in the case when the two shapes compared differ significantly. As a result, automated classification of proteins remains an open issue. We suggested an approach to automated protein classification based on search for structural similarity between protein fragments and functional sites from the PDBSite database. The approach was applied to the classification of the representatives of the main classes of transcription factors. The resulting classification agrees well with the classification obtained by manual analysis.

The proposed workbench has already proven useful in analysis. Further integration with other computational tools would be possible and beneficial.

ACKNOWLEDGMENTS

The work was partially supported by the Russian Foundation for Basic Research (grants Nos. 05-04-49283, 05-07-98012 and 05-04-49141); Development of the Scientific Potential of the Higher School (grant No. N8274); Governmental Contract with the Federal Agency for Science and Technology 'Identification of potential targets for novel medicinal drugs based on reconstructed gene networks', priority direction 'Living system'; The US Civilian Research and Development Foundation for the Independent States of the Former Soviet Union (grants Nos. NO-008-X1 and Rup2-2629-NO-04).

A MARKOV MODEL FOR PROTEIN SEQUENCES

Y. Surya pavan[1], C.K. Mitra[1*], S.M. Bendre[2]
[1] *Department of Biochemistry, University of Hyderabad, Hyderabad 500 046, India, e-mail: ckmsl@uohyd.ernet.in;* [2] *Department of Mathematics and Statistics, University of Hyderabad 500 046, India*
[*] *Corresponding author*

Abstract: The protein primary sequence has information for its folding. Analyzing the interrelationship between the adjacent amino acids and estimating their entropies may be informative. The present work shows that Markov dependencies are clearly evident in the protein primary sequences of various databases studied. The higher-order Markov approximations and their entropy calculations showed that short-range interactions are evident between the neighboring amino acids in the protein primary sequences. Moreover, a strong correlation was observed between the secondary structure elements as expected.

Key words: Markov model; entropy; short-range interactions; secondary structure elements

1. INTRODUCTION

Functionally, proteins are the most diverse of all biological macromolecules. All proteins, whether from the most ancient lines of bacteria or from the most complex highly developed forms of life, are constructed from the same ubiquitous set of 20 amino acids (Lehninger and Nelson, 1993). The three dimensional structure of a protein is uniquely encoded in the primary sequence, and in principle (cf. chaperones), it is possible to predict the most probable three-dimensional structure of a protein based solely on the primary sequence (Anfinsen, 1973). The function of a protein is closely related to its three-dimensional structure, but the prediction of the function from a given structure remains difficult. In reality, this has remained an unsolved problem even today, although several empirical treatments of the problem are available in the literature (Chou and Fasman, 1974a, b) to predict the structure from the primary sequence.

The median length of a protein sequence based on the Swiss-Prot database is ~ 350. However, the distribution of sequence length is quite broad, and sequences smaller than 50–100 residues are quite common. On the larger side, sequences can be as long as 400–500 residues. This again suggests that the final structure is more important than the primary sequence. As an example, there are 20^{350} (~ 10^{455}) possible sequences (assuming a typical length of 350 for a protein), whereas we find less than 10^5 sequences in the living system (e.g., human genome is reported to have only ~ 30 000 genes and most genes code for only one protein). This again suggests that most of the theoretically possible sequences are not biologically meaningful, as they do not meet the essential requirement of a well-defined three-dimensional structure. Therefore, we can conclude that the native sequences are a very special subset (Meeta et al., 1995) of the full set of all possible sequences (most of them will be random without any function).

In this work, we used the principles of information theory, in particular, the concept of entropy, to study the sequences. The entropy of a random variable X with a probability distribution of $p(X)$ will be defined following Shannon method (Cover and Thomas, 1991):

$$H(X) = -E\,(\log p(X)) = -\Sigma\, p(X) \log p(X), \tag{1}$$

which is functionally equivalent to Boltzmann's H-theorem (without the negative sign). Boltzmann showed that in a spontaneous process (collisions), H never increases and can be identified as the entropy of the system, when $p(X)$ is taken as the distribution function of the velocities in an ideal gas.

$$S = -k\,H = k \ln W, \tag{2}$$

where S is the entropy (of a given system) and W is the fraction of all possible arrangements where the given configuration is realized (k is the Boltzmann constant). This statistical mechanical derivation of entropy as a measure of order (or of randomness) is formally equivalent to the classical thermodynamics. A rigorous derivation including quantum statistics must include an additional factor or statistical weight while calculating the probability W. The probability distribution function need not correspond to the equilibrium state of the system, but the equilibrium state can be identified with the most probable probability distribution corresponding to W. This gives us a very powerful but simple technique to measure the degree of order in a given system. The equilibrium state has the most probable distribution and highest entropy. The most ordered system has the lowest probability and the lowest entropy. As a system moves towards the equilibrium, the randomness increases and the entropy also increases. The Shannon definition of entropy only lacks the $-k$ factor required in the classical thermodynamics.

The primary sequence of a protein is directly responsible for the secondary structure. However, the secondary structure is directly responsible for the folded conformation and also for the function of the molecule, and therefore, the primary sequence is only indirectly responsible for the overall three-dimensional structure of the protein. Therefore, it is imperative that analysis may be carried out in a similar fashion, i.e., the primary sequence may be used to predict the secondary structure and only the secondary structure may be used to predict the final folded three-dimensional conformation of the molecule. It is difficult, if not impossible, to predict the tertiary structure starting directly with the primary sequence. On the other hand, various attempts, mostly empirical or semi-empirical, to predict the secondary structure from the primary sequence has been quite successful.

James and Dewey (1997) have demonstrated interesting relations by introducing generalized thermodynamic quantities and relating them to the scaling parameters that are in turn related to fractal dimensions. We have studied the protein sequences looking for the measures of order present in protein sequences that can be similarly generalized. In this paper, we have focused on the Markov model looking for the correlations that can be meaningful in understanding the structure and function of proteins. Similar approach was followed on other non-redundant ASTRAL (http://astral.berkeley.edu) (Brenner et al., 2000) protein databases and on the randomly simulated protein sequences.

2. METHODOLOGY

To obtain the functional significance of pattern of amino acids, we calculated Markov approximations and used them to estimate the entropies of those approximations. A Markov sequence is one in which each term has a direct dependence on the immediate preceding term. This gives rise to a sequence that is a simple or first-order Markov sequence. We can also have higher-order Markov sequences in which each term depends directly on the two, three, or more number of previous terms. In this way, there is always a strong correlation between successive elements of the sequence. However, this is an example of short-range interactions or correlations, as we do not expect that two terms separated by a significant distance to be correlated. We expect short-range interactions to be present in protein sequences and, therefore, also expect Markov dependence. However, it is well known that long-range interactions play a very important role in the overall folding in a protein sequence, and therefore, both short-range and long-range order are expected. Aggregation of residues into a meaningful way has been reported earlier in a different way by using fractal studies (Bonnie and Dewey, 1995).

The frequencies of occurrence of the various amino acid pairs (20 × 20, i.e., 400 entries), triplets (20 × 20 × 20, i.e., 8000 entries), and quadruplets (20 × 20 × 20 × 20, i.e., 160 000 entries) were obtained from Swiss-Prot Protein Sequence Databank (Release 26, 1994). For reasons of convenience, all sequences that are shorter than 256 residues were excluded. No selective filtering of the database was attempted. We feel that screening of the database may introduce a fresh and additional bias (rather than to remove any bias already present). It is difficult to remove any existing bias in the database when the sources of the bias are not well known or completely characterized. The Swiss-Prot database contains a number of fragments (incomplete sequences; approximately 6000) that should be excluded, as the initial starting point is not known. A number of sequences are also very small and very small peptides do not possess a definite three-dimensional structure. Sequences larger than ~ 100 residues are known to have a well-defined structure useful for binding a substrate. However, we must mention that the choice of 256-residue cut-off is purely arbitrary, but our results are not strongly dependent on this (cut-off value). We have also shown that the initial part of the sequences follow a different distribution (overall composition) of amino acids, and this stabilizes only after ~ 50 residues (Mitra and Sen, 2001). Thereafter, the distribution becomes stationary. If the chosen sequences are only 50 residues long, we will be ignoring the stationary distribution. If the chosen sequences are ~ 100 residues long, we shall be observing both the initial and stationary distributions. If the chosen sequences are very long, we shall be seeing only the stationary distributions (but the number of very long sequences is rather smaller). Very short sequences contribute more towards noise than information.

The bias present in the Swiss-Prot Protein Sequence Databank can be attributed to several factors, but it is certain that the database cannot be considered a random sample. The proteins to be selected for sequencing are not chosen at random. Therefore, a complete protein database derived from the genomic data of a given organism may be a better representative sample. However, no such database is available at present. The ASTRAL SCOP (http://astral.berkeley.edu) database suggests a novel idea in selecting a set of non-redundant and, *possibly*, a random representative protein sequence database. We have chosen a set of protein sequences with less than 40 % and 95 % sequence similarity (Brenner et al., 2000) for our reference. However, removing similar sequences does not assure that the database is unbiased. Therefore, the same computations have been repeated with this non-redundant database also.

To compare our results, we simulated 41 408 random sequences using a Monte Carlo method (same as the number being analyzed in Swiss-Prot), having the same amino acid composition as in the database. These random sequences were analyzed in the same way as real sequences.

2.1 Markov approximations on various protein databases

Case I. The symbols are independent and equiprobable. Since there are 20 common amino acids, the Case I approximation will have 20 identical entries to evaluate. This result is independent of any database.

Case II. The symbols are independent. Frequency counts of the various databases were calculated and evaluated. The total entries are 20, one for each amino acid residue.

First-order Markov dependence: The frequencies of the pairs of amino acid counts were taken in to consideration and followed up for evaluation. The total entries are 400 (20×20). This corresponds to a first-order Markov sequence.

Second-order Markov dependence: The frequencies of triplets of amino acids are counted, and the total entries are 8000 ($20 \times 20 \times 20$). This corresponds to a second-order Markov process.

Third-order Markov dependence: The frequencies of quadruplets of amino acids are counted, and the total entries are 160 000 ($20 \times 20 \times 20 \times 20$). As the number of entries is large, the actual frequencies are small and *for a small database, can give rise to significant errors*. For this same reason, higher-order approximations cannot be reliably performed on the available databases.

Apart from the above calculations, we have done studies to check any preferences among amino acids in the database. This was calculated by leaving one amino acid in between. All the Markov approximations are calculated as above by leaving one amino acid in between. The pattern that we followed for calculating the counts for the first-, second-, and third-order approximations are A_iXA_j, $A_iA_jXA_k$, and $A_iA_jA_kXA_l$ respectively, where A_i, A_j, A_k, and A_l are any common residue (labeled by i, j, k, and l) and X is any intervening amino acid.

2.2 Entropies for the Markov approximations

We followed Shannon's limit H for calculating the entropies of Markov approximation (Cover and Thomas, 1991) Eq. (1).

Case I. The amino acids are equiprobable and independent, i.e., $I = \log_2 20$. This is the maximum achievable entropy (4.322 bits per residue).

Case II. The amino acids are independent and the p_i value is the fraction or proportion of individual amino acids:

$$I = -\sum_{i=1}^{20} p_i \log_2 p_i. \tag{3}$$

We can describe the kth order of Markov process as follows for $k = 1, 2, \ldots,$
$Pr(X_{k+1} = x_{k+1} | X_k = x_k, X_{k-1} = x_{k-1}, \ldots, X_1 = x_1) = Pr(X_{k+1} = x_{k+1} | X_k = x_k)$

$$I = -\sum Pr(X_k = x_k) \sum_k Pr(X_{k+1} = x_{k+1} | X_k = x_k) \times \\ \times \log_2 Pr(X_{k+1} = x_{k+1} | X_k = x_k) \tag{4}$$

for all $x_1, x_2, \ldots, x_k, x_{k+1} \in H$

$Pr(X_{k+1} = x_{k+1} | X_k = x_k)$ can be numerically computed from sequence database using the simple formula

$$Pr(X_{k+1} = x_{k+1} | X_k = x_k) = \frac{\text{frequency of } k-\text{tuple } (X_k = x_k)}{\sum_k \text{frequency of } k-\text{tuple}}.$$

This approach was followed for all the databases considered. Note that we have considered only first-, second-, and third-order Markov processes, as the database does not permit accurate determination of higher-order probabilities.

2.3 Calculation of Markov approximations on secondary structure elements

To study the entropy of secondary structure elements, we have used a database, which is a version of the PDBFINDER (ftp.cmbi.kun.nl/pub/molbio/data/pdbfinder2; (Hooft et al., 1996) with the secondary structure obtained based on DSSP program (Kabsch and Sander, 1983). No selective filtering of the database was attempted.

The secondary structure elements obtained from a DSSP program from a protein sequence was described not in terms of the amino acid sequence but by a sequence like

```
ID:     102L
Sequence:
           10          20          30          40          50
MNIFE MLRID EGLRL KIYKD TEGYY TIGIG HLLTK SPSLN AAAKS ELDKA
CCHHH HHHHH HCCEE EEEEC TTSCE EEETT EEEES SSCTT THHHH HHHHH
           60          70          80          90         100
IGRNT NGVIT KDEAE KLFNQ DVDAA VRGIL RNAKL KPVYD SLDAV RRAAL
HTSCC TTBCC HHHHH HHHHH HHHHH HHHHH HCTTH HHHHH HSCHH HHHHH
          110         120         130         140         150
INMVF QMGET GVAGF TNSLR MLQQK RWDEA AVNLA KSRWY NQTPN RAKRV
HHHHH HHHHH HHHTC HHHHH HHHTT CHHHH HHHHH SSHHH HHSHH HHHHH
          160
ITTFR TGTWD AYK
HHHHH HSSSG GGC
```

The above illustration shows the primary sequence of hydrolase (O-glycosyl) with ID 102L, and the line below to the primary sequence represents their respective secondary structure elements.

For our studies, we have considered polypeptide chains of the lengths greater than or equal to 256. This number is a compromise between the typical sequence length and the need of having longer sequences to study dependencies. After trimming (i.e., removing all sequences smaller than 256), the working database of secondary structure contains 19 752 polypeptide chains. The total number of symbols present in this working database is approximately eight millions.

The DSSP program assigns each residue's secondary structure symbol to one of eight classes: α-helix (H), 3_{10} helix (G), β-strand (E), β-bridge (B), coil (C), turn (T), and bend (S). The symbol I (π-helix) was ignored because of very few in number, ~ 30 in the database. We could able to compute the information content of such sequences using a modified form of the Shannon formula. Most of the calculations were been done on Linux Operating System by using C++ language (gcc).

3. RESULTS AND DISCUSSIONS

After trimming the Swiss-Prot database (i.e., removing fragments and short sequences), we found that there are 41 408 protein sequences and a total of 22 408 660 amino acid residues. This working database was used for our further Markov approximation studies (referred to as the working database in the following sections unless explicitly mentioned otherwise).

3.1 Entropy of protein primary sequences on the databases used

The following are the entropy (bits per symbol) values for various databases studied. Table 1 below gives the values of entropies from Case I approximation until third Markov approximation.

Table -1. Entropy (bits per residue) calculated up to the third order in different databases

Entropy (bits per residue)	SWISS-PROT	NRDB 40	NRDB 95	RANDOM
Case I	4.322	4.322	4.322	4.322
Case II	4.185	4.183	4.185	4.185
First-order entropy	4.177	4.176	4.178	4.185
Second-order entropy	4.167	4.155	4.158	4.185
Third-order entropy	4.133	3.904	3.945	4.18

The above table (Table 1) clearly shows that the entropy of proteins, expressed in bits per residue, for the equiprobable distribution is $\log_2 20 = 4.322$. As expected, this value is same for all cases. However, if we introduce the actual probabilities of the 20 different residues as observed in a given database, we see a significant decrease in the entropy. In Case II, we find the entropy to be 4.185 for the Swiss-Prot database. The values are very similar for other databases as well as the abundances of the 20 different amino acids are very similar. As we increase the complexity of the model, we capture more significant patterns of protein sequences, and the conditional uncertainty of the next residue is reduced. The first-order model gave an estimate of 4.177 bits per symbol (Swiss-Prot), while the second-order model gave an estimate of 4.167 bits per symbol (Swiss-Prot). For the third-order model Eq. (4), there is an entropy decrease of 0.186 bits per symbol comparing with the Case I. Although the decrease is not much, this signifies that short-range orders are present and are important. However, this cannot be carried out beyond third order, as the databases (particularly NRDB databases available at the astral site) are not sufficiently large enough.

Although the entropy changes are relatively small, they are nevertheless significant in our opinion. The entropies reported are in bits per residue and in a reasonably large protein, may well add up to a significant contribution. The higher-order entropies could not be calculated, as the databases are not large enough (we may estimate the population size of the non-redundant sequence database to be $1–2 \times 10^5$ sequences). However, the present NRDB databases are too small for an accurate evaluation of the third-order entropies.

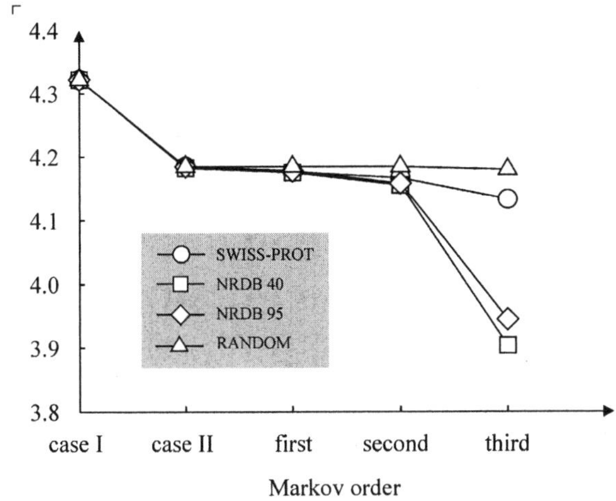

Figure -1. The entropy values (bits per symbol) calculated for Case I to third-order Markov approximations (without gap). The values show that there is a gradual decrease in entropy with more complexity in Markov order approximation.

If we compare the NRDB 40 and NRDB 95 databases, the entropy values obtained seem to be similar with minor differences with respect to Swiss-Prot database up to second-order model (Figure 1). This is somewhat expected as the non-redundant databases are made based on lack of large-scale homology present. Therefore, the preferences that are used for the computations are not altered significantly. Differences were observed in entropy values between Swiss-Prot and non-redundant databases (NRDB 95 and NRDB 40) at the third-order model. This difference might be because of insufficient data in non-redundant databases with respect to Swiss-Prot database. Therefore, the values obtained in the non-redundant databases seem to be unreliable to check the third-order models, mainly because of their much smaller size.

The preference studies for amino acid pairs separated with a single residue (i.e., A_iXA_j, where X is any residue) were computed, and it was found that there are no significant differences between the elements of the transition matrix (as compared to the ungapped pairs). The information content computed from these data (Table 2) is therefore very similar to the first case (Table 1).

The random simulated database was generated using a Monte Carlo technique in which the proportions of the amino acid residues are conserved as to Swiss Prot database composition. The information content of these simulated sequences (Table 1) shows that there is no significant change in the information with increase in the complexity of the model. This behavior is as expected and supports the view that the amino acid residues in the natural sequences show Markov dependency contrary to random simulated sequences.

Table -2. Entropy (per bit) calculated up to the third order in different databases with one gap between the elements of the pairs

Entropy (bits per residue)	SWISS-PROT	NRDB 40	NRDB 95	RANDOM
Case I	4.322	4.322	4.322	4.322
Case II	4.185	4.183	4.185	4.185
First-order entropy	4.177	4.174	4.176	4.185
Second-order entropy	4.167	4.158	4.160	4.185
Third-order entropy	4.136	3.897	3.941	4.18

It is not possible to extend this process to very high order, as the size of the database is a limiting factor. Alternative methods for estimating the entropy of natural protein sequences are therefore required.

3.2 Entropy of secondary structure sequence

The obtained secondary structure database was simplified by DSSP program to a sequence of helixes (H), extended sheets (E), beta sheets (S), and other four symbols (T, C, B and G) for unstructured loops (Table 3).

Table -3. The percent abundance of secondary elements in the database

Symbol	B	C	E	G	H	S	T
Property	β-bridge	Coil	β-strand	3_{10} helix	α-helix	Bend	Turn
Percent abundance	1.375	20.42 2	19.53	4.073	33.48	9.393	11.725

Transition of secondary structure elements. The abundances for various structural elements do not specify how the elements are clustered. A sequence of H is more likely to be found rather than a random distribution. The highest transition probabilities are observed in the transition of H to H (~ 90.8 %), i.e., the persistence of transition of 'H to H' (alpha helix) in the database is very high. Similarly, the occurrence of other highly abundant transition elements are E to E, G to G, T to T, and C to C, which shows that the continuous persistence of these elements is very common in the observed secondary database. In contrast, the less persistent elements are the transition of B to B, i.e., that essentially means that one B is very unlikely to be followed by another B. These characteristics are summarized in the transition matrix that is shown in Table 4.

The maximum entropy for seven states (Case I) is $\log_2 7 = 2.807$ bits. The Case II entropy of the secondary structure is 2.413 bits. However, the first-order (conditioned) entropy is 1.375 bits. This shows that neighboring secondary structure elements are strongly correlated with a difference of 1.038 bits with respect to Case II.

We note the presence of a relatively stronger correlation for the secondary structures. This is not unexpected, as the original amino acid residues have now been categorized into a more meaningful set of secondary structures. However, long-range correlations may still be present and need to be investigated in more details.

Table -4. Transition probability matrix of the secondary structure (in percent) elements

	B	C	E	G	H	S	T
B	2.596	60.229	2.883	1.809	4.746	13.658	14.079
C	4.309	47.782	10.861	3.314	8.407	15.494	9.834
E	0.342	11.789	80.569	0.418	0.543	3.13	3.209
G	0.971	10.012	2.277	70.208	3.243	6.756	6.534
H	0.058	1.778	0.061	0.311	90.855	1.195	5.741
S	2.825	38.386	8.892	1.924	5.497	35.837	6.638
T	1.616	22.895	5.253	1.311	4.672	12.169	52.083

ACKNOWLEDGMENTS

The authors wish to thank Profs. N.A. Kolchanov and Yu.G. Matushkin, Laboratory of Theoretical Genetics, Institute of Cytology and Genetics, Novosibirsk, for various discussions and suggestions. The authors also wish to acknowledge the Department of Science and Technology (DST) for financial support. Y. Surya Pavan thanks the Council of Scientific and Industrial Research (CSIR, India) for a fellowship.

PROTEOME COMPLEXITY MEASURES BASED ON COUNTING OF DOMAIN-TO-PROTEIN LINKS FOR REPLICATIVE AND NON-REPLICATIVE DOMAINS

V.A. Kuznetsov[1*], V.V. Pickalov[2], A.A. Kanapin[3]

[1] *Genome Institute of Singapore, Singapore, e-mail: kuznetsov@gis.a-star.edu.sg;* [2] *Institute of Theoretical and Applied Mechanics, Siberian Branch of the Russian Academy of Sciences, Novosibirsk, 630090, Russia;* [3] *EMBL-EBI Wellcome Trust Genome Campus, Hinxton, CB10 1SD, UK*

[*] *Corresponding author*

Abstract: The entire protein domain set of the proteome of an organism we call the *domainome*. We define the list of domains in domainome, together with the numbers of their occurrences (links to proteins) found in the proteome to be the *domain-to-protein linkage profile* (DPLP). We estimated the DPLP of the proteomes of the 156 complete genomes represented in the InterPro database. This work presents several quantitative measures of the complexity of a proteome based on the DPLP. For each of the 156 studied genomes, we found two large sets of domains: D1, the domains that are not replicated within any protein of the proteome and D2, the domains that occur two or more times in at least one protein of the proteome. Statistics of the observed domain-to-protein links (DPLs) for set D1 and set D2 do not exhibit simple 'scale-free network' properties: for D1, the distribution of DPLs in proteome follows the *Generalized Discrete Pareto* function and for D2, the distribution of DPLs in proteome follows the *inversed gamma* probability function. Dynamical range of DPLs for D1 domains is larger than for D2 domains, and this range correlates with biological complexity of organism. D1 and D2 sets exhibit significant differences of molecular functions of the corresponding proteins, biological processes, and cellular components. The statistical distributions of the number of DPLs in the proteome and the estimates of the differences between the DPLPs for pairs of organisms are used as measures of relative biological complexity of the organisms. In particular, we show quantitatively the greater domain composition complexity of the human proteins relative to that of a mouse or a rat.

Key words: proteome complexity; evolution; domain-to-protein links; skew distributions

1. INTRODUCTION

There is a long history of support of general notion of the overall evolution trends towards increases in size, diversity, and complexity of organism's molecules. However, there are no consensus quantitative measures of the relative or absolute measures of the structural complexity of organism's molecules and its functional diversity. At organism level, the proteome complexity should be characterized by the list of all proteins in a given proteome and their dynamic interactions. This information may be obtained through structural annotation of genome sequences. However, the total number of putative protein sequences, based on complete genome sequence data for a given organism, can only approximately be predicted due to alternative splicing, post-transcription modifications, alternative starts of translation, RNA editing, etc. Our knowledge on dynamical interaction networks of proteins in an organism is also essentially incomplete. Thus, we need more tractable and practical approaches for describing biological complexity.

Proteins contain short structural and/or functional 'blocks', sequences called 'structural motifs' (of length of 5–30 aa) and 'structural domains' (of length of 30–250 aa); those 'signatures' are seen repeatedly in many proteins of species whose genome have been examined (Kuznetsov, 2003b). Proteins complexity can evolve through domains innovation, replication, composition, accretion, shuffling. However, there are a very limited number of domains in nature. Based on non-redundant entries of InterPro signatures, the estimated number of such sequence classes in nature ranges around 10,000 or so (Kuznetsov, 2002; 2003b; Kuznetsov, unpublished, 2004).

In this work, all such classes of sequences will be called, for brevity, *domains*. Domains are essential to biological function(s) of the protein in which they occur and serve as evolutionarily conserved 'building blocks' for the forming of proteins. The domains correspond to specific sequences of DNA within genes that have been evolutionarily conserved. DNA corresponding to a domain may occur multiple times in a given protein-coding gene (repeat-containing protein gene(s)) and/or in many different protein-coding genes (non-repeating multi-domain protein-coding gene) within a given genome and across genomes of many species. It was suggested that the repetitive sequences in proteins associated with recent evolution events than non-repetitive protein sequences (Kajava, 2001). We will attempt to determine the measures of the biological complexity based on analysis of repeating and non-repeating domains in a proteome.

The entire protein domain set of proteome of an organism we call the *domainome*. If a domain D_i occurs in the protein P_j, we say this constitutes a *domain-to-protein link* (DPL). The list of domains together with the numbers of occurrences of each domain (links) found in the proteome of an organism is called the *domain-to-protein linkage profile* (DPLP) of the proteome. The DPLP should allow us to characterize DPL network for each organism. Currently, we do not know all domains and proteins in a given organism. We can study the DPLP of proteome by observing sample DPLP in representative proteome (i.e., in available protein subset that was specified for the organism). However, several protein domain/motif databases provide large enough samples of DPLPs for a variety of organisms. We can analyze the samples of domain-to-protein linkage information in fully sequenced genome organisms to categorize their proteome complexities. This work classifies domains based on different mechanisms of domain–domain interactions within protein and presents several quantitative measures of the complexity of a proteome based on the observed DPLPs that appear to be appropriate and consistent.

2. PROTEIN DOMAIN DATABASE ANALYZER

Our information about DPLPs of the 156 fully-sequenced archaeal, bacterial, and eukaryotic genomes was obtained from the InterPro database (www.ebi.ac.uk/interpro; Mulder et al., 2003); about 10^6 non-redundant protein sequences were obtained from the from the Universal Protein Resource (UniProt) Database (http://www.ebi.ac.uk/). InterPro (Release 7.2) covers 78 % for UniProt protein sequences, in which 11 % of the sequences are not annotated (with GO term: protein unknown function).

We developed a Protein Domain Database Analyzer (PDDA) program, which retrieves data from InterPro database and loads data into local MySQL database (Kuznetsov et al., 2002a). The MySQL database consists of an $N_{tot} \times L$ table, where $N_{tot} = 9609$ domains and $L = 156$ organisms. Each row corresponds to an InterPro domain and each column corresponds to an organism. The (a, P)th entry of the table is the number of occurrences of the domain a in the proteome P. Information of this table was analyzed by PDDA data mining tools, which include statistical and graphical functions, logical functions, descriptive statistics, and correlation analysis.

3. ABSOLUTE AND RELATIVE MEASURES OF PROTEOME COMPLEXITY BASED ON ANALYSIS OF DOMAIN-TO-LINK PROFILES

Let $A = \{a_1, a_2, \ldots, a_N\}$ be the set of observed domains contained in the P proteins of an organism. Let $B = \{b_1, b_2, \ldots, b_P\}$ be the set of proteins in the representative proteome. We define the $N \times P$ adjacency matrix $C' = [c'_{ij}]$ ($I = 1, 2, ..., N$; $j = 1, 2, ..., P$) as follows: the value $c'_{ij} = k$, if protein b_j contains domain a_i exactly k times. Note that $k \in \{0, 1, 2, ...\}$. The value c'_{ij} is the number of DPLs for the (domain, protein) pair $\{a_i, b_j\}$. Let $m'_i = C'_{i*} = \sum_{j=1}^{P} c'_{ij}$ denote the number of occurrences of the domain a_i in all proteins of the representative proteome; $i = 1, 2, ..., N$. We call the value m_i the number of *redundant DPLs* of the domain a_i. Let M' denote the total number of DPLs in the representative proteome of the organism. $M' = \sum_{i=1}^{N} \sum_{j=1}^{P} c'_{ij}$. The value M' is called the redundant connectivity number of the (domain, protein) network. Note that all occurrences of a domain in the proteins are counted.

We also define the adjacency matrix C, where $c_{ij} = 1$, if protein b_j contains domain a_i at least once, and we define $c_{ij} = 0$ otherwise. This matrix reflects the non-redundant structure of the DPLs in the (domain, protein) network. This matrix counts each (domain, protein) link only once. Let $m_i = \sum_{j=1}^{P} c_{ij}$. We call the value m_i the number of *non-redundant DPLs* of the domain a_i. Let $M = \sum_{i=1}^{N} \sum_{j=1}^{P} c_{ij}$. Note that $M \leq M'$. We call the matrix C the non-redundant adjacency matrix, and we call M the non-redundant connectivity number of the (domain, protein) network of proteome. The values M and M' were used as the non-redundant and redundant measures of the proteome complexity of an organism (Kuznetsov, 2002; 2003b).

The DPLPs also allow us to characterize the proteome complexity of the organism. Let the random variable X' denote the number of occurrences of the domain a_i in all proteins of the representative proteome. We call this number as the number of *redundant* DPLs of a random domain of the proteome. We can define the domain occurrence probability function (DOPF) $p'_l = Pr(X' = l)$, where $l = 1, 2, ...$ The value p'_l is the probability that a random domain occurs exactly l times within the proteome.

If a protein contains the given domain at least once, we can count the number of such non-redundant occurrences (non-redundant DPLs) in the (domain, protein) network for the given domain. Let the random variable X denote the number of *non-redundant* DPLs of a random domain of the proteome. We also define the DOPF $p_h = Pr(X = h)$, where $h = 1, 2, ...$ The value p_h is the probability that a random domain occurs exactly h times within the proteome.

4. MEASURES OF RELATIVE PROTEOME COMPLEXITY

To quantify the complexity associated with multiple occurrences of the domain a_i in an organism, we define the redundancy value

$$\delta m_i = m'_i - m_i . \tag{1}$$

To quantify the relative complexity of the domain a_i that redundantly occurs $m_i'^{(s)}$ times in the organism S and which redundantly occurs $m'^{(k)}_i$ times in the organism K, we can define the difference of redundant complexity value for the domain a_i

$$\delta m'^{(s,k)}_i = m'^{(s)}_i - m'^{(k)}_i .$$

By omitting the prim symbol in the above notations, we can also introduce the difference of non-redundant complexity value of the domain a_i

$$\delta m^{(s,k)}_i = m^{(s)}_i - m^{(k)}_i .$$

To characterize the relative redundant complexity of the organism S versus the organism K, we use the empirical bivariate distributions of the random values $(\mu'^{(s,k)}_i, \delta m'^{(s,k)}_i)$, where $\mu'^{(s,k)}_i = (m'^{(s,k)}_i + m'^{(s,k)}_i)/2$ is the mean value of the numbers of the links and $i = 1, 2, \ldots, N^{(s,k)}$; $N^{(s,k)}$ is the total number of domains occurring in DPLPs of the organisms S and K. By omitting the prim symbol in the above notations, we obtain the empirical bivariate distribution of the random values $(\mu^{(s,k)}_i,\ \delta m^{(s,k)}_i)$, which we can use as the relative non-redundant complexity measure of the organisms S and K.

5. RESULTS AND DISCUSSION

5.1 Non-redundant and redundant domains counts

Our analysis shows that the observed number of non-redundant domain-to-protein links for individual domains strongly correlates with the total number of organism's proteins, which is often used as a simple measure of biological complexity. About 50 of the 9609 observed domains (for instance, zinc-finger C2H2 domain, ankyrin, and cytochrome *c* heme-binding site) can be used as a linear measure of the number of proteins not only for prokaryotic organisms, but also for eukaryotic organisms (Figure 1*a*). However, the total number of observed non-redundant links, M, can provide more accurate goodness of fit to the number of proteins in 152 archaeal and bacterial representative proteomes and also in several available eukaryotic proteomes, including yeast, fly, and worm (Figure 1*b*). In this case, an extrapolation of the regression line allows us to improve the InterPro/UniProt's estimates of the number of proteins in the human, mouse and *A. thaliana* organisms. Based on our estimates, the total numbers of proteins in human, mouse, and *A. thaliana* are 33 845, 32 196, and 28 470, respectively. (UniProt/InterPro's data for rat was essentially incomplete and not used in this analysis.) Figures 1*c*, *d* show the empirical DOPFs for the non-redundant and redundant domains counts in the sample human proteome, specifying the proportions of distinct proteins containing 1, 2, ... domains in the sample proteome. Note that for human and for other organisms, the frequency distribution of redundant DPLs has a much longer tail than the frequency distribution for non-redundant DPLs due to the fact that we count every occurrence of the random domain in the sample proteome. Even with their large differences in proteome complexity of archaeal, bacterial, and eukaryotic organisms, all demonstrate similar skewed long-tail (power law-like) DOPFs and all data sets can fit accurately using the Generalized Discrete Pareto (GDP) probability distribution (Kuznetsov, 2002; Kuznetsov et al., 2002):

$$f(m) = Pr(X = m) = \zeta_J^{-1} \frac{1}{(m+b)^{k+1}}, \tag{2}$$

where the $f(m)$ is the probability that a randomly chosen domain occurs m times in the entire domain set (which we call domainome) of an organism. The function f involves two unknown parameters, k and b, where $k > 0$ and $b > 1$; the normalization factor ζ_J is the generalized Riemann zeta-function value: $\zeta_J = \sum_{j=1}^{J} 1/(j+b)^{k+1}$. Note that J, the maximum observed expression level, is a *sample-size* dependent quantity $J = J(M)$. The shape of the empirical DOPFs, the parameter J, and estimated parameters of the best-fit probability function by Eq. 2 correlate on the total number of DPLs M and M' (Kuznetsov, 2003b). We also observed that the difference between the redundant and non-redundant frequency distributions for simpler organisms is much smaller than for more complex organisms (Kuznetsov, 2003b, Kuznetsov et al., 2002).

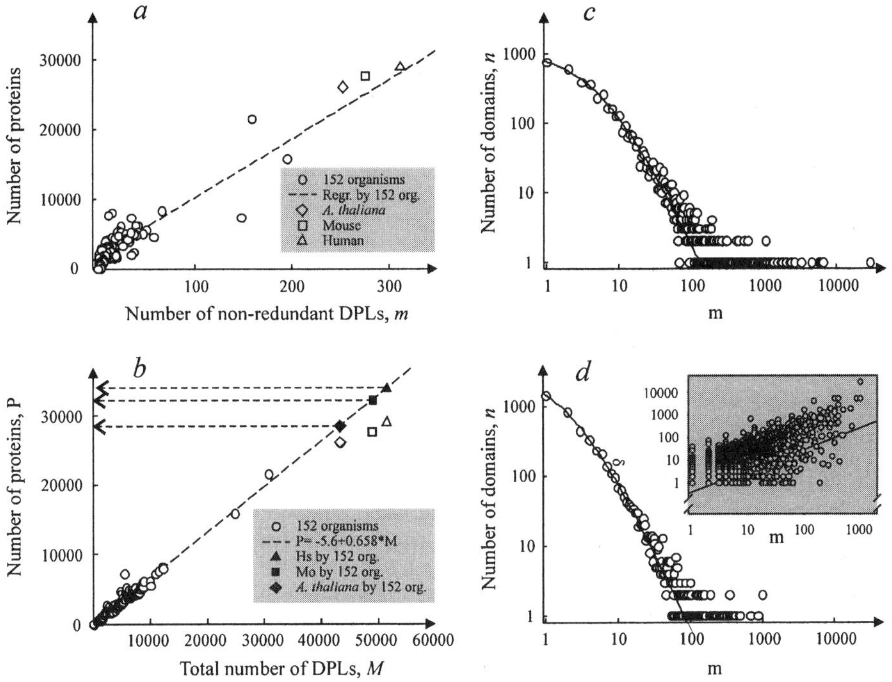

Figure -1. Measures of proteome complexity. The total number of proteins correlates with (*a*) non-redundant DPLs of zinc-finger C2H2 domain ($r = 0.92$) and (*b*) the total number of DPLs in representative domainome ($r = 0.97$). The regression lines on panels (*a*) and (*b*) were calculated based on data of 152 fully sequenced genomes, excluding human, mouse, rat, and *A. thaliana*. Δ,□,◇: InterPro/UniProt estimates of the total number of proteins for human, mouse, and *A. thaliana*, respectively (April, 2004). Arrows on panel (*b*) indicate the numbers of proteins for human, mouse, and *A. thaliana* estimated by using an extrapolation of the regression line of the 152 data points. Skewed frequency distributions of the number of links in the redundant (•) and non-redundant (∘) sets of the human representative domainome. (*c*) GDP best fits to data (•) at k = 1.22+/–0.079, b = 5.58+/–0.358. (*d*) GDP best fit to data (∘) at k = 1.26+/–0.063, b = 2.29+/–0.136; window plot on (*d*): differences between the number of redundant domain-to-protein links and the number of non-redundant domain-to-protein links in the human representative proteome. M = 51 445; M' = 209 595; N = 4761; mean (m) = 10.8+/–0.61; mean (m') = 44+/–7.1.

We observed that when a domain is more common in the entire proteome world, then that domain has a bigger chance to appear multiple times in a protein of any organism. Window plot in Figure 1*d* shows a strong positive trend ($r = 0.88$) of the increment of multiple usage of a domain in the human proteome when the number of non-redundant occurrences of domains increases.

5.2 Identification of two novel sets of domains

For all of the 156 studied organisms, we found two disjointed subsets of domains: the domains that were not multiplied within any protein of the proteome and the domains that occurred two or more times in at least one protein of a proteome. These two subsets of domains we called D1 and D2, for briefly. There are 5068 (53 %) of the 9609 domains in set D1 and 4541 (47 %) of the 9609 domains in set D2. These subsets have not been characterized in the literature. In pooled domain set of the 156 organisms, the empirical probability distribution for set D2 displays a single-peak skewed functional form with a peak at $m = 3$ (Figure 2). However, the frequency distributions of domain-to-protein links for set D1 is represented by the multi-peak skewed distribution having at least three peaks at $m = 2, 8, 14$. Thus, set D1 might be represented by mixture of three or more domain subsets with different molecular properties. Interestingly, for individual organisms, the frequency distributions of DPLs in the domain sets D1 and D2 are presented by single-peak skewed distributions (Figure 3).

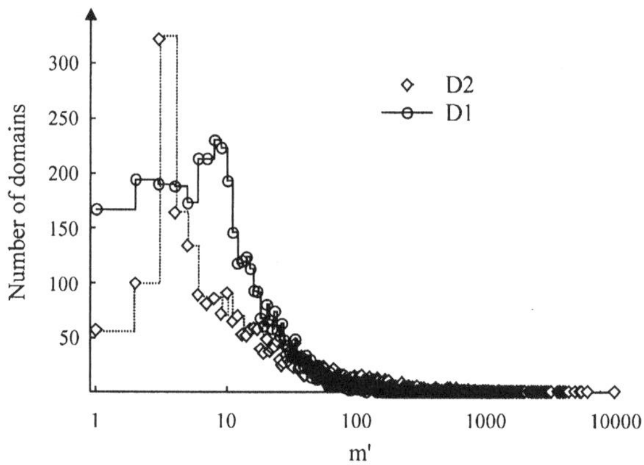

Figure -2. The frequency distributions of two distinct domain subsets. D1: 5068 domains that were not replicated in any protein of the 156 proteomes; D2: 4571 domains replicated in a protein of the 156 organisms. Average number of DPLs: 42+/–1 and 146+/–7 for D1 and D2, respectively.

Panels *a–c* in Figure 4 show the relationships between the numbers of non-redundant and the numbers of redundant DPLs counted for the *T. volcanium*, *E. coli*, and human representative proteomes. These panels demonstrate typical patterns of the bivariate frequency distributions of the random values (m, m') ($(m, m') = \{ (m_1, m'_1), \ldots, (m_{N(s)}, m'_{N(s)}) \}$) corresponding to the numbers of redundant and non-redundant occurrences of domains in the organism *S*.

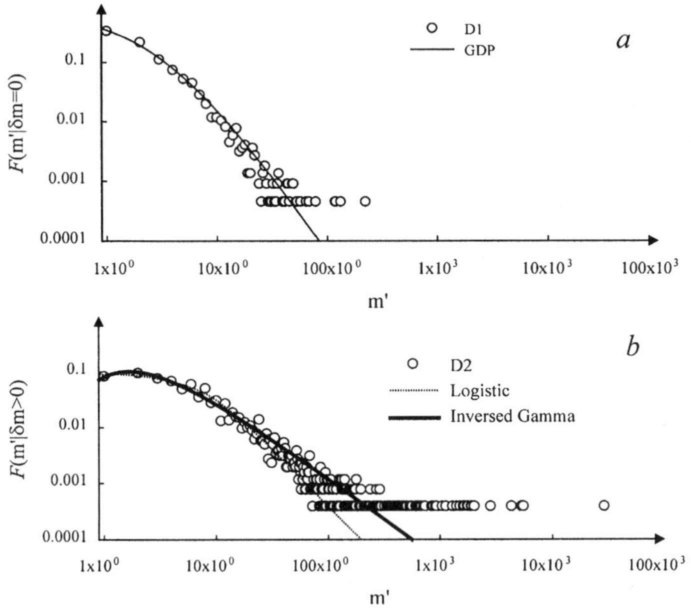

Figure -3. Best-fit probability distributions of the number of DPLs for D1 and D2 domain sets for human. (*a*) GDP (at parameters $k = 1.66 \pm 0.269$, $b = 3.00 \pm 0.559$; $R^2 = 0.9971$) fits to the empirical DOPF for set D1. (*b*) Dotted line: logistic probability function $f(m') = a/(1+(m'/c)^h)$ with parameters $a = 0.0946 +/- 0.0019$, $c = 6.50 \pm 0.200$, $h = 1.92 \pm 0.0544$) fits to data D2. Solid line: inversed gamma probability function $f(m') = a * x_0^k \exp(-x_0/m') * (1/m')^{h+1}$ (Evans et al., 1993) with parameters $a = 0.62 \pm 0.016$, $k = 0.44 \pm 0.026$, $x_0 = 2.41 \pm 0.075$ fits better to data D2.

In Figure 4*a–d*, the points along the diagonal belong to subset D1; all other points (up diagonal) belong to class D2.

All panels in Figure 4 show a preferential distribution of points along the diagonal; the spread of points in the orthogonal direction to the diagonal is smaller. The extreme cases (*T. volcanium* and human proteomes) clearly display the increased complexity of the human proteome due to the $i = 1, 2, \ldots, N^{(s,\,k)}$ increased number of proteins, which preferentially combining different domains together. Panel *d* in Figure 4, showing the observed values (m, m') for the 156 organisms is by a factor of 10 larger then any specific organism. We found that these distribution patterns are typical of the studied archaeal, bacterial, and eukaryotic organisms. Additionally, Table 1 shows that for the seven most abundant short structural motifs in the human, the ratio (*m′/m*) is much bigger than the factor 10 (from 17 to 226).

We found that ~ 45–55 % of eukaryotic proteins containing two or more known domains and multi-domain proteins (m > 2) much oftener appear in organisms with the larger number of genes. Figures 4*a–d* imply that increasing biological complexity in nature is mostly associated with formation of new multi-domain proteins combining different domains rather than with the increase in protein complexity due to replication of domain(s) within proteins in which domains have been already present. However, the rate of creation of new multi-domain protein families may be very diverse; in particular, the higher abundant domains of D2 set might create new diverged protein families much faster than others (Table 1).

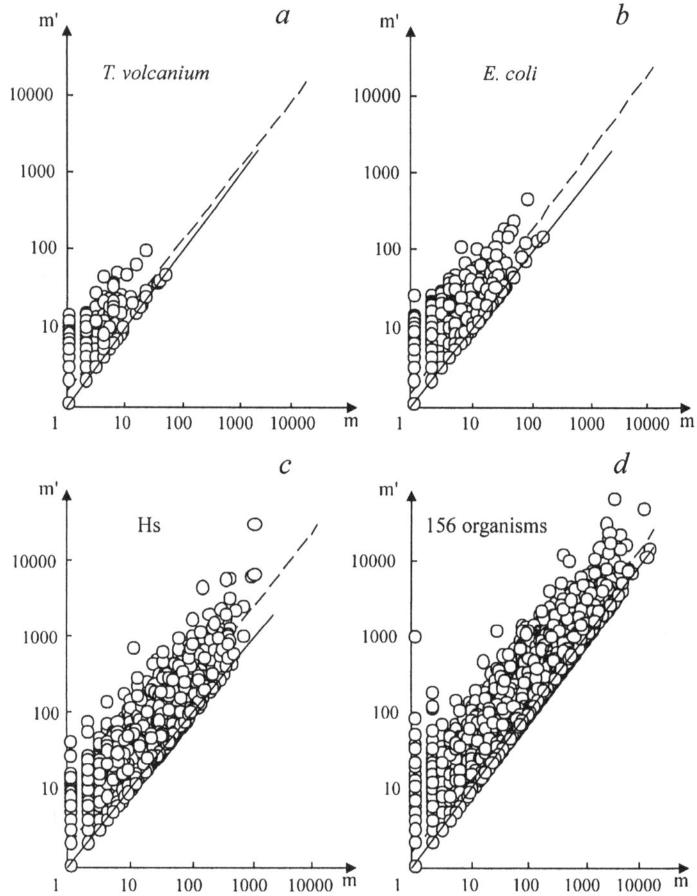

Figure -4. The empirical distributions of the numbers of redundant links versus the numbers of non-redundant links counted for (*a*) *T. volcanium*, (*b*) *E. coli*, (*c*) human representative proteomes, and (*d*) pooled data of 156 organisms. Discontinued line: linear regression; solid line: diagonal.

Thus, there are at least two distinguishing processes of domains spread in nature: integration of two or more different domains in a new protein (forming non-redundant domain-to-protein links) and multiplication of a domain within the same protein (forming redundant DPLs).

Table -1. Abundance of mostly abundant short (length >25) structural motifs from D1 and D2 sets

D1 (length >25 aa)	m	D2 (length >25 aa)	m′
Tyrosine protein kinase, active site	132	Zn-finger, C2H2 type	29816
Endoplasmic reticulum targeting sequence	64	EGF-like domain	3208
Aminoacyl-tRNA synthetase, class I	57	Leucine-rich repeat	2300
ATP-dependent helicase, DEAD-box	44	Olfactory receptor	2270
ATP-dependent helicase, DEAH-box	41	IQ calmodulin-binding region	681
N-6 Adenine-specific DNA methylase	32	Aspartic acid and asparagine hydroxylation site	632
Phosphopantetheine attachment site	30	Zn-finger, C2H2 subtype	566

5.3 Comparison of proteome complexities

Figure 5 demonstrates how the proteome complexity of pairs of organisms can be compared. This figure shows the relative proteome complexity measure for a human versus a mouse, a rat and *A. thaliana*, respectively. For example, panel A shows the bivariate distributions of differences of the number of redundant links for the human and mouse organisms multiplied by factor 0.5 with respect to average number of the redundant links of the human and mouse organisms. This distribution is highly asymmetric about axes *x*: the number of positive differences (abundant for a human) is greater than the number of negative differences (abundant for a mouse), and the positive highly abundant values occur more often in the human sample than in the mouse sample. Similar asymmetric trend we observed for our human–mouse comparison in the case of redundant links (Figure 5*b*). This indicates that the human organism reuses domains more frequently in the same protein and in different proteins and invents more diverse multi-domain proteins than the mouse organism. This analysis suggests that even though the numbers of non-redundant protein-coding genes in a human (~ 33 850 proteins) and a mouse (~ 32 200 proteins) are approximately similar and even though the total number of InterPro domains are also similar in these organisms (~ 5600 for a human and ~ 5100 for a mouse (Kuznetsov, 2003b)), the proteome complexity and diversity of a significant fraction of multi-domain proteins in a human is higher than in a mouse. A large asymmetry in the bivariate distribution of the values ($\mu_i^{(s,k)}, \delta m_i^{(s,k)}$) around axes *x* we observed for a human versus a rat (Figure 5*c*). However, the human and *A. thaliana* proteomes show similar proteome complexity by our criteria (Figure 5*d*). This is not a surprise, because even though the number of protein-coding genes in human is larger than the number of protein-coding genes in *A. thaliana*, ~ 27 000, (Kuznetsov, 2003b), relatively recent massive (and imperfect) genome duplication events in *A. thaliana* might have dramatically increased the protein repertoire due to recombination events, which perhaps increased the slice variants and domain shuffling in proteins. Diversity of proteins might be also increased due to recent evolutionary appearing of a large number of new domains due to global genome duplication event. Indeed, we observed 614 unique entries of the 9609 InterPro entries, which was specific of *A. thaliana*, while only 52 and 34 of the 9606 entries was specific of human and mouse, respectively.

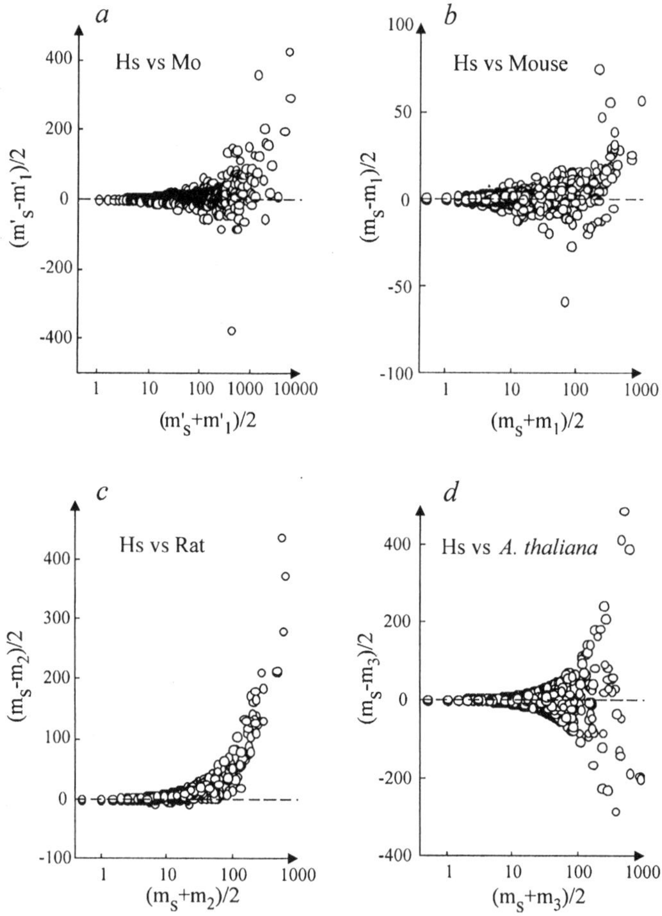

Figure -5. The relative complexity of proteome of a human versus a mouse, a rat (*c*) and *A. thaliana*, respectively. (*a*) The distribution of $(\mu'^{(s,k)}_i, \delta m'^{(s,k)}_i)$; symbol *s* indicates the human organism, $k = 1$ indicates the mouse organism; (*b–d*) the distributions of $(\mu_i^{(s,k)}, \delta m_i^{(s,k)})$; $k = 2, 3, 4$ indicate mouse, rat, and *A. thaliana*, respectively.

These 'organism-specific' domains mostly appear as singletons of domains D1 in the proteome (not presented). Such large number of the *A. thaliana* specific domains should be unlikely associated with erroneous annotation of domains. In the contexts of diversity of domains in *A. thaliana* proteins, we noted that a proportion of gene ontology (GO) terms 'response to stress', associated with D2 set, is much higher in comparison to human, mouse, or rat data sets. (We estimated the proportion by formula: number of D2 terms/(number of D1 terms + number of D2 terms).

Thus, this study identifies two large sets of domains, D1 and D2, based on different types of domain–domain links.

Statistics of the observed domain-to-protein links for D1 and D2 sets follow to distinct skewed distribution function (GDP function and inversed gamma probability (IGP) function, respectively), suggesting different molecular mechanisms of domain–domain interactions in the corresponding proteins. Importantly, both empirical DOPFs do not show simple power law (the 'scale-free' network property) and, thus, forms scale-dependent network. These findings consisting of our early results (Kuznetsov, 2001, 2003a, 2003b) suggested that the scale-free models might not be applicable in understanding the evolution of biological systems and many other evolving interconnected systems.

InterPro GO-slim analysis of the distribution of D1 and D2 sets among proteins of the 156 organisms reveals significant asymmetry in the distribution by several common molecular functions, biological process, and cellular components; in particular, domains of D1 often represent the proteins of unknown function(s) with single domains (that perhaps occurred in evolutionarily recent proteins). D2 domains dominate in 'signal transducer activity', 'cell communication', 'nuclear acid binding', and in many other GO-slim categories (Kuznetsov et al., 2005, submitted). Interestingly, a higher abundant of domains of D2 set often represent more evolutionary recently diverged protein families, while a higher abundant domains of D1 often represent the families of the proteins derived in the Last Universal Common Ancestor (LUCA) or earlier (see examples in Table 1). Some unique, 'organism-specific' InterPro domains of D1 appear to be unlikely associated with erroneous annotation of domains, and these domains might be considered in context of origin of novel domains.

PART 3

COMPUTATIONAL SYSTEM BIOLOGY

A SOFTWARE ARCHITECTURE FOR DEVELOPMENTAL MODELING IN PLANTS: THE COMPUTABLE PLANT PROJECT

V. Gor[1], B.E. Shapiro[1], H. Jönsson[2], M. Heisler[3], G.V. Reddy[3], E.M. Meyerowitz[3], E. Mjolsness[4*]

[1] *Jet Propulsion Laboratory, California Institute of Technology, Pasadena, CA 91109, USA;* [2] *Department of Theoretical Physics, Complex Systems Division, Lund University, Lund, Sweden;* [3] *Division of Biology, California Institute of Technology, Pasadena, CA 91125, USA;* [4] *Institute for Genomics and Bioinformatics; Bren School of Information and Computer Sciences, University of California, Irvine CA 92607, USA, e-mail: emj@uci.edu*

[*] *Corresponding author*

Abstract: We present the software architecture of the Computable Plant Project, a multidisciplinary computationally based approach to the study of plant development. *Arabidopsis thaliana* is used as a model organism, and shoot apical meristem (SAM) development as a model process. SAMs are the plant tissues where regulated cell division and differentiation lead to plant parts such as flowers and leaves. We are using green fluorescent proteins to mark specific cell types and acquire time series of three-dimensional images via laser scanning confocal microscopy. To support this, we have developed an interoperable architecture for experiment design that involves automated code generation, computational modeling, and image analysis. Automated image analysis, model fitting, and code generation allow us to explore alternative hypothesis *in silico* and guide *in vivo* experimental design. These predictions are tested using standard techniques, such as inducible mutants and altered hormone gradients. The present paper focuses on the automated code generation architecture.

Key words: Arabidopsis; Cellerator; correspondence; Delaunay triangulation; meristem; SAM; SBML; softassign; Voronoi diagram

1. INTRODUCTION

Scientists who probe the functionality of dynamic developmental systems often express their models mathematically; to make precise system-specific predictions; these models are typically encoded with high-level computer languages and standard support libraries and solved numerically. However, high-level languages and libraries typically trade efficiency for generality, and thus may not be appropriate for large hybrid dynamical systems. They also typically lack state-of-the-art technologies in such computationally intensive areas as model optimization and fitting. Finally, custom designed systems are rarely interoperable, making it difficult for researchers to disseminate models.

We have developed an architecture aimed at production-scale model inference. We generate simulation code from models specified in biological and/or mathematical language. Other computational tools are used to analyze expression imagery and other data sources, and the simulator combined with nonlinear optimization is used to fit the models to the experimental observations. Key elements include a *mathematical framework* combining transcriptional regulation, signal transduction, and dynamical mechanical models; a *model generation package* (Cellerator) based on a computer algebra representation; *extensions to SBML* (Systems Biology Modeling Language), an exchangeable model representation format, to include dynamic objects and relationships; *a C++ code generator* to translate SBML into highly efficient simulation modules; *a simulation engine* including standard numerical solvers and plot capability; *a nonlinear optimizer*; and ad hoc *image processing* and *data mining* tools. This architecture is capable of simulating processes, such as intercellular signaling, cell cycling, cell birth and death, dynamic cellular geometry, changing topology of neighborhood relationships, and the interactions of mechanical stresses.

2. METHODS AND ALGORITHMS

Models are input in Systems Biology Markup Language (SBML), an XML-based language for exchanging biological models. SBML is currently supported by more than fifty 75 different software packages used by biological modelers and has become the *de facto* standard for exchanging models among the systems biology community (Hucka et al., 2003; Finney and Hucka, 2003). The modeling interface is provided by *Cellerator* (Figure 1; Shapiro et al., 2003), which allows users to specify models in an arrow-based biochem-ical notation and translates them automatically to differential equations using a variety of different schemes. *Cellerator* produces extended SBML Level 2 code utilizing *MathSBML* (Shapiro et al., 2004). SBML encoded models are parsed into internal data structures with a *libSBML*-based parser (Finney et al., 2005).

Several extensions to SBML have been proposed and will likely be adopted in SBML Level 3 (Finney et al., 2004). In particular, SBML Level 2 does not support spatially-dependent models where each biological entity is *individually defined and enumerated* and further, does not provide any easy way to describe dynamic geometry and variable size models resulting from cell birth, death, and differentiation. Therefore, we have adopted (Finney et al., 2003) to describe dynamic topology and connectivity in terms of arrays and have extended *Cellerator*, *MathSBML*, and *libSBML* accordingly.

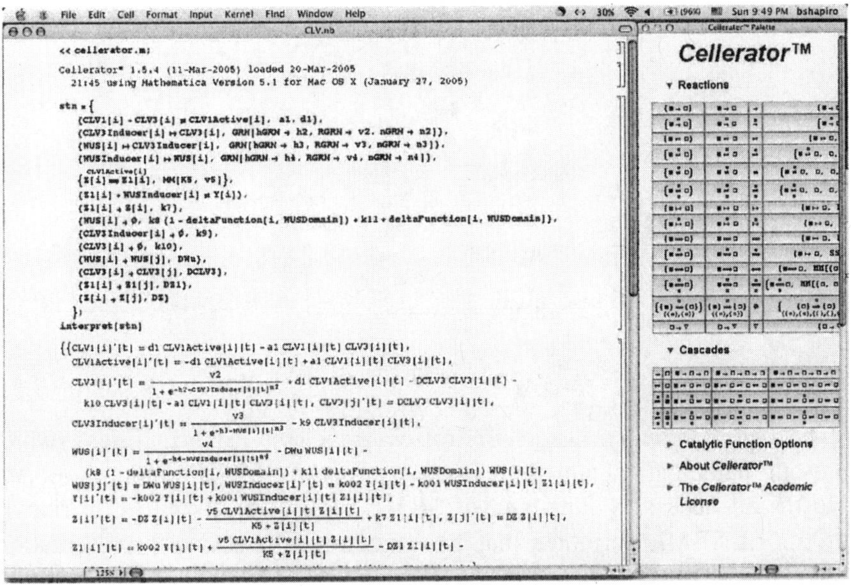

Figure -1. Cellerator screen shot. Signal transduction networks are built in the Mathematica notebook on the left by clicking on the palette on right. The interpret command converts the reactions to differential equations. Array indices are used to indicate different cells.

The *automatic code generator* is central to the architecture. It consists of an *inferencer*, a *rule segmenter* and *optimizer*, and *application code writer* modules (Figure 2). It queries the parser for SBML structures and produces efficient C++ application code. The resulting C++ code is then compiled into object code optimized for the desired application. The first two modules of the automatic code generator—the *inferencer* and *rule segmenter*—are pre-processors. They are called once for each SBML model, independent of the application software to be generated. The *inferencer* receives parsed SBML structures from the parser and infers element attributes given the element name. This reflects the inverse relationships between SBML elements and their attributes. For example, the extended SBML has a `parameter` attribute `foreach` that indicates the compartment; the *inferencer* creates a list of inferred elements, such as the list of parameters in each compartment.

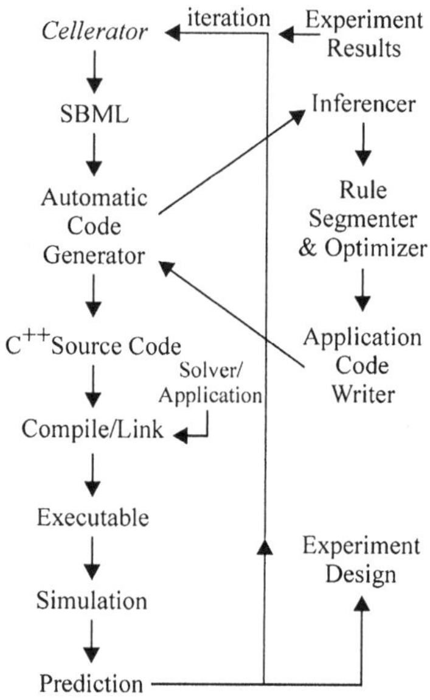

Figure -2. System architecture.

The *rule segmenter* and optimizer translates SBML rules (which represent mathematical equations using a subset of MATHML) into C++ and performs all necessary renaming of SBML model objects into C variables. Portions of SBML formulas that have no immediate C++ representation, such as the MATHML function sum (which sums a formula over an index), are broken up into sub-rules with intermediate variables; these are later translated into loops or other appropriate control and data structures. Future enhancements will include formula optimization. Identical portions of the formula will be separated into intermediate rules that are only executed once; scalar formulas inside loops will be pre-evaluated outside of the loop. The renaming function completes the work of this module. For example, individual array elements are referenced by index with an SBML model utilizing the MATHML selector operator; this is replaced by the appropriate C array reference such as `name[j]`.

The *application code writer* takes as input the C++ model representation generated by the *rule segmenter and inferencer*, along with an application request, chosen from a menu of available applications. The output is application source code that can be compiled and linked with the chosen application. *The application code writer* consists of a three-level library. The top level contains all of the application-dependent code. This *application level software* is the high-level code that is updated as new applications are added. Applications that exist or are being developed include various forward developmental simulators including genetic regulatory network (GRN) temporal synthesis; 4th and 5th order Runge–Kutta differential equation solvers; and optimizers, such as Lam–Delosme simulated annealing. In addition, this top level includes overloaded routines that originate at the second level thereby allowing the top level to access this lower level functionality.

The second level, *SBML level software*, contains all processes that are not application dependent. This library has entry points for accessing all SBML attributes and elements. The third, and lowest level, is *the utility library*, which contains common operations, such as vector algebra and memory maintenance.

3. RESULTS AND DISCUSSION

The SAM (Figure 3) is the plant tissue where regulated cell division and differentiation lead to plant parts, such as flowers and leaves. We are using green fluorescent proteins to mark specific cell types and acquire time series of three-dimensional images via laser scanning confocal microscopy. The three-dimensional reconstruction starts from 'stacks' of horizontal sections (Figure 4).

Such sections are combined to produce four-dimensional visualizations (three spatial dimensions plus time) using various programs we have developed.

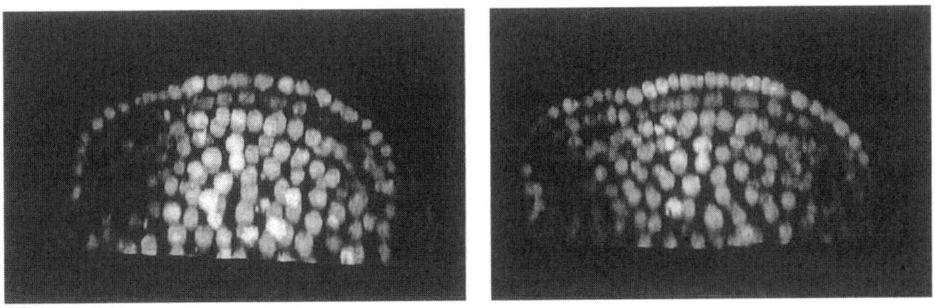

Figure -3. Vertical SAM cross-sections at two different times showing nuclear-localized GFP expressed from a ubiquitous promoter. A flower primordium is emerging at the upper left.

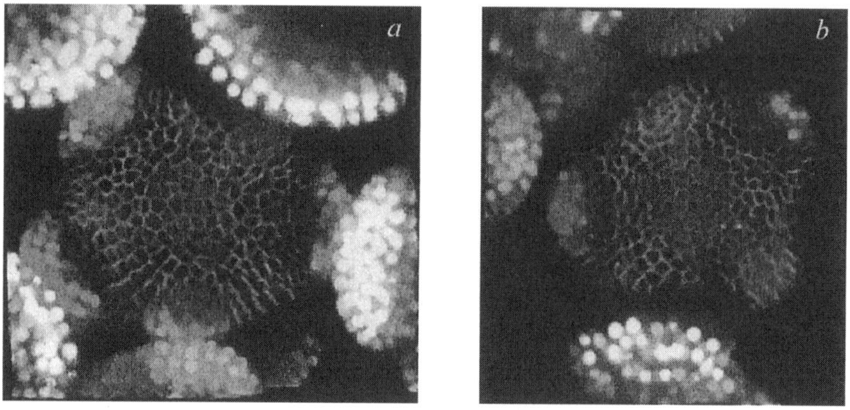

Figure -4. Horizontal SAM cross-sections, showing pPIN1:PIN1GFP (expressed in the cell membranes) in combination with pFIL:dsREDN7 (nuclear, seen here primarily in the primordia) at two time points 33 hours apart; also illustrating the budding of new floral meristems (*a*) initial view; (*b*) final view.

In any 3D image stack, there is a correspondence problem: which cells in one image correspond to which cells in the adjacent cross-section? Cell membranes that are transverse to the image are clearly visible, but it is possible to miss nearly horizontal walls that lie between sections and must be inferred. With a time-course of 3D stacks the formation of floral meristems, cell growth, displacement, and division all complicate the problem. Nuclear locations are determined using a 3D gradient descent algorithm based on image intensity.

Automated extraction of cell walls (or cell membranes) is more complicated, as we have discussed. It is possible to estimate their locations using the Voronoi diagram (also called a Dirichlet tessellation) of the nuclear centers; nearest-neighbor links are then given by the corresponding Delaunay triangulation (Figure 5). The Voronoi 'cell' defined by any nuclear center **p** is the polygon that contains all of the points that are closer to **p** then to any other nuclear center **q**; the Voronoi diagram is the collection of all such Voronoi cells. The Delaunay triangulation is the dual of the Voronoi diagram, namely, the collection of lines drawn from each nucleus to all of its nearest neighbors. The walls of the Voronoi diagram are the perpendicular bisectors of the Delaunay triangulation. This principle that cell walls are equidistant from nuclei is beautifully reflected by the watershed transform (Vincent and Soille, 1991), an image segmentation algorithm based on mathematical morphology (Sternberg, 1986). The watershed transform is used in combination with Voronoi diagrams for detecting cell walls in images. When the walls are visible, they are detected form the gradient of image data; and when the walls are not visible, they are inferred by the Voronoi diagram (Figure 6). Voronoi diagrams and Delaunay triangulations are computed with Qhull (open source software, www.qhull.org).

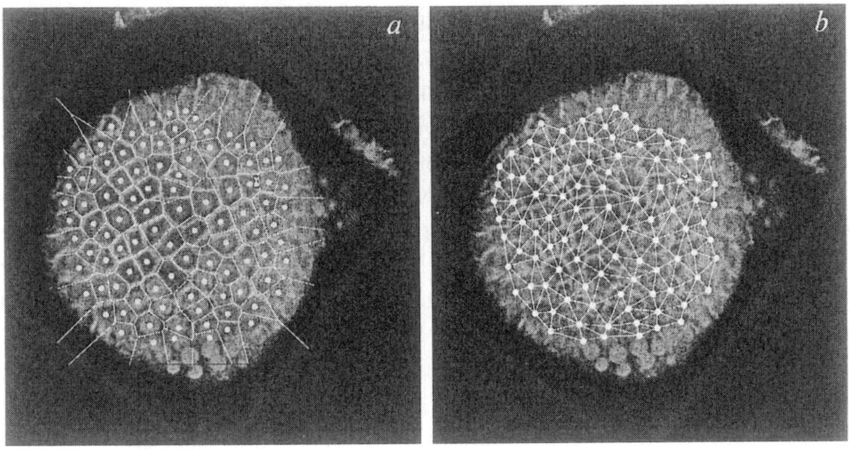

Figure -5. Meristem cross-section stained to show cell membranes and nuclei, superimposed with manually tagged cell-centers and the corresponding Voronoi diagram (*a*) and Delaunay triangulation (*b*). A small number of cell centers were left unmarked; the corresponding Voronoi cells were added to the adjacent cells.

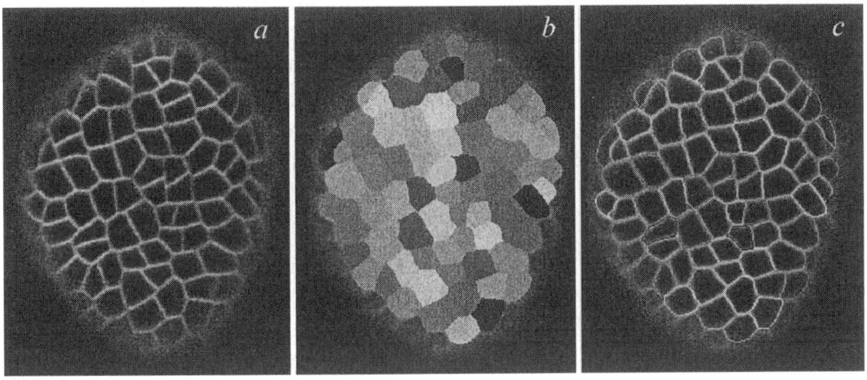

Figure -6. Raw image (*a*), segmentation of image into cells (*b*), and cell walls determined with a watershed algorithm (*c*).

Determining which cells in one 3D image correspond to which cells in the next image is a case of the classic correspondence problem (Post, 1947); many solutions to both point-matching and graph-matching correspondence problems have been published to this extremely difficult problem. Recent work is based on joint estimation of correspondence and spatial mappings via optimization of an energy function (Gold et al., 1996). The general framework uses the method of deterministic annealing in conjunction with the softassign algorithm and clocked objectives to produce an optimizing network and a corresponding energy function (Koslowsky and Yuille, 1994; Gold et al., 1996; Mjolsness and Miranker, 1998). The specific energy function used for cell tracking determines cell correspondence, while estimating the mapping functions, such as affine and thin-plate spline transformations (Chui and Rangarajan, 2000) and cell growth and division history.

We are using this simulation environment to extend and enhance our previously reported developmental simulations of the shoot apical meristem (SAM; Jönsson et al., 2003; 2005; Mjolsness et al., 2004). Our working hypothesis is that SAM development can be described by the differential expression of key regulatory proteins, such as CLV1 (a receptor kinase), CLV3 (thought to be the CLV1 ligand), WUS (a transcription factor negatively regulated by CLV1), and a layer-1 specific protein (L1SP). For example, activation of CLV1 might be described by the reactions

$$CLV1 + CLV3 \rightleftarrows CLV1^*$$
$$CLV3 \rightarrow \varnothing$$

The dependence of CLV1 and CLV3 on WUS, perhaps through a hypothetical diffusible intermediary (CLV3I1), has been inferred from experiments. A second diffusive signal is postulated to originate from L1SP and diffuses into the rest of the meristem via messenger CLV3I2. CLV3 is turned on only if the sum CLV3I1 + CLV3I2 exceeds threshold.

$$WUSI \mapsto WUS$$
$$WUS \mapsto CLV3I1$$
$$L1SP \mapsto CLV3I2$$
$$CLV3I1 + CLV3I2 \mapsto CLV3$$

where WUSI is a hypothetical WUS inducer that originates either in or below the corpus. The expression $A \mapsto B$ is the Cellerator notation for genetic regulation (Table 1). Inhibitory feedback is provided by a proposed entity Z that sequesters activated CLV1, and when activated, sequesters or removes WUSI:

$$Z + CLV1^* \rightleftarrows Z1$$
$$Z1 + WUSI \rightleftarrows Z2$$

Additionally, an unknown diffusible messenger Y creates a surface-specific expression pattern for L1SP, which is itself inhibited by STEM, a hypothetical gene expressed only in the lowest meristem layer:

$$Y \mapsto L1SP$$
$$STEM \mapsto Y$$

Table -1. A selection of typical *Cellerator* arrows

Arrow	Description
$p_1A_1 + p_2A_2 + \cdots \to q_1B_1 + q_2B_2 + \cdots$ $p_1A_1 + p_2A_2 + \cdots \rightleftarrows q_1B_1 + q_2B_2 + \cdots$	Law of mass action, one-way and reversible; optional stoichiometry
$S \overset{E}{\rightleftarrows} P$	Enzymatic mass action, same as $S + E \rightleftarrows SE \rightleftarrows P + E$
$S \underset{F}{\overset{E}{\rightleftarrows}} P$	Reversible enzymatic mass action $S + E \rightleftarrows SE \rightleftarrows P + E$ $P + F \rightleftarrows PF \rightleftarrows S + F$
$S \underset{F}{\overset{E}{\rightleftharpoons}} P$	Reversible enzymatic mass action $S + E \rightleftarrows SE \rightleftarrows PE \rightleftarrows P + E$ $P + F \rightleftarrows PF \rightleftarrows SF \rightleftarrows S + F$
$S \Rightarrow P,\ S \overset{E}{\Rightarrow} P,\ S \underset{F}{\overset{E}{\Rightarrow}} P$	Michaelis–Menten kinetics;one-way and reversible
$S \overset{E}{\mapsto} P$	Conversion of A to B, facilitated by E,via Hill function
$A \mapsto B$	Regulation of B (A unaffected) by Hill function, sigmoid, NHCA, or S-System
$\{S_1, S_2, \ldots\} \underset{\{\{A_1, A_2, \ldots\}, \{I_1, I_2, \ldots\}\}}{\overset{E}{\Rightarrow}} \{P_1, P_2, \ldots\}$	Generalized MWC with multiple substrates, products, activators, and inhibitors
$\{S_1, S_2, \ldots\} \underset{\{\{A_1, A_2, \ldots\}, \{I_1, I_2, \ldots\}, \{\{Q_1, Q_2, \ldots\}, \{R_1, R_2, \ldots\}}{\overset{E}{\Rightarrow}} \{P_1, P_2, \ldots\}$	Generalized MWC with competitive inhibition

Here, the first reaction is activating, and the second is inhibitory; both genetic regulation and inhibition are modeled by Hill functions with different parameters. To maintain homeostasis, we include the reactions

$$CLV3, CLV3I1, WUS \to \varnothing$$

and describe diffusion using a simple compartmental approach. A Cellerator model for a single cell that is very similar to this one is illustrated in Figure 1. A two-dimensional 133-cell Cellerator implementation has 5422 reactions and 1596 differential equations.

The computable plant architecture provides a systematic, highly automated technique for predictive model generation. The approach combines computer–algebraic representations of biological and mathematical models to produce efficient and problem-specific simulation code. This code can be immediately linked with a menu of external solvers and quantitative predictions generated from the resulting simulations. This architecture is scalable and directly applicable to large-scale developmental systems, such as the SAM. The use of extended SBML ensures that models will be interoperable, reusable, and readable by others. Novel to this approach are connections to external solvers by way of automatic code generation and the ability to interpret and solve any biological developmental or cellular process via automatic generation of mathematical and computational tools. Thus, no labor is expended writing and debugging problem-specific code, allowing researchers to spend more time on the wet bench. Further details can be found at the project web-site, http://www.computableplant.org.

ACKNOWLEDGMENTS

This work was supported by the United States National Science Foundation (NSF) under Frontiers in Integrative Biological Research (FIBR) grant No. EF-0330786. HJ was supported, in part, by the Knut and Alice Wallenberg Foundation through Swegene. Portions of the research described in this paper were performed at the California Institute of Technology.

PREDICTION AND ALIGNMENT OF METABOLIC PATHWAYS

M. Chen[1,2], R. Hofestaedt[1*]
[1] Bioinformatics/Medical Informatics, Faculty of Technology, Bielefeld University, 33501, Bielefeld, Germany, e-mail: ralf.hofestaedt@uni-bielefeld.de; [2] College of Life Science, Zhejiang University, Hangzhou, 310029, China
[*] Corresponding author

Abstract: Prediction and reconstruction of metabolic pathways from rudimentary molecular data is a fundamental way to understand the logic of life. Comparing metabolic pathways are important to identify conservation and variations among different biology systems. Alignment is a strong indicator of the biologically significant relationship. In this article, a definition of metabolic pathway is presented. An alignment algorithm and a computational system are developed to reveal the similarities between metabolic pathways. A web-based program for prediction and alignment of metabolic pathway, PathAligner, is available at http://bibiserv.techfak.uni-bielefeld.de/pathaligner.

Key words: Metabolic pathways; prediction; alignment; pathAligner

1. INTRODUCTION

Today, a huge amount of molecular data about different organisms has been accumulated and systematically stored in specific databases (Collado-Vides and Hofestaedt, 2002). This rapid accumulation of biological data provides the possibility of studying metabolic pathways systematically. Predicting metabolic pathways from current biological data, reconstructing metabolic pathways from some rudimentary components, such as genes, gene sequences, proteins, protein sequences, and other biological molecules, and aligning metabolic pathways with each other are some major tasks in current bioinformatics.

Several approaches have attempted to reconstruct metabolic pathways either via genome sequence comparison (Mushegian et al., 1996; Bono et al.,

1998), genome annotation data parsing (Goesmann et al., 2002), annotated whole genome sequence assembly (Overbeek et al., 1997; Selkov et al., 1997; Nakao et al., 1999; Covert et al., 2001; Kim et al., 2001), enzyme assignment (van Helden et al., 2000), and enzyme EC numbering (Ogata et al., 1998). These previous approaches and the existing metabolic pathway databases have a number of limitations in metabolic pathway prediction. Predicting each gene function based solely on sequence similarity searches often fails to reconstruct cellular functions with all necessary components. Some approaches do not contain comprehensive information about metabolic pathways, such as physical and chemical properties of the enzymes that involved; some are not fully computer-aided. The individual database search process requires too much human intervention, and the quality of annotation largely depends on the knowledge and work behavior of human experts.

Metabolic pathway alignment represents an important tool for comparative analysis of metabolism. Although researches on genomic sequence alignment have been intensively conducted, metabolic pathway alignment so far is less studied. Several approaches to metabolic pathway alignment are already made in the past years. Forst and Schulten (1999; 2001) extended the DNA sequence alignment methods to define distances between metabolic pathways by combining sequence information of the involved genes. Dandekar et al. (1999) aligned metabolic pathways by counting the presence/absence of enzymes and their encoding genomic sequences. Tohsato et al. (2000) presented a method for the alignment of reaction similarity of EC numbers. Forst et al. (2001) extended the conventional phylogenetic analysis of individual elements in different organisms to the organisms' metabolic networks. Liao et al. (2002) developed a computational method to compare organisms based on the whole metabolic pathway analysis.

In this article, we present a conceptual framework and web-based computational system that will aid in the prediction and alignment of metabolic pathways.

2. METHODS AND ALGORITHMS

2.1 Formal definition of metabolic pathways

A biochemical pathway is defined by Mavrovouniotis (1995) as an abstraction of a subset of intricate networks in the soup of interacting biomolecules. A prevailing definition is that a metabolic pathway is a special case of a metabolic network with distinct start and end points, initial and terminal vertices, respectively, and a unique path between them, i.e., a directed reaction graph with substrates as vertices and arcs denoting enzymatic reactions (Forst and Schulten, 1999). Traditionally, metabolic pathways have been defined in the context of their historical discovery, often named after key molecules (e.g., 'glycolysis', 'urea cycle', 'pentose phosphate pathway', 'citric acid cycle', and so on). The basic strategy to represent and compute pathways is the reactant–product binary relation. Properties of the pathway that rely upon the integration of two or more input molecules, unrelated output molecules, and feedback effects are ignored.

We consider that a metabolic pathway is a subset of reactions that describe the biochemical conversion of a given reactant to its desired end product.

Let $M = \{m_1, \ldots, m_n\}$ be a set of metabolites in cells. Let $e_i : M \rightarrow M$ be a function for enzymatic reactions taking place in the cells.

The fact that e_i is a function from a set of substrates S ($S \subseteq M$) into a set of products P ($P \subseteq M$). It can be written as follows:

$$e_i : S \rightarrow P.$$

For all $m_1, m_2, m_3 \in M$, the following property holds:

$$e_1(m_1) = m_2 \text{ and } e_2(m_2) = m_3 \Rightarrow e_2(e_1(m_1)) = m_3.$$

Obviously, e_1 and e_2 are two successive enzymatic reactions.

Letting $e_1(m_1) = m_2$, $e_2(m_2) = m_3$, $\ldots$, $e_k(m_k) = m_{k+1}$; we define $e_1e_2 \ldots e_k(m_1) = e_k(e_{k-1} \ldots e_1(m_1)) = m_{k+1}$. A new proposed definition of the metabolic pathway is presented and discussed in the following paragraphs.

Definition 1. *Given $e_i : M \rightarrow M$, a metabolic pathway is defined as a subset of successive enzymatic reactions $P = e_1e_2 \ldots e_k$.*

Each enzymatic reaction e_i ($1 \le i \le k$) is catalyzed by a certain enzyme that is denoted as a unique EC number. The EC number is expressed with a four-level hierarchical scheme that was developed by the International Union of Biochemistry and Molecular Biology (IUBMB; Webb, 1992). The four-digit EC number, $d_1.d_2.d_3.d_4$ represents a sub-sub-subclass indication of biochemical reaction. For instance, arginase is numbered by EC 3.5.3.1, which indicates that the enzyme is a hydrolase (EC 3.*.*.*), acts on the 'carbon–nitrogen bonds, other than peptide bonds' (sub-class EC 3.5.*.*) in linear amidines (sub-sub-class EC 3.5.3.*). Thus, we can adapt the EC number as a unique name for the responding enzyme catalyzed reaction.

Example 1. $e_{3.5.3.1}$ means the biochemical reaction that is catalyzed by the enzyme 3.5.3.1., which catalyzes arginine into urea,

$$e_{3.5.3.1}(\text{arginine}) = \text{urea},$$

and

$$e_{2.1.3.3}e_{6.3.4.5}e_{4.3.2.1}e_{3.5.3.1}(\text{carbamoyl-P}) = \text{urea}$$

indicates that a metabolic pathway $e_{2.1.3.3}e_{6.3.4.5}e_{4.3.2.1}e_{3.5.3.1}$ starts from the enzymatic reaction 2.1.3.3 (symbolized as '1'in Figure 1) with carbamoyl-P and L-ornithine as reactants and after a series of reactions—6.3.4.5 ('2')→4.3.2.1 ('3')→3.5.3.1 ('4')—results in L-ornithine and urea as products.

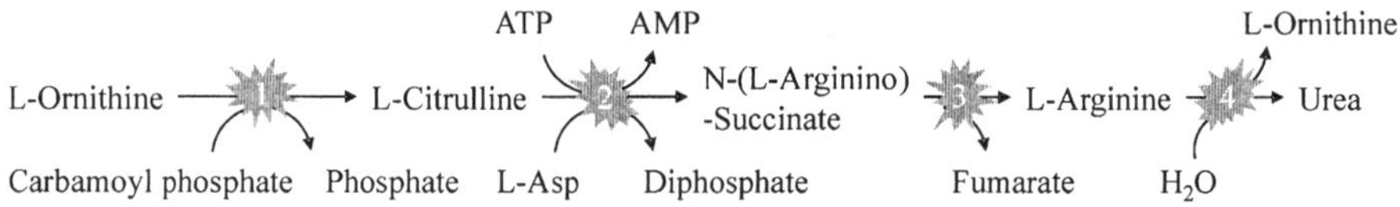

Figure -1. Indication of the pathway e2.1.3.3e6.3.4.5e4.3.2.1e3.5.3.1 as specified above.

2.2 Metabolic pathway prediction

In order to predict metabolic pathways, we develop a system, which exploits the advantages of the Internet to automatically extract metabolic information from multiple heterogeneous molecular biology databases ranging from genomics to metabolism and to represent the functional data with an easy-to-use Web interface. The problem studied here can be started formally as follows.

A 4-tuple (M_0, E_0, G, S) is called a query Q, which describes the components of a rudimentary metabolic pathway:

$M_0 = \{m_1, m_2, ..., m_i\}$is a subset of M;

E_0 is a subset of E;

$G = \{g_1, g_2, ..., g_j\}$ is a set of genes; and

$S = \{s_1, s_2, ..., s_k\}$ is a set of nucleotide or amino acid sequences.

Given a query Q, when the elements of M_0, G, and S are queried to the remote data sources, then sets of associated EC numbers are extracted:

$M_0 \Rightarrow \boldsymbol{E}_{M0} = \{E_{m1}, E_{m2}, E_{m3}, ..., E_{mi}\}$;

$G \Rightarrow \boldsymbol{E}_G = \{E_{g1}, E_{g2}, E_{g3}, ..., E_{gj}\}$; and

$S \Rightarrow \boldsymbol{E}_S = \{E_{s1}, E_{s2}, E_{s3}, ..., E_{sk}\}$,

where E_{mi} is a set of EC numbers related to the m_i ($m_i \in M_0$). E_{gj} is a set of EC numbers related to the g_j ($g_j \in G$), and E_{sk} is a set of EC numbers related to the s_k ($s_k \in S$).

Example 2. Given $Q = (M_0, E_0, G)$, where $E_0 = \{6.3.4.5\}$, $M_0 = \{L\text{-arginine}\}$, $G = \{\text{ASL, OTC}\}$, then $\boldsymbol{E}_{M0} = \{(1.13.12.1,\ 1.14.13.39,\ 2.1.4.1,\ 2.1.4.2,\ 2.3.1.109,\ 2.4.2.31,\ 2.7.3.3,\ 3.5.3.1,\ 3.5.3.6,\ 4.1.1.19,\ 5.1.1.9,\ 6.1.1.19,\ 6.3.2.24)\}$ and $\boldsymbol{E}_G = \{(4.3.2.1), (2.1.3.3)\}$.

Letting
$\boldsymbol{E}'_{M0} = \{E_{m1} \times E_{m2} \times E_{m3} \times \ldots \times E_{mi}\}$,
$\boldsymbol{E}'_{G} = \{E_{g1} \times E_{g2} \times E_{g3} \times \ldots \times E_{gj}\}$, and
$\boldsymbol{E}'_{S} = \{E_{s1} \times E_{s2} \times E_{s3} \times \ldots \times E_{sk}\}$,
we define $\boldsymbol{E}'_{M0} + \boldsymbol{E}'_{G} + \boldsymbol{E}'_{S} = \{E'_{mi} \cup E'_{gj} \cup E'_{sk}\}$,
where $E'_{mi} \in \boldsymbol{E}'_{M0}$, $E'_{gj} \in \boldsymbol{E}'_{G}$; and $E'_{sk} \in \boldsymbol{E}'_{S}$.

Letting $\boldsymbol{P}_c = \boldsymbol{E}'_{M0} + \boldsymbol{E}'_{G} + \boldsymbol{E}'_{S}$, we define $\boldsymbol{P}_c + E_0 = \{P \cup E_0 : P \in \boldsymbol{P}_c\}$; then we have

$\boldsymbol{P}_c + E_0 = \{E_1, E_2, \ldots E_p\}$, which is a set of EC sets, where $E_i = \{e_{i1}, e_{i2}, \ldots, e_{ik} \mid 1 \leq i \leq p\}$.

Continuing **Example 2**, now we have a set of combinatorial EC number sequences:

$\boldsymbol{E}'_{M0} + \boldsymbol{E}'_{G} + \boldsymbol{E}'_{S}$ = {(1.13.12.1, 4.3.2.1, 2.1.3.3), (1.14.13.39, 4.3.2.1, 2.1.3.3), …, (6.3.2.24, 4.3.2.1, 2.1.3.3)} and

$\boldsymbol{P}_c + E_0$ = {(6.3.4.5, 1.13.12.1, 4.3.2.1, 2.1.3.3), (6.3.4.5, 1.14.13.39, 4.3.2.1, 2.1.3.3), …, (6.3.4.5, 6.3.2.24, 4.3.2.1, 2.1.3.3)}.

E_i is then searched for against the pool database to find the metabolic pathway with the highest similarity score.

2.3 Metabolic pathway alignments

In order to score the similarity (percent identity) between two metabolic pathways, we define the similarity function.

Definition 2. *Let E be a finite set of e functions, an edit operation is an ordered pair* $(\alpha, \beta) \in (E \cup \{\varepsilon\}) \times (E \cup \{\varepsilon\}) \setminus \{(\varepsilon, \varepsilon)\}$.

Here, α and β denote four-digit EC strings of enzymatic reaction function, e.g., $\alpha = e_{1.1.1.1}$ and $\beta = e_{2.3.4.5}$, ε denotes the empty string for the null function. However, if $\alpha \neq \varepsilon$ and $\beta \neq \varepsilon$, then the edit operation (α, β) is identified with a pair of enzymatic reaction function.

An edit operation (α, β) is written as $\alpha \rightarrow \beta$ (we can simply write α, β as EC numbers). There are three kinds of edit operations:

$\alpha \rightarrow \varepsilon$ denotes the deletion of the enzymatic reaction function α,

$\varepsilon \rightarrow \beta$ denotes the insertion of the enzymatic reaction function β, and

$\alpha \rightarrow \beta$ denotes the replacement of the enzymatic reaction function α by the enzymatic reaction function β.

Note that $\varepsilon \rightarrow \varepsilon$ never happens.

Definition 3. *Let* $E_1 = e_1 e_2 \ldots e_m$ *and* $E_2 = e_1' e_2' \ldots e_n'$ *be two metabolic pathways, an alignment of* E_1 *and* E_2 *is a pair sequence*

$$(\alpha_1 \rightarrow \beta_1, \ldots, \alpha_h \rightarrow \beta_h)$$

of edit operations such that $E_1' = \alpha_1, \ldots, \alpha_h$ and $E_2' = \beta_1, \ldots, \beta_h$.

Example 3. Given two metabolic pathways $e_{4.3.2.1}e_{6.3.4.5}e_{2.1.3.3}$ and $e_{4.3.2.1}e_{6.3.4.5}e_{2.6.1.1}e_{2.3.1.35}$, then one of their alignments is A = (4.3.2.1→4.3.2.1, 6.3.4.5→6.3.4.5, ε→2.6.1.1, 2.1.3.3→2.3.1.35), which is written as follows, one over the other:

$$\begin{pmatrix} 4.3.2.1 & 6.3.4.5 & \varepsilon & 2.1.3.3 \\ 4.3.2.1 & 6.3.4.5 & 2.6.1.1 & 2.3.1.35 \end{pmatrix}$$

Definition 4. *A similarity function σ assigns to each edit operation* (α, β) *a nonnegative real number. The similarity* σ (α, ε) *and* σ (ε, β) *of the deletion operation* (α, ε) *and insertion operation* (ε, β) *is* 0. *For all replacement operations* (α, β) α ≠ ε, β ≠ ε, *say,* $\alpha = d_1.d_2.d_3.d_4$ *and* $\beta = d_1'.d_2'.d_3'.d_4'$, *then the similarity function* σ (α, β) *is defined by*

$$\sigma(\alpha, \beta) = \begin{cases} 0, \textit{ if } (d_1 \neq d_1'); \\ 0.25, \textit{ if } (d_1 = d_1' \textit{ and } d_2 \neq d_2'); \\ 0.5, \textit{ if } (d_1 = d_1' \textit{ and } d_2 = d_2' \textit{ and } d_3 \neq d_3'); \\ 0.75, \textit{ if } (d_1 = d_1' \textit{ and } d_2 = d_2' \textit{ and } d_3 = d_3' \textit{ and } d_4 \neq d_4'); \\ 1, \textit{ if } (d_1 = d_1' \textit{ and } d_2 = d_2' \textit{ and } d_3 = d_3' \textit{ and } d_4 = d_4', \textit{ i.e. } \alpha = \beta). \end{cases}$$

The definition does not exclude the possibility that d_4, $d_3.d_4$, and $d_2.d_3.d_4$ can be respectively expressed as wild card symbols *, *.* and *.*.*, which means no clear classification of the enzyme.

However, single pair of EC string comparison just means to measure how different EC strings are. Often, it is additionally of interest to analyze the total difference between the two strings into a collection of individual elementary differences. The most important mode of such analyses is an alignment of the pathways. The function σ can be extended to alignments in a straightforward way: the similarity σ (*A*) of an alignment $A = (\alpha_1 \to \beta_1, \ldots, \alpha_h \to \beta_h)$ is the sum of the similarities of the edit operations constituting *A:*

$$\sigma(A) = \sum_{i=1}^{h} \sigma(\alpha_i \to \beta_i).$$

Example 4. The similarity of the alignment A_5 in **Example 3** is
$\sigma(A_5) = \sigma(4.3.2.1 \to 4.3.2.1) + \sigma(6.3.4.5 \to 6.3.4.5) + \sigma(\varepsilon \to 2.6.1.1)$
$+ \sigma(2.1.3.3 \to 2.3.1.35) = 1 + 1 + 0 + 0.25 = 2.25.$

An alignment scoring scheme, Score(E_1, E_2), of two metabolic pathways is the minimal mean similarity of their alignment:

$$\text{Score}(E_1, E_2) = \frac{1}{\max(m,n)} \sigma(A),$$

where *m*, *n* are the lengths of pathways.

Then, the alignment score of **Example 3** and **Example 4** is

$$\text{Score}(E_1, E_2) = \frac{1}{4}(1+1+0+0.25) = 0.55.$$

The scoring scheme is generic. We restrict the alignment algorithm to perform the removal of unmatched elements from both ends of the pathway.

2.4 Alignment algorithms

In general, we align metabolic pathways one above the other. Given two metabolic pathways P_1 and P_2, the implemented algorithm is shown below.

Example 5. Let P_1 = {2.7.4.14, 3.1.3.5, 3.5.4.5, 2.4.2.3} and P_2 = {2.7.4.14, 3.2.2.10, 3.5.4.1, 2.4.2.3, 3.5.4.5} be two pathways. The processes and results of alignment will be:

Step 1: *Finding all* EC *numbers with same 4-level hierarchical numbers from both sides:*

{2.7.4.14, 3.1.3.5, 3.5.4.5, 2.4.2.3}
{2.7.4.14, 3.2.2.10, 3.5.4.1, 2.4.2.3, 3.5.4.5}.

Step 2: *Finding all* EC *numbers with same 3-level hierarchical numbers in each sub-sequence:*

{2.7.4.14, 3.1.3.5, 3.5.4.5, 2.4.2.3}
{2.7.4.14, 3.2.2.10, 3.5.4.1, 2.4.2.3, 3.5.4.5}.

Step 3: *Finding all* EC *numbers with same 2-level hierarchical numbers in each sub-sub-sequence:*

{2.7.4.14, 3.1.3.5, 3.5.4.5, 2.4.2.3}
{2.7.4.14, 3.2.2.10, 3.5.4.1, 2.4.2.3, 3.5.4.5}.

Step 4: *Finding all* EC *numbers with same 1-level hierarchical numbers in each sub-sub-sub-sequence.*

{2.7.4.14, 3.1.3.5, 3.5.4.5, 2.4.2.3}
{2.7.4.14, 3.2.2.10, 3.5.4.1, 2.4.2.3, 3.5.4.5}.

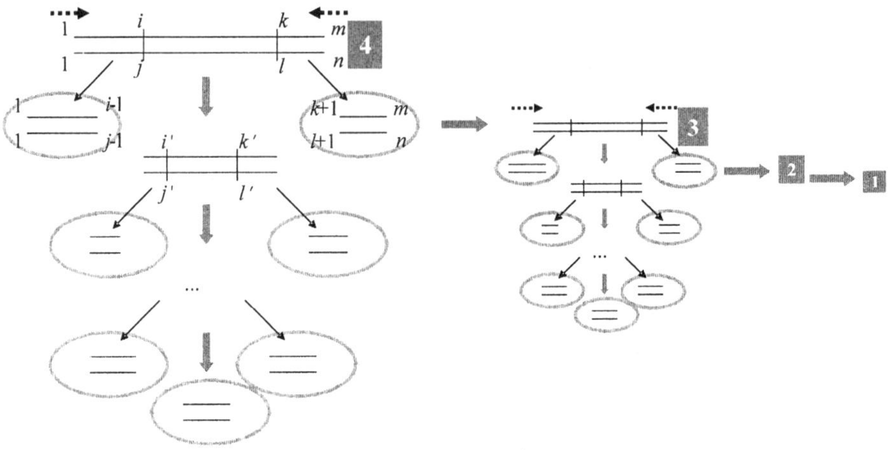

Figure -2. Overview of the alignment algorithm based on a four-hierarchical process. (Step 1: 4-level) Initialize the set of unaligned EC number sequences, their lengths, and score value. Starting from both ends towards the middle, align one sequence to another and attempt to find all EC numbers with same 4-level hierarchical numbers. Score the similarities. Recall the alignment positions where EC numbers are identical and cut the sequences into more subsequences by removing the identical EC numbers. (Step 2: 3-level) Each pair of sub-sequences (inside ellipses) is initialized to begin a new round of 3-level hierarchical EC number matching until all the pairs of sub-sequences are aligned. A similarity score is calculated afterwards. (Step 3: 2-level) Apply the same rule again, find the similarities of the rest unaligned sub-sub-sequences based on 2-level hierarchical EC number matching. (Step 4: 1-level) The sub-sub-sub-sequences on 1-level matching, if any.

3. IMPLEMENTATION

3.1 PathAligner system architecture

PathAligner is a Web-based biological information retrieval system with one main purpose designed for the prediction/alignment of metabolic pathways. The PathAligner system contains a PathModeler and a pathway alignment tool (Figure 3). PathModeler consists of four parts. The first part is a database-mining tool that pulls out the potential metabolic relationships from various databases based on the queried rudimentary components, such as metabolites, genes, sequences, etc. It allows easy access to distributed heterogeneous biological resources through a simple interface. The relationships are then organized, recombined, and queried against metabolic pathway database to retrieve a metabolic pathway result. Genetic and metabolic information involved in the retrieved pathway is extracted and displayed in the second part. In the third part, the retrieved metabolic information is visualized using an interactive graph display module. Finally, an XML data file that contains the basic information about metabolic and regulatory network as well as their kinetic values is formed for further modeling and analysis.

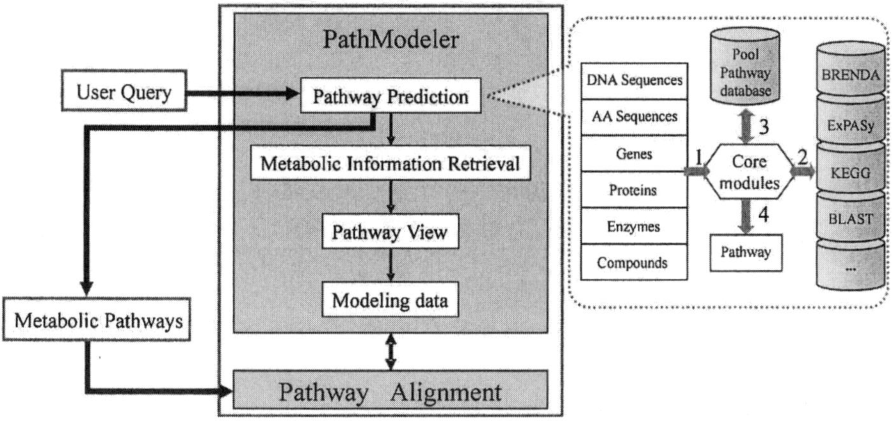

Figure -3. The concept of PathAligner system.

The dotted box in the right side of Figure 3 shows a more detailed architecture of the pathway prediction. By introducing a remote access communication architecture, it allows different distributed heterogeneous biological resources communicating through the same common easy-to-use Web interface and enables researchers to perform efficient and effective biological data retrieval and metabolic pathway reconstruction. It allows the user to enter rudimentary elements that are then run against the remote databases. The processes of the core modules are for distributing and locating the responding database system to answer user's queries via the Web interface. Internet's mechanisms support and maintain the communication between Web browsers and database shells. We have also constructed a pool pathway database of the known metabolic reactions from several online databases, such as KEGG and MetaCyc. It contains metabolic pathways with EC numbers according to our definition of pathway. The pool pathway database is used to identify the most possible pathway corresponding to the rudimentary pathway. The system is completely transparent to the users.

3.2 PathAligner Web interface

PathAligner is available at http://bibiserv.techfak.uni-bielefeld.de/ pathaligner (Figure 4).

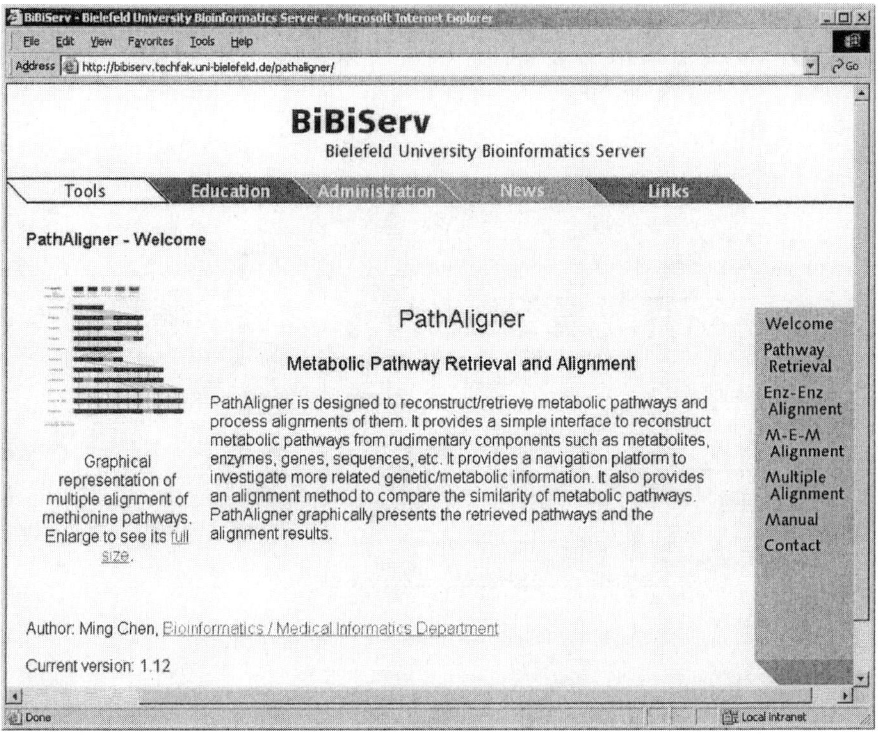

Figure -4. A screenshot of PathAligner.

Besides '*Pathway Retrieval*', there are three other Web-based alignment interfaces that are implemented: '*E-E Alignment*', '*M-E-M Alignment*', and '*Multiple Alignment*'. '*E-E Alignment*' uses the basic algorithm to align two linear metabolic pathways (represented as EC number sequences). User can also align any such metabolic pathway against our pool database to find a list of hits. '*M-E-M Alignment*' considers the differences in metabolites in the two pathways, which are presented as 'Metabolite-EC number-Metabolite' patterns of sequence. It is possible to pick up two such pathways and align them to identify whether they are alternative pathways or partially are. '*Multiple Alignment*' allows the alignment of more than two metabolic pathways.

4. CONCLUSION

We presented a new approach to study the problem of metabolic pathway prediction/alignment. Metabolic pathway defined as a linear reaction sequence is practical for our alignment algorithm. Our algorithm calculates the hierarchical similarities of EC numbers mapping from both ends of the sequences. The algorithm has been successfully implemented into the PathAligner system.

PathAligner is such a Web-based computational system that focuses the efforts on extracting metabolic information from biological databases via the Internet, reconstructing metabolic pathways from diverse rudimentary components, embracing alignment algorithms to integrate diverse biological data more easily, and handling the dynamic nature (continual updating or revisions) of biological data. It provides an easy-to-use interface to retrieve, display, and manipulate metabolic information. There is no problem with the data out of date, because it queries databases remotely instead of locally. Although the procedure might take a little while retrieving data, the accessibility and update are guaranteed, since those databases are global-oriented and maintained by their reputed institutes. It significantly reduces the effort and difficulty involved in data integration and analysis. The retrieved metabolic pathways can be aligned to determine their similarities and distances.

PathAligner provides a heuristic method to analyze the similarity and distance of different metabolic pathways by measuring the functional similarity of enzymatic reactions (EC numbers). The alignment algorithm differs from classical string alignments as well as from previous approaches.

ACKNOWLEDGMENT

This work was supported by the German Research Council (DFG).

GENERAL PRINCIPLES OF ORGANIZATION AND LAWS OF FUNCTIONING IN GOVERNING GENE NETWORKS

R.N. Tchuraev
Department of Physicochemical Biology and Epigenetics, Ufa Research Center, Russian Academy of Sciences, 69 Prosp. Oktyabrya, Ufa, 450054, Russia, e-mail: tchuraev@anrb.ru

Abstract: General principles of organization and laws of activity dynamics have been revealed in intracellular governing eukaryotic gene networks. Following from the hypothesis of *self-differentiation*, a theorem was proved on the existence of *metastable* states in gene networks under *envastat*. A hypothesis was proposed to explain an increase in orderliness during ontogenesis.

Key words: governing gene networks; cellular automata and cell ensembles; metastability

1. INTRODUCTION

In the past 30 years, many attempts have been made to investigate the dynamics of large multicomponent molecular genetic governing systems by means of mathematical and computer modeling using various mathematical theories (Tchuraev and Ratner, 1973; Kauffman, 1974; Glass, 1975; Ratner and Tchuraev, 1978; Tchuraev, 1991; McAdams and Shapiro, 1995; Kolchanov et al., 1998). Dynamics elucidation problems by a predetermined structure have already been solved for a number of real subnetworks (Kananyan et al., 1980; Tchuraev and Galimzyanov, 2003; Kolchanov et al., 2005). The current problem now is to elucidate the principles of organization and laws of functioning in cellular governing gene networks common to quite different eukaryotic organisms. Knowledge of general principles and laws of functioning in governing gene networks makes possible the elaboration of efficient algorithms based on them to solve the problems on analysis and synthesis of real gene networks.

The paper presents the results obtained in attempting to develop a theory of gene networks and, hence, to get rigorously proved statements essential to the understanding of biological aspects of the mechanisms by which hereditary information is stored, encoded, and transmitted, and the way it is realized in ontogenetic processes of self-reproduced multicellular organisms.

2. FROM EQUATIONS OF ACTIVITY TO CELLULAR AUTOMATA

The *objects* of the theory in question are intracellular governing subsystems whose function is to govern rapid metabolic and slow ontogenetic processes.

As the elements of a cellular governing subsystem, we consider gene blocks G_j, i.e., gene j taken in combination with the mechanisms of transcription, processing, transport, and deposition of its final product. Signal transduction from one G-block to another is accomplished by regulatory molecules of different specificity. Five postulates were accepted for them in the context of microapproach (Tchuraev, 1975, 1998), which enabled us to derive general-form equations of activity dynamics for governing (g) eukaryotic (e) gene networks $S_e^g(G)$ represented as a finite oriented graph with a set of gene blocks $G = \{G_1, \ldots, G_i, \ldots, G_N\}$:

$$\mathbf{\Gamma}(t) = \boldsymbol{F}\{\boldsymbol{f}\,[\mathbf{\Gamma}(t-\tau), \boldsymbol{E}(t-\tau)]\}, \qquad (1)$$

where $\mathbf{\Gamma}(t) = \langle \gamma_1(t), \gamma_2(t), \ldots, \gamma_j(t), \ldots, \gamma_N(t)\rangle$ is the γ-vector of the activities of all elements in the network $S_e^g(G)$; in this case, $\gamma = \gamma(t)$ is the binary value; t is the discrete time; $\boldsymbol{F}$ is the column dimension $N \times 1$ whose elements are Boolean functions ('*composition*' of logic structures); $\boldsymbol{f} = \| f_{ij} \|$ is the matrix dimension $J^j \times N$, where each element f_{ij} is a restrictedly determined operator that connects the internal variable υ_{ij} to a sequence of input signals e_{ij} entering via the ith input channel of the jth gene block; $\tau = \max \tau_{ij}$ is the maximum delay of output signals among all G-blocks affecting the given one (G_i); and $\boldsymbol{E}(t-\tau) = \langle e_1(t-\tau_1), e_2(t-\tau_2), \ldots, e_h(t-\tau_h), \ldots, e_H(t-\tau_H)\rangle$ is the word of length H; in this case H is the number of input channels in the network $S_e^g(G)$.

The magnitude of the γ-vector can be experimentally observed by noting either presence or absence of specific gene products at a given instant of time. Hence, $\mathbf{\Gamma}(t)$ is *the observable behavior of the governing gene network.* Following from the Kobrinsky–Trakhtenbrot theorem (1962), it may be stated (1) that *if stationary sequences of input signals enter via all inputs of any finite gene subnetwork, the latter is either at rest or in periodic regime. The number of points of the rest and periodic regimes are finite values.*

It is known that the restrictedly determined operator with a finite weight realized by the network $S_e^g(G)$ can be represented as a finite automaton; we call it *the cellular automaton* $A_e^g(G)$, in which the governing cellular network $S_e^g(G)$ represents its internal structure.

3. PRINCIPLES OF ORGANIZATION IN GOVERNING GENE NETWORKS

Since each specific regulatory substance in the cell is a product of gene blocks, we accept the following *structural postulate*:

For each element G_j of the network $S_e^g(G)$ there is such an element G_i that at least one of its outputs is connected to the input of the element G_j.

Based on the structural postulate and finiteness of elements in the network $S_e^g(G)$, statement (2) can be made in the following way.

Each gene block (element) of the governing gene network $S_e^g(G)$ adheres to at least one governing loop (oriented cycle), i.e., it either enters the oriented cycle or is connected to this cycle with a signal circuit.

Previously, we have established that some governing loops have the properties of dynamic memory elements, which are simply exemplified with genetic triggers (bistable memory modules). Moreover, a number of such elements represent *the epigenes* capable of encoding, storing, and transmitting a part of hereditary information in a series of successive generations without genomic DNA primary succession (Tchuraev, 1975, 1997, 2005; Tchuraev *et al.*, 2000).

In any event, eukaryotic cellular governing gene networks are organized in *a modular fashion,* i.e., any eukaryotic gene network S_e^g can be represented as a network of blocks (subnetworks), where each block at a higher level of complexity is built of the blocks from the previous level of complexity, both levels functioning as a whole (tinkering). As follows from this principle, governing gene networks $S_e^g(G)$ *can be transformed into* $\tilde{S}_e^g(G)$ *whose elements will consist of the following modules: genetic triggers (bistable memory modules), oscillators, and delay logical combinators.* Let us take an example of transformation (Figures 1 and 2).

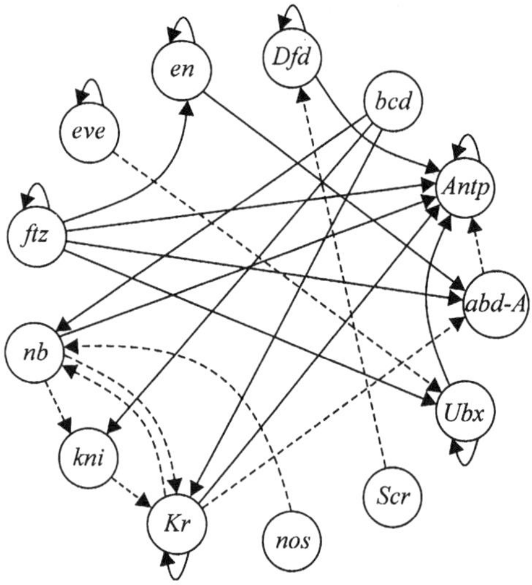

Figure -1. Gene subnetwork S_e^g (Dr) controlling *Drosophila melanogaster* early ontogenesis. This fragment consists of 13 genes: *bicoid* (*bcd*) and *nanos* (*nos*) of the maternal coordinate class; *hunchback* (*hb*), *knirps* (*kni*), and *Krüppel* (*Kr*) of the gap class; *even-skipped* (*eve*) and *fushi tarazu* (*ftz*) of the pair-rule class; *engrailed* (*en*) of the segment-polarity class; and *abdominal-A* (*abd-A*), *Antennapedia* (*Antp*), *Deformed* (*Dfd*), *Sex combs reduced* (*Scr*), and *Ultrabithorax* (*Ubx*) of the homeotic class. The negative regulatory bond among genes (transcription repression or translation inhibition) is denoted with a dotted line; the positive bond (transcription activation), with a solid line.

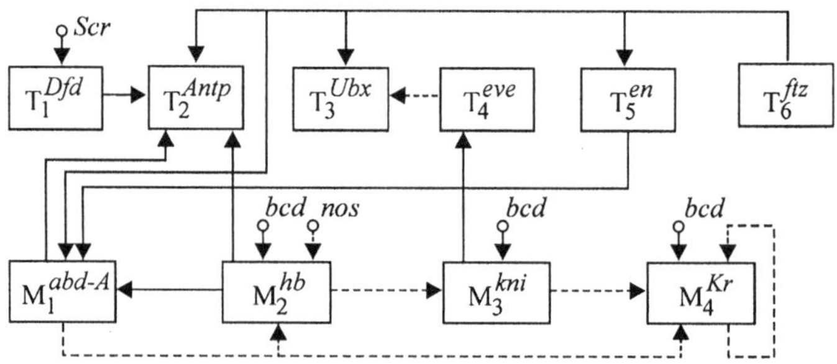

Figure -2. Transformed Gene Network S_e^g (Dr): T, genetic triggers (bistable memory modules) and M, delay logical combinators.

There is a time hierarchy in the structures of cellular governing gene networks S_e^g (*G*) that ensures gene sequential switching during ontogenesis (Tchuraev, Galimzyanov, 2003).

As follows from statement (2), there should be feedbacks in gene networks $S_e^g(G)$. Does the existence of such feedbacks exclude a hierarchic principle in the organization of cellular governing gene networks $S_e^g(G)$? No, it apparently does not, if, as applied to networks $S_e^g(G)$, the term *heterarchy* (McCulloch, 1945) will mean the existence of feedbacks that connect the output channels of gene blocks at different hierarchic levels to the input channels of gene blocks of a higher hierarchic rank. Summing up, the following statement can be made: *any cellular governing gene networks are organized on the principles of both hierarchy and heterarchy.* In addition, a correlation is found between the structure of governing gene network $S_e^g(G)$ and schematic blocks of the ontogenetic hereditary program that realizes the inherited algorithm χ (***principle of correlation between the structure and the ontogenetic function***). An investigation of this correlation (representation) is a rather nontrivial task.

It is known that complicated multimeric complexes, agregulons, whose specificity depends on composition, may serve as carriers of molecular signals (Jacob, 1993). It is easily believed that using n number of different monomers, the q number of different regulatory multimers can be formed:

$$q = \frac{n!}{(n-p)!p!}, \tag{2}$$

where p is the level of multimers. Thus, *there are combinatorial modules in eukaryotic governing networks, and their function consists in generating a large number of different signals from a small number of molecular signals. The combinatorial modules realize a* ***combinatorial principle of gene expression control.***

4. CELLULAR AUTOMATA AND THEIR ENSEMBLES

As was already noted, each eukaryotic cell can be related to cellular automaton A_e^g, in which the governing cellular network $S_e^g(G)$ represents its internal structure. Such cellular automata and their ensembles may be investigated even without going into details of their internal structure, through observing only input and output signals. Thus, in the general form, a canonical description of cellular automaton is as follows:

The cellular automaton A_e^g is described with five symbols (**E**,**∇**, **Ω**, **Φ**, and **Ψ**), where **E** and **∇** are the input and output alphabets, **Ω** is the set of internal memory Ξ states, and **Φ** and **Ψ** denote the transition and output functions, respectively.

The input alphabet **E** of the automaton A_e^g with n_1 number of input channels constitutes a set of corteges (words of length n_1): $\mathbf{E} = \{\mathbf{e}\}$, where $\mathbf{e} = \left\langle \varepsilon_1(\mathrm{t}), \varepsilon_2(\mathrm{t}), ..., \varepsilon_l(\mathrm{t}), ..., \varepsilon_{\mathrm{n}_1}(\mathrm{t}) \right\rangle$, $l = 1, n_1$, and elements ε of the cortege **e** are the binary values.

The output alphabet $\mathbf{\nabla} = \{\mathbf{\Gamma}\}$ of the automaton A_e^g constitutes a set of all possible words **Γ** of the length *N*:

$$\mathbf{\nabla} = \{\Gamma_1, \Gamma_2, \ldots\Gamma_j, \ldots \Gamma_{2^N}\},$$

where $\mathbf{\Gamma} = \mathbf{\Gamma}(t) = \langle\gamma_1(t), \gamma_2(t), \ldots, \gamma_J(t), \ldots, \gamma_N(t)\rangle - \gamma$ is the vector of gene activities in the governing gene network.

The set of states **Ω** of the memory Ξ in the automaton A_e^g is

$$\mathbf{\Omega} = \{\omega_0, \omega_1, \ldots \omega_m, \ldots \omega_M\}, m = \overline{(0,M)}.$$

The *transition* **Φ** and output **Ψ** functions in the automaton A_e^g have the form:

$$\boldsymbol{\omega}(t+1) = \mathbf{\Phi}[\boldsymbol{\omega}(t), \mathbf{e}(t+1)], \qquad \mathbf{\Gamma}(t+1) = \mathbf{\Psi}[\boldsymbol{\omega}(t+1)].$$

The cellular automaton A_e^g described in such a way is the model of the governing gene network in the most general form. Figure 3 represents simple example of this class of automata.

At each discrete instant of time t, it is possible to note the *observable* values or the activities of all genes in the governing gene network S_e^g, i.e., for the output channels of the automaton A_e^g, we consider the output channels of all its elements, not only those unconnected to other elements of the network. Such a representation of output symbols in the cellular automaton is motivated by the possibility to have experimentally observable patterns of gene activities in the governing gene network judging, for example, by the presence (or absence) of primary transcripts.

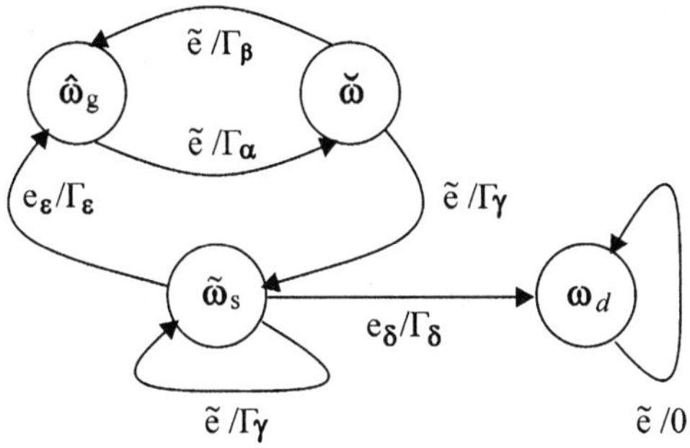

Figure -3. Example of Moor's Diagram – graph of transitions for minimum cellular automata. $\hat{\omega}_g$ is the unstable state corresponding to germ-line; $\breve{\omega}$ is the metastable state; $\tilde{\omega}_s$ is the stable stationary state corresponding to somatic cells; ω_d – blind state, bringing to apoptosis; $\tilde{e}$ is the neutral input signal; $\mathbf{e}_\delta$ is the signal of apoptosis subprogram starting; $\mathbf{e}_\varepsilon$ is the signal to regeneration; ω_d is the irredundant state reducing to apoptosis; launching apoptosis subprogram; $\mathbf{\Gamma}_\alpha$, $\mathbf{\Gamma}_\beta$, $\mathbf{\Gamma}_\gamma$, $\mathbf{\Gamma}_\delta$ and $\mathbf{\Gamma}_\varepsilon$ are the vectors of gene block activities corresponding to subprograms.

It should be noted that with the account for the internal structure of the automaton A_e^g, each state $\boldsymbol{\omega} \in \boldsymbol{\Omega}$ will constitute a complicated composition of ministates in each element in the network S_e^g, the set of states $\boldsymbol{\Omega}$ being the Descartian product of ordered sets of these states N:

$$\boldsymbol{\Omega} = \mathcal{D}_{\otimes}{}_{i=1}^{N} Q_i, \tag{3}$$

where the symbol $\mathcal{D}_\otimes$ denotes the Descartian product and Q_i is the set of states in the ith element of the network S_e^g, in this case $i = \overline{(1, N)}$.

The act of reduplication, introduced for the first time by Tchuraev (1991), has the form:

$$A_e^g \to {}'A_e^g \cup {}''A_e^g,$$

where ${}'A_e^g$ and ${}''A_e^g$ are the copies of the parent automaton A_e^g. The networks S_e^g, ${}'S_e^g$, and ${}''S_e^g$ 'parent' and 'daughter' automata are isomorphic.

During the reduplication of the automata A_e^g, there occurs a sequential formation of *the cellular automata ensemble* $\mathcal{A}_x$ corresponding to the individual x that evolves from a zygote into a reproductive form. A natural requirement for the ensemble $\mathcal{A}_x$ is its capability of being self-reproduced.

Generalized cellular governing gene networks (GGN) in eukaryotes have three fundamental properties:

(a) during the cell sequential divisions, there must be generative cells in which the GGN comes back to their initial state (*'from zygote to zygote'*);

(b) after a series of the initial cell (zygote) divisions, they should be able to give rise to several 'somatic' lines (*ability for divergent determination and differentiation*); and

(c) some functional states of the GGN should be preserved in a series of cell sequential divisions (*stability of determinate states*).

Of course, these three properties should also be inherent in cellular automata (elements of the cellular automata ensemble $\mathcal{A}_x$) corresponding to cellular governing networks in the model of the individual x. These properties are accepted as premises in our theoretical model.

During successive cellular automata reduplications, the automata ensemble should become heterogeneous instead of homogeneous, i.e., *divergent determination* should take place. It is hardly possible that divergent determination results from external environmental influences. Therefore, it would be rightful to perform a mental experiment when a multicellular individual x, which should evolve from a zygote, is placed under conditions of a total absence of external signals, i.e., all external factors are supposed to be neutral (*envastat* conditions). Let us take *the hypothesis on capacity for self-differentiation.*

It is assumed that a clone formed from a zygote in a neutral medium is capable of being self-differentiated.

Or, speaking in terms of the model, the cellular automata ensemble $\mathcal{A}_x$ evolving in a neutral medium becomes heterogeneous (when a pair of automata appears in distinguishable states) instead of homogeneous (when all cellular automata are in similar states) at some instant of time t_{μ}^{k} (when divergent determination takes place).

For two-dimensional cellular automata, Moore (1962) introduced a honeycomb neighborhood ('universe' defined by six positions, one of which is the causality principle. In Moore's formal language, the cell ensemble $\mathcal{A}_x$ is put into correspondence with the honeycomb-like block x and its properties (a–c) are formalized. In the context of this formalism, as a result of the formalized hypothesis of self-differentiation, it is easy to prove a theorem on the existence of specific *metastable states*.

Let us formulate **the theorem on the existence of metastable states**.

If the cell ensemble $\mathcal{A}_x$ evolving in a neutral environment is capable of being self-differentiated, then, in the multitude of states $\mathbf{\Omega}$ *of internal* *memory* Ξ *in any cellular automaton* A_e^g *involved in the particular ensemble* ($A_e^g \in \mathcal{A}_x$), *there would be at least one metastable state* $\breve{\omega}$, *such that:*

$$A_e^g / \breve{\omega} \rightarrow {}' A_e^g / \omega_\upsilon \cup {}'' A_e^g / \omega_p,$$

where ω_υ and ω_p are the distinguishable states.

Since the states of cellular automata represent functional states of the intracellular gene network, we can assert, in the context of the hypothesis on self-differentiation, that *the mechanism of primary divergent determination is in fact the intracellular molecular genetic one.*

The existence of metastable states imposes constraints on the structures of governing gene networks. We have studied the behavior of some simplest networks with a metastable state (Tchuraev, 1980).

5. LAWS OF ACTIVITY DYNAMICS FOR CELLULAR AUTOMATA

Now let us write down the laws of cellular automata behavior.

The *first law* of the activity dynamics has the following form for the observable *cellular* automaton behavior:

$$\boldsymbol{\Gamma}(t) = \boldsymbol{\Psi}\{\boldsymbol{\Phi}\,[\boldsymbol{\omega}(t-1), \mathbf{e}(t)]\} = \tilde{\boldsymbol{\Gamma}}, \qquad \text{if } \boldsymbol{\omega}(t-1) = \tilde{\boldsymbol{\omega}}, \mathbf{e}(t) = \tilde{\mathbf{e}},$$

where $\boldsymbol{\Gamma}(t) = \langle \gamma_1(t), \gamma_2(t), \ldots, \gamma_J(t), \ldots, \gamma_N(t) \rangle$ is the γ-vector of the activities of gene blocks in the governing gene network (subnetwork) S_e^c (G), $\boldsymbol{\Psi}$ is the single-valued function of the state $\boldsymbol{\omega}$, $\boldsymbol{\Phi}$ is the transition function being not necessarily single-valued at some points, $\tilde{\boldsymbol{\omega}}$ is the stationary state, and $\tilde{\mathbf{e}}$ is the neutral input signal.

The *second law* of the dynamics of activities is as follows:

If $\mathbf{e}(t) = \tilde{\mathbf{e}}$ *and* $\boldsymbol{\omega} \neq \tilde{\boldsymbol{\omega}}$, *the operator function* $\boldsymbol{\Phi}$ *(and accordingly* $\boldsymbol{\Psi}$*) are periodic.*

Thus, in the neutral environment (envastat conditions), any cellular automaton and associated intracellular governing network are either in one of the states of rest or function in one of possible periodic regimes. These laws are semi-trivial and resemble Newton's laws of mechanics.

The *third law* of the dynamics of activities is as follows:

If $\mathbf{e}(t) \neq \tilde{\mathbf{e}}$, $\omega \neq \tilde{\omega}$,

$$\Gamma(t) = \Psi\{\Phi[\omega_{i\alpha}(t-1), \mathbf{e}(t)] = \begin{cases} \Gamma_\alpha \\ \Gamma, \text{ if } \mathbf{e} \neq \mathbf{e}_\alpha \end{cases},$$

where $\omega_{i\alpha}$ is the state competent to the signal $\mathbf{e}_\alpha$. This establishes the relation of mutual specificity between states and signals. The existence of metastable states results in the most nontrivial fourth law of the dynamics of activities.

The *fourth law* of the dynamics of activities is as follows:

$$\Gamma(t) = \Psi\{\Phi\,[\omega(t-1), \mathbf{e}(t)]\} = \begin{cases} \Gamma_\upsilon \\ \Gamma_p, \text{ if } \mathbf{e} \neq \mathbf{e}_\alpha \end{cases}$$

where designations have the same meaning as in the expressions given above.

This expression implies the existence of such cell division during ontogenesis when, in a neutral extracellular medium, the daughter cells will differ in the activity of at least one gene (divergent determination).

6. WHENCE ORDER APPEARS DURING ONTOGENESIS

It seems likely that the theory in question suggests an answer to the long-time question: whence the order appears during ontogenesis?

If excluding the involvement of some 'vital force' entelechy) and tautological speculations *à la* Schrödinger about 'feeding on negentropy', we can make the following tentative conclusion (hypothesis):

In the hereditary algorithm χ realized within the hereditary ontogenetic program (HOP), the recursive order determines step-by-step fulfillment of the operation, which is possible during ontogenesis through an exponentially growing quantity of information processing machines (cellular automata A_e^g) with their internal memory increasing capacity that enables to remember all stages necessary in the HOP execution resulting in the organism's self-reproduction.

As was already noted (Tchuraev, 2000; 2005), the entire hereditary information organized in the HOP is stored in the common hereditary memory (CHM). Consequently, this memory also stores the entire information kept in the message (instructions) β of the inherited algorithm χ;

in this case, the starting word of the algorithm χ is the CHM at the instant of time t_0^k. Hence, the CHM initial state is the starting functional states of the activities of all genes in the CHM elements, i.e., the initial value of the γ-vector in the entire network **S**. The activity of the entire cellular gene network $\mathbf{S}_e$ is dictated by the activity of the governing subnetwork $S_e^g \subset \mathbf{S}_e$. Values of γ-vectors of the network S_e^g activity at a molecular level depend not only on the affiliation strength between regulatory protein complexes and DNA sites, but also on the current quantities of extragenomic regulatory molecules and their complexes. The most important factors are accounted for in the mathematical model of genetic element of the governing gene networks constructed in terms of GTM formalism (Tchuraev, 1991, 1993; Tchuraev, Galimzyanov, 2003) in the context of microapproach. It should be noted that transitions from micro to meso and then macroapproaches are by no means trivial.

7. CONCLUSION

Now let us give the summing-up. Just as *brain properties are not the total of neural cell properties, but present those of the integral neural network, so the storage, transmission, processing, and realization of the hereditary information* involve the properties of gene and epigene networks that govern ontogenesis processes of self-reproduction.

ACKNOWLEDGMENTS

I am much grateful to my colleagues I.G. Migranova and A.V. Galimzyanov for valuable discussion and A.P. Maslova for technical assistance.

FROM GRADIENTS TO STRIPES: A LOGICAL ANALYSIS OF *DROSOPHILA* SEGMENTATION GENETIC NETWORK

D. Thieffry[1], C. Chaouiya[1], L. Sánchez[2]
[1] *Institut de Biologie du Développement de Marseille, France, e-mail: {thieffry, chaouiya}@ibdm.univ-mrs.fr;* [2] *Centro de Investigaciones Biológicas, Madrid*
Corresponding author

Abstract: We present a qualitative approach to gene network modelling, together with a computational implementation (GINsim software). Applied to the *Drosophila* segmentation network, our logical analysis leads to the delineation of the crucial interactions and regulatory circuits involved in the main differentiation decisions at the basis of the segmentation process. The resulting logical models can be further used to perform *in silico* perturbations and thereby suggest new experiments. GINsim software is available at the url http://gin.univ-mrs.fr/GINsim. Supplementary materiel and the models referenced in the paper can be downloaded from the url http://gin.univ-mrs.fr/GINsim/Models.

Key words: *Drosophila* development; segmentation; dynamical modelling; regulatory circuit

1. INTRODUCTION

The early embryogenesis of *Drosophila melanogaster* is one of the most extensively studied developmental processes in higher organisms. Saturated mutagenesis followed by careful screening of mutant phenotypes has led to the identification of the key regulatory genes controlling the formation of segments along the anterior–posterior axis of the embryo, prefiguring the specific arrangement of body structures, first in the larva and later in the adult fly (see e.g. Rivera-Pomar and Jäckle, 1996). The setting of segmentation involves dozens of genes, expressed either maternally during oogenesis or in the zygotic syncytium. These genes form a hierarchical genetic network

(Figure 1). On the basis of mutant phenotypes, developmental geneticists distinguish four gene classes, each responsible for a step in the processing of the initial gradients of maternal products, ultimately leading to specific and robust stripes of zygotic gene expression.

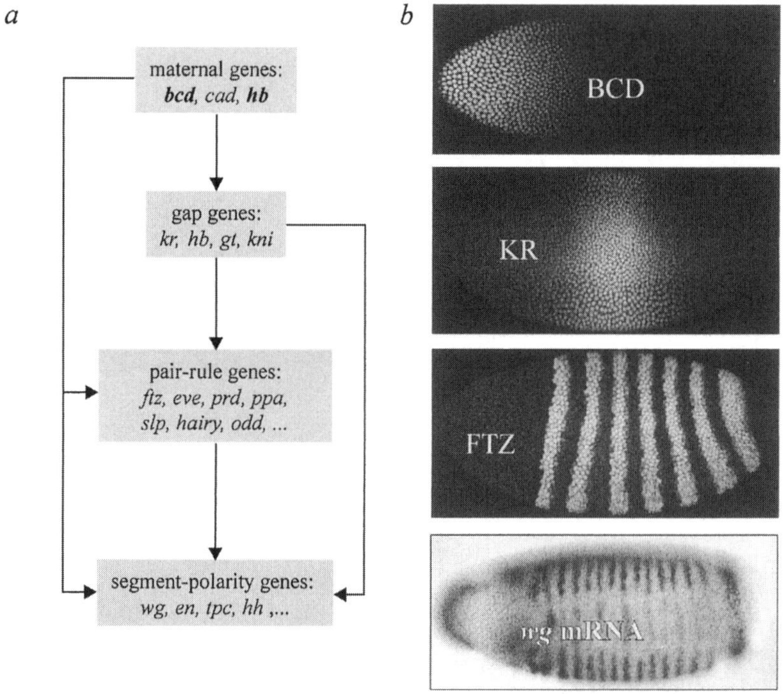

Figure -1. Overview of the Drosophila segmentation genetic system. (*a*) the segmentation genes are organized into four main classes according to mutant phenotypic characteristics and to their expression patterns. (*b*) a typical expression patterns for one gene of each segmentation class. The three top images were extracted from the FlyEx database and reveal proteins (in white) with specific antibodies; the last image comes from the BDGP database and reveals *wingless* mRNA (black).

Early after fertilization, *maternal* regulatory proteins become gradually distributed along the anterior–posterior axis. Together, these maternal products define different functional inputs on the *gap* genes. The *gap* genes are consequently differentially activated along the trunk of the embryo. Furthermore, cross-regulations (predominantly cross-inhibitions) among *gap* genes amplify these initial differences, ultimately leading to well-differentiated expression domains. At a later stage, the maternal and gap products act together on the *pair-rule* genes, leading to a further refinement of the segmented gene expression pattern. Finally, altogether, these genes control the expression of the *segment polarity* genes, which ultimately determine the number of segments that will be formed.

We progressively model the different cross-regulatory modules involved in the control of Drosophila segmentation using the generalized logical formalism, initially developed by R. Thomas and collaborators in Brussels (see Section 2). To foster and ease this approach, we have developed a Java software suite enabling the definition, analysis, and simulation of logical regulatory graphs (section 3). In section 4, logical models, analyses, and simulations concerning the first two cross-regulatory modules are reviewed. These results are discussed and further prospects are proposed in the fifth and last section.

2. LOGICAL MODELING, ANALYSIS, AND SIMULATION OF REGULATORY NETWORKS

Our approach to the modeling and analysis of regulatory networks relies on the logical formalism previously developed by R. Thomas (1991). It is based on the definition of (i) a logical regulatory graph, which describes the regulatory interactions between genes and (or via) their products, and (ii) state transition graphs, which represent the qualitative dynamical behavior of such regulatory graph for given initial states. Hereafter, we provide a brief description of these two types of graphs (more details can be found in Chaouiya et al., 2003).

A regulatory graph is a labelled directed graph where nodes represent genes (or, more generally, regulatory components) and arcs (directed edges) represent interactions between genes. A discrete variable is further associated to each node, representing the current qualitative gene expression level. Each arc (regulatory interaction) is defined by its source and its target and is labelled by an integer interval (Figure 2). An interaction is operating when the current level of expression of its source gene belongs to the related interval.

At this point, we defined the regulatory structure, encompassing the nodes, and their interactions together with the definition of the corresponding expression levels. We have now to define the rules governing the dynamics of the network. This is accomplished through the specification of a logical function attached to each gene. This logical function allows the qualitative specification of the effects of any combination of incoming interactions via the assignment of specific values to each relevant case, each corresponding to a 'logical parameter' ('K's, see Figure 2). Per default, these parameters are assigned zero values, which amounts to single out the combinations of interactions playing an essential role in the generation of dynamical properties consistent with the available wild type and mutant data.

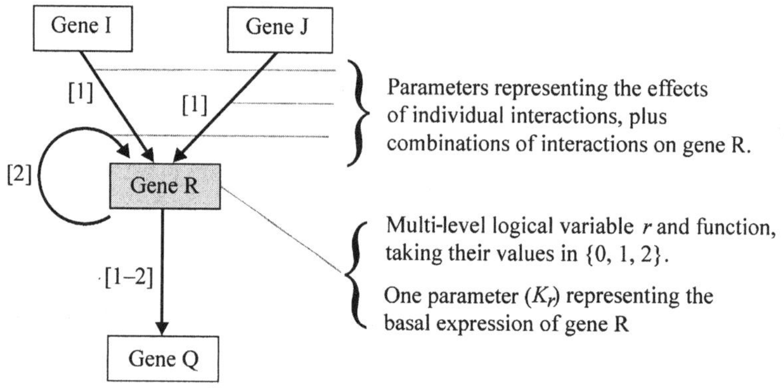

Figure -2. A simple regulatory network: gene R has two regulators (I and J) and is also autoregulated. The associated *logical* variable r can take three values corresponding to the three different qualitative levels of expressions (0, 1, and 2). The autoregulation is operating when r equals two. For the three interactions upon R, eight logical parameters are defined, one for each possible combination of the incoming interaction. In particular, one parameter denotes R basal expression (when no incoming interaction is operating).

An interaction is called an *activation* (resp. a *repression* or an *inhibition*), when its effect on the targeted gene tends to be positive (resp. negative). Note, however, that effective activatory or inhibitory effects generally depend on the presence or absence of cofactors. Indeed, one gene is often the target of several interactions. The types of interactions (activation *versus* inhibition) are implicit in our formal definition of regulatory graphs, but we will use them in the applications below, as biologists very often refer to them.

The (discrete) dynamics of the system can then be represented in terms of *state transition graphs*, where vertices represent *expression* or *activity states* of the system (i.e., n-tuples giving the expression levels of the n genes) and arcs represent *transitions* between these states. For each state of the system, one can determine a set of interactions operating upon each gene. It is then straightforward to determine whether a gene tends to change its level value; indeed, it suffices to consider the relevant logical parameter: if it is higher than the current level of the gene, this gene will tend to increase its expression level, whereas, if the logical parameter is lower, the gene will tend to decrease its expression level (otherwise, in the case of equality, there is no updating call on this gene expression level).

In most applications, the state transition graphs are generated either on the basis of a fully synchronous assumption (all updating calls are then executed simultaneously) or on the basis of a fully asynchronous approach (all updating calls are then considered independently). Whatever the synchronous or asynchronous assumptions, only the elementary transitions (i.e., increase or decrease in level values by unity, at most) are usually considered.

3. ginSIM: A SOFTWARE SUITE FOR THE LOGICAL MODELING, SIMULATION, AND ANALYSIS OF REGULATORY NETWORKS

GINsim is a software suite enabling the definition, simulation, and analysis of regulatory graphs based on the logical formalism introduced in the previous section (Chaouiya et al., 2003). It has been developed as a series of Java classes encompassing four main modules: a user interface, a model parser, a core simulator, and a graph analysis toolbox. Note that GINsim plug-in architecture facilitates the implementation of new modules, in particular, to extend the graph analysis toolbox.

We have developed an interface to ease the interaction between the user and different modules of GINsim. The *Graph Editor* allows the specification of the regulatory graphs as well as the visualization of the corresponding state transition graphs. Similar to a simple drawing software, it has been developed using *JGraph*, the open source Swing Java library for visualization of graphs. The internal graph structures were implemented using the free Java graph library *JGraphT*, which includes graph-theoretical objects and algorithms. The editor further allows the definition of the maximal level of each node (the default value is unity) and of the logical parameters (default values are set to zero).

Given a fully parameterized regulatory graph, the user can then trigger the simulation (i.e., the computation of the state transition graph). A dedicated interface allows defining the set of initial state(s) and choosing between various options: synchronous versus asynchronous updating, limitation of the number of states generated, absolute priorities among genes, etc.

GINsim is available at http://gin.univ-mrs.fr/GINsim. This web site further proposes a tutorial and a collection of models, among which are the *Drosophila* segmentation networks presented in the sequel.

4. LOGICAL MODELING OF *DROSOPHILA* SEGMENTATION NETWORK

4.1 From maternal gradients to specific combinations of gap products

Upon the function of the maternal regulatory products (*Bicoid*, *Hunchback*, and *Caudal*), the *gap* genes (*giant*, *hunchback*, *Krüppel*, and *knirps*) count among the first genes to be expressed along *Drosophila* embryonic development. All these genes encode transcription factors able to bind (often cooperatively) specific short DNA sequences. The action of the maternal products and the expression of the *gap* genes occur when the embryo still forms a single cell (syncytium), although the number of nuclei rapidly increases through a dozen rounds of synchronous divisions. As shown in Figure 3, the *gap* genes extensively inhibit each other's expression. As a result, the *gap* genes quickly set a differentiated pattern along the anterior–posterior axis of the embryo.

On the basis of an extensive analysis of published data, it proved possible to derive most of the qualitative information needed to define a consistent logical model for the regulation of the *gap* genes acting in the trunk of *Drosophila* embryo. On the one hand, we have represented the most important maternal gradients in terms of multilevel logical variables (quaternary variables for *Bicoid* and *Hunchback* regulatory products and a ternary variable for *Caudal*). With respect to the *gap* genes, we were led to use binary variables for *Giant* and *Knirps*, but a ternary variable for *Krüppel*. Note that the gene *hunchback* plays a double role here, as it is expressed both maternally and in the zygote. However, the concentration of its functional regulatory product is represented by a single (quaternary) variable.

In parallel, we defined the rules (logical parameters) directing the expression of each *gap* gene using available information about the wild type and mutant phenotypes. This led us to define a limited number of situations enabling the expression of each *gap* gene. For the parameter values selected and for the initial states corresponding to the different maternal inputs present along the anterior–posterior axis of the trunk of *Drosophila* embryo, the system can reach exactly four different stable states, characterized by the expression of different combinations of *gap* genes, and arranged in a specific order, as illustrated at the bottom of Figure 3. In each of the corresponding regions of the embryo, a unique stable state is thus selected. Our model of the *gap* gene cross-regulatory module thus accounts for the specification of four different expression domains, which are arranged in a specific order along the trunk of the embryo under the influence of the maternal regulatory gradients.

4.2 The pair-rule module: from broad expression to sharp stripes

Under the combined action of the maternal and *gap* regulatory products, the *pair-rule* genes start to be expressed in different places along the anterior–posterior axis, forming alternating stripes of expression. First, only a subset of *pair-rule* genes (*h*, *run*, *eve*, *ftz*, and *odd*) starts to be expressed, mostly in relatively broad domains. Next, other *pair-rule* genes (*prd*, *ppa*, and *slp*) become activated. Most of these genes further show a progressive refinement of their expression pattern from broad domain to sharp stripes.

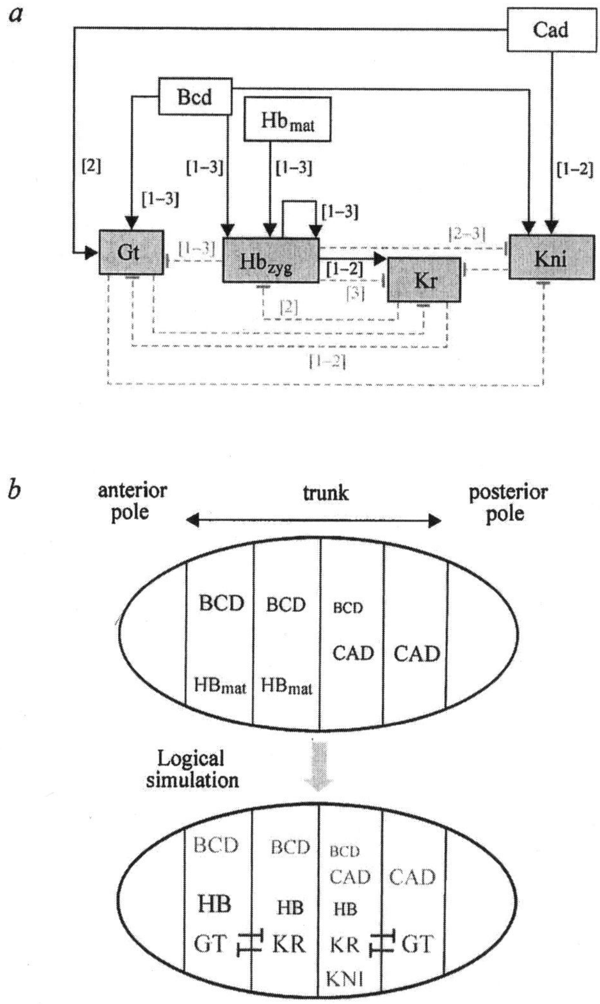

Figure -3. Regulatory graph and qualitative simulation of the *gap* module. (*a*) documented interactions among *maternal* and *gap* genes. Activations are represented by solid arrows and inhibitions, by dotted T-ending arcs. Qualitative intervals associated with each regulatory interaction are given in brackets. In the case of regulatory interactions originating from Boolean elements (Gt and Kni), the intervals ([1]) are omitted. (*b*) qualitative simulations of the wild type *gap* module. In the upper embryo, the initial concentrations of the maternal products are indicated. In the lower embryo, the resulting *gap* expression patterns are schematized. The symbol sizes represent the relative amounts of maternal or *gap* products. Abbreviations: Bcd, Hb, Cad, Gt, Kr, and Kni stands for Bicoid, Hunchback, Caudal, Krüppel and Knirps transcription factors (for more details, see Sánchez and Thieffry, 2001).

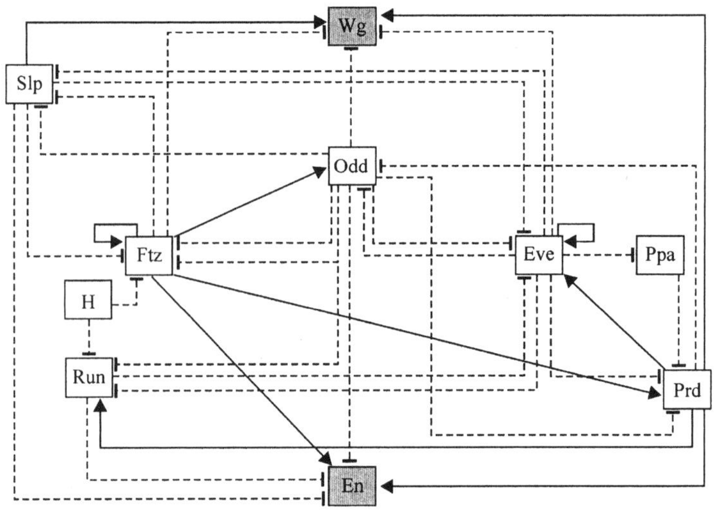

Figure -4. Logical model for the *pair-rule* module: cross-regulation between the main *pair-rule* genes and on the segment polarity genes *engrailed* (*en*) and *wingless* (*wg*; for more details, see Sánchez and Thieffry, 2003).

On the basis of our analysis of published wild type and mutant data, we defined a logical model encompassing the main *pair-rule* genes and their cross-regulations. As in the case of the *gap* cross-regulations, the *pair-rule* genes mainly inhibit each other expression (Figure 4). Taken in isolation, for proper parameter values, this system has four stable states characterized by (i) a high expression of *eve* and *prd*; (ii) a low expression of *eve* and *prd* together with a high expression of *ppm*, *run,* and *ftz*; (iii) a low expression of *prd* together with a high expression of *ppa*, *run,* and *slp;* and (iv) a high expression of *ppa* and *odd* (see supplementary material for the values corresponding to these states). In each of these states, one can single out a specific gene representing the status of the whole system. In what follows, we will refer to these different situations in terms of (i) *eve*, (ii) *ftz*, (iii) *prd/slp*, and (iv) *odd* expression 'modes'. Under the action of upstream segmentation genes, these modes are organized according to a specific anterior–posterior order, which is repeated fourteen times in the trunk.

The simulation of the detailed kinetics of the *pair-rule* gene expression in each embryonic region goes well beyond the scope of this review. In fact, in each of these regions, it is very difficult to identify the dynamical pathway(s) followed in reality. In any case, we are particularly interested in the asymptotical expression states, which will subsequently control expression of the segment polarity genes, and thereby, the location of segmental borders. In this respect, to insure that a unique mode is selected in each of these regions, we were led to further define the three priority rules between alternative dynamical pathways:

(i) the first rule states that *odd* inhibition by *eve* overcomes the reciprocal inhibition; (ii) the second rule states that the activation of *run* by the maternal–gap signal overcomes its inhibition by *eve*; and (iii) finally, inhibitions of *ftz* and *run* by *h* (*hairy*) dominate the regulation of these genes.

Developmental geneticists have extensively studied the effects of the differentiated expression of *pair-rule* genes on two early expressed segment polarity genes, *engrailed* (*en*) and *wingless* (*wg*). We have thus defined the rules of activation of these genes by the *pair-rule* genes. In short, *wg* becomes expressed as a result of the *slp* expression mode, whereas *en* becomes expressed in the regions characterized by the *eve* and *ftz* modes. In the regions dominated by the *odd* mode, neither *wg* nor *en* can be expressed. As a result, at the onset of cellularization, the trunk of the embryo is characterized by a succession of juxtaposed (one-cell wide) *wg* and *en* stripes, separated by broader (*c.a.* two cells) stripes lacking both regulatory products. All *wg* expression stripes seemingly involve the same regulatory mechanism (*slp pair-rule* expression mode), whereas *en* expression stripes involve two different regulatory mechanisms depending on the position of the stripe (*en* even-numbered stripes are activated by the *ftz pair-rule* mode, while the odd-numbered stripes are activated by the *eve* mode). This process results in the specification of a segmental border at the interface between the *en* and *wg* expressing cells, now fully separated by cellular membranes. This last step involves intercellular interactions via external signaling, signal recognition by specific receptors, and, finally, signal transduction leading to transcriptional regulation of specific segment polarity genes in the nuclei of these cells. We are presently working on the logical modeling of this intercellular regulatory network.

4.3 Positive feedback circuits and cell differentiation

Beyond the representation of cellular differentiation in terms of different stable states, the logical approach offers the means to obtain insights in the roles of specific regulatory structures found at the origin of specific dynamical properties. In particular, as well emphasized by R. Thomas, it can be shown that positive circuits (involving an even number of inhibitions) are needed to generate alternative attractors (e.g., stable logical states), whereas negative circuits (involving an odd number of inhibitions) are necessary to generate stable state transition cycles (i.e., sustained gene expression oscillations; Thomas et al., 1995).

Both cross-regulatory modules encompass various feedback regulatory circuits, of various lengths and signs (up to 51 circuits involving from 1 to 7 elements in the case of the *pair-rule* module). For proper parameter values, when a circuit does generate the typical dynamical properties associated with its class (positive *versus* negative), one says that it is *functional*. In the context of the logical formalism, it is possible to identify the *functional* circuits for a specific set of parameter values or yet to compute the constraints on parameter values enabling the corresponding dynamical properties.

Here, we limit ourselves to discussing the results of such computation for the *gap* and *pair-rule* cross-regulatory modules. In the first case, this analysis emphasizes the role of a single positive feedback circuit made of the mutually inhibitory interactions between *giant* and *Krüppel*. As shown in Figure 3 (bottom), this positive circuit ensures non-overlapping expression domains for the two cross-inhibiting genes. Strikingly, this effect occurs at two places: around the middle of first half and around the middle of the second half of the trunk of the embryo. As we shall see in the following subsection, forcing the level of any of the two genes involved in this circuit (i.e., using a loss-of-function mutant or forcing the ectopic expression of one of these genes), some of the stable states of the *gap* system are lost, leading to deep perturbations of the *gap* gene expression patterns.

In summary, our feedback circuit analysis leads to the delineation of a single positive circuit playing a crucial role in the setting of well-differentiated *gap* gene expression patterns. Now, this does not mean that the other genetic interactions encompassed by our model do not play any significant dynamical roles. Although they do not form functional feedback circuits, these interactions further couple the four *gap* genes expression domains. For example, it can be shown that the posterior limit of *knirps* expression domain is controlled by Giant, whereas its anterior boundary is essentially set by Hunchback, despite the absence of functional circuit involving *knirps*.

The situation becomes much more complex in the case of the *pair-rule* cross-regulatory module. However, among the 51 regulatory circuits found in the *pair-rule* regulatory graph, only 7 of these circuits are found (partly) functional, including 6 positive circuits. Three of these positive circuits involve *eve* (i.e., *eve* autoregulation together with *eve/run* and *eve/slp* cross-inhibitory positive circuits). Altogether, these circuits enable the occurrence of three different stable levels of *eve* expression involved in different stable states. Similarly, two positive circuits include *ftz* (i.e., made of *slp/ftz* and *prd/odd/ftz* circular regulatory sequences, respectively). The last functional positive circuit involves *prd* and *odd*. As shown in supplementary material, a loss-of-function affecting any gene of these positive circuits leads to the partial loss of the system multistability property, thereby reducing the number of different stable states and deeply perturbing the global gene expression pattern.

In each of the two analyzed models, corresponding to the *gap* and to the *pair-rule* cross-regulatory modules, we found one three-element negative circuit partly functional. However, the biological significance of these circuits remains unclear, as we could not find any report referring to the occurrence of oscillatory gene expression for these systems.

4.4 *In silico* genetic experiments

The logical modeling approach shows its full power when it is turned towards the simulations of various types of perturbations. The reduced number of values associated with each logical variable, function, or parameter greatly eases the delineation of perturbed situations.

For example, the simulation of complete loss-of-function situations simply consists in blocking the values of the corresponding variable(s) to zero. The results of simulations of maternal, *gap*, and *pair-rule* perturbations are provided in the supplementary material. In all the simulations performed, we have obtained results qualitatively agreeing with the published mutant data.

Developmental geneticists have also described the outcome of experiments consisting in expressing a given *maternal*, *gap*, or *pair-rule* gene ectopically, most often at a constant level in the whole trunk of the embryo. These perturbations can also be easily simulated in the context of the logical formalism, as they correspond to situations where the level of the variable associated with a given gene is lower-bounded at a (non-zero) value.

Finally, it is also possible to simulate more local perturbation, i.e., affecting a single arc in the regulatory graph. This would correspond, for example, to the disruption of enough binding sites for a given regulatory product in the promoter and enhancer regions controlling the expression of a specific target gene. On can perform such perturbations by changing the values of the corresponding parameter(s) in the logical model.

Wrapping up, on the basis of the definition of a logical model (i.e., the specification of a regulatory graph with the corresponding logical parameters), it is possible to simulate various types of perturbations *in silico*. In the first step, the simulation of well-studied perturbations can be performed to check the consistency of the model properties with the published data. Next, the simulation of new types of perturbations (in particular, multiple simultaneous perturbations) can lead to the delineation of interesting or unexpected properties and thereby nourish the design of new experiments.

5. CONCLUSIONS AND PROSPECTS

In this paper, we reviewed our recent work on the logical modeling of the genetic system controlling the onset of the anterior–posterior gene expression pattern during the early embryonic development of the fly *Drosophila melanogaster.* This pattern constitutes the first marks of body segmentation. Focusing on the three first classes of genes involved in this process, namely, the *maternal*, *gap*, and *pair-rule* genes, we have shown how relatively simple discrete models can be defined, which yet enable a qualitative reproduction of the main (pro)-cellular states occurring along the trunk of the embryo. In addition, our logical analysis leads to the delineation of the crucial interactions and regulatory circuits involved in the main differentiation decisions at the basis of the segmentation process. Finally, we have indicated how various types of perturbations can be simulated, first, to check the consistency of the behavior of logical models with the published data and next, to explore new situations *in silico*. This last point is likely the most original aspect of our work when compared to model analyses performed by other groups (see, in particular, Jaeger et al. (2004) for an interesting approach aiming at reverse-engineer a generic differential *gap* gene model and the paper by von Dassow et al. (2000) for an ODE model analysis of the segment polarity module).

We are presently working on the modeling of the segment polarity network. This requires proper formal representations for two new aspects. On the one hand, cellular membranes are formed at that stage, calling for the modelling of intercellular relationships. On the other hand, the segment polarity network involves several types of regulatory interactions beyond transcriptional activations or inhibitions, in particular, regulations at a post-translational level (signal–receptor interactions and proteolysis). Tentatively, our approach consists in encapsulating the regulatory network present in each cell, while taking into account the different input combinations coming from neighboring cells, and then, to simulate its behavior for different input combinations. At least in the case of the stable states, it is then relatively easy to infer multicellular stable states from the knowledge of the stable states corresponding to each input combination.

Now, the identification of the final multicellular stable state(s) depends on a proper integration of the whole segmentation network. In this respect, we still have to specify precisely how the maternal and gap regulatory products control the expression of each *pair-rule* gene stripe along the embryo. Indeed, up to now, we have only delineated general principles for the organization of the *pair-rule* gene expression pattern at the level of a generic segment. Finally, as we face more and more complex regulatory networks, we are led to consider new methodological developments including formal and valuable graph compaction methods as well as constraint or logical programming and model checking approaches.

ACKNOWLEDGMENTS

This work was supported by the grant BMC2002-02858, awarded to L. Sánchez by the *D.G.I.C.Y.T.*, Ministerio de Ciencia y Tecnología of Spain. D. Thieffry and C. Chaouiya acknowledge the support of the Programme inter-EPST Bio-informatique, France. Our collaboration was supported by a CSIC-CNRS cooperation grant (2004-2005).

SELF-OSCILLATIONS IN HYPOTHETICAL GENE NETWORKS

V.A. Likhoshvai[1*], V.V. Kogai[2], S.I. Fadeev[2]

[1] Institute of Cytology and Genetics, Siberian Branch of the Russian Academy of Sciences, prosp. Lavrentieva 10, Novosibirsk, 630090, Russia, e-mail: likho@bionet.nsc.ru; [2] Institute of Mathematics, Siberian Branch of the Russian Academy of Sciences, Novosibirsk, 630090, Russia

[*] Corresponding author

Abstract: Detection of cyclic modes in gene networks is an important problem in analysis of patterns of gene network structure and function. Consequently, study of the special class of ordinary and delay differential equations modeling gene network regulatory circuits is fairly topical. This work presents an efficient method for numerical study of self-oscillations described by autonomous systems of special differential equations modeling class one symmetrical canonical hypothetical gene network (Likhoshvai et al., 2003).

Key words: genetic systems; modeling; delay equation; differential autonomous systems

1. INTRODUCTION

Gene networks (GN) are structurally complex spatial objects composed of hundreds of elements of various natures and complexities, namely, genes and their regulatory regions, RNAs and proteins encoded by these genes, low-molecular-weight compounds, various complexes between enzymes and their targets, etc. The core of GN is regulatory circuits—genes and proteins whose expressions are mutually regulated. Their presence confers on GN a unique ability to respond adequately to changes in external conditions. Thus, detection of possible operation modes of regulatory circuits is an important problem of the theory of gene networks. To approach solution of this problem, it is necessary to investigate systematically the operation patterns of regulatory circuits of various structures. A constructive step in this direction is separation of a finite set of standard elements from natural GN, formalization of the rules for assembling theoretical objects (mathematical models) describing regulatory circuits from these elements, and a systematic analysis of their properties for revealing general biologically significant regularities (Likhoshvai et al., 2003).

We describe in this work an efficient method for searching for cyclic operation modes of symmetrical hypothetical gene networks (HGN) represented by autonomous systems of n differential equations:

$$\frac{dx_i}{dt} = \alpha/(1+\beta z_i) - x_i,\ z_i = \sum_{j=1}^{k-1} x^{\gamma}_{\sigma(i-j)},\ \sigma(i-j) = \begin{cases} i-j, \text{ if } j<i \\ n+i-j, \text{ if } j \geq \mathrm{i} \end{cases} \quad i=\overline{1,n}, \quad (1)$$

where α, β, γ are positive parameters and $1 < k \leq n$. The right parts of the system (1) represent the first class of the regulatory relations considered in (Fadeev and Likhoshvai, 2003). It is easy to demonstrate that the phase trajectories of system (1) coming from the hypercube Ξ with the edge α remain in Ξ. Hereinafter, system (1) will be referred to as the model $M(n, k)$. These HGN are the simplest mathematical objects by their definition.

Properties of symmetrical HGN are detailed in (Fadeev, Likhoshvai, 2003). In particular, according to the hypothesis named (n, k)-criterion, formulated for symmetrical HGN, if the greatest common divisor d of numbers n and k equals k, then there exist $\overline{\alpha} > 0$ and $\overline{\gamma} > 1$ such that at $\alpha > \overline{\alpha}$, and $\gamma > \overline{\gamma}$, the autonomous system (1) has k asymptotically stable stationary solutions. If $d \neq k$, than there exist d stable limit cycles, while the stable stationary solutions are absent. For sufficiently large α and γ, the total number of stationary solutions for the model $M(n, k)$, both stable and unstable, amounts to $2^d - 1$.

The essence of an approach to numerical study of self-oscillations of general autonomous systems depending on the model's parameters is as follows (Kogai and Fadeev, 2001; Kogai, 2002; Fadeev and Kogai, 2004). Let us assume that at a fixed value of model's parameter, for example, fixed α in system (1), the initial data of Cauchy problem for autonomous system of equations are such that solution of the system starting from a certain time point approaches sufficiently well the established self-oscillation mode with a certain period. In this case, the possibility appears to make the description of self-oscillations more precise basing on the boundary value problem for the considered autonomous system with given conditions of periodicity and transversality. Here, the Newton's method is used, where the integration result of Cauchy problem is taken as an initial approximation. The resulting solution of boundary value problem is the starting point in parameter continuation method, allowing for an efficient numerical study of self-oscillations depending on parameter.

The boundary value problem describing self-oscillations of autonomous system (1), if they do exist, may be represented as

$$0 \le s \le 1, \qquad dx_i/ds = T(\alpha/(1+\beta z_i) - x_i), \quad dT/ds = 0, \quad i = 1, 2, \ldots, n. \tag{2}$$

$$x_i(0) = x_i(1), \tag{3}$$

$$\alpha/(1+\beta z_1) - x_1 = 0 \text{ at } s = 0. \tag{4}$$

Here, T is the period to be determined; equality (3), the condition of solution periodicity; and equality (4), one of the possible variants of transversality condition, when the derivative of the first component at $s = 0$ should be equal to zero. As determination of the dependence of boundary value problem solution on parameter, describing self-oscillations by parameter continuation method, is unconnected with the problem of stability, the method may find both stable and unstable periodic solutions, if the latter take place in the mathematical model considered. Later, their stability can be verified using, for example, the algorithm for calculation of the maximum eigenvalue of monodromy matrix.

Biological significance of the results described here consists in that this numerical method for search for cyclic modes is a necessary stage in developing the methods for analysis of regulatory circuits of an arbitrary structure.

2. RESULTS

2.1 Numerical study of self-oscillations of autonomous systems

Boundary value problem (2)–(4) is a special case of a more general problem of search for periodic vector function $x(s)$, $x(s) = x(s + 1)$, with the components $x_1(s), x_2(s), \ldots, x_n(s)$, giving solution of the boundary value problem depending on scalar parameter q of the autonomous system of equations with a sufficiently smooth right parts in a domain of definition:

$$0 \le s \le 1, \qquad dx/ds = T f(x,q), \quad \frac{dT}{ds} = 0, \tag{5}$$

$$x(0) = x(1)\,,\ dx_1(0)/dt = 0\,. \tag{6}$$

Here, $f(x, q)$ is the vector function of vector argument x and scalar argument q with the components $f_1(x, q)$, $f_2(x, q)$. If we denote as y and F the composite vectors $y = \begin{bmatrix} x \\ T \end{bmatrix}$, $F = \begin{bmatrix} Tf \\ 0 \end{bmatrix}$, it is possible to configure boundary value problem (5) in a 'standard' form:

$$0 \le s \le 1, \qquad dy/ds = F(y,q)\,,\ g(y(0), y(1), q) = 0\,, \tag{7}$$

where the vector function $g(y(0), y(1), q)$ represents boundary conditions (6).

Let $y = y(s, q)$ be the solution of boundary value problem (7) at a certain value of parameter q. Let us use the 'multiple shooting' method for numerical determination of $y(s, q)$ (Kogai and Fadeev, 2001; Kogai, 2002; Fadeev and Kogai, 2004). According to multiple shooting method, the segment [0,1] is partitioned by s into m parts:

$$0 = s_1 < s_2 < \ldots < s_m < s_{m+1} = 1\,. \tag{8}$$

Let us designate as p^k the network value of the vector function $y(s, q)$ at the kth node of network (8): $p^k = y(s_k, q)$, $k = 1, 2, \ldots.\ m + 1$. In each of the segments $[s_k, s_{k+1}]$, $k = 1, 2, \ldots.\ m + 1$, let us consider the following series of Cauchy problems in forms of vector and matrix equations:

$$s \in [s_k, s_{k+1}],\ dy/ds = F(y,q),\ \ y(s_k) = p^k \tag{9}$$

$$dY/ds = F_y(y,q)Y, \quad Y(s_k) = I \tag{10}$$

$$du/ds = F_y(y,q)u + F_q(y,q),\ u(s_k) = 0\,, \tag{11}$$

where I is a unit matrix. Let us designate the solutions of series (9) in each of the segments $[s_k, s_{k+1}]$ as $y(s, p^k, q)$; solutions of series (10), as $Y(s, p^k, q)$; and solutions of series (11), as $u(s, p^k, q)$. Evidently, series (9) gives the solution of boundary value problem (7) provided that boundary conditions and continuity conditions of the vector function $y(s, q)$ at the network nodes are fulfilled:

$$\Phi_1 = g(p^1, p^{m+1}, q) = 0,\ \Phi_2 = y(s_2, p^1, q) - p^2 = 0,$$

$$\Phi_{m+1} = y(s_{m+1}, p^m, q) - \mathrm{p}^{m+1} = 0, \qquad \text{or } \Phi(p, q) = 0, \tag{12}$$

where p and Φ are composite vectors with sizes of N, $N = (n + 1)(m + 1)$. Series of Cauchy problems (10) and (11) allows for forming an $(N \times N)$ matrix of Φ_p derivatives of vector function $\Phi(p, q)$ and column of Φ_p derivatives by parameter q:

$$\Phi_p = \begin{bmatrix} g_a(p^1, p^{m+1}, q) & 0 & \dots & 0 & g_b(p^1, p^{m+1}, q) \\ Y(s_2, p^1, q) & -I & \dots & 0 & 0 \\ 0 & Y(s_3, p^2, q) & \dots & 0 & 0 \\ \dots & \dots & \dots & \dots & \dots \\ 0 & 0 & Y(s_{m+1}, p^m, q) & & -I \end{bmatrix}, \Phi_q = \begin{bmatrix} g_q(p^1, p^{m+1}, q) \\ u(s_2, p^1, q) \\ u(s_3, p^2, q) \\ \dots \\ u(s_{m+1}, p^m, q) \end{bmatrix}.$$

Here, for the sake of convenience, the vector arguments of vector function $g(y(0), y(1), q)$ are designated as $a = y(0)$, $b = y(1)$.

Thus, the numerical study of the solution of boundary value problem (7) by the multiple shooting method is formally reduced to numerical study of solution of the system of nonlinear equations (12) with parameter q, determining the vector function $p = p(q)$. Protocol of the process of parameter continuation is described in (Fadeev, 1985; Fadeev et al., 1988; Holodniok et al., 1991). The number of segment [0,1] partitions by s is determined by an 'acceptable stringency' of Cauchy problems (9)–(11), which is achieved through choosing the length of segment $[s_k, s_{k+1}]$, for example, from the condition $s \in [s_i, s_{i+1}]$, $\max \|F_y(y,q)\| \le D$, $D(s_{i+1} - s_i) \approx 1$. Here, the columns of matricant Y, meeting conditions (10) during iterations by the Newton's method, remain virtually orthogonal at $s = s_{i+1}$ in full compliance with the idea of Godunov's orthogonal sweep method (Godunov, 1994). In the case of large N, it is essential for the efficiency of this method to take into account the structure of matrix Φ_z.

2.2 Delay equations describing self-oscillations of the model *M*(*n*, *k*)

As numerical experiments demonstrate, the periodic solutions of boundary value problem (5), (6) for various values of n and k display a partial symmetry, manifesting itself in that all the variables fall into d equipotent groups where the trajectories are identical with the accuracy to phase deviation. Here, d may take values 1, 2, 3 ... If $d = 1$, the cycle displays a complete symmetry and is called symmetrical. In the general case, let us name the cycle d-symmetrical.

The property of a partial symmetry allows for describing self-oscillations of autonomous system (1) by the boundary value problem for a system of *d* delay equations, which has the following vector representation:

$$0 \le s \le 1, \quad \frac{du}{ds} = T\,G(u(s), u(s-\tau_1), u(s-\tau_2), \ldots, u(s-\tau_k), q) \qquad (13)$$

$$u(0) = u(1), \; du_1(0)/ds = 0, -\tau_j \le s \le 0, \;\; u(s) = u(1+s).$$

Here, τ is the delay vector with the components $\tau_1, \tau_2, \ldots, \tau_k$, specified in fractions of period *T*, and $G(u(s), u(s-\tau), q)$ is determined by the right parts of autonomous system (1). The discrete model of boundary value problem (13) in a form of system of nonlinear equations with parameter *q* may be constructed by the method analogous to (Fadeev, 1990).

2.3 Examples of numerical study of self-oscillations

Described below are several results of numerical study of self-oscillations of the model *M*(*n*, *k*). Figure 1 exemplifies solutions of boundary value problem (5), (6) of the model *M*(6, 4) at specified values of parameters $\alpha = 10$, $\beta = 1$, and $\gamma = 5$ and oscillation period $T = 6.21$. Each group is characterized by its own amplitude. Order of shift of the curves in each group is indicated. Note that shift of the neighbor curves in each group is equal to *T*/3. The system of two equations equivalent to the boundary value problem considered has the following form:

$$0 \le s \le 1, \;\; u_1(0) = u_1(1), \quad u_2(0) = u_2(1), \quad du_1(0)/dt = 0,$$

$$du_1/ds = T\,[\alpha/(1+\beta(u_2^\gamma(s-1/3) + u_1^\gamma(s-1/3) + u_2^\gamma(s-2/3))) - u_1(s)],$$

$$du_2/ds = T\,[\alpha/(1+\beta(u_1^\gamma(s) + u_2^\gamma(s-1/3) + u_1^\gamma(s-1/3))) - u_2(s)].$$

Here, $u_1(s) = x_1(s)$, $u_2(s) = x_2(s)$. Then, according to the curve shifts in Figure 1, $x_3(s) = u_1(s-2/3)$, $x_4(s) = u_2(s-2/3)$, $x_5(s) = u_1(s-1/3)$, $x_6(s) = u_2(s-1/3)$.

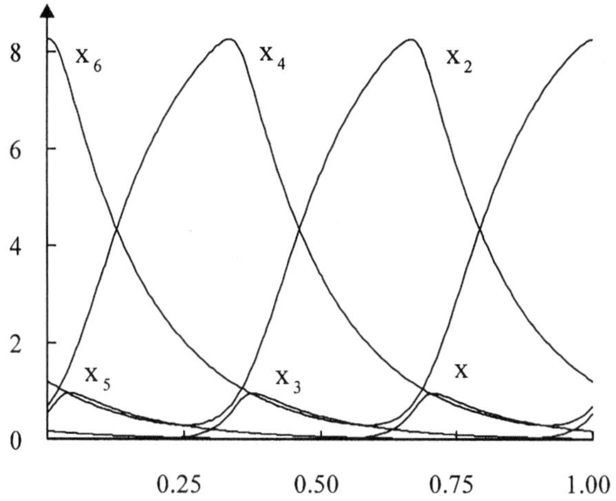

Figure -1. 2-symmetrical self-oscillations of the model *M*(6, 4). The abscissa shows time (conventional unit is equal to the period).

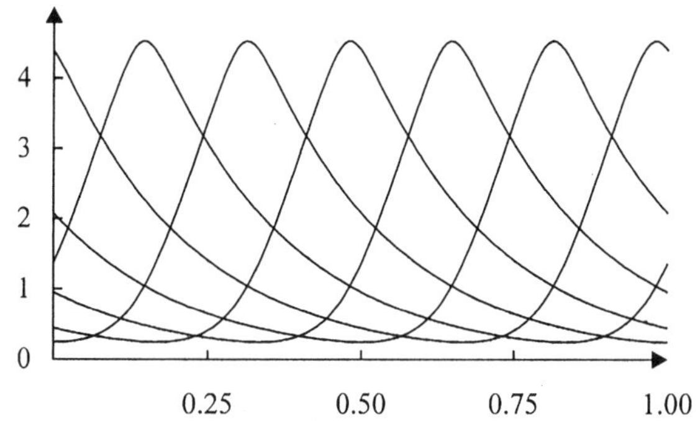

Figure -2. Symmetrical self-oscillations of the model *M*(6, 4) at the parameter values of $\alpha = 10$, $\beta = 1$, and $\gamma = 5$. The abscissa shows time (conventional units).

Another type of self-oscillations of the same model is shown in Figure 2. Here, all the components have the same amplitude. The shift of the neighbor curves is equal to 1/6 period. The corresponding boundary value problem for delay equation has the following form:

$$u(s) = x_1(s),\ 0 \leq 1,\ u(0) = u(1),\ du(0)/dt = 0,$$

$$du/dt = T\ [\alpha/(1+\beta(u^{\gamma}(s-1/6)+u^{\gamma}(s-1/3)+u^{\gamma}(s-1/2))-u(s)].$$

Figure 3 demonstrates the results of numerical study of self-oscillations of the model $M(6, 4)$ depending on parameter α obtained by parameter continuation. The plots marked by (1) represent branches of the dependence plot of period T and amplitude A of oscillations of components of two groups. As is evident from Figure 3, the region of changes in α from 0 to the pivot point at $\alpha = \alpha_1$, where self-oscillations are absent, is followed by the region $\alpha_1 < \alpha < \alpha_2$, where two limit cycles (indicated by (1) and (2)) are present at the same value of α. At $\alpha > \alpha_2$, a limit cycle appears with the components equal in their amplitude (indicated by (3)). The value $\alpha = \alpha_2$, when the amplitudes are zero, corresponds to a singular point on the diagram of stationary solutions of the model $M(6, 4)$. At the singular point, the symmetrical solution i.e., the solution having two identical components, intersects with two partially symmetrical solutions, where even and odd components have their own equal values. In this situation, all the stationary solutions are unstable if $\alpha > \alpha_2$, which explains self-oscillations of the model $M(6, 4)$ at sufficiently large values of α.

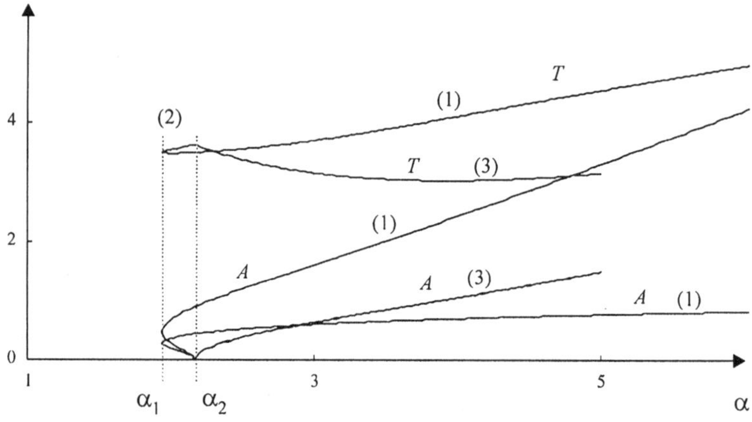

Figure -3. Dependence of self-oscillation period T and amplitude A of the model M(6, 4) on parameter α: $\alpha_1 = 1.95$, $\alpha_2 = 2.17$.

Consider self-oscillations of the model $M(5, 3)$ as another example. According to (n, k)-criterion, here we should anticipate one stable symmetrical limit cycle at sufficiently large values of parameters α and γ. Note that the model $M(5, 3)$ has only one stationary. Starting from certain $\alpha \geq 0$, the stationary solution loses stability switching to self-oscillations.

Figures 4 and 5 show two numerically detected limit symmetrical cycles, each characterized by its own amplitude, period, and order of variables.

The self-oscillations in this system appear at $\alpha > \alpha_1 = 2.44$ and are not connected with Andronov–Hopf bifurcation (Figure 6).

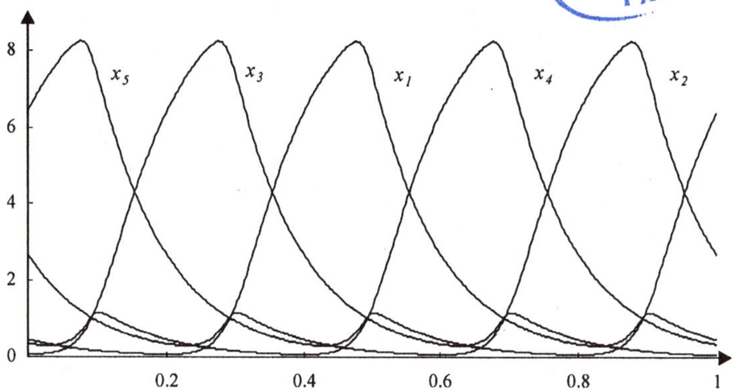

Figure -4. Symmetrical self-oscillations of the model $M(5, 3)$, characterized by similar phase order of variables and equal periods; however, having different amplitudes, $T = 10.35$.

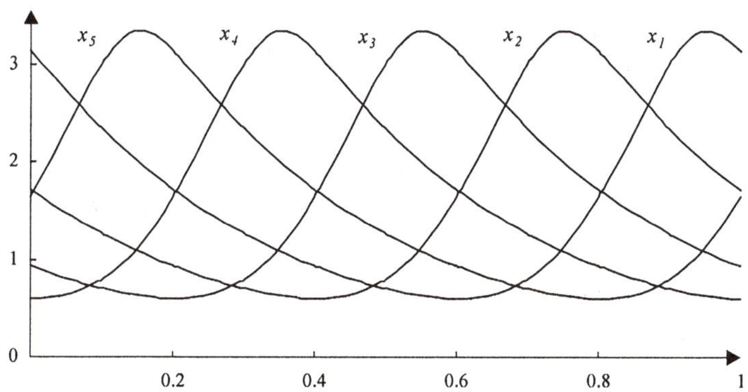

Figure -5. Another form of symmetric self-oscillation of the model $M(5, 3)$ at the same values of parameters as in Figure 4, $T = 3.13$.

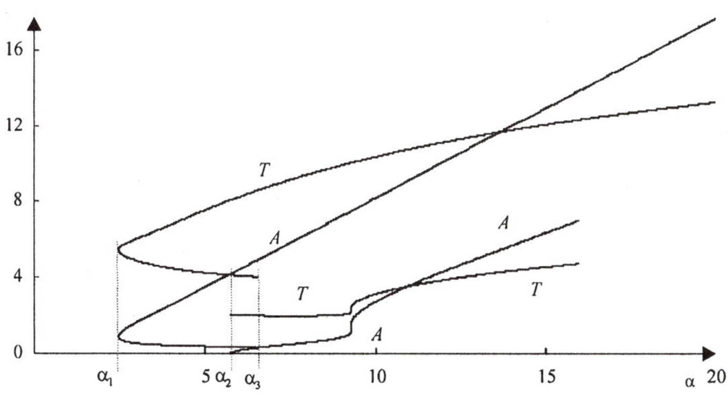

Figure -6. Dependence of oscillation period T and amplitude A of the model $M(5, 3)$ on parameter α.

The following equation corresponds to the curves shown in Figure 4:

$$du/ds = T\,[\alpha/(1+\beta(u^{\gamma}(s-1/5)+u^{\gamma}(s-3/5)) - u(s)].$$

The below equation corresponds to the case shown in Figure 5:

$$du/ds = T\,[\alpha/(1+\beta(u^{\gamma}(s-1/5)+u^{\gamma}(s-2/5)) - u(s)].$$

Thus, two approaches to numerical study of self-oscillations of the model $M(n, k)$ are possible. The results shown in figures were obtained by solving boundary value problem for both the autonomous system and the corresponding system of delay equations.

Note that composition of delay equations requires analysis of plots of the periodic solution components constructed by integration of Cauchy problem for autonomous system (1). Namely, it is necessary to determine the order of components and delay in fractions of period for each group.

Evidently, study of self-oscillations of the model by parameter continuation of boundary value problem solution for system of delayed equations is a more efficient approach, as the problem in this case is reduced to d equations, $d < n$. However, study of the self-oscillations described by the boundary value problem for autonomous system is of general character. In addition, boundary value problem (5), (6) allows known methods for determining stability of periodic solution to be used after such periodic solution is found.

2.4 Determination of stability of self-oscillations

Let us dwell on certain known facts in connection with the problem of stabilities of self-oscillations (Pontryagin, 1961; Bibikov, 1981). Consider the following autonomous system of equations:

$$dx/dt = f(x,q), \tag{14}$$

which at a certain value of parameter q has the T-periodic solution described by the vector function $x = x(t, q)$, $x(t, q) = x(t + T, q)$. Designate as $A(t)$ the matrix of derivatives of the right parts of system (1), $A(t) = f_x(x(t, q), q)$; as $U(t)$, the matricant $dU/dt = A(t)U$, $U(0) = I$. The matrix $M = U(T)$ is called monodromy matrix; its eigenvalues, multipliers.

As is known, the monodromy matrix has a multiplier equal to 1, to which eigenvector $v(T)$ corresponds. According to the Andronov–Witt theorem (Bibikov, 1981), if a multiplier equal to 1 has a repetition factor of 1 and all the rest multipliers lie within a unit circle on a complex plane, the periodic solution of system (14) is stable according to Lyapunov. If at least one of the multipliers lies outside the unit circle, the corresponding periodic solution is unstable. Consequently, determination of periodic solution stability is reduced either to finding of the maximum absolute value of eigenvalue of monodromy matrix (Pontryagin, 1961; Bibikov, 1981) or to dichotomy of monodromy matrix spectrum by a circle of unit radius (Godunov, 1997). As the monodromy matrix always has one eigenvalue equal to 1, it is necessary at first to apply the exhaustion method (Collatz, 1968, Godunov et al., 1988) to the matrix by excluding unit eigenvalue to reduce the dimension of the matrix by one with the spectrum that already does not contain eigenvalue.

The exhaustion method consists in the following. Let A be the $(n \times n)$-matrix with the elements a_{ij}, $i,j = 1, 2, \ldots, n$ and λ, one of its eigenvalues λ, to which corresponds the eigenvector z with the components $z_1, z_2, \ldots, z_n$. Let us also take that the component z_k of vector z is nonzero. Then, $((n-1)\times(n-1))$ is matrix B with the elements $b_{ij} = a_{ij} - z_i a_{kj}/z_k$, $i, j = 1, 2, \ldots, n$, $i \neq k, j \neq k$, that has the same eigenvalues as matrix A except for the excluded eigenvalue λ.

As was noted, the eigenvalue equaling 1, to which corresponds vector v(T), is excluded from monodromy matrix M. Designate as L the matrix got by applying exhaustion method to matrix M and use the power method to calculate the maximum eigenvalue λ_{max} of matrix L assuming that λ_{max} is a real number (Semendyaev, 1943; Faddeev and Faddeeva, 1960; Krylov et al., 1972). In the power method, sequence of vectors $L^k u, k = 1, 2, \ldots, n-1$ with the components $(L^k u)_i$, $i = 1, 2, \ldots, n-1$ is constructed from the specified initial vector u. In this process, for a significantly large values of k, the equation $\lambda_{max}^{(k)} = (L^{k+1}u)_i/(L^k u)_i$ gives an approximate value of λ_{max} that is weakly dependent on the number i.

Figure 7 shows the results of studying the stability of periodic solutions of the model M(6, 4) as the dependences of the maximum absolute value of the eigenvalue of monodromy matrix on parameter α. The values $|\lambda| > 1$ correspond to unstable self-oscillations. From comparison of Figures 3 and 7, it follows that the branch indicated as (1) represents stable self-oscillations; the branch indicated as (2), unstable. The self-oscillations marked (3) are also unstable.

Figure 8 demonstrates dependences of the maximum absolute value of the eigenvalue λ of the monodromy matrices on parameter α corresponding to the model M(5, 3). From comparison of Figures 6 and 8, it follows that the self-oscillations stable for sufficiently large α values retain their stability with decrease in α until the pivot point at $\alpha = \alpha_1$. Then, after the pivot point at $\alpha_1 < \alpha < \alpha_2$, self-oscillations become unstable. In the region $\alpha_2 < \alpha < \alpha_3$, self-oscillations are again stable and loss stability at $\alpha = \alpha_3$.

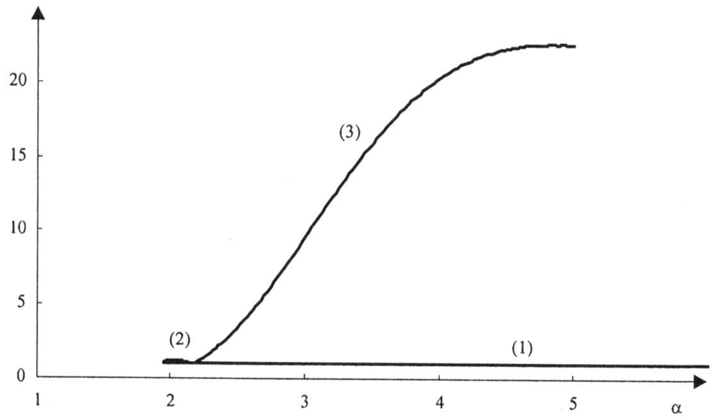

Figure -7. Dependences of the maximum absolute value of eigenvalue λ of the monodromy matrices of the model $M(6, 4)$ on parameter α.

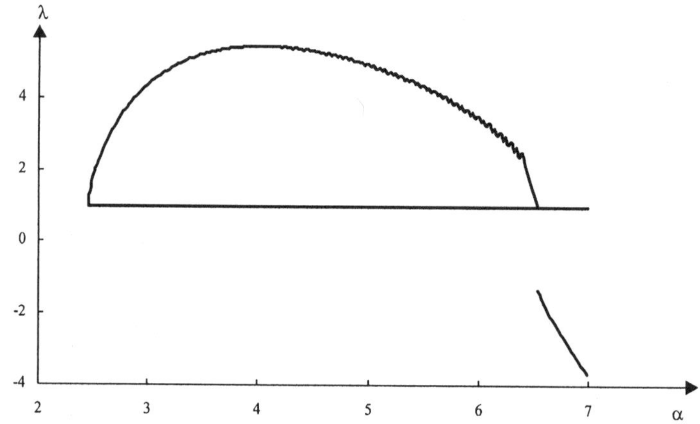

Figure -8. Dependences of the maximum absolute value of eigenvalue λ of monodromy matrices of the model $M(5, 3)$ on parameter α.

Note another approach to studying stability of periodic solutions of autonomous system depending on parameter that consists in determination of solution norm of discrete matrix Lyapunov equation (Godunov, 1997).

3. CONCLUSION

Study of properties of the mathematical models describing hypothetical gene networks is an important stage in studying operation patterns of natural gene networks, which underlie performance of virtually all vital functions of the organisms. The simplest nonstationary types of gene network operation are the cyclic oscillations. Studied in this work are systems (1) that describe the first class of hypothetical gene networks (Likhoshvai et al., 2003). For the systems of this type, an efficient method is developed for search for oscillation modes that is independent of oscillation stability. The method is based on the correspondence between unstable cyclic trajectories of systems of autonomous equations (1) and stable trajectories of the corresponding systems of delay equations of a smaller dimension. In connection with the discovered correspondence, the problem of study of type (1) delay systems arises. Numerical results demonstrate that here the behavior palette may be very rich even in the case of one equation. Let us consider the below equation as an example:

$$dx/dt = \alpha/(1 + x^{\gamma}(t-\tau) + x^{\gamma}(t-a\tau)) - x\,. \qquad (15)$$

It follows from the results shown below that this equation at certain values of parameters α, γ, a, and τ has the trajectories that are identical to cyclic trajectories of system (1) at $k = 3$. However, at $\alpha = 50$, $\gamma = 50$, $a = 25$, and $\tau = 0.0313$, the numerical calculation gives the solution behavior shown in Figure 9. Preliminary analysis demonstrates that the trajectory is neither periodic nor quasi-periodic and possibly, we have a strange attractor. Since equation (15) may be also interpreted as a model describing protein biosynthesis with two negative feedbacks, we may infer that a chaotic protein synthesis is possible already in the simplest biological systems.

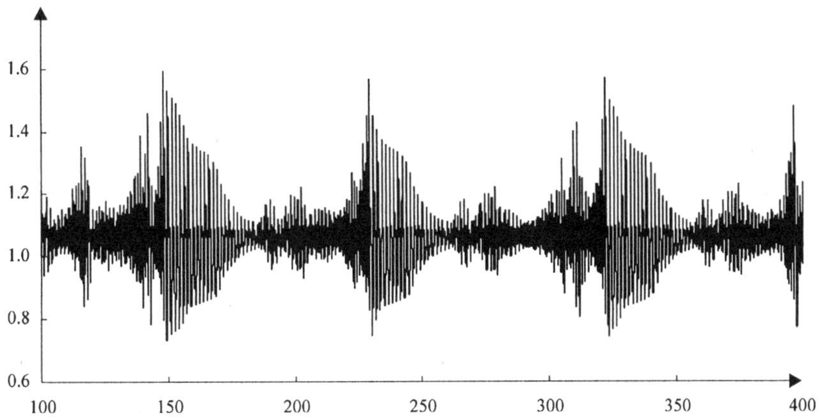

Figure -9. Plot of solution of equation (15) at $\alpha = 50$, $\gamma = 50$, $a = 25$, $\tau = 0.031$, $x(t) = 0.9$, and $t - a\tau \leq 0$: the abscissa, values of variable t; the ordinate, values of function x at point t.

These results may find practical application in synthesis of gene networks with specified properties as well as may be used when solving practical problems in the fields of biotechnology, biotherapy, genetic engineering, and pharmacogenetics.

ACKNOWLEDGMENTS

The work was supported by RFBR (grants Nos. 03-01-00328, 03-04-48506, 03-07-96833, 03-04-48829, 04-01-00458, 05-07-98011, 05-07-98012, and 05-04-08084), Siberian Branch of the Russian Academy of Sciences (integration project No. 119), program of Basic Research of the Presidium of the Russian Academy of Sciences and Branches of RAS (grant No. 10.4), state contract No. 10002-251/P-25/155-270/200404-082, and NSF:FIBR No. EF-0330786.

PERIODIC TRAJECTORIES AND ANDRONOV–HOPF BIFURCATIONS IN MODELS OF GENE NETWORKS

V.P. Golubyatnikov[1, 2*], V.A. Likhoshvai[3, 4], E.P. Volokitin[1], Yu.A. Gaidov[5], A.F. Osipov[5]

[1] *Sobolev Institute of Mathematics, Siberian Branch of the Russian Academy of Sciences, prosp. Koptyuga 4, Novosibirsk, 630090, Russia, e-mail: glbtn@math.nsc.ru;* [2] *Weizmann Institute of Sciences, Rehovot, Izrael;* [3] *Institute of Cytology and Genetics, Siberian Branch of the Russian Academy of Sciences, prosp. Lavrentieva 10, Novosibirsk, 630090, Russia;* [4] *Ugra Institute of Informatics, Hanty-Mansiisk, Russia;* [5] *Novosibirsk State University, ul. Pirogova 2, Novosibirsk, 630090, Russia*

* *Corresponding author*

Abstract: Multistability is an important property of the gene network functioning. The estimate of possible numbers of limit cycles and stationary points of gene networks with various types of regulations is a fundamental problem of the applied mathematics. We prove existence of limit cycles for four classes of the gene network models where the control of protein concentrations is realized at the level of regulation of their stability. We show that the change in regulation from the stage of gene expression activation to the stage of degradation of the synthesis products does not affect the dynamical properties of the gene networks under consideration.

Key words: Dynamical systems; Andronov–Hopf bifurcation; fixed point theorem; negative feedback; regulation of degradation and synthesis of mRNA and proteins

1. INTRODUCTION

Detection of closed trajectories in any particular dynamical system is a hard mathematical problem even in the low-dimensional cases. Here, we consider special dynamical systems as the models of gene networks. We study their periodic trajectories and stationary points. The existence of these regimes is very important from the viewpoint of the gene network design for the needs of biotechnology, biocomputing, and gene therapy (Elowitz and Leibner, 2000; Gardner et al., 2000; Golubyatnikov et al., 2003). We analyze here the gene network models based on the regulation of degradation stages of the synthesis products of the genetic elements in contrast with Golubyatnikov and Makarov (2004), where the regulation is realized at the stages of initiation of mRNA and protein synthesis. We show that in both cases the qualitative properties of the corresponding dynamical systems are similar and depend on the general structure of the gene network rather than on the particular realization of the regulation mechanism.

2. METHODS AND ALGORITHMS

We consider the following dynamical systems as the models of gene networks, introduced by Likhoshvai et al. (2001):

$$\frac{dx_i}{dt}=\frac{\alpha}{1+x_{i-1}^{\gamma}}-x_i,\ \alpha>0,\ \mathrm{i}=1,2,3 \tag{1}$$

$$\frac{dx_i}{dt}=\frac{\alpha}{1+x_{i\text{-}1}^{\gamma}+x_{i\text{-}2}^{\mu}}-x_i,\quad \alpha>0,\ i=1,2,3 \tag{2}$$

$$\frac{dx_i}{dt}=\frac{\alpha}{1+x_{i-1}^{\gamma}\cdot x_{i-2}^{\mu}}-x_i,\quad \alpha>0,\ i=1,2,3 \tag{3}$$

$$\frac{dx_i}{dt}=\frac{\alpha}{(1+x_{i-1}^{\gamma})(1+x_{i-2}^{\mu})}-x_i,\quad \alpha>0,\ i=1,2,3 \tag{4}$$

The regulation processes in these models take place at the stages of mRNA initiation and/or at the stages of the protein synthesis (mechanism **I**). In a similar way, we construct the gene network models with regulation processes at the stages of degradation of the products of synthesis (mechanism **D**). These models are described by the following systems:

$$\frac{dx_i}{dt}=\alpha-x_i(1+x_{i-1}^{\gamma}),\ \alpha>0,\ i=1,2,3 \tag{5}$$

$$\frac{dx_i}{dt} = \alpha - x_i(1 + x_{i-1}^{\gamma} + x_{i-2}^{\mu}),\ \alpha > 0,\ i = 1, 2, 3 \tag{6}$$

$$\frac{dx_i}{dt} = \alpha - x_i(1 + x_{i-1}^{\gamma} \cdot x_{i-2}^{\mu}),\ \alpha > 0,\ i = 1, 2, 3 \tag{7}$$

$$\frac{dx_i}{dt} = \alpha - x_i(1 + x_{i-1}^{\gamma}) \cdot (1 + x_{i-2}^{\mu}),\ \ \alpha > 0,\ i = 1, 2, 3. \tag{8}$$

We assume that $\gamma > \mu > 1$, $i - 1 = 3$, $i - 2 = 2$ for $i = 1$, and $i - 2 = 3$ for $i = 2$. Each of the dynamical systems (1)–(8) is symmetric with respect to the cyclic permutation of the variables $x_3 \to x_1 \to x_2 \to x_3$. Sometimes we shall use notations $x_1 = x$, $x_2 = y$, $x_3 = z$ just for simplicity. All trajectories of these systems eventually enter the cube $Q = [0, \alpha] \times [0, \alpha] \times [0, \alpha] \subset R^3$ and do not leave it. The diagonal $\Delta = \{x_1 = x_2 = x_3\}$ of this cube contains exactly one stationary point $M_*^{(j)}$ of each of these systems. Here, the number j corresponds to the equation number (1)–(8). Linearizations of all these systems in the neighborhood of their diagonal stationary points are described by the matrix

$$A = \begin{pmatrix} -1-s & -p & -q \\ -q & -1-s & -p \\ -p & -q & -1-s \end{pmatrix},\ p,\ q,\ s \geq 0. \tag{9}$$

One of its eigenvalues, $\lambda_1 = -1 - p - q - s$, corresponds to the vector $(1, 1, 1)$, which is parallel to the diagonal Δ. For $p \neq q$, the other eigenvalues λ_2 and λ_3 are complex $\operatorname{Im} \lambda_3 = -\operatorname{Im} \lambda_2 \neq 0$, $2\operatorname{Re} \lambda_{2,3} = p + q - 2 - 2s$.

3. RESULTS AND DISCUSSION

3.1 Existence of closed trajectories

The behavior of the trajectories of the systems (1) and (5) is much simpler than those in the other systems under consideration. Let $r(X)$ be the vector that joins an arbitrary non-diagonal point X with its projection onto the diagonal Δ and let $v(X) = \dfrac{dr(X)}{dt}$. Simple calculations show that for the points X outside the diagonal and the coordinate axes, all coordinates of the vector product $[r(X), v(X)]$ are equal and strictly negative. Hence, we obtain:

Theorem 1. *All trajectories of the systems (1), (5) turn around the diagonal Δ with non-zero angular velocity.*

This theorem implies that there are no stationary points of the dynamical systems (1), (5) outside the diagonal Δ. Linearization of the system (1) in some neighborhood of the stationary point $M_*^{(1)} \in \Delta$ with the coordinates $x_*^{(1)}$ is described by the matrix (9) with $p = 0$, $q = \gamma(x_*^{(1)})^{\gamma+1}\alpha^{-1}$. In a similar way, we can obtain the linearization of the system (5) in some neighborhood of the point $M_*^{(5)} \in \Delta$. For $\alpha(\gamma-2) < \gamma x_*^{(1)}$, the real parts of the eigenvalues $\lambda_{2,3}$ are negative; in this case, there are no attractors in the systems (1), (5) except the stationary points $M_*^{(1)}$, $M_*^{(5)}$. If $\alpha(\gamma-2) > \gamma x_*^{(1)}$, then the points $M_*^{(1)}$ and $M_*^{(5)}$ are unstable and the angular velocities of the trajectories of these two systems are bounded away from the zero outside some neighborhood $U(\Delta)$ of the diagonal Δ and outside some neighborhood W of the coordinate axes. For some positive t_0, each point in $D = Q\backslash(U(\Delta)\cup W)$ makes at least one complete turn around Δ. Let $T_1 = D \cap H_1$ $\{x_1 > x_2 = x_3\}$, $T_2 = D \cap H_2$ $\{x_2 > x_1 = x_3\}$, and $T_3 = D \cap H_3$ $\{x_3 > x_2 = x_1\}$.

According to Theorem 1, the trajectory of each point $M_3 \in M_3$ arrives to T_1 at some moment $t_1(M) < t_0$. Let $\tau_1 : T_3 \to T_1$ be the shift of the points of the set T_3 along their trajectories. At some moment $t_1(M) + t_2(M) < t_0$, the point M_3 arrives to T_2. Let $\tau_2 : T_1 \to T_2$, $\tau_3 : T_2 \to T_3$ be analogous shifts. At some moment $t_1(M) + t_2(M) + t_3(M) < t_0$, the point M_3 returns to T_3 for the first time. Denote by $\varphi_1 : T_1 \to T_3$, $\varphi_2 : T_2 \to T_1$, $\varphi_3 : T_3 \to T_2$ the rotations of the compact contractible sets T_i around Δ by the angle 120°.

Consider the composition $\varphi_1 \circ \tau_1 : T_3 \to T_3$ of continuous mappings τ_1 and φ_1. The topological fixed point theorem implies the existence of at least one point $M_0(x_0, x_0, z_0) \in T_3$, such that $\varphi_1 \circ \tau_1(M_0) = M_0$. In other words, the shift $\tau_1(M_0)$ of this point is obtained by rotation of this point around the diagonal Δ. Since the systems (1) and (5) are symmetric with respect to the cyclic permutation of the variables $x_3 \to x_1 \to x_2 \to x_3$, the composition $\varphi_2 \circ \tau_2 : T_1 \to T_1$ maps the point $\tau_1(M_0)$ to itself as well. Hence, the shift $\tau_2 \circ \tau_1(M_0) = (x_0, z_0, x_0)$ coincides with the result of the rotation $\varphi_2^{-1} \circ \varphi_1^{-1}$ of the point M_0. Finally, the composition $\varphi_3 \circ \tau_3 : T_2 \to T_2$ maps the point $\tau_2 \circ \tau_1(M_0)$ to itself; the total shift $\tau_3 \circ \tau_2 \circ \tau_1(M_0)$ (M_0) coincides with the result of the complete turn $\varphi_3^{-1} \circ \varphi_2^{-1} \circ \varphi_1^{-1}(M_0)$. Hence, we obtain the following:

Theorem 2. *If Re* $\lambda_{2,3} > 0$, *then each of the dynamical systems (1) and (5) has at least one periodic trajectory symmetric with respect to the cyclic permutation of the variables.*

It is worthy to note that the topological fixed point theorem does not ensure the uniqueness or stability of the periodic trajectory in Theorem 2.

Figure 1 shows the cycles of the system (1) at the values of the parameters $\alpha = 3$, $\gamma = 10$ (left) and $\alpha = 3$, $\gamma = 50$ (right). Note that at $\alpha > 1$, $\gamma >> 1$, the function $g(x) = (1 + x^\gamma)^{-1}$ defined on the segment $[0, \alpha]$ is approximated by the piecewise constant function

$\text{sgn}(x) = 1$ for $0 \le x < 1$, $\text{sgn}(1) = 1/2$, $\text{sgn}(x) = 0$ for $1 < x \le \alpha$,

and for large values of γ, the trajectories of the system (1) are approximated by those of the system

$$\frac{d x_1}{d t} = \alpha \cdot \text{sgn}(x_3) - x_1, \frac{d x_2}{d t} = \alpha \cdot \text{sgn}(x_1) - x_2, \frac{d x_3}{d t} = \alpha \cdot \text{sgn}(x_2) - x_3. \quad (10)$$

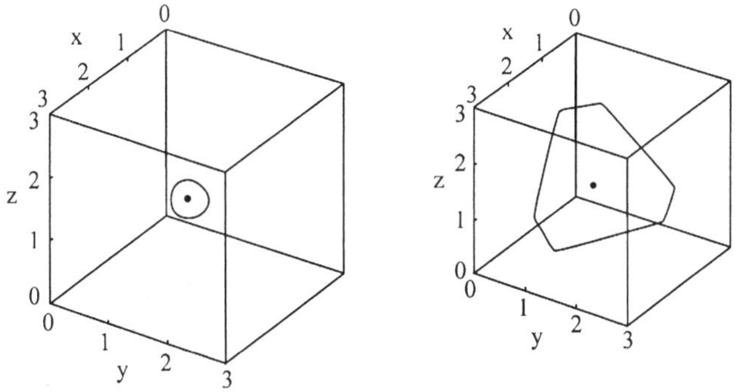

Figure -1. Closed trajectories of system (1).

In the same way, one can approximate systems (2)–(4), which are invariant with respect to the permutations $x_3 \to x_1 \to x_2 \to x_3$ as well.

It is easy to see that their trajectories are piecewise linear and their linear segments are located inside the parallelepipeds composed by the faces of the cube Q and the planes $x_i = 1$, $i = 1, 2, 3$.

As it was shown by Volokitin (2004), system (10) has a unique cycle symmetric with respect to rotations around the diagonal Δ by an angle of 120°. This is a hexagon contained between the two parallel planes orthogonal to Δ. This cycle is stable, and its period T can be expressed as

$T = -\ln z$, where z is the corresponding root of the equation $z^2 + z(1 - \frac{\alpha^2}{\alpha - 1}) + 1 = 0$.

3.2 Andronov–Hopf bifurcation

Some other results on uniqueness and stability of the cycles can be deduced from the Andronov–Hopf bifurcation theorem. For systems (1) and (5), the relation Re $\lambda_{2,\,3} = 0$ is equivalent to $\alpha = \frac{\gamma}{\gamma - 2} \cdot \left(\frac{2}{\gamma - 2}\right)^{1/\gamma}$. Simple calculations show that if $\gamma > 2$, then $\frac{d}{d\alpha}(\mathrm{Re}\,\lambda_{2,3}) > 0$ for both systems (1) and (5). More advanced analysis shows that the Lyapunov parameter v_1 is negative here. Hence, the bifurcation theorem (Kuznetzov, 1995) implies the uniqueness and stability of the cycles that appear in systems (1), (5) at their bifurcation points. It is easy to verify that for any fixed γ and μ, if Re$\lambda_{2,\,3} = 0$ at $\alpha = \alpha_0$, then $\frac{d}{d\alpha}(\mathrm{Re}\,\lambda_{2,3}) > 0$ at $\alpha = \alpha_0$ for each of the systems (2)–(4) and (5)–(8). As above, the Andronov–Hopf theorem implies that for $\alpha = \alpha_0$ sufficiently close to α_0, some small neighborhood of each stationary point $M_*^{(i)}$ contains a periodic trajectory of the corresponding dynamical system.

Figure 2 shows the convergence of the two trajectories of system (2) to its Hopf cycle from two opposite directions along the central manifold of the bifurcation point; this is the 'slow variables' surface.

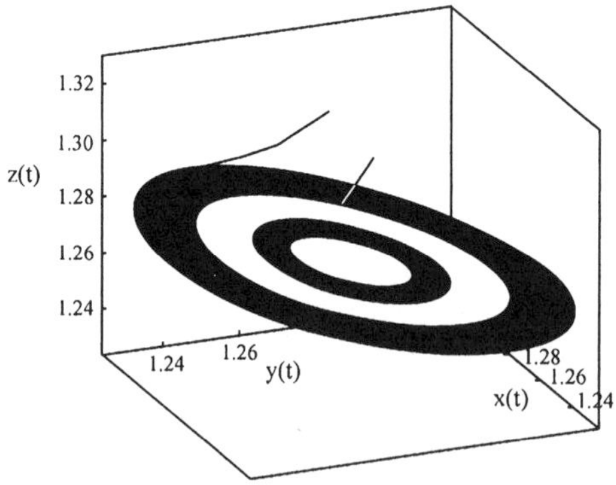

Figure -2. Andronov–Hopf bifurcation in system (2).

The coordinates of the starting points of the interior trajectory are $x = 1.285$, $y = 1.285$, $z = 1.29$; for the exterior trajectory, $x = 1.27$, $y = 1.27$, $z = 1.31$. Here, $\alpha = 5.908$, $\gamma = 2.981$, and $\mu = 2.0$.

Other gene networks models considered here and their higher-dimensional analogues have similar bifurcations. If $\alpha > 2$ and $\gamma + \mu > 4$, then systems (3) and (7) do not have bifurcation cycles.

3.3 Other gene network modeling results

In contrast with systems (1) and (5), the behavior of the trajectories of other dynamical systems considered above is more complicated. If the values of the parameters α, γ, and μ are sufficiently large, then each of the systems (2) and (6) has six stationary points outside the diagonal Δ. Three of them are stable and are contained in some neighborhoods of the vertices $(\alpha, 0, 0)$, $(0, \alpha, 0)$, and $(0, 0, \alpha)$ of the cube Q. Other three points are unstable (hyperbolic). In these cases, there is no bifurcation near the diagonal stationary point. Figure 3 shows a collection of the trajectories of system (2). Here $\alpha = 10.0$, $\gamma = 5.0$, and $\mu = 1.5$. The 'vertices of the curvilinear triangle' are located near the unstable stationary points; trajectory in the left part of the figure is attracted by the stable stationary point in a neighborhood of the vertex (0, 10, 0) of the cube Q. In a similar way, one can observe the phase portrait of system (6).

Here, in Figure 4, $\alpha = 10.0$, $\gamma = 6.0$, and $\mu = 2.0$. The trajectory with the starting point C (1, 0, 0.001) is attracted by the 'triangle' cycle in the center of Figure 4. Another trajectory with the starting point B (2.5, 0.9, 1.8) approaches this cycle, then passes near the unstable stationary point, and, finally, is attracted by the stable stationary point S in a neighborhood of the vertex (0, 0, 10).

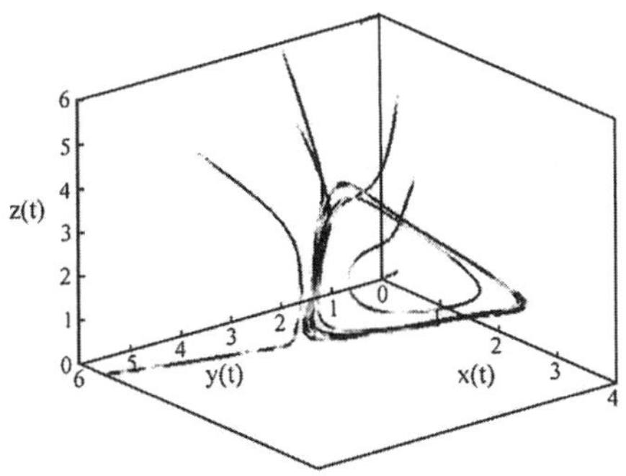

Figure -3. The phase portrait of system (2).

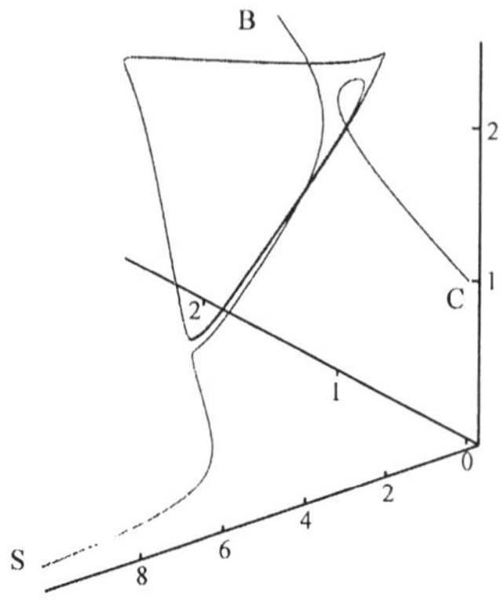

Figure -4. Two trajectories of system (6).

The trajectory of system (3) in the Figure 5 does not have a constant direction of rotation around Δ, although it is attracted by a closed cycle. This direction changes near one of the unstable stationary points. Here, $\alpha = 3.237$, $\gamma = 1.725$, and $\mu = 1.434$.

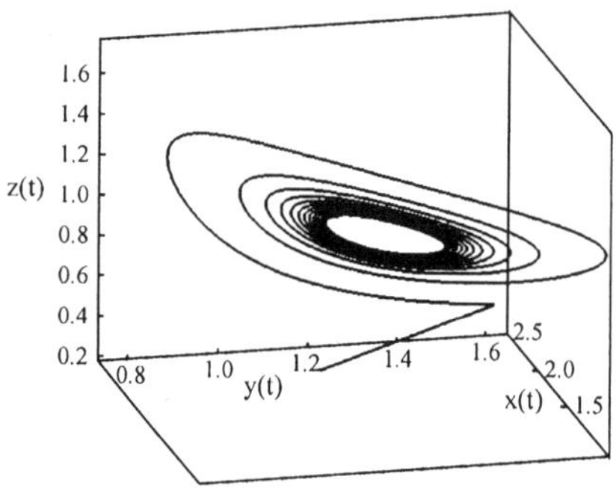

Figure -5. A limit cycle of system (3).

It is easy to verify that for all dynamical systems (1)–(8) and for their higher-dimensional analogues div $X(t) < 0$. Hence, the cubes Q in the positive octants of R^n, $n > 3$, do not contain proper subsets that have positive Lebesgue measure and are invariant with respect to the trajectories of these systems. Therefore, there are no invariant tori in these systems. As in our previous studies, we used in the simulations of gene networks the multiple precision algorithms and the software complex GeneNetSTEP, developed specially for analysis of the chemical kinetic nonlinear dynamical systems.

It was shown that the standard algorithms used in Maple 6 did not give realistic results in the case of system (6) near the point S (Figure 4) and in some other quite simple situations. Thus, the mathematical arguments should not be neglected in the numerical simulations of these natural phenomena.

4. CONCLUSIONS AND FUTURE WORKS

For the considered models of gene networks, we gave the mathematical proofs of (1) the existence of closed trajectories and (2) the stability of the bifurcation cycles.

Comparison of the dynamical systems (1) and (4), (2) and (5), and (3) and (6), corresponding to the transcriptional regulation mechanism I and to the post-transcriptional mechanism D, respectively, shows that in both cases, the qualitative properties of their phase portraits are similar and determined by the general structure of the models, i.e., by the gene network graphs. Notwithstanding relative simplicity of these systems, we have seen the hierarchical principles in the design of the gene network structure: here, the main role belongs to the gene network graph (Likhoshvai et al., 2001). Further precision of the properties of these networks can be obtained by means of the appropriate choice of the regulatory mechanism, and its finer adjustment is realized at the level of parameter variations. Some analogues of the stationary points and the oscillating stationary regimes considered here were observed in the functioning of the natural gene networks (Elowitz and Leibner, 2000; Gardner et al., 2000). As we have seen above, the behavior of the trajectories of the nonlinear dynamical systems near the unstable stationary points does not seem to be regular. Hence, the problem of detection of the unstable stationary points and the unstable cycles, as well as the shapes of the separatrix is very important for the numerical analysis of the gene network models.

It was shown by Likhoshvai et al. (2001) and Volokitin (2004) that for large values of the parameters α, γ, and μ, the multistability property of the dynamical systems approaches to that of the finite automata. In order to investigate the boundary between the dynamical systems and the finite automata, we plan to study in our future work the following deformation problems:

1. Are the cycles of systems (1) and (4) in theorem 2 unique and stable?
2. What happens with the bifurcation cycles when these parameters grow?

Our current task is to describe the multistability properties of the gene network models composed by collections of the lower-dimensional models of the types similar to (1)–(8). In particular, we plan to realize a similar approach to the modeling of gene networks by means of the dynamical systems $\frac{dx_i}{dt} = f(x_{i-1}) - x_i$ in the cases $f(z) = \frac{\alpha z}{1 + z^{\gamma}}$, as in the Glass–Mackey equation (Mackey and Glass, 1977), and $f(z) = \alpha - 2\left|\frac{\alpha - 2z}{2}\right|$ for $0 \le z \le \alpha$ (Schuster, 1984). The dynamical systems of these types describe the models of the gene networks with positive feedbacks for small values of the concentrations. Preliminary studies of their phase portraits show a chaotic behavior of their trajectories in the case $n > 3$.

ACKNOWLEDGMENTS

The work was supported by the leading scientific schools grant No. 311.2003.1 of the President of the Russian Federation; Russian Foundation for Basic Research (grants Nos. 03-04-48829, 04-01-00458, 05-07-98011, 05-07-98012, and 05-04-08084); Siberian Branch of the Russian Academy of Sciences (integration project No. 119); program of Basic Research of the Presidium of the Russian Academy of Sciences and branches of RAS (grant No. 10.4); state contract No. 10002-251/P-25/155-270/200404-082; NSF:FIBR (grant No. EF-0330786); and by Rosi and Max Varon Visiting Professorship at the Weizmann Institute of Sciences. The authors are indebted to V.M. Goldshtein, A.V. Ratushny, and K.V. Storozhuk for helpful discussions and assistance.

ON THE PROBLEM OF SEARCH FOR STATIONARY POINTS IN REGULATORY CIRCUITS OF GENE NETWORKS

V.A. Likhoshvai
Institute of Cytology and Genetics, Siberian Branch of the Russian Academy of Sciences, prosp. Lavrentieva 10, Novosibirsk, 630090, Russia, e-mail: likho@bionet.nsc.ru; Novosibirsk State University, ul. Pirogova 2, Novosibirsk, 630090, Russia
Corresponding author

Abstract: This work considers properties of the genetic automatons modeling regulatory circuits of gene networks. It is proved that the problem of search for stationary points of genetic automatons is equivalent to the problem of search for covering of oriented graphs. For particular cases, a complete description of the structure of stationary points for genetic automatons is given.

Key words: genetic systems; regulatory circuits of gene networks; modeling; discrete methods; stationaries; stationary points

1. INTRODUCTION

Gene networks (GN) are structurally complex spatial objects formed by hundreds of elements of various natures and complexities, namely, genes and their regulatory regions, RNAs and proteins encoded by these genes, low-molecular-weight compounds, various complexes between enzymes and their targets, etc. (Kolchanov et al., 2000). The GN elements are united into a functional system via complex nonlinear biochemical processes of synthesis and degradation of substances (Kolchanov et al., 2002). The GN are open systems whose operation is sustained by continuous inflow of certain substances and energy to the medium and outflow of the resulting products. GN operation may be characterized by temporal trajectories of changes in concentrations of certain substance set of particular GN. The most important property of GN is the ability to change its state (concentrations of substances) in response to alterations in the external and internal conditions. The change in the state is achieved through altering the level of expression of certain gene groups via regulatory substances. The regulatory processes are sequences of molecular events (often, fairly complex and branched) involving frequently (1) numerous substances both coming outside (external signals) and synthesized by the gene network itself (internal signals) and (2) regulatory regions of genes. Thus, the GN core is the genes and the RNAs and proteins they encode, whose expression is mutually regulated. These subnetworks are the regulatory circuits of GN. Study of the properties of regulatory circuits represent the most important problem in the field of bioinformatics related to the research into dynamics of GN operation, as these particular circuits underlie the unique ability of GN to respond adequately to changes in external conditions.

This work develops a discrete approach that makes it possible to draw the information on the presence of stationaries in gene networks directly from analysis of the oriented graphs representing regulatory circuits of gene networks omitting construction and calculation of dynamic models. The results obtained are also an additional source of hypotheses on the properties of the corresponding dynamic models developed by Likhoshvai et al. (2001, 2003, 2004) and Fadeev and Likhoshvai (2003).

2. RESULTS

Let us consider only the part of the gene network that comprises the genetic elements regulating expression efficiencies of other genes or being themselves subjected to control by other genes. Let us name it as the regulatory circuit of the gene network. Let us represent the scheme of activity regulation of gene expression as a bipartite oriented graph $G(U_1, U_2, W_{12}, W_{21})$. In this graph, the first type nodes (U_1) identify proteins; the second type nodes (U_2), independent regulatory mechanisms; edges W_{12}, coming from the first type nodes to second type nodes, indicate that a protein is involved in the corresponding regulatory mechanism; and edges W_{21} identify that the mechanism is acting on particular protein. Let us construct a monopartite oriented graph $G(U_1, W)$ on the nodes of the bipartite graph $G(U_1, U_2, W_{12}, W_{21})$. In this monopartite graph, the edges indicate that a particular protein is involved in the mechanism of expression regulation of the genes encoding the proteins whereto the edges are coming. Let us name the monopartite graph $G(U_1, W)$ as the graph associated with $G(U_1, U_2, W_{12}, W_{21})$.

Let us ascribe to each first type node u_i a non-negative variable x_i, changing within the range $[0, p_i]$, where p_i is the upper limit of changing in the ith variable.

Let us specify a non-negative number q, non-negative integer k, and positive $s_1, \ldots, s_k$. Then, $B_0(y_1, \ldots, y_k|q, s_1, \ldots, s_k)$ will denote the Boolean function defined in $R_+^k = \{(y_1, \ldots, y_k) \mid y_i \geq 0, \quad i = \overline{1,k}\}$, taking the zero value if and only if $y_1^{s_1} \cdot \ldots \cdot y_k^{s_k} \leq q^{\sum_{i=1,k} s_i}$. Otherwise, $B_m(y_1, \ldots, y_k|q, s_1, \ldots, s_k) = 1$. If $k = 0$, $B_0(\varnothing|q, 0) = 0$ by definition. For negating $B_0(y_1, \ldots, y_k|q, s_1, \ldots, s_k)$, let us introduce the designation $B_1(y_1, \ldots, y_k|q, s_1, \ldots, s_k)$.

For B_0, the simplest biological prototype of the functions introduced is the mechanism of a threshold inhibition of transcription activity of genetic element by the multimer composed of s_1 molecules of type y_1, ..., and s_k molecules of type y_k; for B_1, the mechanism of the corresponding threshold activation. In the general case, $s_1, \ldots, s_k$ are real numbers called Hill coefficients, which specify the degree of nonlinearity of involvement of a corresponding effector in the regulatory mechanism.

Let i be the number of the first type nodes. The set of node numbers wherefrom the edges in oriented graph $G\ (U_1, W)$ come to u_i are designated as D_i. Let l_i edges be directed to the ith node from the second type nodes, and let them have the numbers $\sigma_i(j)$, $j = 1, \ldots, l_i$. Let us designate as $R_{i,j}$ the set of numbers of the first type nodes wherefrom the edges are directed to the second type node with the number $\sigma_i(j)$. If the second type node is rooted, $R_{i,j} = \varnothing$ by definition. Let us associate the integer $\delta_{i,j} \in \{0,1\}$ and the function $B_{\delta_{i,j}}$ with each second type node and construct the Boolean function

$$\Omega_i(x_j \mid j \in D_i) = \\ = \bigvee_{j:\delta_{i,j}=0} B_{\delta_{i,j}}(x_r \mid r \in R_{i,j}, q_{i,j}, s_{i,j,1}, \ldots, s_{i,j,l_{i,j}}) \vee (\bigwedge_{j:\delta_{i,j}=1} B_{\delta_{i,j}}(x_r \mid r \in R_{i,j}, q_{i,j}), \quad (1)$$

which describes the total mechanism of activity regulation of the ith genetic element.

2.1 Definition of genetic automaton

Let us associate with each first type node the three non-negative numbers α_i, δ_i, p_i and the variable x_i, possessing integral values in the range $[0, p_i]$. Let us name the function $G_p(x_1, \ldots, x_n) = (x_1^+, \ldots, x_n^+)$, where

$$x_i^+ = \begin{cases} \max(0, x_i - \delta_i)\ ,\ \Omega_i(x_j \mid j \in D_j) = 1 \\ \min(p_i, x_i + \alpha_i)\ ,\ \Omega_i(x_j \mid j \in D_j) = 0 \end{cases} \quad (2)$$

the *genetic automaton* (G automaton) constructed on the bipartite graph $G(U_1, U_2, W_{12}, W_{21})$.

The sequence of points $X^0 = (x_1^0, ..., x_n^0), X^1 = (x_1^1, ..., x_n^1) = G(x_1^0, ..., x_n^0)$, $X^2 = G^2(x_1^0, ..., x_n^0), ..., X^k = G^k(x_1^0, ..., x_n^0)$, ... will be named as the trajectory of G automaton (starting from the point X^0). Let us name one act of the automaton operation as step. Evidently, due to a finiteness of the space of value vectors, any trajectory of genetic automaton after a finite number of steps becomes cyclic. Let us name the minimal number of the non-repeated steps of trajectory as the cycle length. If the cycle length equals unity, the cycle is called the stationary point. The problem stated is to describe the stationary points of genetic automatons. The biological significance of this problem lies in that the stationary points of genetic automatons correspond to the stationaries of gene networks with the corresponding structures of their regulatory circuits.

Lemma 1. If $X^0 = (x_1^0, ..., x_n^0)$ is the stationary point of genetic automaton, than for each i, $x_i^0 = 0$ or p_i.

Lemma 1 is valid because if $x_i^0 \neq 0$ and p_i, than the following results from (2): x_i^1 increases if $\Omega_i(x_j | j \in D_j) = 0$, and x_i^1 decreases if $\Omega_i(x_j | j \in D_j) = 1$.

The following reasoning demonstrates that the stationary points do not depend on the values of parameters $q_{i,j}, s_{i,j,1}, ..., s_{i,j,l_{i,j}}$ α_i, δ_i, p_i of the automaton, i.e., are determined only by the structure of bipartite oriented graph.

Let us assume that the genetic automaton has stationary points at a certain specified set of parameters $q_{i,j}, s_{i,j,1}, ..., s_{i,j,l_{i,j}}$ α_i, δ_i, p_i. Then, by virtue of Lemma 1, $x_i^0 = 0$ or p_i for any number i. Consequently, this point is a stationary point for the genetic automaton with arbitrary threshold values $q_{i,j} < p_i$. In particular, we can take all $q_{i,j} = 0$. Evidently, at zero threshold values, the values of parameters α, δ, and p also can be taken arbitrary positive. Let us take them equaling unity, as this decreases maximally the value range of automaton. Then, it is evident that the stoichiometric coefficients (s) may also be taken equal to unity.

Thus, we demonstrated that all the genetic automatons constructed on a fixed bipartite graph had the same stationary points. This means that when solving the problem of search for stationary points, it is possible to limit to consideration of the double-valued genetic automaton.

Definition. *The totality V of the first type nodes is called g-base of (bipartite associated) oriented graph if and only if (i) in the case the node* u_i *is rooted in the associated oriented graph, the node necessarily lies in V; (ii) in the case the node* u_i *is from V and not rooted in the associated oriented graph, then (1) in any not empty* $R_{i,j}$ *such that* $\delta_{i,j} = 0$, $\exists$ *the node with number* $U_1 \backslash V$ *or (2) either there does not exist* $R_{i,j}$ *such that* $\delta_{i,j} = 1$ *or there always exist at least one* $R_{i,j}$ *for which* $\delta_{i,j} = 1$ *such that any node with number* $R_{i,j}$ *belongs to V; and (iii) for each node* u_i *from* $U_1 \backslash V$ *either* $\exists$ *a non-empty* $R_{i,j}$ *for which* $\delta_{i,j} = 1$ *such that a certain node with number* $R_{i,j}$ *belongs to* $U_1 \backslash V$ *or* $\exists$ *a non-empty* $R_{i,j}$ *for which* $\delta_{i,j} = 0$ *such that all the nodes with numbers from* $R_{i,j}$ *belong to V.*

Here is the biological interpretation of the definition. All the nodes from V correspond to active genes, while all the nodes from the complement, to inactive genes. Consequently, all the constitutively expressed genes are members of the set of active genes, which is specified by condition (i). However, to be expressed, a regulated genetic element should be either activated or have a nonzero basal activity level. For a certain activation mechanism to function, it is necessary that all its activators are synthesized, i.e., that condition (ii2) is fulfilled, whereas all the inhibition mechanisms are switched off, i.e., condition (ii1) is fulfilled. On the contrary, all the genetic elements whose corresponding nodes lie in $U_1 \backslash V$ will be inactive only when condition (iii) is fulfilled. This reasoning evidently leads to Lemma 2.

Lemma 2. Let the genetic automaton be constructed on the bipartite oriented graph $G(U_1, U_2, W_{12}, W_{21})$. Then, any g-base V in $G(U_1, U_2, W_{12}, W_{21})$ generates in genetic automaton a stationary point of the following type: $x_i = 1$ if $x_i \in V$ and $x_i = 0$ otherwise.

The opposite statement is also true. For any point of the genetic automaton constructed on the prespecified bipartite oriented graph $G(U_1, U_2, W_{12}, W_{21})$, the subset of first type nodes with nonzero values of the variables x_i is the g-base. Thus, the problem of description of stationary points of gene network regulatory circuits is reduced to the problem of search for special coverings of oriented graphs.

3. DISCUSSION

Study of the general operation patterns of natural gene networks is extremely complex due to the unique structure of each gene network. Study of hypothetical constructions created according to certain prespecified rules may considerably help in solving of this intricate problem. For example, genetic elements and regulatory mechanisms may be taken as building units to construct hypothetical gene networks (Likhoshvai et al., 2001; 2003; 2004; Fadeev and Likhoshvai,

2003). The assumption that regulatory mechanisms represent a certain combination of independent events composed of the threshold-type interactions between some complexes with target sites leads us to description of gene network regulatory circuits by genetic automatons. This result demonstrates that the assumption on a threshold-type action of the regulatory mechanisms gives the inference that the presence of stationaries is completely determined by the structure–function organization of gene network regulatory circuits, represented by the bipartite oriented graph $G(U_1, U_2, W_{12}, W_{21})$ and is independent of the parameters of the genetic automaton model. The obtained description of stationaries of genetic automatons in terms of g-bases gives the theoretical solution for the problem of search for stationary points of arbitrary regulatory circuits of gene networks. For this purpose, it is necessary to construct the corresponding bipartite oriented graph and find all its g-bases. In this process, structures of all the stationary points are determined simultaneously. Practical application of this result requires construction of an algorithm searching for g-bases and development of the corresponding software tools; however, this is a technical problem.

The result obtained put forth an important theoretical problem of search for convenient criteria for defining g-bases. In the general case, this problem is a generalization of the problem of describing 1-bases of oriented graphs (see Harari, 1973, for definition of 1-bases). Indeed, when considering only the regulatory circuits where all the regulatory mechanisms are controlled by homomultimeres and belong to the negative action type, the g-bases appear identical to 1-bases. As the problem of defining 1-bases stands for already 25 years and yet has no solution, it is unlikely that simple searching criteria for g-bases can be developed for the general case. Most probably, such criteria may be constructed for certain classes of oriented graphs. For example, of considerable interest initially is the description of g-bases of the simplest class of oriented graphs permitting a group of rotations by the specified commutations $(1, 2 \ldots n)$, where n is the number of nodes in the oriented graphs.

ACKNOWLEDGMENTS

The work was supported by RFBR (grants Nos. 03-01-00328, 03-04-48506, 03-07-96833, 03-04-48829, 04-01-00458, 05-07-98011, 05-07-98012, and 05-04-08084), Siberian Branch of the Russian Academy of Sciences (integration project No. 119), the Program of Basic Research of the Presidium of the Russian Academy of Sciences and branches of RAS (grant No. 10.4), state contract No. 10002-251/P-25/155-270/200404-082, and NSF:FIBR grant No. EF-0330786.

MODELING OF GENE EXPRESSION BY THE DELAY EQUATION

V.A. Likhoshvai[1, 3*], G.V. Demidenko[2], S.I. Fadeev[2]
[1] *Institute of Cytology and Genetics, Siberian Branch of the Russian Academy of Sciences, prosp. Lavrentieva 10, Novosibirsk, 630090, Russia, e-mail: likho@bionet.nsc.ru;* [2] *Sobolev Institute of Mathematics, Siberian Branch of the Russian Academy of Sciences, Novosibirsk, Russia;* [3] *Ugra Research Institute of Information Technologies, Khanty-Mansyisk, Russia*
[*] *Corresponding author*

Abstract: We consider an autonomous system of differential equations modeling a substance synthesis without branching. We prove the theorem on passage to the limit from the system to a delay differential equation for unlimited increase in number of intermediate stages.

Key words: genetic systems; modeling; delay equation; differential autonomous systems

1. INTRODUCTION

Study of the regularities in functioning of gene networks is one of the pivotal problems of the post-genomic molecular biology and genetics. With this aim in view, the methods of computer and mathematical modeling are widely used (Thomas et al., 1995; Edwards and Glass, 2000; Elowitz and Leibler, 2000; Gardner et al., 2000; Likhoshvai et al., 2003). Specifics of gene networks as an object of mathematical and computer study lies in the fact that they belong to superlarge systems where the flows of substance and energy are mediated by synthesis of many dozens and hundreds of thousands intermediate forms of DNA, RNA, and proteins. Syntheses of these substances are executed by fundamental and multistage processes of replication, transcription, and translation. Accounting for intermediate stages of DNA, RNA, and protein syntheses gives rise to the systems of differential equations with huge dimensionality. Thus, a demand arises to decrease dimensionality of gene network models without the loss of their adequacy.

One of the approaches that could be used in modeling is dividing the processes into rapid and slow with subsequent reduction of the differential equation systems on the basis of Tikhonov's theorem (Tikhonov, 1952). The second approach is based on elimination from the model of some variables due to these or that considerations that emerge from specificity of the modeled system and/or from the essence of the task to be solved. As a rule, these considerations are poorly described by exact methods and are based on semi-intuitive considerations. For example, under modeling, a group of subsequent processes is frequently substituted by delay parameter that equals numerically the summarized duration of the processes. In this study, we give the exact grounding of adequacy of this approach application for reducing the model in the class of substance synthesis without branching. This class of models frequently appears as a constituent part of more general models of gene networks, because they help to describe the processes of replication, transcription, translation, and enzyme chain reactions. In this work, we prove the theorem on the limiting transition from the system of autonomous equations to differential system with delay, under tending to infinity of the number of equations describing the intermediate stages of synthesis. We also constructed the functions and biases appearing during transition from the finite models to delay equation.

2. RESULTS

We consider a mathematical model of irreversible multistage substance synthesis process without branching:

$$\begin{cases} \dfrac{dy_1}{dt} = -\dfrac{n-1}{\tau} y_1 + f(y_n), \\ \dfrac{dy_i}{dt} = \dfrac{n-1}{\tau}(y_{i-1} - y_i), \qquad i = \overline{1, n-1}, \\ \dfrac{dy_n}{dt} = \dfrac{n-1}{\tau} y_{n-1} - \theta\, y_n \, . \end{cases} \tag{1}$$

where y_i is the concentration of intermediate stage of the protein synthesis; y_n, concentration of final product of the synthesis; $f(y_n)$, a function describing the regulatory mechanism of initiation of the substance synthesis; $\tau > 0$, the total time needed to proceed via the stages from the 1st condition to the nth condition; and $\theta > 0$, the constant of dissipation rate of the final product from reaction mixture.

In a real situation, substance synthesis process can have hundreds of thousands of intermediate stages; i.e., the system (1) can have a great many equations. However, solving the Cauchy problem for (1), we cannot decrease the dimension of the system (1), since biologists need principally the synthesis product, i.e., the last element $y_n(t)$. Therefore, we need to find the last element $y_n(t)$ of the solution to the system (1). As follows from the system (1), we cannot neglect any equations, i.e., we cannot reduce this system to a system of smaller dimension. Hence, for example, if n equals several million, then solving the system (1) is impossible by using computer, and we confront with the problem of 'high dimension'.

We point out a method in order to find effectively the approximate values of the last element $y_n(t)$ of the solution to the Cauchy problem for the system (1) with initial conditions

$$y_j|_{t=0} = y_j^0, \quad j = 1, \ldots, n, \tag{2}$$

in the case of great dimension n.

The idea of our method is to solve the Cauchy problem for the delay equation

$$\frac{dx}{dt} = -\theta x + f(x(t-\tau)) \tag{3}$$

instead of the Cauchy problem (1), (2). We prove that for sufficiently large n, values of $y_n(t)$ will be closely approximated by values of the solution $x(t)$; moreover, for $|y_n(t) - x(t)|$, we obtain estimates from which we see that the larger n the exacter the approximation.

We sketch our method below.

3. SKETCH OF THE METHOD

For simplicity, we suppose that the initial conditions (2) are trivial. Assume that the function $f(z)$ in (1) satisfies the Lipschitz condition. Then, the Cauchy problem (1), (2) is uniquely solvable. Enlarging the system, we consider a sequence of problems of the form (1), (2). We need only the last component of solution to each problem. As a result, we obtain the sequence $\{y_n(t)\}$. From the next theorem it follows that the sequence converges and its limit is a solution to the equation (3).

Theorem 1. *Suppose that the function $f(z)$ satisfies the Lipschitz condition:*

$$|f(z_1)-f(z_2)|\leq L|z_1-z_2|,\ z_1,z_2\in R. \tag{4}$$

Let $T>\tau$ be defined by the inequality

$$L\left(\frac{1-e^{-\theta T}}{\theta}\right)<1. \tag{5}$$

Then, the sequence $\{y_n(t)\}$ is uniformly convergent on the segment $[0,T]$:

$$y_n(t)\to x(t),\qquad n\to\infty,$$

and the limit function $x(t)$ satisfies the identity

$$\frac{dx(t)}{dt}\equiv-\theta x(t)+f(x(t-\tau)),\ t>\tau. \tag{6}$$

By Theorem 1, from the theoretical standpoint, we obtain a method for approximately solving the problem of substance synthesis. Namely, instead of solving a problem for the system (1) with a very large dimension, it is sufficient to solve the corresponding problem for the delay differential equation (3). However, the method is effective, if exact estimates for the approximation exist. Hence, for solving practical problems, it is necessary to answer the following question.

How can we estimate the difference $|y_n(t)-x(t)|$ for sufficiently large n?

We give the answer to the question in the following theorem.

Theorem 2. *Suppose that the function $f(z)$ satisfies the Lipschitz condition (4) and*

$$\sup_{z\in R}|f(z)|=G<\infty$$

Let the number T be defined by (5). Then, the following estimate holds:

$$\max_{t\in[0,T]}|y_n(t)-x(t)|\leq\left(1-L\frac{1-e^{-\theta t}}{\theta}\right)^{-1}I_n,\ n>>1, \tag{7}$$

where

$$I_n = G\left(A_n \frac{1-e^{-\theta T}}{\theta} + n^{-1/4} \frac{1}{\left(1-\frac{\theta\tau}{n-1}\right)^{n-1}} \left(3\frac{1-e^{-\theta T}}{\theta} + 8\tau\right)\right),$$

$$A_n = e^{\theta\tau} - \left(1-\frac{\theta\tau}{n-1}\right)^{1-n}.$$

The expression in the right-hand side (7) is unwieldy. However, using the estimate, we can approximate the concentration $y_n(t)$ by the function $x(t)$ with guaranteed accuracy, and from Theorem 2, we have the asymptotic equality

$$|\, y_n(t) - x(t) \,| = O\left(n^{-1/4}\right),\ n \to \infty,\ t \in [0,T]. \tag{8}$$

Sketch a proof of Theorem 1, 2.

Using the Cauchy formula, taking into account the trivial initial conditions (3), we obtain from (1) the integral equation for the function $y_n(t)$:

$$y_n(t) = \int_0^t \Psi_n(t-s) f(y_n(s)) ds, \tag{9}$$

where

$$\psi_n(t) = \frac{e^{-\theta t}}{\left(1-\frac{\tau\theta}{n-1}\right)^{n-1}} S_n(t), \tag{10}$$

$$S_n(t) = 1 - e^{-\omega t} \sum_{k=0}^{n-2} \frac{(\omega t)^k}{k!},\ \omega = \frac{n-1}{\tau} - \theta. \tag{11}$$

By (9) we have

$$y_{n+1}(t) - y_n(t) = \int_0^t \left(\psi_{n+1}(t-s) - \psi_n(t-s)\right) f(y_{n+1}(s)) ds +$$

$$+\int_0^t \psi_n(t-s)\big(f(y_{n+1}(s))-f(y_n(s))\big)\,ds$$

for any *n*, *l*. Taking into account the representation and properties of the kernel (10), we show that the sequence $\{y_n(t)\}$ is fundamental in the space $C[0,T]$, where *T* is defined by (5). Then, by completeness of the space $C[0,T]$, the sequence $\{y_n(t)\}$ converges to a function $x(t)$. Further, we show that we can proceed to the limit in (9). As a result, we have the identity (6) and the estimate (7).

The crucial point in the proof of convergence of the sequence $\{y_n(t)\}$ is the proof of convergence of the sequence $\{S_n(t)\}$, given by (11). From the next two theorems, it follows that the convergence holds

$$S_n(t)\to u(t-\tau),\ n\to\infty, \tag{12}$$

where $u(t-\tau)$ is the Heaviside function:
$u(t-\tau)=0$ for $t<\tau$, $\qquad u(t-\tau)=1$ for $t>\tau$.

Theorem 3. *Let* $t=\tau p,\ p>1, n_p=\left[\dfrac{p\theta\tau}{p-1}\right]+1$. *Then, the estimate*

$$\left|S_{n+1}(t)-1\right|<\frac{1}{\sqrt{2\pi(n-1)}\left(p\left(1-\dfrac{\theta\tau}{n}\right)-1\right)}\left(\frac{p}{e^{p-1}}\right)^n e^{\theta p\tau}\left(1-\frac{\theta\tau}{n}\right)^n \tag{13}$$

holds for $n\ge n_p$.

Theorem 4. *Let* $t=\tau/p,\ p>1$. *Then, the estimate*

$$\left|S_{n+1}(t)\right|<\frac{1}{\sqrt{2\pi n}\left(1-\dfrac{1}{p}\left(1-\dfrac{\theta\tau}{n}\right)\right)}\left(e^{1-\frac{1}{p}\left(1-\frac{\theta\tau}{n}\right)}\frac{1}{p}\left(1-\frac{\theta\tau}{n}\right)\right)^n \tag{14}$$

holds for $n>\theta\tau$.

Sketch a proof of Theorem 3, 4.

Represent the function $S_{n+1}(\tau p)$ in the form $S_{n+1}(\tau p)=1-F_n$, where

$$F_n = e^{-q_n}\sum_{k=0}^{n-1}\frac{q_n^k}{k!}, \; q_n = np_n, \; p_n = p\left(1-\frac{\theta\tau}{n}\right).$$

Obviously, to prove (13), it is necessary to establish the estimate

$$F_n < \frac{1}{\sqrt{2\pi(n-1)}(p_n-1)}\left(\frac{p}{e^{p-1}}\right)^n e^{\theta p\tau}\left(1-\frac{\theta\tau}{n}\right)^n, \; n \geq n_p. \tag{15}$$

Note that the inequalities hold

$$\frac{q_n^k}{k!} < p_n^{-(n-k-1)}\frac{q_n^{n-1}}{(n-1)!}, \; k < n-1.$$

for $n > \theta\tau$. Then,

$$F_n < e^{-q_n}\frac{q_n^{n-1}}{(n-1)!}\sum_{k=0}^{n-1}p_n^{-(n-k-1)} = e^{-q_n}\frac{q_n^{n-1}}{(n-1)!}\frac{1-p_n^{-n}}{1-p_n^{-1}}.$$

Since $p_n > 1$ for $n \geq n_p$, hence

$$F_n < e^{-q_n}\frac{q_n^{n-1}}{(n-1)!}\frac{p_n}{p_n-1}. \tag{16}$$

Further, we use the Stirling inequality

$$\sqrt{2\pi m}\left(\frac{m}{e}\right)^m < m! < \sqrt{2\pi m}\left(\frac{m}{e}\right)^{\mathrm{m}}\left(1+\frac{1}{4m}\right). \tag{17}$$

Consequently, by the inequality from (16), we have

$$F_n < e^{-q_n}q_n^{n-1}\frac{p_n}{p_n-1}\frac{1}{\sqrt{2\pi(n-1)}}\left(\frac{n}{n-1}\right)^{n-1}$$

$$= \frac{1}{\sqrt{2\pi(n-1)}\,(p_n - 1)} e^{-(p-1)n+\theta p\tau-1} \left(\frac{n}{n-1}\right)^{n-1} p_n^n.$$

Since $\left(\frac{n}{n-1}\right)^{n-1} < e$, then we obtain (15) and (13).

To prove (14), we introduce the following notation

$$\sigma_n = \frac{1}{p}\left(1 - \frac{\theta\tau}{n}\right), \qquad \rho_n = n\sigma_n .$$

Using the notation, rewrite the function $S_{n+1}(\tau/p)$, $p > 1$ as follows:

$$S_{n+1}\left(\frac{\tau}{p}\right) = 1 - e^{-\rho_n} \sum_{k=0}^{n-1} \frac{\rho_n^k}{k!} = e^{-\rho_n} \sum_{k=0}^{\infty} \frac{\rho_n^k}{k!} - e^{-\rho_n} \sum_{k=0}^{n-1} \frac{\rho_n^k}{k!} = e^{-\rho_n} \sum_{l=0}^{\infty} \frac{\rho_n^{n+l}}{(n+l)!}. \tag{18}$$

Note that the following inequalities

$$\frac{\rho_n^{n+l}}{(n+l)!} < \left(\frac{p}{1 - \frac{\theta\tau}{n}}\right)^{-l} \frac{\rho_n^n}{n!}, \qquad l \geq 1, \text{ for } n > \theta\tau .$$

Then, from (18), we obtain

$$S_{n+1}\left(\frac{\tau}{p}\right) = e^{-\rho_n} \frac{\rho_n^n}{n!} \sum_{l=0}^{\infty} \left(\frac{\rho_n^n}{n!}\right)^{-1} \frac{\rho_n^{n+l}}{(n+l)!} <$$

$$< e^{-\rho_n} \frac{\rho_n^n}{n!} \sum_{l=0}^{\infty} \left(\frac{p}{1 - \frac{\theta\tau}{n}}\right)^{-l} = e^{-\rho_n} \frac{\rho_n^n}{n!} \left(\frac{p}{p - 1 + \frac{\theta\tau}{n}}\right) \text{ for } n > \theta\tau .$$

By the Stirling inequality (17), it is easy to derive (14):

$$S_{n+1}\left(\frac{\tau}{p}\right) < e^{\rho_n}\rho_n^n \left(\frac{p}{p-1+\dfrac{\theta\tau}{n}}\right)\frac{1}{\sqrt{2\pi n}}\left(\frac{e}{n}\right)^n = \frac{1}{\sqrt{2\pi n}(1-\sigma_n)}\left(e^{1-\sigma_n}\sigma_n\right)^n.$$

Note that from (13) and (14), it follows that the convergence (12) is uniform on the segments

$$[0,\tau-\varepsilon],\ [\tau+\varepsilon,T]$$

for any $\varepsilon > 0$. Then by (9), one can establish uniform convergence of the sequence $\{y_n(t)\}$ on the segment $[0,T]$. Taking into account the convergence

$$\left(1-\frac{\theta\tau}{n}\right)^{n-1} \to e^{-\theta\tau}, \qquad n\to\infty$$

in the equation (9), we can proceed to the limit as $n\to\infty$. Hence, for the limit function $x(t)$, we get the identity

$$x(t) \equiv \int_0^{t-s} e^{-\theta(t-s-\tau)} f(x(s))ds\,, \qquad t>\tau\,.$$

Consequently, differentiating it, we have the identity (6).

As a result, from Theorems 1, 2, we obtained an effective method for approximation of $y_n(t)$. Namely, to find an approximation of $y_n(t)$, it is sufficient to solve the delay differential equation (3). Using Theorem 2 we may study qualitative properties of $y_n(t)$. However, our results hold only on the segment $[0,T]$, where T is defined by (5). Therefore, to research asymptotic stability of substance synthesis process, it is necessary to have analogous results on the half-line $t \geq 0$. In the following theorem, we indicate sufficient conditions under which such results hold.

Theorem 5. *Suppose that the function $f(z)$ satisfies the Lipschitz condition (4). Let $L \geq \theta$, where the parameter θ is defined by a utilization law. Then the sequence $\{y_n(t)\}$ is uniformly convergent on the half-line $[0,\infty)$:*

$$y_n(t) \to \infty\,, \qquad n\to\infty\,,$$

where the function $x(t)$ *is a solution to the equation (3). Moreover, the asymptotic equality (8) holds.*

The theorem can be proven by the scheme given above.

4. DISCUSSION

Systems of equations of the form (1) appear as constituent elements of more general systems of differential equations that model gene networks, because network functioning is based on such fundamental matrix processes as replication, transcription, and translation, which could be referred in a first approximation to irreversible processes composed of a large number of consequent rapidly proceeding intermediate stages. The entry of $y_n(t)$ in the right-hand side of the first equation of the system (1) appears, if there exists a regulation (repression or activation) of the effectiveness of a process by its final substance (product). From this viewpoint, study of the properties of systems of the form (1) and its generalizations is an important task of the theory of gene network modeling. The result obtained gives evidence that if the synthesis has sufficiently large number of linear stages and the rate of each intermediate stage is rather high, then the kinetics of production of the final product is almost independent of the kinetics of inner stages of synthesis. The whole process is determined by the mechanism of regulation of the synthesis initiation (launching the first stage of synthesis) and the value of delay, which equals the average summarized time of duration of all the intermediate stages. In other words, the result obtained in this work estimates the relationships between the micro- and macrolevels of the system's functioning in the case we consider stages of synthesis for a microlevel and the final product for a macrolevel, respectively. These relationships may be expressed as the following statement: *A single stage of synthesis occurring at the microlevel influences less the kinetics of the final product production if lesser is the time it occupies in the whole integrity of subsequently occurring microprocesses.* In the limit at the macrolevel, only one characteristics of microlevel is revealed, namely, the summarized duration of the process of synthesis. With respect to the suggested interpretation of the result obtained, natural questions arise on criticality of linearity conditions and reversibility of intimidate stages, which are necessary for validation of the limited theorem proved. Indeed, in real biological systems, separate stages of DNA, RNA, and protein syntheses are linear and irreversible only in the first approximation. In general, they are nonlinear, because they are represented by integrity of biochemical reactions. Due to the same reasoning, the stages of synthesis lose irreversibility to ever greater extent when we disintegrate them, gradually approximating to the level of elementary biochemical events. Thus, the study of limiting transitions in the systems describing multistage processes under different mechanisms of intermediate stages of synthesis is very important for constructing the theory of gene networks. Justification of the limiting transition makes a theoretical basis for intuitive understanding of the fact that for adequate modeling of processes at the macrolevel, the knowledge on gene network functioning at every microlevel stage is not necessary. In future, we plan to develop the limiting theory towards attenuating conditions set for the system (1).

ACKNOWLEDGMENTS

The work was supported by RFBR (grants Nos. 03-01-00328, 03-04-48506, 03-07-96833, 03-04-48829, 04-01-00458, 05-07-98011, 05-07-98012, and 05-04-08084), Siberian Branch of the Russian Academy of Sciences (integration project No. 119), Program of Basic Research of the Presidium of the Russian Academy of Sciences and branches of RAS (grant No. 10.4), state contract No. 10002-251/P-25/155-270/200404-082, and NSF:FIBR grant No. EF-0330786.

AGNS—A DATABASE ON EXPRESSION OF ARABIDOPSIS GENES

N. Omelyanchuk[1*], V. Mironova[1], A. Poplavsky[1], N. Podkoldny[1], N. Kolchanov[1], E. Mjolsness[2], E. Meyerowitz[3]

[1] *Institute of Cytology and Genetics, Siberian Branch of the Russian Academy of Sciences, prosp. Lavrentieva 10, Novosibirsk, 630090, Russia, e-mail: nadya@bionet.nsc.ru;* [2] *Institute for Genomics and Bioinformatics; Bren School of Information and Computer Sciences, University of California, Irvine CA 92607, USA;* [3] *Division of Biology, California Institute of Technology, Pasadena, CA 91125, USA*

[*] *Corresponding author*

Abstract: AGNS (Arabidopsis GeneNet supplementary database) is an Internet-available resource that provides access to description of the functions of the known Arabidopsis genes at various levels—the levels of mRNA, protein, cell, tissue, and ultimately at the levels of organs and the organism in both wild type and mutant backgrounds. AGNS annotates published papers on gene expression and function and by this way integrates, systematizes, and classifies this heterogeneous, disparate, and scattered information. AGNS consists of three databases—the Expression Database (ED), the Phenotype Database (PD), and the Reference Database (RD)—and two controlled vocabularies. ED describes gene expression in wild type, mutants, and transgenic plants; PD contains information on phenotypic abnormalities in mutant and transgenic plants; and RD includes references to the papers and description of plant growth conditions with an indication of the ecotypes used as control in the experiments. Detailed controlled vocabularies on growth stages and morphology were developed around the annotated data. AGNS makes possible the search for genes expressed in particular organs, at particular stages, for genes whose expression is altered in particular mutants, for alleles causing particular phenotypic abnormalities, and for pleiotropic effect of particular mutations. Navigation table and search tools are used to browse the informational content of the database and controlled vocabularies.

Key words: Arabidopsis; gene expression; phenotype; database; mutant; transgenic plants

1. INTRODUCTION

Completion of sequencing of Arabidopsis genome in 2000 and the planned determination of the functions of all Arabidopsis genes by the year 2010 makes this plant species a model object for informational biology, as this provides the possibility for development of approaches to systematization of the data and integration of the knowledge starting from nucleotide sequences up to the phenotype. At the level of nucleotide sequences and proteins, this problem is solved successfully in several databases, such as GenBank (Benson et al., 2003), SWISS-PROT (Apweiler et al., 2004), and in the specialized databases on the Arabidopsis genome TAIR (Rhee et al., 2003) and MAtDB (Schoof et al., 2004). Regrettably, all these databases contain very scant data related to the detailed description of cell-, tissue-, organ-specific, and temporal expression patterns of Arabidopsis genes as well as to regulation of their expression at the levels of transcription, translation, protein interactions, and phenotype. However, a complete understanding of the function of a gene can be reached only when the information about temporal and spatial gene expression patterns will be associated with its input in the phenotype at the organismal level. This missing information now is available in a fairly large number of publications on stages of organ development, their morphology, and anatomy; expression of genes at the levels of mRNA and protein; the phenotypic abnormalities found in various mutants and transgenic plants; and the effects of mutations or transgenes on expression of other genes in the genome. This wealth of multilayered, heterogeneous, and autonomous data demands integration in a systematized and classified form. The aim of AGNS is to create an Internet-accessible resource accumulating these data from annotations of published papers and thereby, to provide a description of the functions of the Arabidopsis genes at various levels—the levels of mRNA, protein, cell, tissue, and ultimately at the levels of organs and the organism and in different genotypes. Thus, AGNS supplements the already available informational resources related to Arabidopsis with the data either presently absent in these resources or underrepresented and allows individual researchers to access, analyze, and utilize the vast amounts of rapidly accumulating data on gene function in Arabidopsis. The first version of AGNS describes gene function in a network of shoot apical meristem (SAM) development in embryogenesis and at the vegetative phase.

2. METHODS AND ALGORITHMS

The web resource AGNS can be viewed with Netscape Navigator 6.0 (IE Explorer 5.0, Mozilla 1.0) or higher. The web interface of this database is realized using Java Server Faces (JSF 1.1), Sun Microsystems. The JSF technology is the last development for web-enabled Graphical User Interfaces (GUI) in the Enterprise Edition Java stack. Berkeley DBXML v. 2.0.9, providing efficient access to data basing on the modern query language XQuery 1.0, is used for data storage and analysis. The entire system runs under the control of an open source servlet—Jakarta Tomcat 5 container, developed by Apache Software Foundation.

The database is available at http://wwwmgs2.bionet.nsc.ru/mgs/dbases/agns (home site) and http://emj-pc.ics.uci.edu/mgs/dbases/agns (mirror site).

3. RESULTS AND DISCUSSION

3.1 AGNS Expression Database: description of gene expression patterns in wild type and mutant backgrounds

Gene expression may be markedly different depending on the organ, tissue, stage, and action of external factors. The expression is also modified under the effect of mutations in other genes (for example, transcription factors). Information about expression of genes involved in the development of Arabidopsis is accumulating in the AGNS_ED. A unit entry in this database corresponds to a gene. Expression of genes is described in tables consolidating the data in a unified format (Table 1). The first field in each table (block of fields) corresponds to the allele.

The normal expression pattern is described in the tables where 'wild type' is indicated in the field Allele. The field Allele contains reference to the paper wherefrom the information about the expression pattern described in this block was extracted. This reference is linked to the Reference Database, where plant growth conditions and the ecotype taken as the wild type are described and which is linked to PubMed. Altered expression of a gene in plants carrying mutations in other genes is described in the tables where the field Allele specifies the mutant allele (alleles) of genes regulating expression of the gene in question. If a gene has several names, for example, *dcl1* and *caf* or *zll* and *pnh,* the most commonly used gene name is indicated first followed by the second name separated by hyphen.

In the case the allele number in the new gene name is unknown, the allele will be nonetheless identified, as the old name with the corresponding number is indicated. The transgenic constructs are inputted into the database as they are named in the corresponding paper. The second field of this format is the field Experiment, which describes the method used for detecting expression.

Table -1. The format for description of expression patterns in AGNS_ED by an example of CLV3 gene expression

Field name	Example of description of CLV3 gene in wild type	Example of description of CLV3 gene in wus-1 and stm-11 mutants
Allele/Alleles	wild type [Fletcher J.C. et al., 1999] [Brand U. et al., 2002]	wus-1 [Brand U. et al., 2002] stm-11 [Brand U. et al., 2002]
Experiment	mRNA, AR mRNA, GUS	mRNA, GUS
Developmental stage	Heart stage	Mature embryo
Organ	Embryo, the apical domain, the presumptive SAM	Embryo, SAM
Expression level	Present	None
Anomaly		Absent
Comments	CLV3 mRNA expression is first detected in heart stage embryos, in a patch of cells between the developing cotyledons predicted to give rise to the SAM [Fletcher J.C. et al., 1999]	wus-1, all embryos lacked the SAM and failed to express CLV3::GUS [Brand U. et al., 2002] stm-11, most mutant embryos lacked the SAM and did not express the CLV3::GUS reporter. However, in 6 of 86 stm-11 mutant embryos analyzed, weak GUS staining was observed in two to four cells between the cotyledons [Brand U. et al., 2002]

Authors use various methods for detection of expression, determining it according to the presence of the corresponding mRNA or protein. First, the field Experiment contains information about what was detected—mRNA or protein—followed by the method applied to detect expression—*in situ* hybridization with antisense riboprobe (AR), GUS reporter gene (GUS), GFP reporter gene (GFP), blot hybridization (blot), or RT–PCR (RTPCR). The third field of this format indicates the developmental stage where the expression was detected. This field is linked to the controlled vocabulary on developmental stages, where the user can get a detailed description of this particular stage. The fourth field indicates the organ or its part where the expression was detected. If expression is detected in a part of an organ, the information in this field is presented in a hierarchical form, for example, the anther wall is described as flower, whorl 3, stamen, anther, and anther wall. This field is linked to the controlled vocabulary on morphology, providing the user with a detailed description of morphology of the corresponding organ. In the third and fourth fields and, consequently, in the controlled vocabularies, we use several layers of definitions, starting with very basic definitions (hierarchical manner); the fifth field specifies the level of expression (present, low, very low, none, or high). The tables describing expression of the gene in mutant background contain an additional field for indicating the abnormalities or deviations from the wild type (absent, increased, decreased, or ectopic). To ensure all information is accurate, each block of fields is supplemented with comments cited from the corresponding paper. The comments explain why the mutation is included in this block as well as contain (optional) a detailed description of the abnormality (since every bit of data may become useful in the future), the frequency of this abnormality among the plants carrying such mutation and organs of this particular plant at the stage in question, and other quantitative data. The comments have references to the papers from which the information was extracted. Thus, the database is subdivided into two parts: one for describing expression in the wild type and the other, for its change under the effect of mutations or in a transgenic plant. Based on the ED, the following automated queries are provided: (1) gene expression pattern; (2) the genes expressed in certain organs; (3) the genes expressed at the stage queried; and (4) abnormal expression of genes in mutant or transgenic plants.

3.2 AGNS Phenotype Database: description of morphological abnormalities in mutants

The PD provides collection, primary processing, and classification of the data on phenotypic effects of particular mutations during the development of particular organs. A unit entry in AGNS_PD corresponds to a mutant allele. The mutant phenotype is described in the tables organized according to a unified format (Table 2). The field Allele is the first field in each table (block of fields). This field has reference to the publication from which the data on the morphological abnormality described in the block was taken.

If a number of mutations result in a morphological abnormality, the first field is repeated several times. In the case epistatic effects or an additive interaction causing the same abnormality is found, the first field specifies double or triple mutations or combination of mutations with transgenes. This field is linked to the Reference Database, where plant growth conditions and the ecotype taken as the wild type are described and which is linked to PubMed. The second field of this format indicates the plant developmental stage where the abnormality was detected. The majority of abnormalities were described in adult plants and, correspondingly, 'adult phenotype' is indicated in this field. In the case an abnormality described is connected with development of an organ and the field specifies the developmental stage, this field in linked to the controlled vocabulary on developmental stages, providing the user with a detailed description of the stage in question. The third field of the format AGNS_PD indicates the organ or its part that carries the abnormality. This field is linked to the controlled vocabulary on morphology, where the user can find a detailed description of morphology of the corresponding organ. The fourth field describes the abnormality in question (increased or decreased size, increased or decreased number of organs, ectopic organ, and other).

The block of fields ends with comments cited from the paper where this abnormality is described with reference to this paper. If available in the paper, the comments contain a detailed description of the abnormality; penetrance of the trait, i.e., the number of plants carrying the abnormality in the lines homozygous with respect to the mutations in question; and expressivity of the abnormality in plant (for example, its maximal manifestation in basal flowers). The comments also can be used further for a more detailed classification of a particular abnormality with accumulation of new data, for highlighting of new fields in description, or subdividing the abnormalities into groups.

Table -2. The format for description of phenotypic abnormalities in AGNS_PD by an example of the absence or strong reduction of shoot apical meristem in mature embryo not restored at germination

Field name	How it is described in AGNS_PD
Alleles*	cuc2 [Aida M. et al., 1997], cuc2 cuc1/+ [Aida M. et al., 1997], cuc2 cuc1 [Aida M. et al., 1997], stm-1 [Barton M.K. and Poethig R.S., 1993] [Clark S.E. et al., 1996] [Byrne M.E. et al., 2002], stm-1 as-1 knat1-bp [Byrne M.E. et al., 2002], zll-pnh-2 [Lynn K. et al., 1999], zll-3 wus-1 [Moussian B. et al., 2003]
Develop-mental stage	Mature embryo
Develop-mental stage	Seedling
Organ	Primary SAM
Anomaly	Absent or very strongly reduced and is not restored at germination
Comments**	cuc2, in 0.08 % of plants, all these seedlings have cotyledons fused along one side. Some of the heart-type seedlings did not produce shoots (two of 13 heart-type seedlings). These heart-type seedlings lacked a SAM [Aida M. et al., 1997]

* For the sake of brevity, not all the alleles causing this abnormality are listed in this example.
** For the sake of brevity, only one comment is given in this example.

The PD format allows the data extracted from various papers to be pooled, the mutations and transgenes causing the same abnormality to be detected, and epistatic and additive interactions to be described. This makes it possible to define the group of genes responsible for a certain trait and accumulate in one block of fields a detailed description of the contribution of these genes to the abnormality as well as to make conclusions about gene interactions in the case of double or triple mutations. Thus, one table (block of fields) in AGNS_PD lists the members of the gene network responsible for the trait indicated in the field Allele. The PD allows for the following automated queries, which help the user to find the needed information in AGNS: (1) the mutations resulting in phenotypic abnormalities of the organs selected and (2) the phenotypes of the mutants or, in other words, the pleiotropic effect of the queried mutation.

3.3 AGNS controlled vocabularies

As the annotation proceeded, the need arose to build two controlled vocabularies—the vocabulary on developmental stages and the vocabulary on morphology of organs—around the contents of annotated data based on both the data provided by the authors of the papers annotated and specialized publications on plant developmental stages and morphology in the wild type. The majority of papers concern only individual stages or morphology of individual organs and/or only some their aspects or parts. All these data have references to the papers from which they were annotated. Thus, the controlled vocabularies contain information about the available descriptive systems of Arabidopsis morphology and development, which is systematized and compared. The most frequently used names of the stages and organs are highlighted, and their synonyms are given. Each description of stages and organs is accompanied by detailed comments. The vocabularies are supplemented with new research data as they become available. The vocabularies with detailed descriptions allow for establishing the correspondence and equivalence of terms used by different authors when describing the same phenomena, providing a unified terminology in the database, and are a necessary requisite for inputting the new information into this database. The most widely used term is the main term in the database; the rest are given in the vocabularies as synonyms. The cases of different meaning of identically named objects and overlapping meanings of different objects are discussed separately and described in the vocabularies. For example, the three following meanings of the word 'embryo' are met in papers. First is the descendent of the diploid zygote (conceptus in animals, preembryo, or proembryo). Based on the first descendants of the diploid zygote (apical and basal cells), the embryo can be divided into the apical region (descendents of the apical cell) and the basal region (descendents of the basal cell). From the quadrant stage, the apical region can be divided into the apical and central domains. This establishment of the apical–basal axis of polarity in early embryogenesis makes the body plan of the mature plant. Second, in a more traditional meaning, the embryo or embryo proper is only the descendant of the apical cell (one of two cells derived after the first division of the zygote). All names of the stages in early embryo development (one cell embryo, two cell embryo, etc.) imply this definition of the embryo. Third is an intermediate variant—the embryo as a cell mass that will develop into a mature embryo. In this variant, the hypophysis and descendants of the apical cell altogether are called the embryo (as well as embryo proper), because they make up the mature embryo after suspensor senescence. In our database for organ and tissue definitions, we use the term 'embryo' meaning the descendents of the diploid zygote. For stage names in embryo development, we use the traditional names; thus, the embryo in this case is the descendant of the apical cell. Unfortunately, the terms 'shoot apex', 'SAM', and 'apical meristem' are used in scientific papers with broad definitions or interchangeably. In the database, we define SAM as the part of the shoot apex lying distal to the youngest visible primordia (Medford, 1992; Long and Barton, 2000). The lower border of the SAM changes in the course of SAM development and is defined by adaxial margins of bulging or cleaving of the nearest primordia. SAM contains leaf primordia up to and including the P0 stage (Long and Barton, 2000).

Browsing of the informational content of controlled vocabularies at separate pages indicated with the corresponding bookmarks Development (vocabulary on developmental stages of organs) and Morphology (vocabulary on morphology of organs) is provided. The vocabularies can be searched for a detailed description of a certain developmental stage or morphology of a particular organ in the same pages. The informational content is presented as an ontological description. The problem with organ hierarchical ontology is that some organs can be subdivided into different overlapping regions. For example, the SAM can be divided into two outer layers and the inner corpus and into the central, peripheral, and rib zones.

In such cases, both division variants are included concurrently. For this example, two AGNS fields—developmental stages and organs—develop into the ontology and their lexicon is defined in controlled vocabularies, containing detailed descriptions of organs, their parts, and developmental stages. Thus, we build ontology and controlled vocabularies around the annotated data, i.e., the ontology and controlled vocabularies are developed concurrently with filling of the database. When new objects are added to the database, papers on development of these objects are annotated in the controlled vocabularies.

3.4 Reference Database

The Reference Database contains references to papers with a link to PubMed; the data on growth conditions of plants in experiments and the ecotype used as the wild type are added to it. These data are required for studying temperature- or light-sensitive mutations as well as mutations in different ecotypes.

3.5 Informational content of the AGNS

At present, expression of 20 genes and 1 gene family (23 genes) and mutations of 109 genes are described in AGNS through the annotation of 124 papers. The phenotypic abnormalities of mutants and the expression of genes are most comprehensively studied in the shoot apical meristem, flower, and leaf. In the course of data accumulation in the database, the problem arose as to how to represent the entire information volume of the database so that it would be convenient for searching. For this purpose, the navigation table was designed, where the rows represent organs and their hierarchical structure and the columns, the sequence of developmental stages. Such a structure of the navigation table demonstrates appearance of new organs in the course of development of the organism and makes the formal concepts of types and interactions of the objects better defined. In addition, this navigation table underlines that AGNS has a two-dimensional ontology, i.e., describes the gene expression and phenotypic abnormalities in 2D space (organ–stage). For example, description of organs lacks the term 'adult leaf'; however, contains the terms 'leaf' in the list of organs and the term 'adult phenotype' in the list of developmental stages. Such scheme of ontology representation allows surplus description to be avoided. Otherwise, we should describe separately the stage of adult leaf in the vocabulary on developmental stages and the adult leaf in the vocabulary of organs. It is also evident from the table that the change in phenotype with time presents complications in the recording of the expression or abnormality location, as new organs developed at the place of the previous organs, for example, presumptive stamen primordia, stamen primordia, and stamens. In these cases, all the organs are put in one line. A specialized bookmark Compound View is provided in AGNS to open this navigation table.

The table contains figures in the filled cells, which indicate the number of expression patterns (colored blue) or phenotypic abnormalities (colored green) described so far in AGNS for a particular organ at a particular stage. Clicking allows the user to access the description of expression of a gene or possible mutational changes characteristic of a particular organ at a particular stage. When browsing the navigation table, it is apparent that only an insignificant number of the cells represent both expression and phenotypic abnormality. Partially, this results from incompleteness of the database; however, the main reason for this fact is related to the specific features of the annotated material. A considerable part of the phenotypic abnormalities is described for an adult organism, and when expression of genes is described, the main attention is focused on alterations in expression in various organs at various stages of plant development. The navigation tables, dynamically developed now, form the platform for development of this database, since they serve as interface for introducing new fields into the already constructed blocks and for determining positions of the new blocks. The navigation table is used as a platform in the input system to standardize and increase the efficiency of data updating.

3.6 Conclusions

The AGNS database is a detailed annotation of published experiments and observations related to expression of Arabidopsis genes as well as to regulation of gene expression at the levels of transcription, translation, protein interactions, and the phenotype. Automated queries provide an access to the component of the database directly relevant to the user's interest. These data may be helpful to a wide range of researchers in the area of plant genetics and development; furthermore, controlled vocabularies may be used for both explanation and comparison of the data from AGNS and may provide a curated gateway to various descriptions of the stages and synonyms of the Arabidopsis organs. We also hope that experimenters will use our controlled vocabularies with detailed criteria and consistent nomenclature for description of their results and that the database content or its incompleteness will suggest further experiments. The approach, methodology, and software developed for the model plant *Arabidopsis thaliana* can be further applied for describing the phenotype and gene expression in any other organism.

ACKNOWLEDGMENTS

The work was supported in part by the Russian Foundation for Basic Research (grants Nos. 03-04-48506, 05-07-98012 and 03-04-48469), Presidium of the Russian Academy of Sciences (grant No. 10.4), Siberian Branch of the Russian Academy of Sciences (Integration project No. 119), and the US National Science Foundation (FIBR EF-0330786 *Development Modeling and Bioinformatics*).

STUDY OF THE INTERACTIONS BETWEEN VIRAL AND HUMAN GENOMES DURING TRANSFORMATION OF B CELLS WITH EPSTEIN–BARR VIRUS

E. Ananko[1*], D. Oshchepkov[1], E. Nedosekina[1], V. Levitsky[1, 2], I. Lokhova[1], O. Smirnova[1], V. Likhoshvai[1, 2], N. Kolchanov[1, 2]

[1] *Institute of Cytology and Genetics, Siberian Branch of the Russian Academy of Sciences, prosp. Lavrentieva 10, Novosibirsk, 630090, Russia, e-mail: eananko@bionet.nsc.ru;*
[2] *Novosibirsk State University, ul. Pirogova 2, Novosibirsk, 630090, Russia*
[*] *Corresponding author*

Abstract: Epstein–Barr virus (EBV) is a common herpes virus that establishes a life long persistence in the human host and can directly transform B lymphocytes. EBV can drive B cell development and survival in the absence of normal B cell receptor signals. The goal of our work was to define the mechanisms by which EBV regulates B cell fate. We developed new sections of TRRD and GeneNet databases containing the information on the processes occurring during EBV infection and transformation of B cells as well as about signal transduction pathways, regulatory proteins, and genes whose expression changes in transformed B cells. Analysis of the information from these new sections revealed several transcription factors important for regulation of B cell fate by EBV. The samples of corresponding binding sites were constructed, and the methods for recognizing binding sites of these factors were developed. Regulatory regions of genes expressed in B cells and the complete genome of EBV were scanned by these methods. In addition to *CD23*, already known from the literature data, seven most probable target genes for the viral transcription factor EBNA-2 were discovered. According to the study performed, the cellular transcription factors AP1, BSAP, NF-κB, STAT, and c-Myc may activate EBV promoters.

Key words: gene networks; genotype; computer analysis; mathematical model

1. INTRODUCTION

Epstein–Barr virus (EBV) causes infectious mononucleosis and is associated with several human malignancies, such as Burkitt's lymphoma, Hodgkin's disease, AIDS-associated B cell lymphoma, primary CNS non-Hodgkin's lymphoma, gastric adenocarcinoma, X-linked lymphoproliferative syndrome, nasopharyngeal carcinoma, and post-transplant lymphoproliferative disease. EBV can drive B cell development and survival in the absence of normal B cell receptor signals. *In vitro*, EBV transforms B cells into lymphoblastoid cell lines. Several EBV-encoded proteins enable the virus-infected cells to avoid apoptosis (Cohen, 1999).

Different approaches are used to clarify the mechanisms by which EBV regulates the B cell fate. First, changes in the cell phenotype by EBV may indicate what signal transduction pathways will be affected. This could include cell surface receptors and components of signal transduction pathways. Second, some of the cellular proteins could be targets for EBV-encoded proteins. The first protein that can interact with the EBV-encoded EBNA proteins was identified as the e-subunit of human chaperonin TCP-1 complex (Kashuba et al., 1999). The second protein shown to bind EBNA-3 turned out to be the minor subunit of aryl hydrocarbon receptor complex (Kashuba et al., 2000).

One of the key moments in development of pathologies connected with EBV is activation of the virus Wp promoter (Kirby et al., 2000), which leads to expression of EBNA-2 and EBNA-LP proteins. These particular proteins switch on the program of cell transformation through activating several cellular and viral genes. Wp promoter contains a number of binding sites for cellular transcription factors, namely, widely expressed FRS, REBATE (Kirby et al., 2000), NF-κB (Sugano et al., 1997), and BSAP (Pox-5; Kirby et al., 2000) specific of lymphocytes.

Can other viral genes be regulated by cellular transcription factors? Moreover, can EBV-encoded proteins influence transcription of human genes? To answer these questions, experimental data obtained by different approaches were collected in TRRD (Kolchanov et al., 2002) and GeneNet (Ananko et al., 2005) databases. The sections are freely available at http://wwwmgs.bionet.nsc.ru/mgs/gnw/genenet/viewer/ for GeneNet and at http://wwwmgs.bionet.nsc.ru/mgs/gnw/trrd/sections1.shtml for TRRD. The collected information was analyzed, allowing for detection of the human genes that may be potentially regulated by the viral transcription factor EBNA-2. In addition, EBV genome was searched for the presence of target genes for cellular transcription factors, whose activation might assist development of pathologies caused by EBV.

2. METHODS AND ALGORITHMS

For formalized description of interactions between two genomes—EBV and human (B cells)—in hybrid gene networks, GeneNet technology (Kolchanov et al., 1999; Ananko et al., 2005) was applied.

TRRD database (Kolchanov et al., 2002) was used to create the samples of natural binding sites for AP-1, BSAP (Pax-5), c-Myc, EBNA-2, EGR, IRF, ISGF3, KLF, NF-AT, NF-E2, NF-κB, NRF-2, NF-Y, p53, Pu.1, STAT, and USF.

The SITECON method (Oshchepkov et al., 2004a, b) was utilized for recognition of binding sites for c-Myc, EBNA-2, EGR, KLF, NF-AT, NF-E2, NRF-2, NF-Y, p53, Pu.1, STAT, and USF.

Binding sites for AP-1, BSAP/Pax-5, IRF, ISGF3, and NF-κB were recognized by the SiteGA method (Levitsky and Katokhin, 2003; Levitsky et al., 2005).

The mathematical model of the effect of BSAP binding sites on functioning of EBV Wp promoter was constructed using the generalized chemical kinetic approach (Likhoshvai et al., 2001).

3. RESULTS AND DISCUSSION

Two sections of the GeneNet database devoted to hybrid gene networks of interaction between the EBV and human genomes were created: (1) 'EBV infection' contains description of the processes in B cells infected with Epstein–Barr virus and (2) 'EBV transformation' compiles the information about signal transduction pathways and regulatory proteins as well as about the genes and proteins whose expression changes in transformed B cells. Table 1 shows the information content of these two GeneNet sections.

Table -1. GeneNet database, sections 'EBV infection' and 'EBV transformation'

GeneNet section	Genes	Proteins	Relationships	Annotated papers
'EBV infection'	9	19	26	35
'EBV transformation'	29	75	102	117

Three signal transduction pathways playing an essential role in EBV-transformed B cells were described in these sections of GeneNet, namely, (1) Jak/STAT signal transduction pathway (activation of the transcription factors Stat1, Stat3, and IRF-7); (2) activation of NF-κB factor by the viral protein LMP1 via TRADD/TRAF2; and (3) activation of the MEKK/ERK signal transduction pathway via TRADD/TRAF2 (Figure 1). Some other transcription factors are also activated in EBV-transformed B cells, namely AP-1 (Kieser et al., 1997) and ATF2 (Eliopoulos et al., 1999).

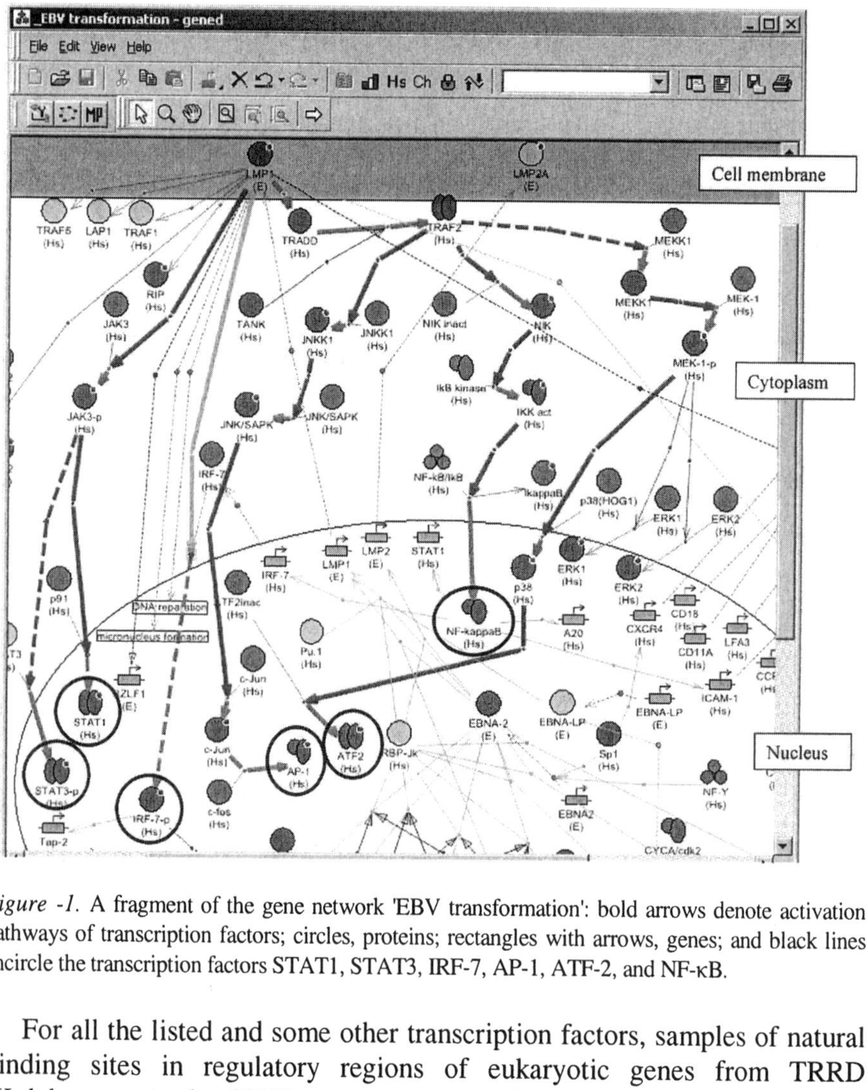

Figure -1. A fragment of the gene network 'EBV transformation': bold arrows denote activation pathways of transcription factors; circles, proteins; rectangles with arrows, genes; and black lines encircle the transcription factors STAT1, STAT3, IRF-7, AP-1, ATF-2, and NF-κB.

For all the listed and some other transcription factors, samples of natural binding sites in regulatory regions of eukaryotic genes from TRRD (Kolchanov et al., 2002) were constructed. Basing on these samples, recognition methods for 17 transcription factors were developed using SITECON (Oshchepkov et al., 2004) and SiteGA (Levitsky et al., 2003) methods. The recognition method was chosen basing on the best ratio of type I and II errors. The main requirement to this method was its ability to predict no more than one site per 10 000 bp due to accidental reasons.

The complete genome of EBV (GenBank entry NC_001345, 172 281 bp) was scanned by the methods developed. Results of recognition for 17 transcription factors and characteristics of the methods used for recognition of each site type are listed in Table 2. The cellular transcription factors activated by the virus (AP1, NF-κB, STAT, and c-Myc) may bind to EBV genome approximately at the same rate as the virus-encoded factor EBNA-2 (Table 2), thereby suggesting that they may activate certain viral genes.

The B cell-specific transcription factor BSAP deserves a special attention. It is known that this factor induces expression of the virus Wp promoter (Kirby et al., 2000). According to our data, it can also regulate the promoters located at 90 051, 125 113, 127 237, and 129 377 positions of the EBV genome.

We may infer that binding sites for c-Myc, KLF family, p53, and NF-AT (Table 2) are presumably important for regulation of some EBV promoters.

The number of binding sites for the EGR family of transcription factors, IRF, ISGF3, NF-E2, NRF-2, NF-Y, Pu.1, and USF in EBV genome does not exceed the background level (Table 2). This suggests that the transcription factors in question are unlikely to play an essential role in regulation of expression of EBV genes.

Table -2. Characteristics of the methods for predicting transcription factors used in the work

Transcription factor	Type I error (underprediction), %	Type II error (overprediction)	Predicted by chance, 1 site per base pairs	Number of predicted sites in complete EBV genome
AP1	78.9	8.37E-05	11 947	26
BSAP	0.0	9.10E-06	109 890	16
c-Myc	27.8	8.80E-05	11 364	44
EGR	86.4	5.00E-05	20 000	9
KLF	85.7	1.00E-04	10 000	29
IRF	34.5	2.07E-05	48 309	4
ISGF3	30.8	2.58E-06	387 597	2
NF-AT	63.6	7.00E-05	14 286	16
NF-E2	62.5	1.00E-05	100 000	4
NF-κB	47.8	6.99E-05	14 306	28
NRF-2	33.3	4.45E-05	22 472	5
NF-Y	74.2	3.30E-05	30 303	7
p53	60.0	2.40E-05	41 667	13
Pu.1	60.9	5.00E-05	20 000	9
STAT	75.0	2.30E-05	43 478	14
USF	88.6	9.80E-06	102 041	5
EBNA-2	25.0	6.82E-05	14 663	25

According to the information available in TRRD and GeneNet databases, Wp promoter of EBV contains binding sites for RF-X, CREB, and NF-κB (Sugano et al., 1997; Kirby et al., 2000) and two sites for BSAP (Figure 2). The B cell–specific transcription factor BSAP, binding to B and D sites (Figure 2), plays the key role in regulation of expression of Wp promoter in B lymphocytes (Kirby et al., 2000). We constructed a mathematical model of the effect of BSAP binding to these sites on the functioning of Wp promoter. Currently, the model contains description of more than 100 dynamic variables.

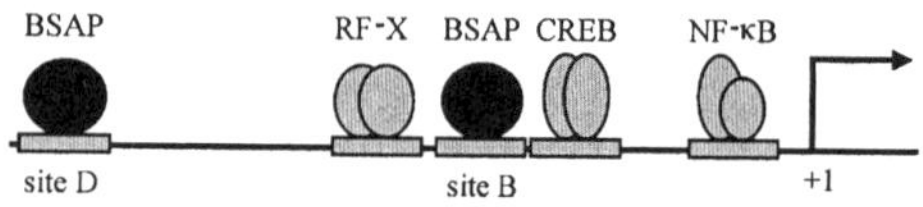

Figure -2. Wp promoter of EBV.

According to the model, mutation in one BSAP binding site reduces Wp expression three–fourfold and mutation in both BSAP binding sites decreases expression five–tenfold. Such drastic decrease is explained by the fact that these binding sites act in a cooperative manner. It is assumed that the distal site D have a stronger effect on the expression compared with the proximal site B (Tierney et al., 2000), and we took into account this hypothesis in the model developed. The results obtained using the constructed model of Wp promoter function are shown in Figure 3.

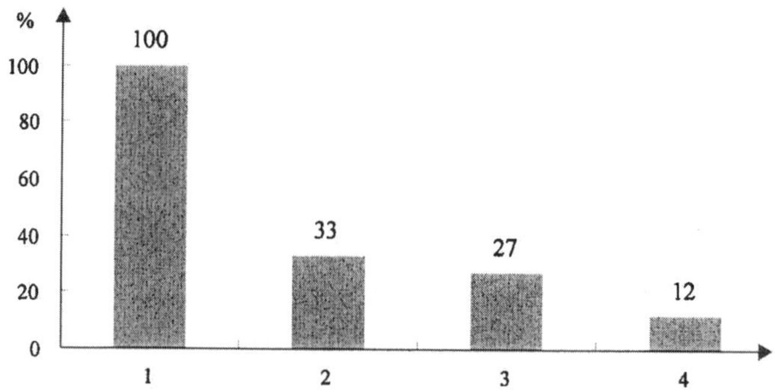

Figure -3. Results of modeling of Wp promoter expression in the case of mutations in binding sites for the transcription factor BSAP: 1 – wild type promoter; 2 – mutated site B; 3 – mutated site D; and 4 – mutations in both sites B and D.

As is evident from analysis of the information accumulated in GeneNet and TRRD databases, the activation of signal transduction pathways caused by viral proteins plays a very important role in transformation of B cells by Epstein–Barr virus. In addition, the virus-encoded transactivators, in particular, EBNA-2, expressed from Wp promoter (Tierney et al., 2000) and capable of stimulating expression of genes of the host cell, are essential for the process in question. Correspondingly, development of methods for recognizing binding sites for the transcription factors activated in EBV-transformed B cells and for searching for potential binding sites of these factors in the genes expressed in B cells is a highest priority task. In addition, search for potential binding sites for EBV-encoded transactivators in regulatory regions of human genes is of great interest.

It is known that EBNA-2 in complex with the cellular transcription factor CBF1 (RBP-Jk) influences transcription of human genes, such as *c-fgr* (Knutson, 1990), *CD21* (Cordier et al., 1990), and *CD23* (Ling et al., 1994). A highly conservative sequence GGGAA is present in the center of all EBNA-2 binding sites in both viral promoters and regulatory regions of human genes. We did not succeed in finding references to recognition methods, consensus sequence, or weight matrices for binding site of the viral protein EBNA-2 in the published data. The consensus obtained basing on the sample of natural EBNA-2 sites from TRRD is shown below:

VSYYGTGGGAAAWHDGT,

where V = G, C, A,
S = G, C,
Y = T, C,
W = A, T,
H = A, C, T,
D = G, A, T.

We constructed a sample of 36 human genes annotated in the TRRD database and related to EBV transformation. To find out which of these genes might be regulated directly by the viral transcription factor EBNA-2, the regulatory regions of the genes expressed in B cells were scanned by the developed method to find the binding sites of this factor (Table 2, last line). The binding sites for EBNA-2 were detected in promoter regions of only 8 genes of the 36 analyzed; note that the vast majority of these genes encode membrane proteins (Table 3).

Table -3. Results of recognition of EBNA-2 in the promoter regions of genes expressed in B cells

Gene	Position relatively to ST	Orientation	Sequence
CD19	–182	Reverse	TACCACGGGAAATGATC
CD23	–170	Direct	CCCTGTGGGAACTTGCT
CD40	–107	Direct	ACTTGTGGGAATGTTCT
CD150	–557	Direct	CTGGGAGGGAATCCACA
GPC	–769	Direct	TGCTTGGGGAACTATAA
HLA-B	–101	Reverse	GGGAGTGGGAAGTGGGG
MIP-1a	–1303	Direct	AGAAATGGGAAATCAAG
LFA3	–931	Direct	GGTTCTGGGAATAGGGT

In addition to the known EBNA-2 binding sites in *CD23* gene (Ling et al., 1994), we with a high probability recognized putative EBNA-2 sites in promoter regions of *CD19*, *CD40*, *CD150*, *GPS*, *HLA-B*, *MIP-1*a, and *LFA3* genes. A complete matching with the consensus in all the sites found is observed only in the central part for the highly conservative pentanucleotide GGGAA with deviations from the obtained consensus in the flanks.

However, all the found sites in their DNA structural characteristics that were used for construction of the recognition method for EBNA-2 binding site (Oshchepkov et al., 2004b) coincided no less than by 85 % with the actual sites of the training sample (data not shown). We believe that the genes listed in Table 3 are directly regulated by EBNA-2.

The EBNA-2 binding sites in *CD21* gene are present only in the silencer localized to the intron downstream of the start of transcription. Note here that EBNA-2 binding sites in introns downstream of the transcription start site were recognized in many genes from our sample (data not shown).

We thus undertook the first step in an integrated search for interactions between the two genomes, virus and human, during transformation of B cells by EBV. Putative target genes of cellular transcription factors were predicted in EBV genome. Of the 36 genes expressed in B cells, 7 potential targets for a direct regulation by the viral transactivator EBNA-2 were detected.

ACKNOWLEDGMENTS

The work was supported in part by the INTAS (project No. 21-2382); Russian Foundation for Basic Research (grants Nos. 03-04-48506, 03-07-90181, 05-04-49111, and 03-04-48469); Russian Academy of Sciences (grant No. 10.4); Siberian Branch of the Russian Academy of Sciences (integration project No. 119); agreement with the Institute of Mathematics, Siberian Branch of the Russian Academy of Sciences under the Russian Foundation for Basic Research grant No. 03-01-00328 'Mathematical Modeling and Analysis of Gene Networks'; the state contract with the Federal Agency for Science and Innovations; 'Identification of Promising Targets for Action of New Drugs Basing on Reconstruction of Gene Networks' within the priority direction; 'Living Systems'; CRDF (grant No. Y2-B-08-02); and NATO (grant No. LST.CLG.979816). Authors are grateful to G.B. Chirikova for assistance in translation.

PROBING GENE EXPRESSION: SEQUENCE-SPECIFIC HYBRIDIZATION ON MICROARRAYS

H. Binder

Interdisciplinary Centre for Bioinformatics, University of Leipzig, Haertelstr. 16-18, D-04107 Leipzig, Germany, e-mail: www.izbi.de, binder@izbi.uni-leipzig.de

Abstract: *Background:* DNA microarrays are routinely used to monitor the transcript levels of thousands of genes simultaneously. However, the array design, hybridization conditions, and oligodeoxyribonucleotide probe sequence influence the performance of the DNA microarray platform and must be considered by data analysis. *Results:* We analyzed the signal intensities of GeneChip microarrays in terms of a microscopic binding model. It considers specific and non-specific transcripts, which both compete for duplex formation with perfect match (*PM*) and mismatch (*MM*) oligonucleotide probes. Intensity simulations enable us to judge the accuracy and precision of gene expression measures. The accuracy of the estimated fold changes ranks according to $PM - MM > PM > MM$, whereas the precision decreases with $PM \geq MM > PM - MM$, where $PM - MM$ denotes the respective intensity difference. *Conclusions:* *MM* probes possess the potency to correct the intensity of the respective PM probe for the non-specific background. The middle base related bias of the *MM* intensity must, however, be considered by improved algorithms of data analysis. Moreover, the knowledge of base pair interactions suggests substituting the complementary mismatches on GeneChips by alternative rules of *MM* design.

Key words: DNA/RNA duplex stability; perfect match and mismatch probes; gene expression

1. INTRODUCTION

The gene chip microarray technology empowers researchers in the collection of large-scale data on gene expression. The method is based on the selectivity of the hybridization reaction between the target RNA transcribed from the gene of interest and the complementary DNA probes grafted on the chip. The formation of probe/target duplexes is a complex process governed by an intricate interplay between several effects, such as binding and saturation, surface electrostatics, and non-equilibrium thermodynamics (Vainrub and Pettitt, 2002; Hekstra et al., 2003; Held et al., 2003; Naef and Magnasco, 2003; Zhang et al., 2003; Halperin et al., 2004). The proper interpretation of microarray data in terms of gene expression requires a detailed understanding of the hybridization mechanism at the level of base pairings at different concentrations of target RNA.

A typical GeneChip microarray, such as the human genome HG-U133 chip, consists of nearly 500 000 probe spots on an area of about 1.5 square centimeters. Each spot is formed by 25-meric DNA oligonucleotides of (almost) one sequence, which are grafted with the 3'-end on the glass support. The sequence of these perfect match (*PM*) probes corresponds to a 25-meric fragment in the consensus sequence of the target gene. The DNA oligomers are expected to capture the complementary messenger RNA by sequence-specific duplex formation, and in this way, to probe its abundance. The amount of bound RNA is detected using fluorescent labels. Consequently, each probe spot gives rise to an intensity value, which, in an ideal case, is directly related to the concentration of target RNA.

The sample solution represents a complex mixture of RNA fragments of different lengths and sequences. Consequently, the total RNA concentration can be split into two fractions, namely, that of the target RNA, which is specific (*S*) for a given probe, and that of the non-specific (*NS*) RNA fragments, i.e. $c_{\mathrm{RNA}} = c_{\mathrm{RNA}}^{S} + c_{\mathrm{RNA}}^{NS}$. Unfortunately, the latter RNA also can possess a non-negligible affinity for duplex formation with the probe oligomers. This *NS* hybridization is problematic for chip analysis, because it adds a 'chemical' background intensity, which is not related to the expression degree of the target gene. To deal with this problem, each *PM* probe is paired with the so-called mismatch probe (*MM*) on microarrays of the GeneChip-type (Affymetrix, 2001). The *MM* sequence is identical with that of the respective *PM* probe except for the base in the middle of the oligomer, which is replaced with its complement to prevent *S* hybridization. In this way, the *MM* probe is intended to measure the amount of *NS* hybridization and, thus, to provide a correction of the *PM* intensity for the chemical background.

The lower binding affinity of the *MM* probes predicts a systematically smaller spot intensity, if compared with that of the respective *PM*, i.e., $I^{MM} < I^{PM}$. Figure 1 correlates the *MM* with the *PM* intensities of a typical GeneChip microarray in a logarithmic scale. More than 40 % of the data points are found above the diagonal referring to so-called 'bright' *MM* with a larger fluorescence intensity compared with their *PM* counterpart (Naef et al., 2002). This result implies that the conventional hybridization theory is simply inadequate, and particularly, that the basic mechanism of *MM* hybridization is not understood yet. As a consequence, many algorithms of gene expression analysis simply ignore *MM* intensity data or they are considered in an empirical fashion to exclude the 'bad' probes from the analysis (see Irizarry et al., 2003, for an overview).

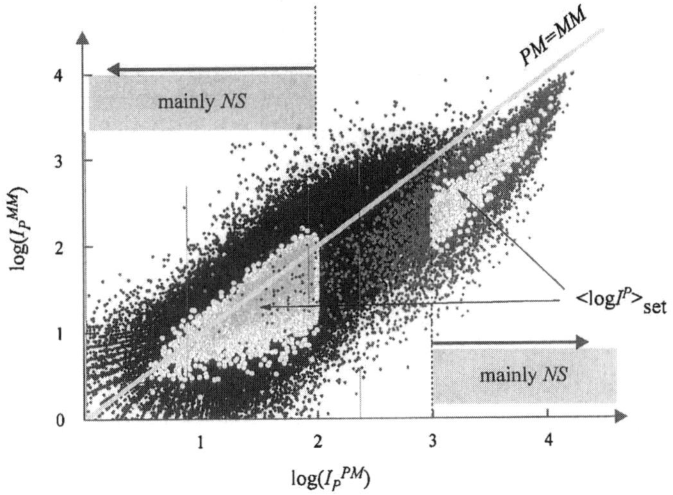

Figure -1. PM/MM intensity correlation plot of the probe pairs taken from the probe sets that meet the condition $<\log I^{PM}>$set < 2 (left data cloud, the open symbols showing the set averages) and $<\log I^{PM}>$set > 3 (right data cloud). The former and latter data refer to predominantly non-specifically (*NS*) and specifically (*S*) hybridized probes, respectively. Note the small amount of bright *MM* in the *S* subset (< 5 %) and the high amount in the *NS* subset (> 40 %).

This study deals with the basic issues of the GeneChip technology, such as the systematic effects of the probe sequence and of matched and mismatched base pairings on the signal intensity, which at present are still unsolved. This sequence-specific view is expected to improve the data analysis as well as chip design.

2. DATA

We have used microarray data from a calibration experiment provided by Affymetrix (http://www.affymetrix.com/support/technical/sample_data/datasets.affx). In this experiment, specific transcripts of the 42 genes referring to 462 probes were titrated in definite concentrations onto a series of chips in three replicates to study the relation between the ('spiked-in') RNA concentration and the intensity of the respective 'spiked-in' probe. For a more detailed description of the HG U133 Latin Square data set, see, for example, the paper by Binder and Preibisch (2005). The Affymetrix technology uses fragmented biotin-labeled RNA for hybridization, which is obtained by a reverse transcription of the extracted RNA into cDNA (mRNA→cDNA) and the subsequent *in vitro* transcription (cDNA→cRNA), biotinylation, and fragmentation.

3. RESULTS AND DISCUSSION

3.1 Probe intensities at specific and at non-specific hybridization

Besides the *PM/MM* pairs, the GeneChip technology uses a second redundancy of the probe design to get independent estimates of the expression degree of each gene. Usually, 11 *PM/MM* probe pairs referring to different regions of the same gene and, thus, to a common concentration value of the target RNA are collected into the so-called probe sets. The set-averaged mean log intensity provides a rough measure of the concentration of *S* transcript according to $<\log I^P>_{\text{set}} \infty [\log c_{\text{RNA}}^S + Z^{\text{set}}]$, where Z^{set} is a set-specific constant, which scatters with a standard deviation of ~ ± 0.5 about its chip average (Binder et al., 2004).

Figure 1 selects two subsets of the probe intensities meeting the conditions of $< \log I^{PM}>_{\text{set}} < 2$ and $<\log I^{PM}>_{\text{set}} > 3$, respectively. The former one includes the probes referring to relatively small concentrations of *S* transcripts and, thus, to the limiting case of dominating NS hybridization. The respective data cloud nearly symmetrically spreads about the diagonal with a relatively large fraction of bright *MM* ($I^{MM} > I^{PM}$) of more than 40 %. On the contrary, the data cloud formed by the second ensemble of probes is clearly shifted away from the diagonal with a tiny fraction of bright *MM* of less than 5 %. These probes correspond to the limiting case of dominating *S* hybridization with a relatively high concentration of *S* transcripts. Hence, the effect of bright *MM* is related to *NS* hybridization. The *S* hybridization nearly exclusively produces bright *PM*, $I^{PM++} > I^{MM}$, as expected from the hybridization theory.

3.2 Binding model of duplex formation

The fluorescence intensity per probe spot can be described by (Binder et al., 2004)

$$I^P \approx F_{\text{chip}} \cdot c_{\text{RNA}} \cdot K^{P,S} \cdot \left[x^S + (1 - x^S) \cdot r^P\right] \cdot S^P, \qquad (1)$$

if one neglects the optical background. The binding 'strength' (or affinity) of the DNA probe for duplex formation with RNA is characterized by the binding constants of *S* and *NS* hybridizations, $K^{P,h}$ ($h = S, NS$; $r^P = K^{P,NS}/K^{P,S}$ denotes their ratio) and the saturation term

$$S^P = \left(1 + K^{P,S} \cdot c_{\text{RNA}} \cdot \left[x^S + (1 - x^S) \cdot r^P\right]\right)^{-1}.$$

The fraction of target RNA is $x^S = c_{\text{RNA}}{}^S/c_{\text{RNA}}$, and the fraction of NS RNA, involving other sequences than the intended target, is $x^{NS} = (1 - x^S)$. The chip specific constant F_{chip} specifies the detection 'strength' of the technique. It includes, besides other factors, the amount of labeling.

3.3 PM/MM trajectories of individual probes

Each probe is characterized by a '*PM/MM* trajectory', which describes the intensity change upon increasing the content of *S* transcripts ($0 \leq x^S \leq 1$) in the $\log I^{MM}$ versus $\log I^{PM}$ correlation plot. Figure 2 shows the experimental intensity data of six selected probes together with fits by means of Eq. (1) (compare curves and symbols). The trajectory, typically, 'starts' near the diagonal line in the absence of *S* transcripts (i.e., $I^{PM} \approx I^{MM}$ for $x^S = 0$), 'moves' towards bright *PM* (i.e., $I^{PM} > I^{MM}$) with increasing x^S and, finally, the trajectory returns back in direction of the diagonal at high x^S values owing to saturation.

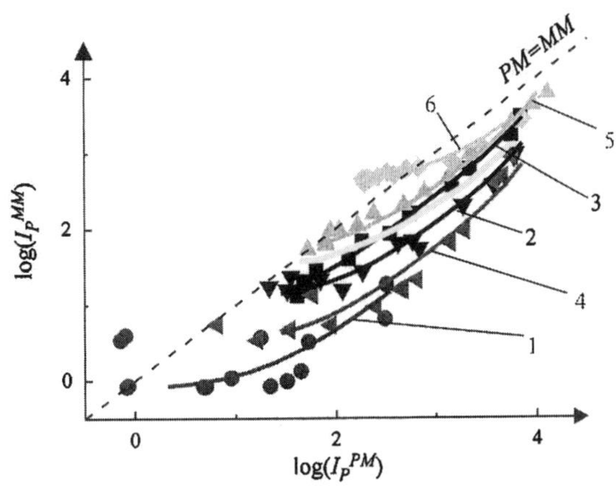

Figure -2. PM/MM trajectories of selected spiked-in probes (see Table 1 for probe no. and sequence). The curves are calculated using Eq. (1) and the parameters listed in Table 1.

Each *PM/MM* trajectory is characterized by four model parameters: the affinity constant for *S* binding, $K^{P,S}$, and the effective affinity ratio r^P for the $P = PM$ and MM probes (Table 1). It turns out that the S binding constants of the PM exceed that of the MM by a factor of between about two and twenty. Therefore, the PM intensity of all the considered probes is distinctly higher than that of the respective MM probe at larger S transcript concentrations. The relation between PM and MM intensities, however, is more heterogeneous in the limit of dominating NS hybridization. The trajectories can start on both sides of the diagonal line in Figure 2. This result indicates that the affinity of the PM probes for NS transcripts is either higher or lower compared with that of the respective MM.

Table -1. Binding constants of the selected probes (see trajectories in Figure 2) and the mean *PM/MM* trajectories (see Figure 3)

	PM sequence	No. of bases		*PM*		*MM*	
No.	Selected probes	C	A	$\log K^{PM,S}$	$\log r^{PM}$	$\log K^{MM,S}$	$\log r^{MM}$
1	TATAATCTTTTA**T**ACAGTGT CTTAC	4	7	−1.3	−2.7	−2.7	−1.5
2	GAGGATTCATCT**T**GCACAT CTGAGA	5	7	0.3	−3.0	−0.7	−2.3
3	GACAGGTCCTTT**T**CGATGGT ACATA	5	6	0.3	−2.7	−0.3	−2.8
4	GCACAAGTTTTT**C**TACACTC AGTGT	6	6	0.3	−3.0	−1.0	−2.3
5	GTGATGCTCAAT**G**GATCCC GCAGTA	7	6	0.7	−3.0	0.2	−2.5
6	TAGGCCATTTGG**A**CTCTGCC TTCAA	7	5	0.0	−1.8	−0.4	−1.1
	Middle base averages (PM)			$\log K_B^{PM,S}$	$\log r_B^{PM}$	$\log K_B^{MM,S}$	$\log r_B^{MM}$
B=	**A**			−0.15	−2.45	−0.7	−1.8
	T			−0.05	−2.45	−0.8	−2.05
	G			0.0	−2.45	−0.8	−1.5
	C			+0.20	−2.45	−0.9	−1.75
	Standard deviation			0.14	0.0	0.08	0.23
	Total mean			$\log K_0^{PM,S}$	$\log r_0^{PM}$	$\log K_0^{MM,S}$	$\log r_0^{MM}$
				0.0	−2.45	−0.8	−1.8

3.4 Mean *PM/MM* trajectories

The *PM/MM* trajectories of individual probes are well described by the suggested binding isotherms (Eq. (1)). To generalize these results in terms of mean trajectories, we calculated the 'total' average over the log intensities of all the 462 spiked-in probes using also all the three available replicates at each concentration $\langle \log I^P \rangle \equiv \langle \log I^P \rangle_{\text{sp-in}}$ ($P = PM$, MM) as well as partial averages over subsets of probes with the common middle base B = A,T,G,C at position $k = 13$ of their sequence, $\log I_B^P \equiv \langle \log I_p^P \rangle_B$. The respective trajectories characterize the average intensity relation between the *PM* and *MM* probes (see symbols in Figure 3). In addition, the mean intensities are well approximated by Eq. (1) (see lines) where the probe-specific binding constants are substituted with effective values. They can be interpreted as log averages over the considered ensemble of probes (i.e., $\log K^{P,h} \rightarrow \log K_0^{P,h} \approx \langle \log K^{P,h} \rangle_{\text{sp-in}}$ and $\log K^{P,h} \rightarrow \log K_B^{P,h} \approx \langle \log K^{P,h} \rangle_B$). The mean binding constant of the *PM* probes for target RNA, $K_0^{PM,S}$, exceeds that of the *MM* almost by one order of magnitude (Table 1). On the other hand, the mean binding constant of the probes for non-specific binding is by two–three orders of magnitude weaker than that for specific binding ($\log r_0^{PM} = -2.45$, $\log r_0^{MM} = -1.8$).

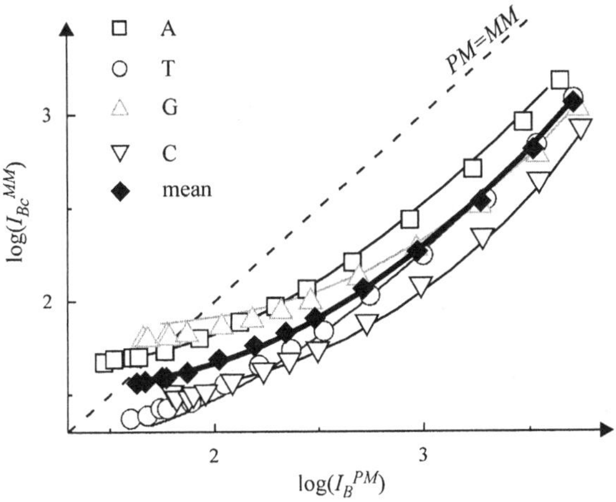

Figure -3. Mean *PM/MM* trajectories averaged over all spiked-in probe pairs with a common *PM* middle base and their total mean.

The middle-base specific *PM/MM* trajectories diverge in a systematic fashion from each other. For example, the trajectories of the purine middle bases $B = $ A,G start in the range of bright *MM* (i.e., $I^{PM} < I^{MM}$) at small intensities (and dominating *NS* hybridization) in contrast to that of the pyrimidines $B = $ T,C. The trajectories of G and T, however, merge with increasing x^S (at higher intensities). This behavior indicates that *S* and *NS* RNAs are binding differently to the probes as a function of their middle base. Note that the mean *S* binding constant of the *PM* is decreasing according to C > T ≈ G > A (see $\log K_B^{PM,S}$ in Table 1) in contrast to that of the *MM*, which is almost constant.

The data shown in Figure 3 are averaged over the limited ensemble of the 462 spiked-in probes. To generalize these results for the whole set of 250 000 *PM* and *MM* probes of the chip, we correlate the intensities of the

PM, which possesses a common middle base with their paired *MM* probe intensities (and complementary middle base, see Figure 4). The data cloud for B = A is clearly shifted towards the bright *MM* compared with that for T. The same tendency was obtained for G and C (not shown here). The systematic trend due to the different middle bases can be filtered out more clearly, if one calculates running averages over 1000 subsequent probes along the axes (see lines in Figures 4 and 5). Thus, it turns out that the curves referring to the whole set of probes show virtually the same features as the respective curves for the spiked-in probes (compare Figures 5 and 3).

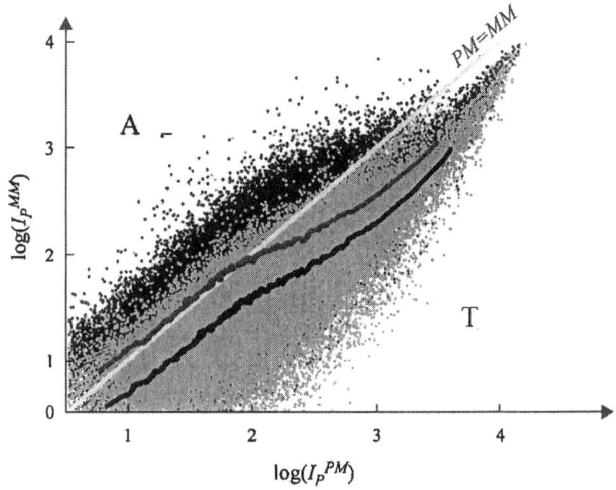

Figure -4. PM/MM correlation plot of probe pairs with *PM* middle bases A and T. Both data clouds are shifted in vertical direction to each other. The lines are running averages through the respective clouds.

3.5 Sequence specific binding: single-base model and probe sensitivity

The observed intensities are functions of the affinity for DNA/RNA duplex formation, which, in turn, depends on the sequences of the 25-meric probe and of the bound RNA fragments. Note that the trajectories of most of the selected probes in Figure 2 and Table 1 are shifting systematically towards higher intensities (and $K^{P,S}$) with increasing C and decreasing A contents (see columns 'No. of bases' in Table 1). For a more detailed description, we used the positional-dependent single-base (*SB*) model, which approximates the deviation of the probe intensity from its set average by a sum of base-specific terms according to

$$Y^P = \log I^P - \left\langle \log I^P \right\rangle_{set} \approx \sum_{k=-N_{out}+1}^{N_b+N_{out}} \sigma_k^P(\xi_k^P), \; P = PM, MM, \tag{2}$$

where ξ_k^P is the base (A, T, G, or C) at position k of the probe sequence taken from the target gene. Eq. (2) defines the sensitivity of the probe, Y^P, which, in a first-order approximation, characterizes its ability to detect a certain amount of RNA independently of the experimental conditions given by the chip specific factor and the total RNA concentration, F_{chip} and c_{RNA}, respectively (see Eq.(1)).

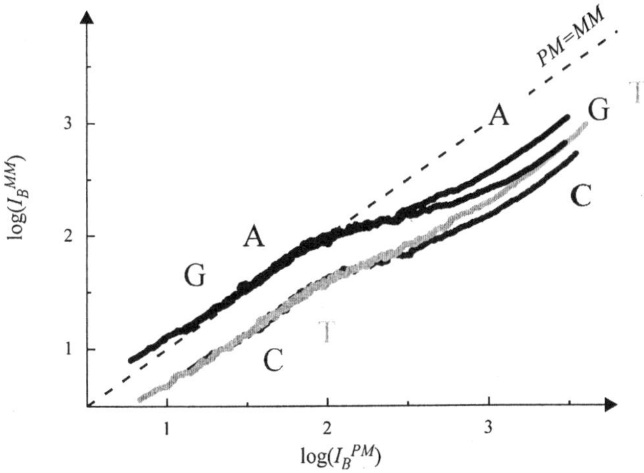

Figure -5. Running averages for all four *PM* middle bases. Note the correspondence with the middle-base averages over the spiked-in probes (Figure 3).

Note that it is by the DNA probe sequence only that we identify the SB sensitivity. It consequently refers to matched and/or to mismatched pairings with the RNA in the respective duplex. Moreover, the length of the RNA fragments typically exceeds the length of the 25-meric probes. Hence, the bases that dangle outside of the target sequence also can affect the binding affinity, because they modify the propensity of the RNA fragments for intramolecular folding. In addition, the fluorescently labeled bases outside of the target region also contribute to the measured fluorescence intensity. The model, therefore, considers the next $N_{out} = 20$ bases, which precede and follow the probe sequence of $N_b = 25$ nucleotides in the sequence of the target gene.

The sensitivity coefficients of the *SB* model, $\sigma_k^P(B)$, were determined by means of multiple linear regression of the Y^P values of selected subsets of *PM* and *MM* probes referring predominantly to *S* and *NS* hybridization. In accordance with our previous results, we collect all probe pairs of the chip meeting the condition $<\log I^{PM}>_{set} > 3$ and $<\log I^{PM}>_{set} < 2$ into the former and latter subset, respectively (Figure 1).

The shapes of the sensitivity profiles of the *PM* probes of both subsets and of the *NS* hybridized *MM* probes are very similar (Figure 6). In particular, the profiles for B = C,A show a typical parabola-like shape within the region of the probe sequence ($1 \leq k \leq 25$). They show the maximum and the minimum in the middle of the sequence, respectively, whereas the sensitivity contributions for B = T,G change almost monotonously (see also Binder et al., 2003, 2005; Mei et al., 2003; Naef and Magnasco, 2003).

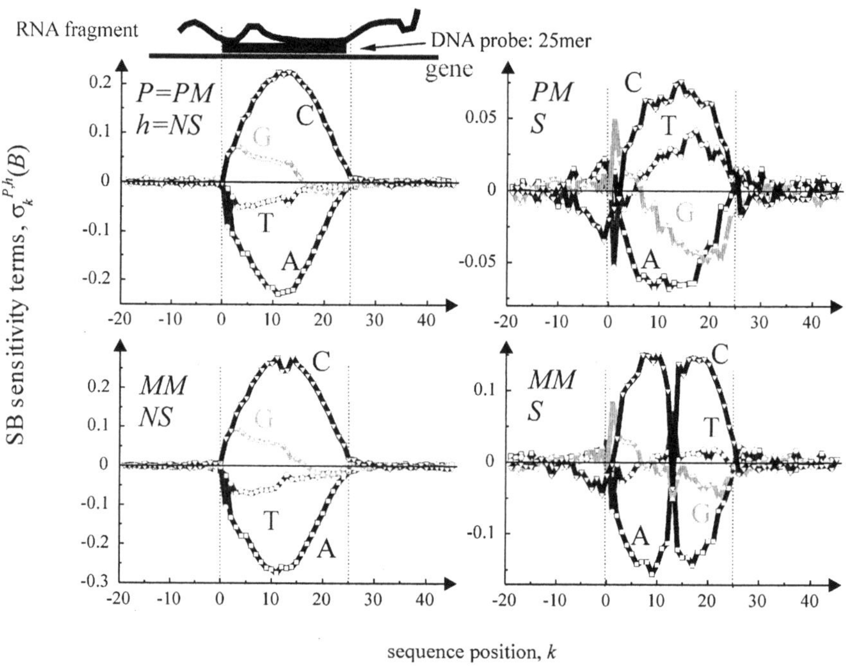

Figure -6. Single-base sensitivity profiles of *PM* and *MM* probes in the limit of specific (*S*) and non-specific (*NS*) hybridizations. The profiles consider 65 positions and extend to 20 bases before and after the 25-meric probe sequence (see the cartoon). Note the 'dent' in the middle of the *MM* S profiles for B = C and A.

In contrast, the profiles for B = A,C of the *S* hybridized *MM* distinctly differ in the middle of the sequence from the other profiles considered. Namely, the sensitivity contribution of the middle base markedly drops to tiny values near zero. Hence, the mismatched middle base of the *MM* probes on the average provides only a weak base-specific contribution to the probe intensity within the limit of *S* hybridization. On the other hand, the remaining sequence positions at $k \neq 13$ show similar sensitivity profiles for the *PM* and *MM* probes under all conditions.

Finally, the small sensitivity contributions outside of the target region at $k < 1$ and $k > 25$ indicate that these positions only weakly contribute to the probe intensity in a base-specific fashion.

3.6 Base pairings in probe/target duplexes

We analyzed the *PM* and *MM* probe intensities using two approaches: first, by averaging over all the probes with a common middle base and analysis of the respective *PM/MM* trajectories in terms of the binding model and, second, by the fit of the probe sensitivities by the sum of *SB* terms, which explicitly extract the relative contribution of the middle base to duplex stability. Both independent approaches are complementing each other. Note that both the middle base-specific binding strength and the respective *SB* sensitivity term characterize the effective interactions of the middle base in the RNA/DNA oligonucleotide duplexes, i.e., $\log K_B^{P,h} \approx \sigma_{13}^{P,h}(B)$.

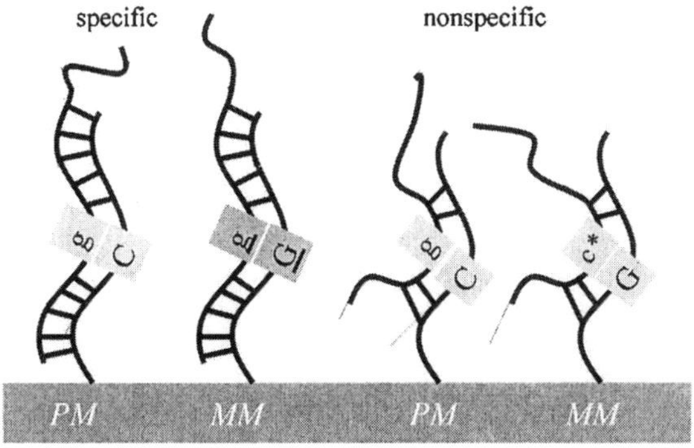

Figure -7. Base pairings in the middle of duplexes between DNA probes and RNA fragments. The *NS* duplexes are stabilized by a smaller number of *WC* pairings compared with the *S* duplexes. The middle base of the *MM* forms an *SC* (self-complementary) pairing upon *S* hybridization. Note the reversal of the *WC* pair in the *NS* duplexes of the *PM* and *MM*

The results give rise to the following interpretation in terms of the base pairings that stabilize the DNA/RNA duplexes (Figure 7). The *PM* probes 'per definition' form exclusively Watson–Crick (*WC*) pairs with the complementary sequence of the target RNA. The central *WC* pair of the *PM*, $B \bullet b^c$ (lower case letter refer to RNA; the superscript denotes the complement, e.g., C•g) is replaced with the respective self-complementary (*SC*) pair, $\underline{B}^c \bullet \underline{b}^c$ (e.g. $\underline{G} \bullet \underline{g}$) in the respective *MM*/target duplex. The shift of the trajectories into the range of bright *PM* at dominating *S* hybridization indicates that the *SC* pairing of the *MM* is considerably weaker than the respective WC pairing of the *PM*. The middle base averaged binding constants,

$K_B^{P,S}$, reflect the relative strength of the respective *WC* pairing. Its values reveal a purine–pyrimidine asymmetry according to C•g > G•c* ≈ T•a > A•u* (the asterisk denotes labeling).

On the other hand, the '*NS* background' represents a mixture of RNA fragments with a broad distribution of base compositions, which enables the formation of a sufficient number of *WC* pairings, which stabilize the *NS* duplexes. The middle bases on the average are assumed to form *WC* pairings. This reverses the direction for each *PM/MM* pair: $B{\bullet}b^c$ for the *PM* becomes $B^c{\bullet}b$ for the *MM*. The probe pairs split into two fractions with purine (A,G) middle bases of the *PM* and preferentially bright *MM* ($I^{MM} > I^{PM}$) and with pyrimidines (C,T) in the middle and the reverse intensity relation ($I^{PM} > I^{MM}$) due to the purine/pyrimidine asymmetry of interaction strengths.

3.7 Simulated intensity data

To illustrate the effect of the probe sequence on the intensity, we used a synthetic, randomly generated 'target gene' of 3000 nucleotide bases. The intensity of all the possible PM and MM probes was calculated by means of the following equations adapted from Eqs. (1) and (2)

$$
\begin{aligned}
I_p^P &\approx K_0^{P,S} \cdot \left[x^S \cdot 10^{Y_p^{P,S}} + (1-x^S) \cdot r_0^P \right] \cdot S_p^P \quad \text{with} \quad P = PM, MM; \\
Y_p^{P,h} &= \sum_{k=p-12}^{p+13} \sigma_k^{P,h}(\xi_k^P), \quad h = NS, S; \qquad (3) \\
S_p^P &= \left(1 + c_{\mathrm{RNA}} \cdot \left[x^S \cdot 10^{Y_p^{P,S}} + (1-x^S) \cdot r_0^P \right] \right)^{-1}.
\end{aligned}
$$

The *PM* probe sequence refers to a sliding window of 25 positions, which moves along the gene sequence, ξ_p ($p = 1, \ldots 3000$). For the respective MM sequence, the middle base at $k = p$ was replaced with its complement. The model parameters, namely, the total binding constants and the *SB* sensitivity contributions, were taken from the fits of the mean *PM/MM* trajectories and from the fits of the *SB* model to the experimental sensitivity data (see above).

Figure 8 shows the calculated intensities as a function of sequence position. The *PM* and *MM* intensities are correlated in Figure 9. Note that this correlation plot shows essentially the same characteristic features as the plot of the experimental data (compare with Figure 1). In particular, the data shift towards 'bright' *PM* with increasing x^S and, finally, they turn back to the diagonal line owing to saturation.

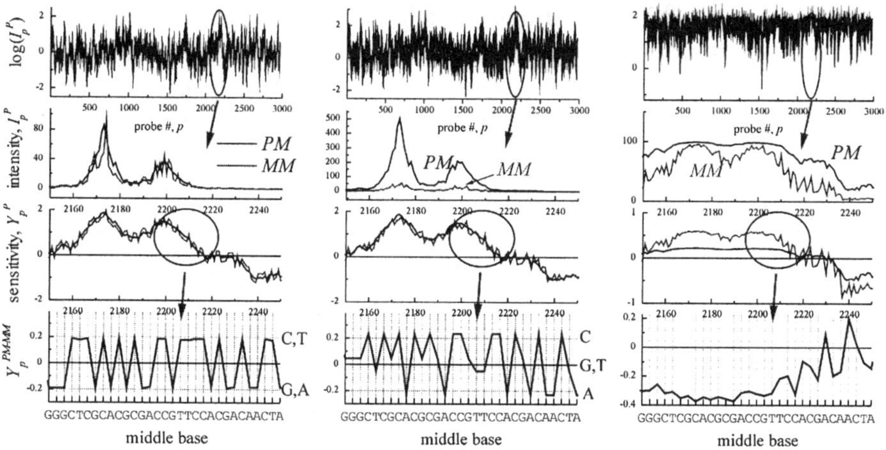

Figure -8. Simulated *PM* and *MM* intensity data, the respective sensitivities, and the difference $Y^{PM-MM} = Y^{PM} - Y^{MM}$ (from top to bottom). The left and the middle panels refer to non-specific hybridization ($x^S = 0$) and to an *NS* + *S* mixture with a fraction of specific transcripts of $x^S = 0.03$, respectively, without considering saturation. The right panel considers saturation. Note the different scattering patterns of Y^{PM-MM} as a function of the middle base and of saturation.

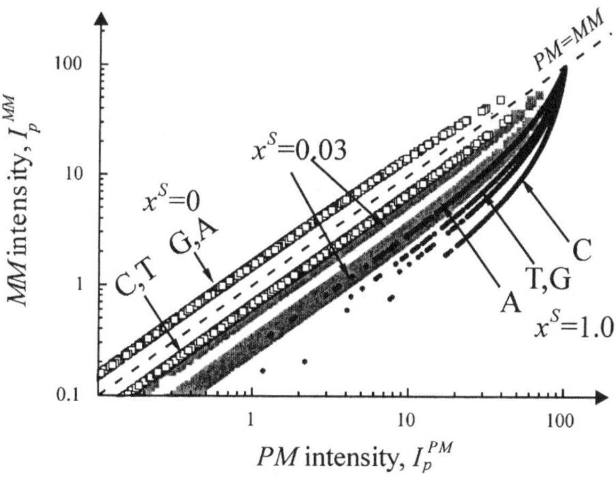

Figure -9. PM/MM correlation plot of the calculated intensities referring to three *NS* + *S* mixtures of different composition (see x^S values within the figure). Compare the simulated with the experimental data shown in Figure 1. The branches refer to the probe pairs with a common middle base of the *PM* (see figure).

Let us at first neglect saturation ($S_p^P = 1$). In this special case, the simulated *PM* and *MM* intensity data vary by about four orders of magnitude due to differences in their sequence within the limits of *NS* ($x^S = 0$, left panel of Figure 8) and *S* hybridizations (middle panel of Figure 8). Note that the neighboring probes with the indices p and $p + 1$ are shifted by only one base each to another. Both the intensity and the sensitivity of the probes smoothly change along the target gene as a consequence.

The comparison of the respective Y_p^{PM} courses shows that the sensitivity of the *PM* is invariant for changes of the fraction of specific transcripts. This property reflects the constant, i.e., middle base independent ratio of the *S* and *NS* binding constants, i.e., $r_B^{PM} \approx \text{const}$ (Table 1). The MM sensitivity reveals a more complex behavior. First, the main course of Y_p^{MM} changes parallel to that of Y_p^{PM}, because both sequences are identical for all positions $k \neq 13$. Second, the individual *MM* values scatter however about the *PM* sensitivities due to their complementary middle bases (see Y^{PM-MM}). Third and the most interesting, the scattering pattern is different for *NS* and *S* hybridized *MM* owing to the different sensitivities of the middle base (see also Figure 6). Note that the relative binding strength for *NS* binding of the *MM*, $\log r_B^{MM}$, distinctly varies upon change of the middle base, giving rise to the maximum standard deviation among the data considered (Table 1).

The intensity data are strongly modified by the saturation of the probes with bound transcripts (Figure 8, right panel). This effect especially decreases the peak values of the *PM* intensity accompanied by a marked drop in the respective sensitivity. The effect of saturation also smoothes out the scattering of the *MM* sensitivity in the range of high intensities (see Y^{PM-MM} in Figure 8).

3.8 Differential expression: accuracy and precision

The basic application of the GeneChip technology intends to estimate the level of differential gene expression in terms of the change in the RNA transcript concentration between different samples, e.g., between the sample of interest and an appropriately chosen reference. The respective ratio of target concentration, $R_{\text{true}} \equiv x^S(\text{samp})/x^S(\text{ref})$, defines the 'true' fold change, which the analysis algorithm aims to extract from the probe intensities. In the simplest approach, the intensities themselves provide the apparent fold changes in terms of the ratio $R_p^P \equiv I_p^P(\text{samp})/I_p^P(\text{ref})$ with $P = PM$, MM, and $PM - MM$ for *PM*-only, *MM*-only, and $I_p^{PM-MM} = I_p^{PM} - I_p^{MM}$ difference estimates, respectively.

Our intensity simulation enables judging the accuracy and precision of the apparent fold change by direct comparison with the true value. Note that R_p^P varies as a function of the probe sequence for a fixed R_{true}. In our notation, the precision specifies this variability in terms of the standard deviation, $SD(R_p^{P}*)$, of the relative apparent fold change $R_p^{P}* = R_p^P/R_{\text{true}}$ for all probes of the generated test gene (see the previous section). On the other hand, the accuracy reflects the consistency between the true and apparent fold changes in terms of $R^{P*} = \langle R_p^{P*} \rangle_{gene}$, the averaged relative fold change. Ideally, $SD(R_p^{P*})$ and R^{P*} adopt values near zero and unity, respectively.

Figure 10 and Table 2 compare special situations referring to a 'true' two- and fourfold concentration change ($R_{true} = 2$ and 4). It clearly turns out that the $PM - MM$ intensity difference provides the best accuracy with the R^* near unity. The subtraction of the *MM* intensity obviously provides a suitable correction of the *PM* data for the chemical background caused by non-specific hybridization. On the other hand, the $PM - MM$ data are behaving relatively noisy giving rise to, by far, the worst precision. Saturation decreases both the accuracy and the precision (see '+ sat' in Table 2).

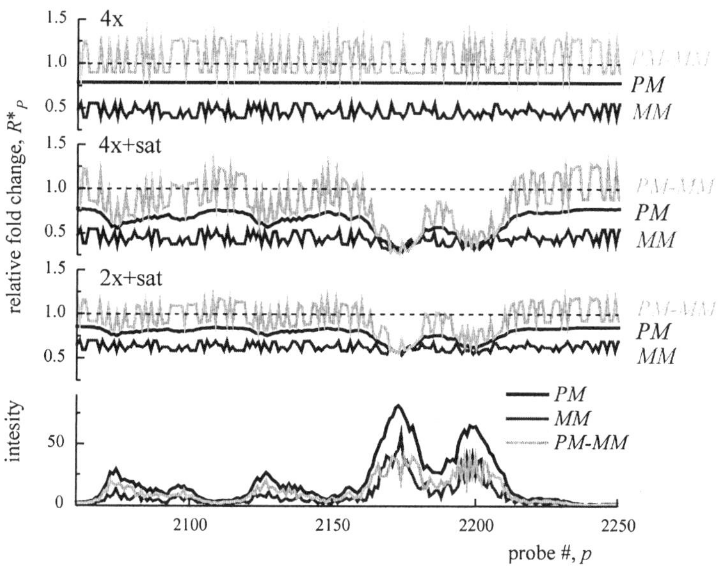

Figure -10. Simulated fold changes in *PM*, *MM*, and $PM - MM$ intensity measures. The data are normalized with respect to the 'true' fold change in 4*x* and 2*x*, i.e., $R^* = R/R_{true}$ (ideally equalling 1). See Table 2 for assignments. The accuracy ('agreement with unity') ranks according to $PM - MM > PM > MM$, whereas the precision ('scattering width about the mean') decreases with $PM > MM > PM - MM$.

In summary, the accuracy of the estimated fold changes ranks according to $PM - MM > PM > MM$, whereas the precision decreases with $PM \geq MM > PM - MM$. The former result can be simply explained by the decreasing relative contribution of non-specific hybridization to the total signal intensity, which is minimal for $PM - MM$ and maximal for *MM*. The latter trend is caused by the variability of the *MM* sensitivity owing to the changing affinity of the *MM* middle base in *S* and *NS* duplexes.

Table -2. Accuracy (R^*) and precision (SD) of *PM*, *MM* and *PM-MM* intensity measures for fold changes of gene expression (R_{true}). Saturation is neglected in one of the $4x$ samples and considered in the '+sat' samples. The fraction of specific transcripts is $x_S = 0.03$ in the reference. See text

	PM		*MM*		*PM-MM*	
R_{true}	R^*	SD	R^*	SD	R^*	SD
4x	0.79	0.0	0.47	0.06	1.08	0.19
4x+sat	0.70	0.09	0.45	0.06	0.95	0.23
2x+sat	0.82	0.04	0.64	0.04	1.00	0.14

The potential accuracy advantage of the analysis algorithm using the $PM - MM$ difference is opposed by its low precision. Instead, a *PM*-only algorithm for extracting differential expression measures seems to afford a suited compromise between the accuracy and the precision in agreement with recent results (see Irizarry et al., 2003, and references cited therein).

4. CONCLUSIONS: CONSEQUENCES FOR DATA ANALYSIS AND CHIP DESIGN

NS hybridization considerably complicates the analysis of microarrays, because it adds a background intensity not related to expression degree of the gene of interest. The probes with mismatched base pairings possess the potency to estimate the background level and, in this way, to correct the intensity of the respective *PM* probe. We found that the intensity of the complementary *MM* however introduced a systematic source of variation relative to the intensity of the respective *PM* probe owing to different base pairings in the NS duplexes. In consequence, the naive correction of the *PM* signal by subtracting the *MM* intensity decreases the precision of expression measures. Our results imply improved algorithms of data analysis that explicitly consider the middle base–related bias of the *MM* intensities to reduce their systematic variability. Moreover, the knowledge of base pair interactions suggests substituting the complementary mismatches on GeneChips by alternative rules of *MM* design.

ACKNOWLEDGMENTS

The work was supported by the DFG (grant No. BIZ 6-1/2). I thank P. Stadler and M. Loeffler for the stimulating discussion and T. Kirsten and St. Preibisch for their help in practical questions.

DETERMINATION OF THE DEVELOPMENTAL AGE OF A *DROSOPHILA* EMBRYO FROM CONFOCAL IMAGES OF ITS SEGMENTATION GENE EXPRESSION PATTERNS

E. Myasnikova[1*], A. Samsonova[2], S. Surkova[1], M. Samsonova[1], J. Reinitz[3]
[1] St. Petersburg State Polytechnic University, 29, Politehnicheskaya, St. Petersburg, 195257, Russia, e-mail: myasnikova@spbcas.ru; [2] European Bioinformatics Institute, Wellcome Trust Genome Campus, Cambridge, CB10 1SD, UK; [3] University at Stony Brook, Stony Brook, NY,11794, USA
[*] Corresponding author

Abstract: In the paper, we address the problem of the temporal characterization of *Drosophila* embryos. We have developed a method for automated staging of an embryo on the basis of a confocal image of its gene expression pattern. Phases of spectral Fourier coefficients were used as the features characterizing temporal changes in expression patterns. The age detection is implemented by applying support vector regression, which is a machine learning method for creating regression functions of arbitrary type from a set of training data. The training set is composed of embryos for which the precise developmental age was determined by measuring the degree of membrane invagination. Testing the quality of regression on the training set showed a good prediction accuracy.

Key words: gene expression; embryo staging; *Drosophila*; support vector regression

1. INTRODUCTION

One of the primary patterning decisions required in fruit fly *Drosophila* is the division of a major body axis into serially repeated units or segments. Immediately following fertilization and egg deposition, the newly formed zygotic nucleus starts to divide. By cleavage cycle 10, the embryo becomes a hollow ellipsoid of nuclei, which are not separated by cell membranes, called the syncytial blastoderm. In this paper, we focus on cleavage cycle 14A, which lasts approximately from 130 to 180 minutes after fertilization. It is characterized by determination of embryonic segments and cellularization of the blastoderm, as the membranes invaginate and surround nuclei (Foe et al., 1983).

The initial determination of segments is a consequence of the expression of about 16 genes, which are mainly transcription factors. Most of these genes are zygotic and are expressed in patterns that become more spatially refined over time. The most of the gain in resolution happens during cycle 14A. Of particular importance are members of the 'gap' and 'pair-rule' classes of segmentation genes (Akam, 1987; Ingham, 1988).

To provide full spatiotemporal information about expression of these genes, the data must be obtained at cellular resolution in space and at temporal resolution that is close to the characteristic time for changes in gene expression. In our experiments, such data are obtained by means of immunofluorescence histochemistry and confocal scanning microscopy.

One of the important procedures of data processing is temporal characterization of embryos. The problem of embryo age detection arises, as gene expression data have been acquired from fixed embryos for which a precise developmental time was not known. The experimental methods for detection of embryo age are time consuming and expensive. However, the expression patterns themselves can be used to determine developmental time in cleavage cycle 14A. We staged the embryos on the basis of an expression pattern of the pair-rule gene *even-skipped* (*eve*), which is highly dynamic during cycle 14A. This is accomplished by standardizing the *eve* expression pattern against the developmental time determined experimentally by measuring membrane invagination *in vivo.* The degree of membrane invagination as a function of time gives a standard curve that makes it possible to assign a precise developmental age to fixed embryos (Merrill et al., 1988).

We have already reported the method for automated assignment of expression-based age to an embryo belonging to the late part of cycle 14A (Myasnikova et al., 2002). The method was developed for the data presented in terms of nuclear location, and we restricted ourselves to the data extracted from the central 10 % horizontal strip running in an A–P direction on the midline of an embryo. Moreover, the method application was limited only to those embryos in which the full set of seven well-defined *eve* stripes was already formed. We have generalized the method to any raw image of gene expression obtained directly from the microscope and presented in a raster format.

The development of the method can be subdivided into two major stages. The first stage is the extraction of characteristic expression features of embryos of different age, and the second is standardization against morphological data. In the present paper, we pay particular attention to the first problem. Basing on our results (Aizenberg et al., 2002), we use the phases of Fourier spectral coefficients as the features characterizing temporal changes in expression patterns. The feature extraction method includes image preprocessing, extraction of Fourier coefficients from the standardized image, and transformation of spectral phases to the form suitable for the regression. The solution of the temporal characterization problem is based on support vector (SV) regression and is described in detail by Myasnikova et al. (2002).

2. METHODS AND ALGORITHMS

2.1 Dataset

In our experiments, gene expression was measured using fluorescence tagged antibodies as described by Kosman et al. (1998). For each embryo, a 1024×1024 pixel confocal image with 8 bits of fluorescence data in each of the three channels was obtained. At present, our dataset contains confocal scans of about 1400 embryos, of which 809 are wild type and belong to cycle 14A. Of these, all are stained for the pair-rule segmentation gene *even-skipped* (*eve*) and two other genes that vary among the dataset. Cycle 14A is about one hour long and is characterized by a rapid transition of the expression patterns of pair-rule genes, which culminates in the formation of seven stripes. We selected embryos for scanning without regard for age, so we expect our dataset to be uniformly distributed in time. As an initial step of temporal characterization, the embryos were divided by visual inspection of the pair-rule gene expression patterns into eight temporal equivalence classes (Myasnikova et al., 2001).

In this study, only images of *eve* expression pattern are considered. The patterns of *eve* gene are particularly important for temporal characterization, since they are highly dynamic and each embryo is stained for this gene. Overall, 120 of these embryos were rephotographed in Nomarski optics to visualize the morphology of the blastoderm. These data were used to measure the degree of membrane invagination for the embryos by means of manual separation of the region of interest and by texture analysis. Using the standard curve giving membrane invagination as a function of developmental time (Merrill et al., 1988), the precise developmental age of each embryo was determined. For each embryo, 20 measurements were made, and the age was computed as an average value over all measurements. The mean measurement error was about 2.5 minutes. The ages turned out to be distributed uniformly over the range from 20 to 60 minutes from the onset of cleavage cycle 14A; according to the preliminary classification, the embryos belong to six temporal classes from 3 to 8. Representative images of embryos from classes 3 and 8 are shown in Figure 1. The 120 rephotographed and standardized embryos are used as a training set for temporal analysis.

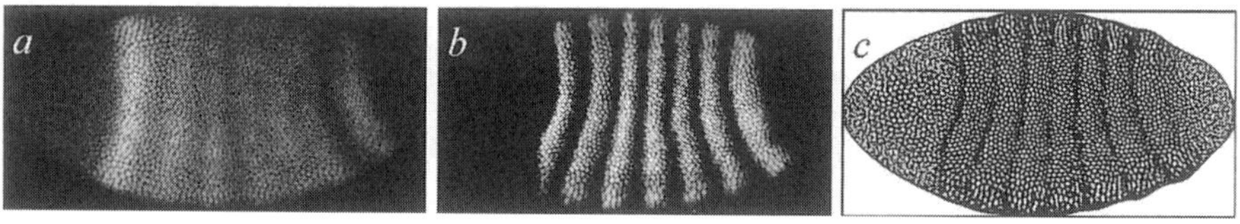

Figure -1. Confocal images of *eve* expression patterns in embryos belonging to (*a*) early and (*b*) late parts of cleavage cycle 14A. According to visual classification, they are attributed to classes 3 and 8, respectively. Small white spots correspond to nuclei with the fluorescence intensity proportional to gene expression. (*c*) Binary mask of the image (*b*).

2.2 Preprocessing of confocal images

Prior to feature extraction, confocal images are preprocessed to bring the raw images to a unified standard form. The preprocessing procedure is implemented in several steps.

Segmentation of images. In an image, there are areas arising from *internuclear* space that do not contain any information about the gene expression. It is necessary to distinguish between these areas and the so-called *nonexpressing* areas, i.e., *intranuclear* space where a given gene is not expressed but where a nonzero fluorescence is induced by the nonspecific background staining. To exclude the internuclear areas, an image is subjected to the segmentation procedure.

The segmentation of an image is preceded by the preliminary filtering. To denoise the images, we apply the multivalued nonlinear filter (MVF), introduced by Aizenberg et al. (2000). Next, the sliding histogram equalization filter is applied to amplify the image details. Finally, the image is thresholded by the following rule: every pixel whose brightness is lower than a given threshold is replaced with 0 and every pixel larger than or equal to the threshold is replaced with 255. The end result of the segmentation procedure is a binary image in which contiguous groups of 'on' pixels, separated from other groups by 'off' pixels, define the intranuclear regions (Figure 1*c*). Now, the data on gene expression is read off the 'on' regions of the segmented image.

Background removal. Images of expression patterns usually contain a nonspecific background signal varying considerably among the dataset. The presence of background prevents the data obtained in different experiments from being combined and averaged. To bring the data to the normalized form with zero background and to remove distortions of gene expression patterns, the background removal procedure is applied (Myasnikova et al., 2005). The main idea of the method is to approximate the background signal by a broad paraboloid from the nonexpressing areas of an embryo and then, to rescale the whole image by this paraboloid. Background is removed from the image by a linear mapping of intensity that transforms the fluorescence at or below background level to zero and transforms the maximum fluorescence to itself. As a result, those regions of the expression pattern where *eve* gene is not expressed are mapped to zero, and all the expression domains are rescaled to a unified form.

Filtering. To denoise the normalized image, the mean filter with a 3×3 window is applied.

Resizing. To facilitate the feature extraction, all the images are brought to the same size of 256×256 pixels.

2.3 Extraction of characteristic features

To be able to detect the age of an embryo on the basis of knowledge about its *eve* expression pattern, it is necessary to present the pattern in terms of a small number of parameters that characterize well the temporal changes in *eve* expression domains. It has been shown in our previous study (Aizenberg et al., 2002) that the frequency domain representation of the images is very important for their classification into discrete temporal classes. Hence, it may be used to detect the characteristic features that mark the development of expression patterns over time. The Fourier spectrum is extracted from two-dimensional images by means of the fast Fourier transform (FFT). The FFT algorithm works the most efficiently, if an image is presented by a quadratic matrix with the dimensions of a power of two. For this reason, our images are resized to 256×256 pixels.

The Fourier spectrum of a two-dimensional function $f(x, y)$ is defined by the discrete Fourier transform estimated on the two-dimensional lattice of integers. Suppose that we have N^2 sampled values $f_{kl} = f(x_k, y_l, x_k = k\Delta_x,$ $y_l = l\Delta_y, k, l = 0, \ldots, N-1$. Discrete Fourier transform of the N^2 points f_{kl} is given by a two-dimensional array

$$F_{nm} = \sum_{k=0}^{N-1} \sum_{l=0}^{N-1} f_{kl} \exp\left[2\pi i \left(\frac{kn}{N} + \frac{lm}{N}\right)\right]. \tag{1}$$

Phases of the spectral coefficients are given by phase(F_{nm}) = arctan (Im(F_{nm})/Re(F_{nm})), where Im(·) and Re(·) are the imaginary and real parts of a complex number, respectively.

It was shown by Oppenheim and Lim (1981) that a spectral phase mostly contained information about an object presented by a signal, while the spectral amplitude contained more information about the signal behavior, presence of noise or blur (if any), etc. Thus, with a view of detecting the characteristic features, it makes sense to consider only the spectral phases ignoring amplitudes. One more observation made in the course of classification is that for different classes of discrete signals, the sets of quarter to half low frequency spectral coefficients are very close for the signals belonging to the same temporal class and, hence, are important for temporal characterization. Extraction of the phases from Fourier spectral coefficients is organized according to the frequency ordering. We start from the lowest ones and then proceed according to the so-called 'zigzag' rule up to the frequency 8. The total number of parameters extracted in such a way is 84, but due to the symmetry of the spectrum, only 42 of them are pairwise different.

2.4 Training set

At the next stage, we consider the group of 120 embryos for which the precise developmental age was determined by measuring membrane invagination. These embryos are used as training data for creating the regression function with the characteristic features used as independent variables. However, the spectral phases cannot be directly involved into the regression analysis for two reasons: first, phases are periodic values with a period equal to 2π and, second, the number of independent parameters is too big as compared with the size of the training set.

Periodicity. To treat periodicity, we apply the following algorithm. Let p_{kj}, $k = 1, \ldots, 42, j = 1, \ldots, N$, be the array of spectral phases taking values in the interval $[0,2\pi]$, where $N = 120$ is the size of the training set. Each vector $\{p_{kj}\}_{j=1}^{N}$, composed of the values of kth parameter, is arranged in increasing order. Denote ranked values of the phases $p^{(j)}$ (omitting the parameter index k), so that $p^{(j)} \leq p^{(j+1)}$, and choose such a rank j^* that the transform

$$\tilde{p}^{(j)} = \begin{cases} p^{(j)} & j \leq j^* \\ p^{(j)} - 2\pi & j > j^* \end{cases}, \quad j = 1, \ldots, N \tag{2}$$

minimizes $\left|\tilde{p}^{(N)} - \tilde{p}^{(1)}\right|$, the maximal in absolute value pairwise difference between the values of the parameter over the training set. As a result, for each parameter, the standard range of values is defined as $[\tilde{p}^{(1)}, \tilde{p}^{(N)}]$. Finally, the transformed parameter values $\tilde{p}^{(j)}$ are rearranged back to their initial order.

Principal component analysis. The problem of high dimensionality of the feature space often arises in the regression prediction. If the training set is of a small size relative to the dimension of the feature vector, this may cause the so-called overfitting effect, which, in turn, may cause unreliable age prediction and lack of robustness with respect to changes in the algorithm parameters. As the spectral phases are strongly correlated, and hence, the feature set is redundant, it is possible to reduce the dimension of the feature vector using the principal component analysis (PCA).

To reduce the data dimension by PCA, several correlated variables are linearly combined into one factor so as to maximize the variance of the 'new' variable (factor), while minimizing the variance orthogonal to the new variable. After the first factor, which maximizes the overall variance of the data, has been extracted, we define another factor that contains the maximum amount of the remaining variability, and so on. In this manner, the consecutive factors are extracted, which are orthogonal to one another, and hence, independent. This procedure implies that the new factors are linear combinations of the initial variables. Thus, applying the PCA, we reduce the set of spectral phases to a small set of size L that still contains most of the information of the full set.

2.5 Construction of regression function

Each embryo of the training set is now characterized by a multidimensional vector containing as components the value of developmental age together with parameters of gene expression patterns. The regression function for the age prediction is created from the set of the training data applying the support vector regression (SVR) (Schölkopf and Smola, 2002). SVR is a statistical method in pattern recognition theory, which is more flexible compared to classic regression, because it allows for the use of loss functions of various types. The SVR algorithm as applied to embryo age detection from quantitative data on gene expression is described in more detail by Myasnikova et al. (2002); here, we will give just a brief statement of the main ideas of the method.

Suppose we are given the training data represented by N observations (embryos). Each observation consists of a pair: a vector of L characteristic features $\mathbf{p}_i = (p_{1i}, \ldots, p_{iL})$, $i = 1, \ldots, N$, and the associated value A_i (an embryo age), given to us by a trusted source (membrane invagination). In linear ε-SVR (Vapnik, 1995), the goal is to find a function $f(\mathbf{p}) = (\mathbf{w}, \mathbf{p}) + b$, that minimizes the regularized empirical risk functional

$$R_{\mathrm{reg}}[f] = \frac{1}{N}\sum_{i=1}^{N} | A_i - f(\mathbf{p}_i) |_\varepsilon + \frac{1}{2}\|\mathbf{w}\|^2 \tag{3}$$

with the ε-insensitive loss function $|\xi|_\varepsilon = \begin{cases} 0 & \text{if } |\xi| \le \varepsilon \\ |\xi| - \varepsilon & \text{otherwise} \end{cases},$

and the regularization term $\frac{1}{2}\|\mathbf{w}\|^2$.

The problem can be represented as a convex optimization problem, which is solved more easily in its dual formulation through utilizing Lagrange multipliers.

2.6 Age prediction

To determine the expression-based age of an embryo not belonging to the training set, the image of its *eve* expression pattern is subjected to the same preprocessing and feature extraction procedures as the training data.

First, a raw image is segmented and normalized to remove the background signal; then, filtered and resized to the standard form. Characteristic features are extracted from a standardized image using FFT (Eq. (1)). Then, the spectral phases are brought to the standard range defined by Eq. (2). If a phase cannot be transformed into the standard interval, it means that it is impossible to determine the age of this embryo from our training set. This usually happens when a precise age of an embryo is less than the age of the earliest embryo in the training set or if the *eve* expression pattern has, for some reason, an irregular shape. At the next step, the PCA is applied to reduce the number of features to the regression number, L, required in SV.

Finally, the embryo age is defined by substituting the vector of parameters into the regression function $f(\mathbf{p})$ as independent variables and computing the value of the function with the regression coefficients estimated from the training set.

3. RESULTS

Characteristic features were extracted from 120 images of embryos belonging to the training set for which the precise developmental age was experimentally determined. All the images were brought to the standard normalized and resized form. The phases of low frequency Fourier coefficients up to the frequency 8 were extracted using FFT. The total number of these parameters were 84, but due to the symmetry of the spectrum, only 42 were pairwise different. For each parameter, the standard range was determined.

Applying the PCA, we have reduced the parameter set to five significant uncorrelated variables, which contain 60 % of information originally contained in the full set of 42 parameters. The multiple correlation between all the five factors involved in the model and the precise developmental age is 94 %. Even the first factor alone, which captures the overall variance of the data, shows 91 % correlation with the measured ages. The regression function $f(\mathbf{p})$ was constructed from the training set. The results of regression estimation are shown in Figure 2. The quality of regression is characterized by the minimal value of the empirical risk functional (Eq. (3)) with no regularization term, $R_{\text{emp}} = \frac{1}{N}\sum_{i=1}^{N} | A_i - f(\mathbf{p}_i) |_{\varepsilon}$, in other words, the average ε-insensitive deviation between the observed and predicted ages. For our training set, the best fit is achieved at $R_{\text{emp}} = 2.2$.

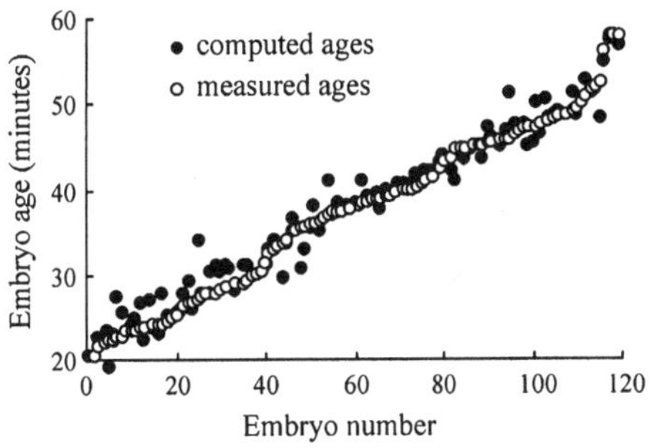

Figure -2. Embryo ages (measured in minutes from the onset of cycle 14A) computed for the training set using the regression function (black) and measured in experiment (white).

To cross-validate the accuracy of prediction, we apply the leave-one-out test excluding one by one a single item from the training set and predicting the age for the excluded embryo. As a criterion of the quality of prediction, the risk functional is used with the entries computed for the excluded items

$$R_{\text{test}} = \frac{1}{N}\sum_{i=1}^{N} \left| A_i - f_i(\mathbf{p}_i) \right|_{\varepsilon}, \tag{4}$$

where f_i are the different functions each time newly estimated for the training set with the exclusion of the ith embryo. The value of the criterion defined by Eq. (4) is equal to 2.4, and the results of testing are visualized in Figure 3.

Taking into account that the ages vary over a range of 20 to 60 minutes from the onset of cycle 14A, the mean error value a little greater than 2 minutes is small enough to permit one to conclude that the predictions are reliable. In addition, the error value is comparable to the error of membrane measurement estimated as 2.5 minutes.

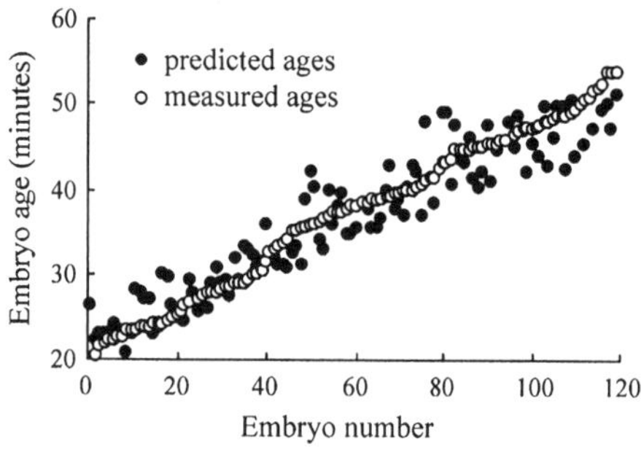

Figure -3. Embryo ages predicted for the members of the training set using the leave-one-out cross-validation (black) and the same experimental data as in Figure 2 (white).

Next, we used the regression function $f(\mathbf{p})$ for the prediction of the expression-based age of embryos not included into the training set. The results are displayed in Figure 4 for 426 embryos.

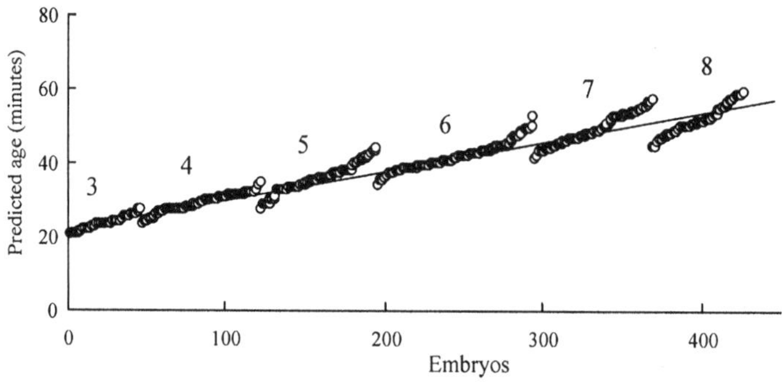

Figure -4. Predicted values of ages of the embryos not belonging to the training set. The data are grouped according to the predefined temporal classes 3 to 8.

Their developmental ages were not experimentally measured, but in our previous work (Myasnikova et al., 2001), all these embryos were assigned to one of six temporal classes (from 3 to 8) on the basis of visual inspection of their gene expression patterns, in particular, that of *eve*. The developmental age of all these embryos was greater than 20 minutes, and hence, the spectral phases extracted from their images were within the standard range defined over the training set. It is clear that the visual classification is not exact and, thus, can serve only as an indirect criterion of the quality of age prediction. Figure 4 shows a good correlation with the predefined temporal classes; the *t*-test confirms the statistically significant pairwise discrimination between all the temporal classes at 0.05 confidence level. However, it is evident from Figure 4 that the prediction results for the embryos belonging to late temporal classes are less accurate than for the early ones, which is true, in particular, for the embryos from temporal classes 7 and 8.

4. DISCUSSION

In this paper, we address the problem of temporal resolution of segmentation gene expression patterns by providing a new method for their temporal characterization. We have already reported the method for the detection of developmental age of embryos belonging to the late part of cycle 14A (Myasnikova et al., 2002). The method allowed us to assign the age only to the embryos in which the full set of seven *eve* stripes had been already formed, and for the same reason, we could not use all the embryos for which the precise developmental age was experimentally measured, as training data. Use of the spectral phases as characteristic features allowed us to generalize the method to any image of *eve* expression pattern presented in a raster format.

The regression function for age prediction was constructed using the SV regression method. To make it possible to consider the spectral phases as independent regression variables, we transformed them to five nonperiodic uncorrelated factors. It is particularly remarkable that the multiple correlation between the expression-based features and the experimentally determined developmental age is as high as 94 %. Such a strong correlation between the characteristics originating from different independent sources confirms the validity of the choice of spectral phases as the features characterizing temporal changes in *eve* expression pattern. In addition, the high linear correlation validates the linearity of the regression model and explains the good prediction results. The accuracy of prediction on the training set is estimated at 2.4 minutes, which is of the same order as the error of experimental method. The prediction of ages for embryos not included into the training set, i.e., for which we do not possess any information about their precise developmental age, shows a good correlation with the predefined temporal classes. However, the prediction for embryos belonging to late temporal classes, though statistically significant, is less accurate than for early ones. A possible way to reduce the prediction error is to improve the quality of experimental data. When standardizing against the membrane invagination, we are currently limited to published experiments (Merrill et al., 1988). Now, we are undertaking the acquisition of our own *in vivo* membrane observations, which we anticipate will provide a more precise standard of age. The exploration of the method on improved experimental data will be the subject of the future work.

ACKNOWLEDGMENTS

This work is supported by NIH (grant No. RR07801) and GAP (award No. RBO-1286). We thank Dr. Igor Aizenberg for key contributions to this work.

PART 4

BIOMOLECULAR DATA AND PROCESSES ANALYSIS

TOPICAL CLUSTERING OF BIOMEDICAL ABSTRACTS BY SELF-ORGANIZING MAPS

M. Fattore, P. Arrigo*
ISMAC, via De Marini 6 16149 Genova, Italy, e-mail: arrigo@ge.ismac.cnr.it
* *Corresponding author*

Abstract: One of the major challenges in the post-genomic era is the speed-up of the process of identification of molecules involved in a specific disease (molecular targets). Even if the experimental procedure has greatly enhanced the analytical capability, the textual data analysis still plays a central role in the experimental activity design or in the data collection. The extraction of useful information from published papers is still strongly dependent on the human expertise in the selection and retrieval of the relevant papers. The search for abstracts in the MEDLINE or PubMed databases is a common activity for researcher. Often, the navigation in textual databases is not simple, and in many cases, the user can retrieve only a list of abstracts without any kind of additional information about the relatedness of the abstract content with the submitted query. In the last decade, the application of natural language processing tools has acquired some relevance in bioinformatics field. The possibility to retrieve and organize the textual information according to the specific topics allows the user to select and analyze only a reduced set of papers. In our work, we present the application of a document clustering system founded on self-organizing maps to reorganize in a hierarchical way the cluster of abstracts retrieved by a PubMed query. The system is available at http://www.biocomp.ge.ismac.cnr.it.

Key words: text mining; conceptual clustering; self-organizing maps; latent semantic analysis

1. INTRODUCTION

The biomedical literature is a major repository of scientific knowledge. In many cases, the published papers constitute the main source of data for a large number of biological databases. In molecular biology, the data warehousing task is very complex and often performed manually. The personnel involved in this task spend a lot of time for visual analysis of the texts in order to collect the data for a specific repository. In addition, the accompanying information of a submitted nucleotide or protein sequence includes the bibliographic references. The availability of web-based technologies has enhanced the possibility to access in a fast and simple way the MEDLINE or other bibliographic databases (Sable, 2000).

The exploitation of published papers is one of collateral effects of genome projects. However, as the volume of literature increases, the burden of data warehousing increases too. The application of Natural Language Processing (NLP) tools can help the analysis of biomedical textual information. The text mining has acquired great relevance after the completion of the rough sequencing of many genomes. The capability to define the real biological functionality depends on the possibility to integrate multiple heterogeneous data sources, either molecular or cellular. The discovery process needs two different steps: the hypothesis generation and its validation (Weeber, 2003). This task is strongly dependent on the efficiency of database interoperability and data-mining integration. The different fields of bioinformatics and computational biology allow for obtaining some highly specialized knowledge. In order to obtain a more general knowledge discovery, we need to perform a knowledge fusion. This task is very complex, because many data are not really reliable and the different bioinformatics tools not always communicate in an efficient way. In a data integration perspective, the text-mining procedures could be a good contact point among different kind of data and programs (Altman, 2003; Raychauduri, 2003).

In order to improve the capability to analyze heterogeneous information, a Natural Language Processing approach can efficiently support the process of scientific hypothesis formulation. The generation of a hypothesis in biomedical field is now supported by specialized thesauruses, such as the Unified Medical Language System (McCray, 1998) or, more recently, by the development of specialized ontologies (Guarino, 1995; Baker, 1999). In the last decade, the relevance of NLP application to biomedical field has rapidly increased its relevance; many computer programs (Nenadic, 2002) have been designed to extract various molecular biological findings from Medline abstracts or full-text articles. One of the great challenges for NLP is the decoding of the semantic structure; this is essential for the concept formation process.

The textual knowledge discovery offers powerful methods to support the knowledge discovery and the construction of topic maps and ontologies. The challenge is to manage the increasing volume, complexity, and specialization of knowledge expressed in this literature.

Although information retrieval or text searching is useful, it is not sufficient to find specific facts and relations. Information extraction methods are evolving to extract automatically specific, fine-grained terms corresponding to the names of entities referred to in the text and the relationships that connect these terms. Many tools for text mining are founded on clustering methods (Chen, 1998), because the biomedical abstracts are not always organized into topic classes; these algorithms allow partitioning of the data set into clusters that can be analyzed separately.

2. METHODS

The aim of our work is to investigate the possibility to identify automatically the rough semantic structure. The proposed system try to extract the hierarchical (semantic) structure embedded in a user-defined set of retrieved documents.

The latent semantic analysis (Derweester, 1990) is a well-established approach for investigation of the conceptual relations in texts. The majority of the other available methods uses, as support information, the conceptual structure of NCBI's Medical Subject Headings (MeSH) terms; in our work, we do not use these data and consider only the textual content of the abstracts. Our approach is founded on the following considerations:

1. An initial PubMed query is normally quite general. Often, the user must refine it by using other specific keywords. The selection of relevant abstract is the result of an iterative search procedure that requires a continuous user feedback.
2. The retrieved abstract list does not give information about the real relatedness of the content with the submitted query.
3. The retrieved list does not allow for founding the terms that have some dependency with the query.

The common way to retrieve a set of abstracts is founded on the submission of a query. It is constituted by a single term or a logical combination of several words. The result is a simple list of bibliographic references. The content information only depends on the query terms. This kind of report does not allow the user to obtain any information about the conceptual relatedness of the abstracts in the list with the original query. The discovery of the semantic structure in the retrieved document set could be a useful tool for the researcher in order to speed up the screening of the literature.

In order to achieve this aim, the extraction of the latent semantic structure is essential. Using a predefined semantic or ontological structure is not always able to face the variability of words used to describe the same concept. The aim of the latent semantic analysis (LSA) is the identification of high-order structures, the semantics, taking into account the intrinsic document variability. These techniques are commonly based on the application of statistical methods to estimate the semantic similarity by using a specific parameter—the Latent Semantic Indexing (LSI; Derweester, 1990). Recently, some authors have applied machine-learning approaches, such as kernel methods, for LSA or unsupervised algorithm for text classification (Kohonen, 2000). These methods allow for obtaining a good document classification, but they are not yet well focused on the extraction of semantic structure. In our paper, we propose a combination of self-organizing maps (SOM; Kohonen 1999) with agglomerative clustering to obtain a conceptual classification of PubMed abstracts. Our work is addressed to the automatic detection of the identification of a rough semantic structure that can be compared with existing specialized ontologies. The conceptual clustering is an inductive machine-learning approach (Michalsky and Stepp, 1983; Fisher, 1987) that allows obtaining a semantic description starting from a set of unlabelled samples (learning by experience). The process requires two steps: the first is called aggregation phase; the second is defined as characterization phase. In the aggregation, the samples must be grouped; this implies the application of clustering methods. In the characterization phase, the description of each cluster must be obtained. The semantic similarity among different groups is subsequently evaluated by a conceptual distance. The system proposed here applies the approach of Self-Organizing Maps (SOM) for the aggregation phase and a single link clustering method for the hierarchy construction (Jain, 1999).

In our paper, we use a set-based distance as a conceptual distance. The system workflow is briefly sketched in Figure 1, and the main steps are listed below:

1. Query submission and document retrieval;
2. Document file cleaning and splitting;
3. Linguistic pre-processing;
4. SOM classification;
5. Hierarchical clustering of SOM-defined groups;
6. Textual characterization of hierarchical cluster; and
7. Visualization.

Each operation will be described in detail in the following paragraphs.

The system operates at the intermediate level between the NCBI PubMed database and the end user. The starting point is the submission of a query to PubMed according to the syntactical rules defined by NCBI. In order to ensure the maximal standardization of the system, we process the abstracts in XML format. This choice allows us to modify easily the input data, because the XML is currently a worldwide standard. For instance, we can easily adapt the system to analyze the annotation of biosequences or to classify other documents retrieved from other bibliographic databases, such as INSPEC or CAS (Chemical Abstract Services).

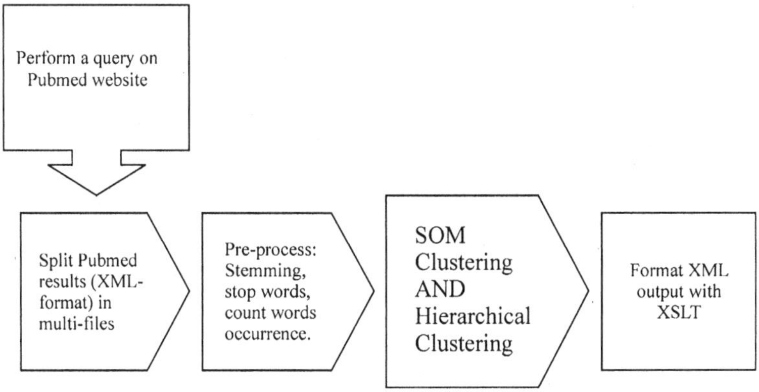

Figure -1. The flowchart of the document clustering method.

The abstracts are retrieved in a single file; this file is locally downloaded and it is subsequently split into a set of single abstract files. In many cases, the abstract of a paper is not available in the PubMed database; remove these entries from the retrieved set. The number of downloaded documents is restricted according to the NCBI rules (no more than 5000 abstracts for each query).

The program performs the linguistic pre-processing only on the 'abstract' field of XML document.

The linguistic pre-processing phase is very complex and plays a critical role in the subsequent conceptual classification step. In our application, at the moment, we do not use any thesaurus and do not apply any word sense disambiguation system. The word sense disambiguation is a step in a typical Natural Language Process. A word can have different meanings; the correct meaning depends on the context. The word sense disambiguation indicates the procedure needed to assign the correct meaning to a specific word in a specific context. We have decided to apply a complete 'blind' analysis, because we want to investigate the capability of an unsupervised neural classifier system to autonomously find a rough semantic structure without the support of any additional information.

The linguistic pre-processing requires making the following operations:

1. Text segmentation and 'stopword' removal;
2. Stemming and word count;

3. Word sense disambiguation;
4. Anchor term selection; and
5. Document matrix creation.

First of all, we performed the abstract segmentation in order to obtain all the words.

We eliminated the irrelevant terms, the 'stopwords' (articles, conjunctions, pronouns), that can affect the frequency count. For our application, we created with the support of a molecular biologist and a linguistic expert, a specific 'stopword' list that contained a larger set of words that were not relevant for biomedical literature.

On the 'cleaned' segmented text, we applied a stemming procedure based on the Porter's algorithm (Porter, 1980) to remove the prefix and suffix from a word, in order to obtain its root.

On the basis of the retrieved documents, we created a local dictionary of words. In NLP field, a word is defined as 'anchor', if it characterizes a document. The selection of anchor terms allows the reduction of the dimensionality of the document matrix. In order to select the 'anchor' terms, we compute for each token the Term Frequency/Inverse Document Frequency (TFIDF; Salton, 1983). There are different approaches to compute the *TFIDF*; here, we applied a more standard formula given below:

$$TFIDF(w) = f_w \log(N/\#(w)). \tag{1}$$

Here, f_w indicates the occurrence frequency of the word w and the parameter N stands for the total number of abstracts in the retrieved set. The $\#(w)$ indicates the number of abstracts that contain a specific term w. In order to obtain a reduced set of terms, we introduced a threshold value. This constrain allowed us to reduce the dimensionality of the Vector Space Representation (VSR) of an abstract. The current threshold was selected by using the following heuristic approach. We select the words that has a *TFIDF* that lies in the range defined below as

$$[|N|/10] < TFIDF(w) < [|N| - |N|/4]. \tag{2}$$

Here, $|N|$ indicates the cardinality of the set of retrieved documents (total number of PubMed abstracts) and $TFIDF(w)$ is the value computed by Eq. (1).

The resulting set of terms is subsequently used to create the VSR of each document. The VSR is a well-established method to convert the textual document into a more useful format for machine-learning applications. In the VSR, each document is represented by a d-dimensional feature vector $v = \{t_1, \ldots, t_d\}$. Each vector location represents the *TFIDF* of a specific term

(t_d). This conversion originates a *Nxd* document matrix (***M***). This matrix constitutes the training set for the Neural Network classifiers. Taking into account the Conceptual Clustering perspective, we applied self-organizing maps (Kohonen, 1999) instead the commonly used clustering methods. The Document matrix, obtained by linguistic analysis, is the training set for the SOM. In the past, we applied the SOM for the biosequence analysis (Arrigo et al., 1991); now, we are using this method for conceptual clustering. For this application, we use a 2D-lattice. The network topology is computed automatically according the dimension of the training set. One of the major difficulties in the application of SOM is the possibility to extract the representative patterns for each cluster.

The classification phase originates a set of clusters, represented by the activated computational elements (nodes). These elements constitute the set $C^0 = \{c_1, \ldots, c_k\}$. This set defines the lowest level of the hierarchy (zero level is identified by the superscript 0). Taking into account the weight vectors at the end of the classification phase, we extract the representative term vectors from each activated element of the set C^0. Each representative is subsequently binarized. We obtain a new set of binary patterns (B) with the cardinality $|C^0|$. In order to convert the original floating point vectors into binary, we applied the following rule:

$$f_d = \begin{cases} 1, & \text{if} \quad w(f_d) > \theta \\ 0, & \text{otherwise.} \end{cases} \tag{3}$$

Here, f_d is the value of binarized feature and $w(f_d)$ is the original floating point value of the feature d. The value θ defines the fixed binarization threshold. This parameter reflects the relative relevance of each feature. The binary vectors represent the combination of a high relevant term for each level of C^0 cluster. An agglomerative clustering procedure is applied on the C^0 set. We selected a single-link clustering approach. The single-link agglomerative clustering is founded on the iterated calculation of pairwise similarity; in our case, we create a similarity matrix for the C^0 clusters. Many similarity measures are suitable for agglomerative clustering. For our application, we selected a set-based measure, the Tanimoto similarity. The main advantage of this measure, with respect to the Hamming distance, is that it overcomes the problem of different vector lengths. For a binary case, the Tanimoto similarity is expressed in the following way:

$$S_T = (n_{x \cap y})/(n_x + n_y - n_{x \cap y}). \tag{4}$$

The Hamming distance gives the distance taking into account the differences between two vectors. Instead, the S_T takes into account the 'rate' between the shared and not-shared features between the vectors. The system performs the agglomerative phase taking into account the previously defined similarity measure. Two C^0 clusters are merged together, if S_T is over a predefined threshold (the default value is 75 %). The merging phase proceeds until the previous condition is satisfied. We obtain the set C^1 of high-level clusters. The agglomerative procedure is iterated on the high-level cluster set until the above-described similarity condition is satisfied.

The software prototype is a client–server system. It is implemented by using JAVA under LINUX RED HAT version 9.0 operating system.

The current version of the output visualization is written in HTML format. The HTML output allows the user to click on the MEDLINE_ID code to view the requested abstract. In addition, the user can add one of the displayed terms to the query string in order to refine the original query; this ensures obtaining an interactive abstract selection procedure. A prototype of the system is currently available at http://biocomp.ge.ismac.cnr.it/.

Figures 2, 3 show the input form and the output visualization of our text-mining system.

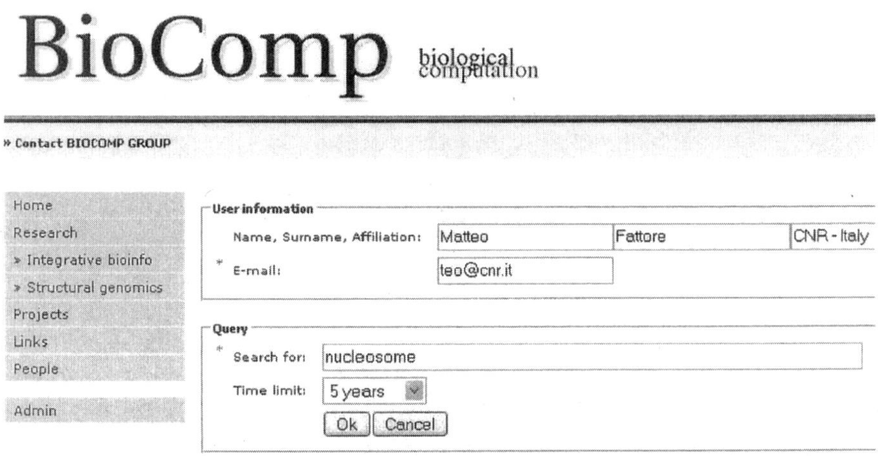

Figure -2. A screenshot of the input form.

3. RESULTS AND DISCUSSION

As it was previously explained in the paper, our aim is the development and implementation of a textual mining service on the web than can allow the user to retrieve and organize the PubMed abstracts in a *conceptually* homogeneous way. The proposed system seems to be able to point out dependencies of terms embedded in the abstracts. The identification of this kind of correlations permits to obtain a rough conceptual organization of the documents. As an example, we performed the conceptual classification of a very general query based on the term '*nucleosome*' (2601 abstracts). The box below shows the set of rough conjunctive rules that involve the combination of more relevant anchor terms. The user's query is the antecedent of the rule; the arrow identifies the implication; and the terms on the right are the consequent.

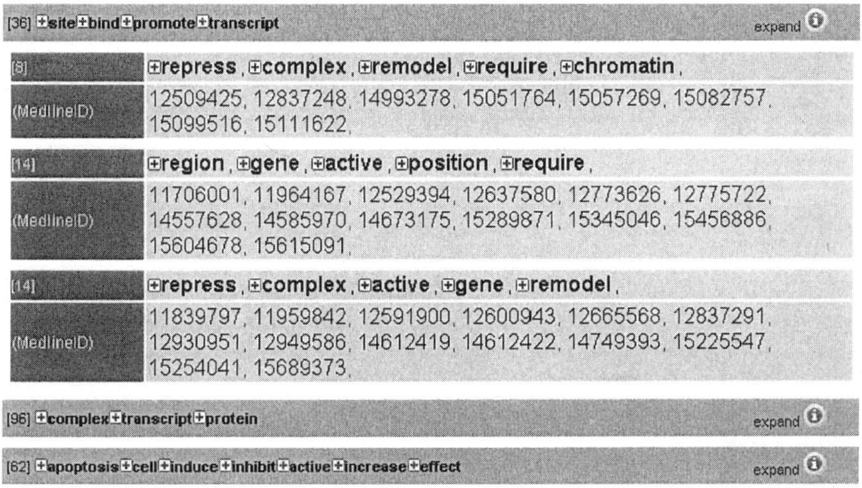

Figure -3. A screenshot of an output. The image shows the two-level clusters (dark grey bars indicate high-level clusters; light grey bars, low-level clusters). In each bar, the main relevant term is shown. The cross near each term allows the user to add a new term to the query. The retrieved abstracts are represented by their PMID code.

Rough first-level conjunctive forma for 'Nucleosome' query (2601 abstracts)

1 Nucleosome→ [promote & transcription & active & region]
2 Nucleosome→ [Chromatin]
3 Nucleosome→ [promote]
4 Nucleosome→ [remodel & chromatin & complex]
5 Nucleosome→ [DNA & core & histone & structure]
6 Nucleosome→ [Assembly & chromatin & histone & complex]
7 Nucleosome→ [Apoptosis & cell & increase & inhibit & induce & active]
8 Nucleosome→ [bind & access & DNA & target]

The current version of the prototype allows obtaining only the first-level rough rules. In order to complete the conceptual agglomerative procedure, we are implementing the possibility to extract the rules that combine conjunctive and disjunctive forms. The results presented here highlight the potentiality of self-organization in the NLP.

ACKNOWLEDGMENTS

This research was supported by the project 'Oncology over Internet'. This project was funded by the Italian Ministry of Education, University, and Research (MIUR) for the period 2002–2005.

SOFTWARE FOR ANALYSIS OF GENE REGULATORY SEQUENCES BY KNOWLEDGE DISCOVERY METHODS

E.E. Vityaev[1, 3*], T.I. Shipilov[2], M.A. Pozdnyakov[1], O.V. Vishnevsky[1, 2], A.L. Proscura[1], Yu.L. Orlov[1], P. Arrigo[4]

[1] Institute of Cytology and Genetics, Siberian Branch of the Russian Academy of Sciences, prosp. Lavrentieva 10, Novosibirsk, 630090, Russia; [2] Novosibirsk State University, ul. Pirogova 2, Novosibirsk, 630090, Russia; [3] Sobolev Institute of Mathematics, Siberian Branch of the Russian Academy of Sciences, prosp. Koptyuga 4, Novosibirsk, 630090, Russia, e-mail: vityaev@math.nsc.ru; [4] ISMAC, via De Marini 6, 16149 Genova, Italy

[*] Corresponding author

Abstract: Application of knowledge discovery techniques to search for and analysis of regularities in the context signals in DNA sequences involved in transcription regulation is described. The software developed allows for interactive construction and visualization of complex signals for samples of DNA regulatory sequences and evaluation of their statistical significance. The complex signals in DNA comprise simple context signals (oligonucleotides), the signals with specified localizations in promoters, predicted and experimentally detected transcription factor binding sites, and other user-specified signals. The regularities are statistically assessed by first-order logic with probabilistic estimates. The computer tool developed was tested using samples of nucleotide sequences of eukaryotic gene promoters extracted from databases according to tissue specificity, type of gene regulation, or joint expression in a functional system. A set of conservative context signals relating the nucleotide sequences and gene functional class was found. The software is available by request to the corresponding author.

Key words: eukaryotic promoter; recognition; transcription factor binding sites; knowledge discovery; data mining

1. INTRODUCTION

Analysis of gene regulatory sequences is a challenge to the theory of data mining and machine learning. Promoters in eukaryotic genomes act as the molecular 'switches' that turn genes on and off (Zhang, 1998). They contain transcription factor binding sites (TFBS)—short stretches of DNA sufficiently conserved to provide a specific recognition by the corresponding protein. The presence and location of transcription factor binding sites in regulatory regions of genes correspond to the tissue- and stage-specific features of gene expression in an organism. The problem arises to both determine the TFBS in a DNA sequence and predict location of promoter, which determines regulation of gene expression, according to the localization pattern of particular TFBS (Qiu, 2003).

The problem of TFBS prediction is computationally difficult due to a tremendous diversity of the experimentally detected protein-binding nucleotide sequences compiled in databases (Zhang, 1998; Ohler and Niemann, 2001). Several machine learning approaches to TFBS prediction are developed, including statistics, Markov models, and neural network predictors (Ohler and Niemann, 2001; Vityaev et al., 2001; 2002; Thijs et al., 2002).

Positional weight matrix is a traditional model for predicting binding sites (Chen et al., 1995). Use of the predicted TFBS as signals in nucleotide sequence forms the background for forecast of gene promoters and further determination of the gene function connected with the manner how its transcription is regulated (Vityaev et al., 2001; Kolchanov et al., 2003).

Although many computational methods consider identification of individual transcription factor binding sites, very few of them focus on the analysis of mutual location and interactions between these sites (Cartharius et al., 2005). For example, several approaches are developed to search for clusters of sites in promoters (Zheng et al., 2003; Yu et al., 2004).

Nonetheless, such software is able to give only a general estimate on the degree of randomness in localization of a given number of sites; however, these programs take into account neither the overall diversity of specific features of site localizations nor other characteristics potentially important for regulation of gene transcription. In particular, the latter signals include polytracts, the regions capable of forming noncanonical DNA structures, and sites of preferred nucleosome formation (Orlov et al., this issue).

The main goal of this work is to predict the gene function by using a set of integrated methods for recognition of regulatory elements and transcription factor binding sites. The computer program developed searches for the regularities in locations of binding sites in a regulatory region. For this purpose, the set of context characteristics (specific features) and potential (predicted) TFBS should be first detected for the contrast training sample of promoter sequences. The features considered include computationally predicted binding sites in the sequence analyzed (predicted basing on the consensus, weight matrices, and homology to known sites stored with databases), specific oligonucleotides, low complexity regions, and several other characteristics.

A distinctive feature of our approach is use of specific feature patterns describing a subgroup of the training set (Vityaev et al., 2001). The search patterns for regularities are constructed in the first-order logic augmented by probabilistic estimates. The program is written in C++ and supplied with a user-friendly interface.

We analyzed sets of promoter nucleotide sequences extracted from TRRD and EMBL databases according to tissue-specificity and type of gene regulation. A set of regularities relating the nucleotide sequences and gene functional class were found.

2. SYSTEMS AND METHODS

An interactive system ExpertDiscovery was developed; it allows the user (an expert biologist) to construct hierarchically the complex signals, visualize locations of these signals in DNA sequences, and determine statistical parameters of the signals in a contrast data sample analyzed. Complex signals are determined recursively basing on the primary signals. The following signals may represent the primary signal *S:*

1. A potential functional site predicted by the homology (or weight matrix) with annotated sequences in specialized molecular biological databases (Pozdniakov et al., 2001);
2. A context signal, i.e., a sequence of symbols in a 15 single letter–based code;
3. A site with conserved conformational or physicochemical features (i.e., double-helix angle twist or DNA melting temperature; Oshchepkov et al., 2004);
4. A secondary structure element (Z-DNA or RNA hairpin); and
5. A low complexity region (polytracks); (Orlov and Potapov, 2004).

The complex signal is described hierarchically as

- The primary signal itself;
- Orientation of the signal (direct, symmetrical, or inverted);
- Repetition of the signal N times ($2 \leq N_{min} \leq N \leq N_{max}$); the distance between neighbor copies of the signal falls into a user-specified range;
- Occurrence of the complex signal in a certain range relative to transcription start (or the beginning of a phased sequence); and
- An ordered pair of the complex signals S_1 and S_2 with the distance between them varying in a particular range.

The binding sites can be known and annotated in databases or predicted by certain external software. A universal format of signal layout representation is used to input information about sites and any other context signals into the system ExpertDiscovery. An example of formal TFBS presentation as a signal is given in the scheme below:

```
<Matrics_TFBS>
 Signal_Number 272
 <Signal 1>
name 103_AP2_AS00103
TAGAAAGCCCCGGT
   method_name weight_matrics
 </Signal 1>.
```

The system ExpertDiscovery allows the user to

- Input data as a positive and negative samples of DNA sequences;
- Input layout of external signals for the samples in question;
- Specify complex signals in an interactive manner using signal editor and obtain visual pattern of signal localization in DNA sequences and statistical significance of the signal in question;
- Edit any complex signal altering its parameters and involving additional primary signal;
- Detect automatically the statistically significant complex signals specifying beforehand the predicate operations for the search to be performed (for example, all the pairs or all the triplets of primary context signals); and
- Save the project and the relevant data for further expert work.

Figure 1 exemplifies the interface of ExpertDiscovery.

Promoter sequences were extracted from TRRD database (Kolchanov et al., 2002) and divided into several groups according to their tissue specificities (endocrine system genes, cholesterol homeostasis, heat shock response system, interferon-regulated, glucocorticoid-regulated, and cell cycle genes). The negative sets, containing the sequences that are not promoters, were constructed using genomic data (exons and noncoding nonregulatory regions). In addition, randomly generated sequences with the same nucleotide frequencies as in the positive set sequences were used as a negative set.

3. RESULTS AND DISCUSSION

The oligonucleotides specific of promoter regions detected by the program ARGO (Vityaev et al., 2001) and the TFBS predicted by the weight matrices constructed using TRRD database were the primary signals. The predicted sites were involved because the number of experimentally confirmed sites is insufficient for a large-scale statistical analysis. Expert Discovery succeeded in finding numerous regularities for joint presence of context signals in promoter regions.

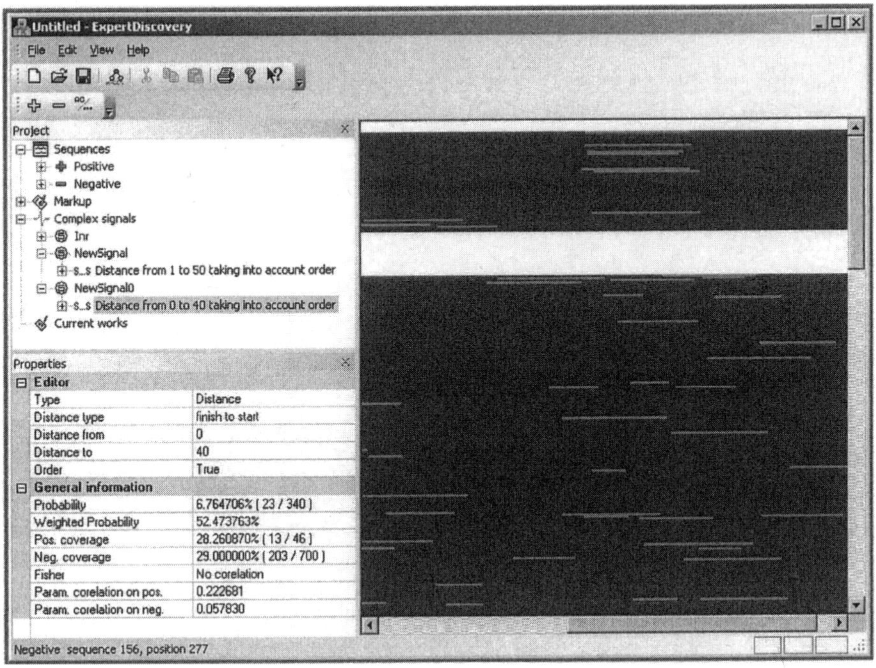

Figure -1. Interface of the system ExpertDiscovery. The right panel represents visualization of TATA box location in promoters of genes of the endocrine system.

The resulting regularities are stored in ExpertDiscovery as complex signals in an IF–THEN form:

$$(A_1\&\ldots\&A_k) \Rightarrow A_0,$$

where the IF part, $A_1\& A_2\&\ldots\&A_k$, consists of true/false logical statements $A_1, \ldots, A_k$ concerning the presence of context features (potential TFBS) in a sequence and the THEN part consists of a single logical statement A_0 concerning promoter class and gene function.

Oligonucleotides in a 15 single letter–based code may, for example, serve as context signals A_k. The record of a complex signal is simultaneously a human-readable forecasting rule, which in text format is accepted also by computer program.

Consider an example of a complex signal (YCTNNYTS, DRVSCAG, WTAWWWR) found by the program in promoters of lipid metabolism genes (Figure 2).

As a rule, complex signals characterize a number of sequences (a subset of sequences from a set); in this case, nine sequences (Figure 2). The probability to get a signal due to a random cause is indicated as well as the sequences wherein the signal is located.

```
Regularity 203
IF YCTNNYTS =  1    (Fisher criterion 0.004058)
AND NDRVSCAG =  1    (Fisher criterion 0.005992)
AND WTAWWWRN =  1    (Fisher criterion 0.020397)
THEN Class =  1   (with frequency 9 / (0 + 9) = 1.000000)
Regularity apply to objects: 2(+) 5(+) 16(+) 17(+) 24(+) 26(+)
31(+) 37(+) 50(+)
```

Figure -2. Signal {YCTNNYTS, NDRVSCAG, WTAWWWRN}. Output of the program.

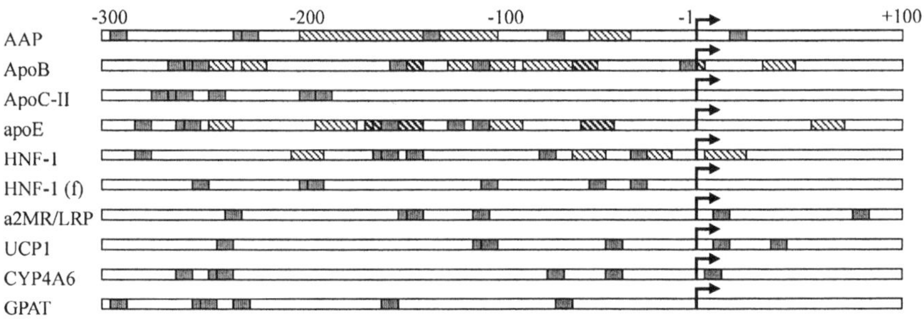

Figure -3. An example of location pattern of potential binding sites (grey rectangles) in gene promoter sequences phased relative to the transcription start (arrow). Genes: human AAP, human ApoB, human apoC-II, human apoE, rat HNF-1, clawed frog HNF-1, human alpha2MR/LRP, rat UCP1, rabbit CYP4A6, and mouse GPAT.

Oligonucleotides may be used for localizing the binding sites in promoters; however, layout of potential binding sites obtained using weight matrices is a more accurate signal. Such layout of the potential TFBS in promoters was made involving the weight matrices constructed using the ArtSite database (Khlebodarova et al., 2005, this issue). An example of a location pattern of potential binding sites in promoters of endocrine system genes is shown in Figure 3. The promoter sequences are aligned relative to the transcription start (position +1 bp), indicated by arrows. Gene names are given to the left.

Potential transcription factor binding sites composing the complex signal are shown as gray rectangles; hatched rectangles indicate those annotated in TRRD databases.The complex signal shown in Figure 3 comprises six potential binding sites, some of which are repeated. The formal signal representation is (239_USF)&(239_USF)&(240_USF)&(323_SPZ1)&(240_USF)&(239_USF).

Here, 239_USF is the weight matrix for USF1 (upstream stimulatory factor 1), determined according SELEX experiment (Khlebodarova et al., 2005, this issue); the protein binds to DNA as a homodimer; 240_USF is a weight matrix for USF1 (matrix for the second half-site); and 323_SPZ1 is the weight matrix for the transcription factor SPZ1 (spermatogenic Zip 1).

Thus, the system ExpertDiscovery helps construction and detection of such complex signals and regularities in promoter regions that cannot be formulated in terms of one of the used approaches (for example, only a group of nucleotides or a pair of binding sites). Functional meaning of the signal could be treated in terms of the transcription factor binding sites or DNA conformational properties. Our study demonstrates that not only pairs of context signals, but also triplets, quadruplets, and larger sets of ordered context signals, which may correspond to groups of jointly functioning transcription factors, are statistically significant.

The regularities found could be analyzed by a molecular biology expert as unique complex signals that are essential for proper promoter functioning. The research suggested that functional promoter modules could be detected by formal models independently of the degree of homology between sequences (Klingenhoff et al., 1999).

This data mining approach is applicable to analysis of the context gene structure at all levels of gene hierarchy: promoter, regulatory regions, and transcription factor binding sites (Liu and Wong, 2003). The algorithm is flexible enough to search for structural patterns that are typical of a whole set of sequences as well as a subset of sequences.

Study of the regulatory regions with the help of contextual signals and predicted transcription factor binding sites may be supplemented with methods of comparative genomics (Dieterich et al., 2005).

ACKNOWLEDGMENTS

The authors are grateful to E. Bedareva, O.A. Podkolodnaya, and N.A. Kolchanov for scientific discussion. This work was supported in part by the Russian Foundation for Basic Research (grants Nos. 05-04-08084,

05-04-49111, 05-07-90185, 05-07-98012, 03-04-48506, 03-07-90181, and 03-04-48469), Siberian Branch of the Russian Academy of Sciences (integration project No. 119), Presidium RAS Program on molecular physicochemical biology (grant No. 10.4), Program for Development of Science Potential of Higher School (grant No. 8319, 308-05), Novosibirsk State University (grant No. 3H-308-05), and NATO (grant No. LST.CLG 979815).

A MATHEMATICAL MODEL OF THE DISCONTINUOUS GENETIC STRUCTURES FIXATION PROCESS

E.Ya. Frisman[1], O.L. Zhdanova[2*]

[1] Complex Analyses of Regional Problems Institute FEB RUS, ul. Sholom-Aleyhem 4, Birobidzhan, 679016, Russia; [2] Institute for Automation and Control Processes FEB RUS, ul. Radio 5, Vladivostok, 690041, Russia, e-mail: axanka@iacp.dvo.ru

[*] *Corresponding author*

Abstract: An integral model of the evolution of a Mendelian one-locus population of diploid organisms with continual allele diversity developing under density-limiting conditions or without density limitation was proposed and analyzed. The model was used to study the mechanism of the appearance of discrete genetic structures, i.e., the fixation of a limited number of alleles. Local resistance of the resultant genetic distributions to homogeneous mutations was demonstrated.

Key words: mathematical model; computer analysis; fitness; genetic distributions; mutations

1. INTRODUCTION

To explain the mechanisms of formation of discrete biological taxa is one of the main problems of evolutionary theory. Although the famous work by Charles Darwin is entitled *On the Origin of Species*, neither this work nor synthetic evolutionary theory, which is based on both Darwinism and contemporary genetics, cannot satisfactorily answer the questions as to why the entire biological diversity is ultimately discrete and why life exists in the form of genetically isolated species, with practically no transitional forms left between them. Moreover, the genetic diversity within a species is also often discrete and strictly limited. The reason is hardly the discreteness of the 'heredity carrier' itself, i.e., DNA consisting of monomers. A protein consists of several hundred amino acids. Mutational variation may yield a vast diversity of molecules of a given protein, with most of these molecules functioning normally (Altukhov, 2003). However, only one form of a given protein is usually fixed in the populations frequently. Two forms of a protein are seldom fixed; three forms, even more seldom; etc. What is the mechanism of fixation of some alleles and loss of others? There are two main hypotheses answering this question: (1) a random loss of alleles because of gene drift and (2) a balanced polymorphism determined by the selective advantage of heterozygotes. Both of them have supporters and opponents (Crow and Kimura, 1971; Lewontin, 1974; Altukhov, 2003), but neither provides a definite solution to the problem. Our study is one more attempt to analyze this issue.

2. METHODS AND ALGORITHMS

Let us consider a large Mendelian panmictic sexless population in which the inheritance of a certain character is determined by one gene with m alleles. The genetic-structure and population dynamics in this population can be described by the following simultaneous equations:

$$\begin{cases} x_{n+1} = \overline{W}_n x_n \\ q_{i,n+1} = q_{i,n}\left(\sum_{j=1}^{m} W_{ij} q_{j,n}\right) / \overline{W}_n, \\ \overline{W}_n = \sum_{i=1}^{m} \sum_{j=1}^{m} W_{ij} q_{i,n} q_{j,n}, \end{cases} \tag{1}$$

where n is the ordinal number of the generation, x_n is the population size, $q_{i,n}$ is the frequency of the ith allele, W_{ij} is the fitness of the genotype ij, and $\overline{W}_n$ is the population mean fitness in the nth generation (see, e.g., Svirezhev and Pasekov, 1982).

If the locus in question has an infinite number of alleles, then it would be reasonable to replace sums by integrals and to go on from allele frequencies to the density of allele frequencies. Thus, the population dynamics is described by a set of simultaneous discrete–continuous equations (by analogy with Tuzinkevich, 1980):

$$\begin{cases} x_{n+1} = \overline{W}_n x_n, \\ q_{n+1}(\tau) = q_n(\tau)\left(\int_0^1 W(\xi,\tau) q_n(\xi) d\xi\right) / \overline{W}_n, \\ \overline{W}_n = \int_0^1 \int_0^1 W(\xi,\tau) q_n(\xi) q_n(\tau) d\xi d\tau. \end{cases} \tag{2}$$

Here, the function $q_n(\tau)$ is the frequency density of allele τ in the population in the nth generation, and τ may be any real number within the interval [0, 1]. In fact, we approximate a finite-dimensional situation by an infinite-dimensional one (Gorban' and Khlebopros, 1988) in order to use all the possibilities of analysis of continuous functions.

2.1 The integral model of a Mendelian one-locus population with density-dependent selection

If density-dependent selection takes place in the population, then fitnesses are decreasing functions of the population size. The exponential dependence of fitness on population size is suitable for analysis. This dependence can be written as follows:

$$W(\xi,\tau,x_n) = \exp(R(\xi,\tau)(1 - x_n / K(\xi,\tau)). \tag{3}$$

In this case, each genotype is characterized by two parameters $R(\xi, \tau)$ and $K(\xi, \tau)$—the Malthusian and the resource parameters, respectively (Evdokimov, 1999).

The presence of the density-dependent component of natural selection has been shown experimentally and found in natural populations. In the early studies in this field, Birch (1955) demonstrated experimentally that the equilibrium frequencies of inversions *Standard* and *Chiricahua* in chromosome 3 of *Drosophila pseudoobscura* strongly depended on the larval density.

We considered the distribution of allele frequencies $q_n(\tau)$ to be correct, if it satisfied the following conditions:

$$\begin{cases} \int_0^1 q_n(\tau)d\tau = 1 \\ q_n(\tau) \geq 0, \forall \tau \in [0,1]. \end{cases} \tag{4}$$

Obviously, the value of population size x_n is correct, if it is simply a non-negative value.

It is possible to obtain the following analytical results (Frisman and Zhdanova, 2003):

Lemma 1. The evolution described by set of simultaneous equations (2) transforms a correct set of genetic composition distributions and of population size into correct ones.

Lemma 2. If $\int_0^1 W(\xi, \tau, x_n) d\xi = \Psi(\tau, x_n) \neq f(x_n)$, where $f(x_n)$ is the function of population size alone and does not depend on τ, then the set (2) has no continuous stationary distributions of allele frequencies density $q_n(\tau)$.

Therefore, it can be expected that the evolution of set (2) will result in the transformation of uniform continuous density distributions of allele frequencies into vastly not uniform ones, provided there is a diversity of fitness values.

In genetic terms, $\int_0^1 W(\xi, \tau, x_n) d\xi$ is the integral sum of the fitnesses of all genotypes containing allele τ. It would be reasonable to assume that this integral is a function of τ in the case of a fixed population size, because there are indeed both advantageous and deleterious alleles in real populations. For example, let allele τ_i be a deleterious dominant mutation; then, at a fixed population size x, the inequality $\int_0^1 W(\xi, \tau_1, x) d\xi < \int_0^1 W(\xi, \tau_2, x) d\xi$, where τ_2 is one of normal alleles, is true. If we assume that $\int_0^1 W(\xi, \tau, x_n) d\xi = f(x_n)$, then this summarized equality of the contributions of all alleles to fitness could only be explained by either a complete absence of deleterious dominant mutations or expression of all these alleles only in the post-reproductive period (like Huntington's chorea in humans), as well as monogenic heterosis for all recessive lethal alleles. However, these assumptions do not agree with facts. For example, one-gene heterosis is very rare in natural populations (Lewontin, 1974; Altukhov, 2003); a proportion of mutations of some loci is eliminated from populations very rapidly (Altukhov, 2003), which corresponds to directional selection where the fitness of the heterozygote is intermediate between the fitnesses of the homozygotes. On the basis of these considerations, we further studied the dynamics of model (2) for genotype fitnesses meeting the condition $\int_0^1 W(\xi, \tau, x_n) d\xi = \Psi(\tau, x_n) \neq f(x_n)$. The dynamics of set (2) was studied numerically. We analyzed the population dynamics and changes in population genetic structure with time for different variants of the initial distributions of the density of allele frequencies and fitness functions.

2.2 The integral model of a Mendelian one-locus population without density limitation

The special case of the integral dynamic model of a one-locus diallelic population with fitnesses independent of the population size, i.e., in the absence of density control, is of special interest. This model is a logical generalization of the classical model of the dynamics of a one-locus diallelic population with constant fitnesses of genotypes (Ratner, 1977; Frisman, 1986), extending it to the case when there is a continual number of alleles of one locus. In the absence of density limitation, the population dynamics is of no interest, because it will either infinitely grow (if $\bar{W} > 1$) or constantly decrease (if $\bar{W} < 1$). Therefore, let us consider separately the dynamics of allele frequency density:

$$q_{n+1}(\tau) = q_n(\tau)\left(\int_0^1 W(\xi,\tau)q_n(\xi)d\xi\right) / \int_0^1\int_0^1 W(\xi,\tau)q_n(\xi)q_n(\tau)d\xi d\tau. \quad (5)$$

As in the previous case, it is easy to show that the evolution described by Eq. (5) transforms correct distributions of allele frequency density into correct ones. In addition, if the genetic space is heterogeneous with respect to the fitnesses $\left(\int_0^1 W(\xi,\tau,x_n)d\xi = \Psi(\tau,x_n) \neq f(x_n)\right)$, then any continuous density distribution of allele frequencies is not stationary.

Thus, the results of our analytical study allow us to expect that evolution (5) will transform the continuous density distributions of allele frequencies density into almost discrete ones, provided that there is a diversity of fitness values $\left(\int_0^1 W(\xi,\tau)d\xi \neq \text{const}\right)$. Then, we analyzed the behavior of model (5) numerically.

2.3 Effect of mutations on the dynamics of the integral model of an unlimited population

Mutation, as well as natural selection, is one of the main factors of evolution, supplying it with elementary material. Therefore, any evolutionary model that does not take into account mutation seems somewhat artificial. Hence, although simulation taking into account mutation is often causing mathematically challenging, we have taken the risk of adding some homogeneous mutations into model (5).

Let mutations occur before selection; then, the allele frequency density in the *n*th generation after mutation takes the form

$$\tilde{q}_n(\tau) = \int_0^1 q_n(\xi)\mu(\xi,\tau)d\xi, \quad (6)$$

where $\mu(\xi,\tau)$ is the probability density of the mutation from ξ to τ. After this, selection takes place:

$$q_{n+1}(\tau) = \tilde{q}_n(\tau)\left(\int_0^1 W(\xi,\tau)\tilde{q}_n(\xi)d\xi\right) / \\ / \int_0^1\int_0^1 W(\xi,\tau)\tilde{q}_n(\xi)\tilde{q}_n(\tau)d\xi d\tau \quad (7)$$

Assume that mutations are homogeneous:

$$\mu(\xi,\tau)=\begin{cases}\delta, & \forall\xi\neq\tau\\ 1-\delta, & \forall\xi=\tau\end{cases},\quad \xi,\tau\in[0,1]. \tag{8}$$

Then, Eq. (6) may be rewritten as

$$\tilde{q}_n(\tau)=(1-\delta)q_n(\tau)+\delta. \tag{9}$$

We performed numerical analysis of the dynamic behavior of the set of simultaneous equations (7), (9) at a fixed value of $\delta = 0.1$.

3. RESULTS AND DISCUSSION

3.1 Stable stationary distributions in the case of density-dependent selection

We analyzed model (2) of the population dynamics and changes in the population genetic structure with time for various choices of the initial distributions of allele 'frequencies' and fitness functions in the form (3). If fitnesses $W(\xi, \tau, x_n)$ depended on both genetic variables (ξ, τ), then the stationary distributions of genetic composition were actually almost discrete. Note that it is possible that only a few or even one of the numerous alleles will be preserved in the course of evolution (Figure 1, panel *a*). There are distributions with two and three peaks (Figures 1 and 2, panels *b*). In these cases, the population size may either vary or remain constant.

3.2 Stationary distributions in the absence of density limitation and the effect of mutations

Introduction of mutations into the model in the cases where the fitness function has the form $W(\xi, \tau) = 0.1 + (\xi-\tau)^2$, $W(\xi, \tau) = 1.6 - (\xi-\tau)^2$, or $W(\xi, \tau) = \xi + \tau$ has practically no effect on the pattern of the distributions (Figures 3, 4, and 5, respectively), except that the mutations slightly 'smeared' the almost discrete distribution.

Figure 6 shows the selection determined by the fitness function $W(\tau, \xi) = 1.5 + 1.5\sin(8\pi\tau\xi)$; in this case, the system is sensitive to the choice of initial conditions even without mutations.

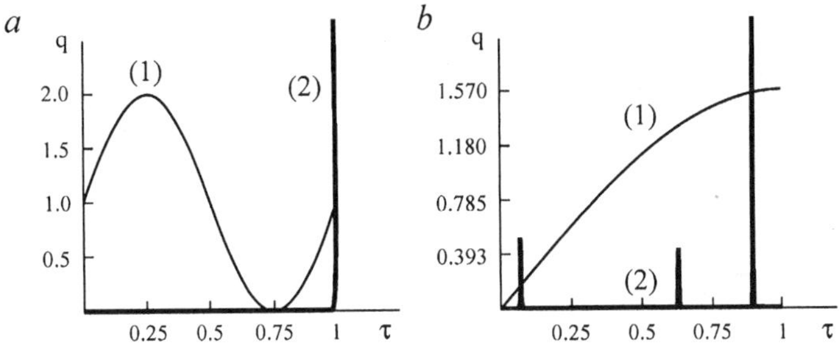

Figure -1. Curves (1, thin line) show the initial density distributions of allele frequencies. Curve (2, bold line) in panel *a* shows the density distribution of allele frequencies after 550 generations of evolution described by the formula $W(\xi, \tau, x_n) = \exp(1 + \xi + \tau - x_n)$ (monomorphism). Curve (2, bold line) in panel *b* shows the density distribution of allele frequencies after 30 050 generations of evolution described by the formula $W(\xi,\ \tau,\ x_n)$ $\exp(1,5 + 1,5\sin(8\pi\xi\tau) - x_n)$; the pattern resembles that of a stable triallelic polymorphism, but two of these alleles are extremely rare.

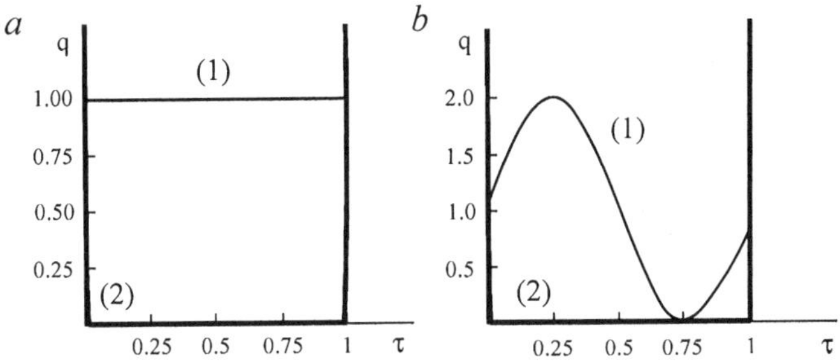

Figure -2. Stable polymorphism of two alleles, allele 0 and allele (1) $W(\zeta,\ \tau\ x_n) = \exp(0.5 + (\tau - \zeta)^2 - x_n)$. Curves (2, bold line) show the density distributions of allele frequencies after 550 generations of evolution in the cases of different initial distributions shown by curves (1, thin line).

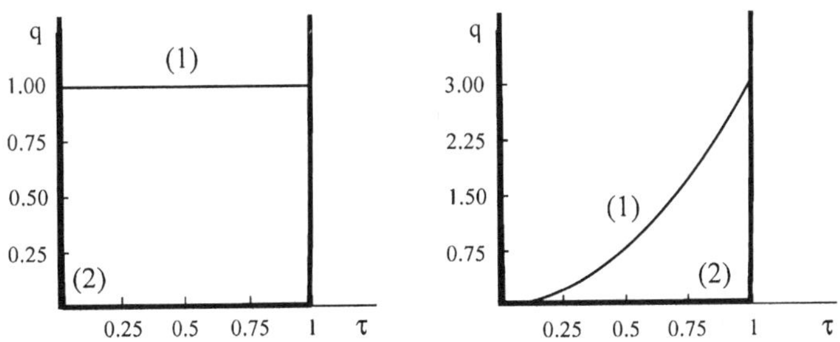

Figure -3. Diallelic polymorphism. $W(\xi,\ \tau) = 0.1 + (\xi - \tau)^2$. Curves (2, bold line) show the density distributions of allele frequencies after 300 generations of evolution in the cases of different initial distributions shown by curves (1, thin line).

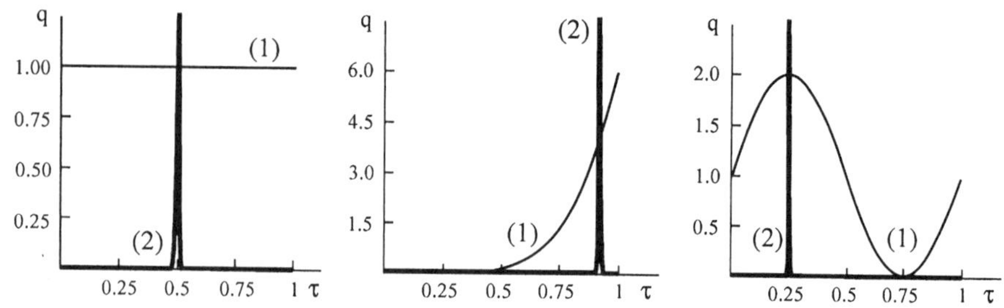

Figure -4. Monomorphism. $W(\xi, \tau) = 1.6 - (\xi-\tau)^2$. Curves (2, bold line) show the density distributions of allele frequencies after 2000 generations of evolution in the cases of different initial distributions shown by curves (1, thin line).

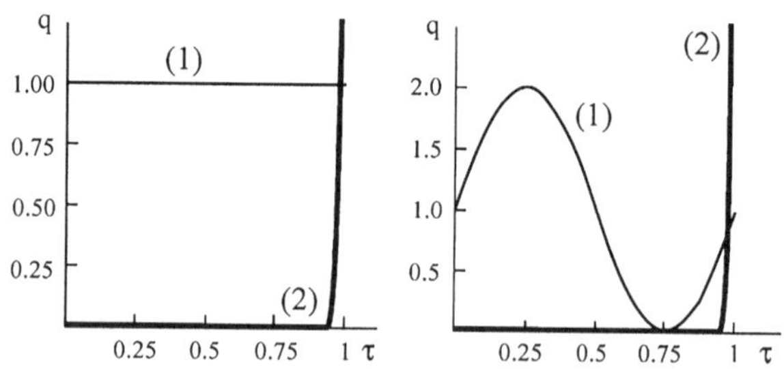

Figure -5. Monomorphism. $W(\xi, \tau) = \xi + \tau$. Curves (2, bold line) show the density distributions of allele frequencies after 300 generations of evolution in the cases of different initial distributions shown by curves (1, thin line).

Mutations can slightly smear the 'discrete' distribution (Figure 6, the top row of plots) and, in addition, increase the size of the peak (Figure 6, the second row from the top) and considerably change the distribution pattern (Figure 6, the third and fourth rows). Note, however, that markedly heterogeneous distributions with small numbers of peaks are still observed. Taking into account that the mutation rate that we set in the model ($\delta = 0.1$) is substantially exaggerated compared with the actual value in natural populations (10^{-5}), the peaks seen in the plots may be considered almost discrete. Note that mutations have increased genetic diversity wherever possible (Figure 6, the second to fourth rows).

3.3 Results and Conclusions

Let us compare the results of our numerical experiment with the existing evolutionary theories and some factual data on the allele frequencies in natural populations.

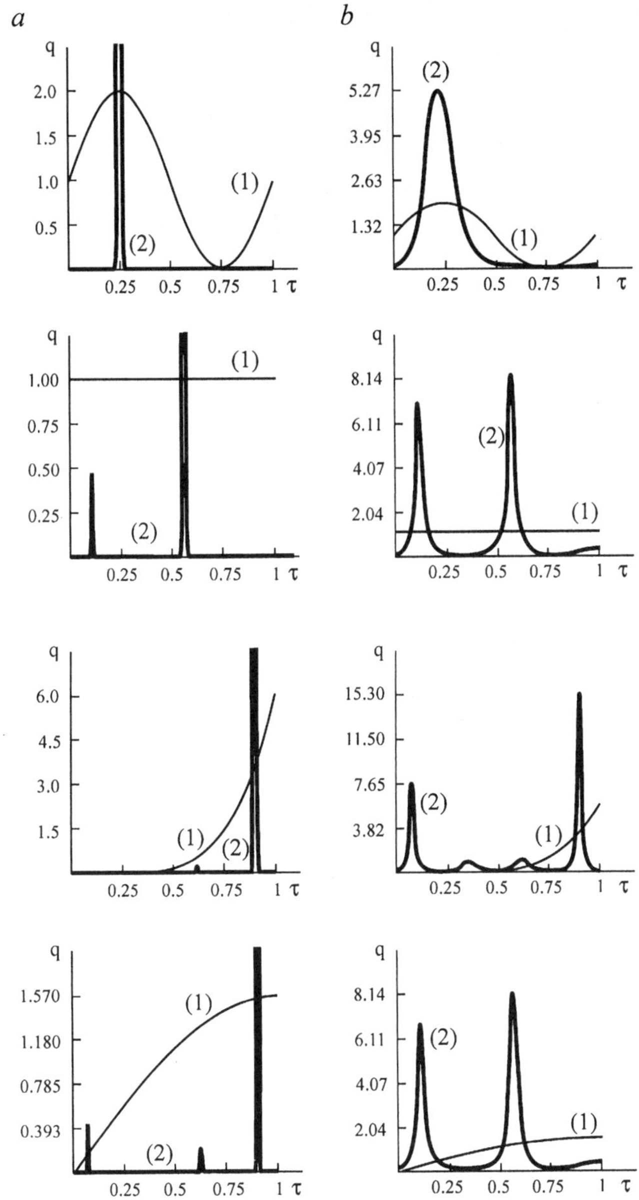

Figure -6. $W(\tau, \xi) = 1.5 + 1.5\sin(8\pi\tau\xi)$. Curves (1, thin line) and (2, bold line) show the initial density distributions of allele frequencies and their distributions after 1500 generations of evolution, respectively. Plots in the left panels correspond to a complete absence of mutations; plots in panels (*b*), to the presence of mutations of type (8). In panels (*b*), scaling was made in order to make the peak q(t) discernible. The initial distributions are described by formulas $q_0(\tau) = 1 + \sin(2\pi\tau)$, $q_0(\tau) = 1$, $q_0(\tau) = 6\tau^5$, and $q_0(\tau)$ $\pi\sin(\pi\tau/2)/2$ in the plots shown in the first, second, third, and fourth rows of panels, respectively.

When specifying the form of the fitness function as $W(\xi, \tau) = 0.1 + (\xi - \tau)^2$, $W(\xi, \tau) = 1.6 - (\xi - \tau)^2$, or $W(\xi, \tau) = \xi + \tau$, we acted in compliance with the so-called 'classical' model of the population genetic structure of species (Altukhov, 2003). The resultant unimodal distributions (Figures 4, 5) may adequately describe the monomorphic loci, which are more typical of the genome than the polymorphic ones (Altukhov, 2003). The diallelic polymorphism obtained in the classical model (Figure 3) looks peculiar, because the amounts of both alleles are equal, which is uncharacteristic of experimental data.

The form of the fitness function $W(\tau, \xi) = 1.5 + 1.5\sin(8\pi\tau\xi)$ corresponds to the so-called 'balance' model of the population genetic structure of species (Altukhov, 2003). In this case, infinitely many genotypes are 'optimal', a small proportion of them being homozygous and others being heterozygous. Note that, although the number of optimal genotypes is infinite in this case, still only a few alleles remain in the population in the course of evolution. The resultant genetic distributions are qualitatively similar to the distributions of the allelic frequencies of some loci in a few populations of *D. pseudoobscura* (as demonstrated by Prakash, Lewontin, and Hubby (1969) and Prakash, Lewontin, and Crumpaker (1973)). For example, 2 out of 12 populations studied appeared to be monomorphic for the *malate dehydrogenase* locus (chromosome 4; Figure 6, the top of panel *a*). Diallelic (Figure 6, panel (*a*), the second row) and triallelic (Figure 6, panels (*a*), the third and fourth rows) polymorphisms were found in eight and two populations, respectively. The malate dehydrogenase locus in chromosome 4 is characterized by a low polymorphism, with the concentration of the most frequent allele being higher than 90 %.

In the distributions of allelic frequencies obtained in our integral model without mutations (Figure 6, column *a*), the polymorphism is even lower, with the frequencies of rare alleles being about several per one tenth of percent. The introduction of mutations into the integral model has yielded higher polymorphisms of three and four alleles (Figure 6, panels (*a*), the second, third, and fourth rows). Qualitatively similar triallelic and tetraallelic polymorphisms for the locus *pt-8* (chromosome 2) are observed in some populations of *D. pseudoobscura*.

Apparently, further modification of the sinusoid function of fitness corresponding to the balance model and selection of a more adequate mutation rate will make it possible to fit the model values of allele frequencies to the experimental ones, as well as to describe the polymorphisms of, e.g., the *esterase 5* and *xanthine dehydrogenase* loci (the X chromosome and chromosome 2 of *D. pseudoobscura*, respectively), which involve more than four alleles.

Thus, the results of our study demonstrate that, 'in the general case', even an infinitely (continually) large diversity of alleles is reduced to a small number of almost discrete alleles in the course of evolution under strictly determined conditions. Even introduction of some equiprobable mutations into the model does not result in stable homogeneous distributions. The mutation process somewhat 'smears' the resultant distributions; however, a few almost discrete peaks are preserved, although both their number and heights may increase. Apparently, this situation will not change even in the case of a large (but finite) number of original alleles. The dynamic equations are such that evolution does not lead to homogeneous distributions of large numbers of alleles. Typically, distributions with small numbers of forms appear. Apparently, this explains why the allelic diversities of many genes are substantially limited in natural populations.

Note that we have proposed the discreet–continuous model and, then, have demonstrated analytically that this model has not any continuous stationary distributions if there is fitness heterogeneity in the space of genotypic classes. Then, we have explored numerically the asymptotic features of the solutions of this model, but the analytical investigation of the asymptotic features of that requires further study.

ACKNOWLEDGMENTS

This study was supported by the Russian Foundation for Basic Research (grants Nos. 03-01-00044-a and 04-04-97000-p).

References

Abe, T., Kanaya, S., Kinouchi, M., Ichiba, Y., Kozuki, T., and Ikemura, T., 2003, Informatics for unveiling hidden genome signatures, *Genome Res.* **13**(4):693–702.

Abnizova, I., Boekhorst, R. te, Walter, K., and Gilks, W.R., 2005, Some statistical properties of regulatory DNA sequences, and their use in predicting regulatory regions in *Drosophila* genome: the fluffy-tail test, *BMC Bioinformatics* **6**:109.

Abnizova, I., Schilstra, M., te Boekhorst, R., and Nehaniv, C.L, 2003, A statistical approach to distinguish between different DNA functional parts, *WSEAS Transactions on Computational Methods* **2**(Issue 4):1188–1196.

Adachi, O., Moonmangmee, D., Toyama, H., Yamada, M., Shinagawa, E., and Matsushita, K., 2003, New developments in oxidative fermentation, Appl. Microbiol. Biotechnol. 60:643–653.

Aerts, S., Van Loo P., Thijs, G., Moreau, Y., and De Moor, B., 2003, Computational detection of cis -regulatory modules, *Bioinformatics* **19**, Suppl 2:II5–II14.

Affymetrix, 2001, Affymetrix microarray suite 5.0., in: *User Guide*. Affymetrix, Inc., Santa Clara, CA.

Ahmad, S., Gromiha, M.M., and Sarai, A., 2003, Real value prediction of solvent accessibility from amino acid sequence, *Proteins* **50**:629–635.

Aizenberg, I., Aizenberg, N., and Vandewalle, J., 2000, *Multi-Valued and Universal Binary Neurons: Theory, Learning, Applications*, Kluwer Publishers, Boston/Dordrecht/London.

Aizenberg, I., Myasnikova, E., Samsonova, M., and Reinitz, J., 2002, Temporal classification of *Drosophila* segmentation gene expression patterns by the multi-valued neural recognition method, *Mathem. Biosci.* **176**:145–159.

Akaike, H., 1974, A new look at the statistical model identification, *IEEE Transactions on Automatic Control* **19**(6):716–723.

Akam, M., 1987, The molecular basis for metameric pattern in the *Drosophila* embryo, *Development* **101**:1–22.

Alexandrov, I., Kazakov, A., Tumeneva, I., Shepelev, V., and Yurov, Y., 2001, Alpha-satellite DNA of primates: old and new families, *Chromosoma* **110**(4):253–266.

Alexandrov, N., and Mironov, A., 1990, Application of a new method of pattern recognition in DNA sequence analysis: a study of *E. coli* promoters, *Nucleic Acids Res.* **18**:1847–1852.

Altman, R.B., et al., 2003, Indexing pharmacogenetics knowledge on the World Wide Web, *Pharmacogenetics* **13**:3–5.

Altschul, S.F., Gish, W., Miller, W., Myers, E.W., and Lipman, D.J., 1990, Basic local alignment search tool, *J. Mol. Biol.* **215**:403–410.

Altschul, S.F., Madden, T.L., Schaffer, A.A., Zhang, J., Zhang, Z., Miller, W., and Lipman, D.J., 1997, Gapped BLAST and PSI-BLAST: a new generation of protein database search programs, *Nucleic Acids Res.* **25**(17):3389–3402.

Altukhov, Yu.P., 2003, *Genetic Processes in Populations*, Akademkniga, Moscow.

Ananko, E.A., Naumochkin, A.N., Fokin, O.N., and Frolov, A.S., 1998, Programs for data input to the transcription regulatory regions database, in: *Proceedings of the First International Conference on Bioinformatics of Genome Regulation and Structure, 1998 August 24 –August 31; (BGRS'98), Novosibirsk.* ICG, Novosibirsk, **1**, pp. 29–32.

Ananko, E.A., Podkolodny, N.L., Stepanenko, I.L., Podkolodnaya, O.A., Rasskazov, D.A., Miginsky, D.S., Likhoshvai, V.A., Ratushny, A.V., Podkolodnaya, N.N., and Kolchanov, N.A., 2005, GeneNet in 2005, *Nucleic Acids Res.* **33**:D425–D427.

Ando, S., Yang, H., Nozaki, N., Okazaki, T., and Yoda, K., 2002, CENP-A, -B, and -C chromatin complex that contains the I-type alpha-satellite array constitutes the prekinetochore in HeLa cells, *Mol. Cell. Biol.* **22**:2229–2241.

Andrade, M.A., Ponting, C.P., Gibson, T.J., and Bork, P., 2000, Homology-based method for identification of protein repeats using statistical significance estimates, *J. Mol. Biol.* **298**:521–537.

Andrea, T.A., and Kalayeh, H. 1991, Applications of neural networks in quantitative structure–activity relationships of dihydrofolate reductase inhibitors, *J. Med. Chem.* **34**:2824–2836.

Andreeva, A., Howorth, D., Brenner, S.E., Hubbard, T.J., Chothia, C., and Murzin, A.G., 2004, SCOP database in 2004: refinements integrate structure and sequence family data, *Nucleic Acids Res.* **32**:D226–D229.

Anezo, C., Vries, A.H. de, Holtje, H.D., Tieleman, D.P., and Marrink, S.J., 2003, Methodological issues in lipid bilayer simulations, *J. Phys. Chem. B.* **107**:9424–9433.

Anfinsen, C.B., 1973, Principles that govern the folding of protein chains, *Science* **181**:223–230.

Antosiewicz, J., Briggs, J.M., Elcock, A.H., Gilson, M.K., and McCammon, J.A., 1996a, Computing the ionization states of proteins with a detailed charge model, *J. Comp. Chem.* **17**:1633–1644.

Antosiewicz, J., McCammon, J.A., and Gilson, M.K., 1994, Prediction of pH-dependent properties of proteins, *J. Mol. Biol.* **238**:415–436.

Antosiewicz, J., McCammon, J.A., and Gilson, M.K., 1996b, The determinants of pKa's in proteins, *Biochemistry* **35**:7819–7833.

Apweiler, R., Bairoch, A., and Wu, C.H., 2004, Protein sequence databases, *Current Opin. in Chem. Biol.* **8**(1):76–80.

Argaman, L., Hershberg, R., Vogel, J., Bejerano, G., Wagner, E.G.H., Margalit, H., and Altuvia, S., 2001, Novel small RNA-encoding genes in the intergenic regions of *Escherichia coli*, *Curr. Biol.* **11**:941–950.

Arrigo, P., Giuliano, F., Scalia, F., Rapallo, A., and Damiani, G., 1991, Identification of a new motif on nucleic acid sequence data using Kohonen's self-organizing map, *Comput. Appl. Biosci.* **7**(3):353–357.

Audic, S., and Claverie, J.M., 1998, Self-identification of protein-coding regions in microbial genomes, *Proc. Natl. Acad. Sci. USA* **95**(17):10026–10031.

Azad, R.K., and Borodovsky, M., 2004, Effect of choice of DNA sequence model structure on gene identification accuracy, *Bioinformatics* **20**:993–1005.

Azbel, Y.M., 1995, Universality in a DNA statistical structure, *Physical Review Letters* **75**:68–171.

Bafna, V., and Huson, D.H., 2000, The conserved exon method for gene finding, in: *Proceedings of the Eighth International Conference on Intelligent Systems for Molecular Biology*; AAAI Press, 3–12.

Bagga, R., Michalowski, S., Sabnis, R., Griffith, J.D., and Emerson, B.M., 2000, HMG I/Y regulates long-range enhancer-dependent transcription on DNA and chromatin by changes in DNA topology, *Nucleic Acids Res.* **28**(13):2541–2550.

Bairoch, A., and Bucher, P., 1994, PROSITE: recent developments, *Nucleic Acids Res.* **22**:3583–3589.

Baker, D., and Sali, A., 2001, Protein structure prediction and structural genomics, *Science* **294**:93–96.

Baker, P.G., et al., 1999, An ontology for bioinformatics applications, *Bioinformatics* **15**(6):510–520.

Baldi, P., 2002, On the convergence of a clustering algorithm for protein-coding regions in microbial genomes, *Bioinformatics* **16**(4):367–371.

Baldi, P., and Brunak, S., 2001, *Bioinformatics: the Machine Learning Approach*, The MIT Press, Cambridge, Massachusetts.

Baldi, P., Brunak, S., Chauvin, Y., and Krogh, A., 1996, Naturally occurring nucleosome positioning signals in human exons and introns, *J. Mol. Biol.* **263**(4):503–510.

Baldi, P., Brunak, S., Frasconi, P., Soda, G., and Pollastri, G., 1999, Exploiting the past and the future in protein secondary structure prediction, *Bioinformatics* **15**:937–946.

Balestrazzi, A., Chini, A., Bernacchia, G., Bracci, A., Luccarini, G., Cella, R., and Carbonera, D., 2000, Carrot cells contain two top1 genes having the coding capacity for two distinct DNA topoisomerases, *J. Exp. Bot.* **51**:1979–1990.

Barron, A., Rissanen, J., and Yu, B., 1998, The minimum description length principle in coding and modelling, *IEEE Trans. Inform. Theory* **44**:2743–2760.

Bartlett, G.J., Porter, C.T., Borkakoti, N., and Thornton, J.M., 2002, Analysis of catalytic residues in enzyme active sites, *J. Mol. Biol.* **324**:105–121.

Bashford, D., and Gerwert, K., 1992, Electrostatic calculations of the pKa values of ionizable groups in bacteriorhodopsin, *J. Mol. Biol.* **224**(2):473–486.

Bashford, D., and Karplus, M., 1991, Multiple-site titration curves of proteins: an analysis of exact and approximate methods for their calculation, *J. Phys. Chem.* **95**:9556–9561.

Batzoglou, S., Pachter, L., Mesirov, J.P., Berger B., and Lander, E.S., 2000, Human and mouse gene structure: comparative analysis and application to exon prediction, *Genome Res.* **10**(7):950–958.

Bava, K.A., Gromiha, M.M., Uedaira, H., Kitajima, K., and Sarai, A., 2004, ProTherm, version 4.0: thermodynamic database for proteins and mutants, *Nucleic Acids Res.* **32**:D120–D121.

Bejerano, G., 2004, Algorithms for variable length Markov chain modeling, *Bioinformatics* **20**(5):788–789.

Belikov, S., Holmqvist, P.H., Astrand, C., and Wrange, O., 2004, Nuclear factor 1 and octamer transcription factor 1 binding preset the chromatin structure of the mouse mammary tumor virus promoter for hormone induction, *J. Biol. Chem.* **279**(48):49857–49867.

Bendall, A.J., and Molloy, P.L., 1994, Base preferences for DNA binding by the bHLH-Zip protein USF: effects of MgCl2 on specificity and comparison with binding of Myc family members, *Nucleic Acids Res.* **22**:2801–2810.

Benos, P.V., Bulyk, M.L., and Stormo, G.D., 2002, Additivity in protein-DNA interactions: how good an approximation is it? *Nucleic Acids Res.* **30**(20):4442–4451.

Benson, D.A., Karsch-Mizrachi, I., Lipman, D.J., Ostell, J., and Wheeler, D.L., 2003, GenBank, *Nucleic Acids Res.* **31**(1):23–27.

Bentley, S.D., Chater, K.F., Cerdeno-Tarraga, A.M., Challis, G.L., Thomson, N.R., James, K.D., Harris, D.E., Quail, M.A., Kieser, H., Harper, D., Bateman, A., Brown, S., Chandra, G., Chen, C.W., Collins, M., Cronin, A., Fraser, A., Goble, A., Hidalgo, J., Hornsby, T., Howarth, S., Huang, C.H., Kieser, T., Larke, L., Murphy, L., Oliver, K., O'Neil, S., Rabbinowitsch, E., Rajandream, M.A., Rutherford, K., Rutter, S., Seeger, K., Saunders, D., Sharp, S., Squares, R., Squares, S., Taylor, K., Warren, T., Wietzorrek, A., Woodward, J., Barrell, B.G., Parkhill, J., and Hopwood, D.A., 2002, Complete genome sequence of the model actinomycete *Streptomyces coelicolor* A3(2), *Nature* **417**:141–147.

Berendsen, H.J.C., Postma, J.P.M., Gunsteren, W.F. van, DiNola, A., and Haak, J.P., 1984, Molecular dynamics with coupling to an external bath, *J. Chem. Phys.* **81**:3684–3690.

Berendsen, H.J.C., Spoel, D. van der, and Drunen R. van, 1995, Methodological issues in lipid bilayer simulations, *Comp. Phys. Comm.* **91**:43–56.

Berezikov, E., Bucheton, A., and Busseau, I., 2000, A search for reverse transcriptase-coding sequences reveals new non-LTR retrotransposons in the genome of *Drosophila melanogaster*, *Genome Biol.* **1**(6):RESEARCH0012.

Berg, O.G., and Hippel, P.H. von, 1987, Selection of DNA binding sites by regulatory proteins I: statistical-mechanical theory and application to operators and promoters, *J. Mol. Biol.* **193**:723–750.

Berg, O.G., and Hippel, P.H. von, 1988, Selection of DNA binding sites by regulatory proteins, *Trends Biochem. Sci.* **13**:207–211.

Berman, B., Nibu, Y., Pfeiffer, B., Tomancak, B., Celniker, S., Rubin, G., Levine, M., and Eisen, M., 2002, Exploiting TFBS clustering to identify CRM involved in pattern formation in *Drosophila* genome, *Proc. Natl. Acad. Sci. USA* **99**(2):757–762.

Berman, H.M., Westbrook, J., Feng, Z., Gilliland, G., Bhat, T.N., Weissig, H., Shindyalov, I.N., and Bourne, P.E., 2000, The protein data bank, *Nucleic Acids Res.* **28**(1):235–242.

Bernaola-Galvan, P., Grosse, I., Carpena, P., Oliver, J.L., Roman-Roldan, R., and Stanley, H.E., 2000, Finding borders between coding and noncoding DNA regions by an entropic segmentation method, *Phys. Rev. Let.* **85**(6):1342–1345.

Besemer, J., and Borodovsky, M., 1999, Heuristic approach to deriving models for gene finding, *Nucleic Acids Res.* **27**:3911–3920.

Besemer, J., Lomsadze, A., and Borodovsky, M., 2001, GeneMarkS: a self-training method for prediction of gene starts in microbial genomes. Implications for finding sequence motifs in regulatory regions, *Nucleic Acids Res.* **29**:2607–2618.

Bestor, T.H., and Coxon, A., 1993, The pros and cons of DNA methylation, *Curr. Biol.* **6**:384–386.

Bibikov, Yu.N., 1981, *Course of Ordinary Differential Equations*, Leningrad University, Leningrad, pp. 232 (in Russian).

Bielawski, J.P., Dunn, K.A., and Yang, Z., 2000, Rates of nucleotide substitution and mammalian nuclear gene evolution. Approximate and maximum-likelihood methods lead to different conclusions, *Genetics* **156**(3):1299–1308.

Biessmann, H., Walter, M.F., Le, D., Chuan, S., and Yao, J.G., 1999, Moose, a new family of LTR-retrotransposons in the mosquito *Anopheles gambiae*, *Insect. Mol. Biol.* **8**(2):201–212.

Billeter, M., 1996, Homeodomain-type DNA recognition, *Prog. Biophys. Mol. Biol.* **66**(3):211–225.

Binder, H., and Preibisch, S., 2005, Specific and non-specific hybridization of oligonucleotide probes on microarrays, *Biophys. J.* (in press) (see http://arxiv.org/ftp/q-bio/papers/0410/0410028.pdf).

Binder, H., Kirsten, T., Loeffler, M., and Stadler, P., 2003, Sequence specific sensitivity of oligonucleotide probes, *Proc. of the German Bioinformatics Conf.* **2**:145–147.

Binder, H., Kirsten, T., Loeffler, M., and Stadler, P., 2004, The sensitivity of microarray oligonucleotide probes – variability and the effect of base composition, *J. Phys. Chem. B.* **108**(46):18003–18014.

Binder, H., Preibisch, S., and Kirsten, T., 2005, Base pair interactions and hybridization isotherms of matched and mismatched oligonucleotide probes on microarrays, http://www.arvix.org/abs/q-bio.BM/0501008.

Binder, S., and Brennicke, A., 2002, Gene expression in plant mitochondria: transcriptional and post-transcriptional control, *Phil. Trans. R. Soc. Lond.* **358**:181–189.

Birch, L.C., 1955, Selection in *Drosophila pseudoobscura* in relation to crowding, *Evol.* **9**(4):160–165.

Birnboim, H.C., Sederoff, R.R., and Paterson, M.C., 1979, Distribution of polypyrimidine. polypurine segments in DNA from diverse organisms, *Eur. J. Biochem.* **98**:301–307.

Black, B.E., Foltz, D.R., Chakravarthy, S., Luger, K., Woods, V.L. Jr., and Cleveland, D.W., 2004, Structural determinants for generating centromeric chromatin, *Nature* **430**(6999):578–582.

Blackwell, T.K., and Weintraub, H., 1990, Differences and similarities in DNA-binding preferences of MyoD and E2A protein complexes revealed by binding site selection, *Science* **250**:1104–1110.

Blattner, F.R., Punkett III, G., Bloch, C.A., Perna, N.T., Burland, V., Riley, M., Collado-Vides, J., Glasner, J.D., Rode, C.K., Mayhew, G.F., Gregor, J., Davis, N.W., Kirkpatric, H.A., Goeden, M.A., Rose, D.J., Mau, B., and Shao, Y., 1997, The complete genome sequence of *Escherichia coli* K12, *Science* **277**:1453–1462.

Blaxter, M.L., De Ley, P., Garey, J.R., Liu, L.X., Scheldeman, P., Vierstraete, A., Vanfleteren, J.R., Mackey, L.Y., Dorris, M., Frisse, L.M., Vida, J.T., and Thomas, W.K., 1998, A molecular evolutionary framework for the phylum Nematoda, *Nature* **392**(6671):71–75.

Boekhorst, R. te, Nehaniv, C., and Abnizova, I., 2005, An adapting sliding window algorithm for inferring DNA functionality from sequence information (under review in *Applied Bioinformatics*).

Boffelli, D., McAuliffe, J., Ovcharenko, D., Lewis, K.D., Ovcharenko, I., Pachter, L., and Rubin, E.M., 2003, Phylogenetic shadowing of primate sequences to find functional regions of the human genome, *Science* **299**:1391–1394.

Bolshoy, A., 2003, DNA sequence analysis linguistic tools: contrast vocabularies, compositional spectra and linguistic complexity, *Appl. Bioinformatics* **2**(2):103–112.

Bolshoy, A., Shapiro, K., Trifonov, E.N., and Ioshikhes, I., 1997, Enhancement of the nucleosomal pattern in sequences of lower complexity, *Nucleic Acids Res.* **25**(16):3248–3254.

Bonnie, J.S., and Dewey, G.T., 1995, Multifractal and decoded walks: applications to protein sequence correlations, *Phys. Rev. E* **52**:6588–6592.

Bono, H., Ogata, H., and Goto, S., 1998, Reconstruction of amino acid biosynthesis pathways from the complete genome sequence, *Genome Res.* **8**:203–210.

Bormann, E.R., Eikmanns, B.J., and Sahm, H., 1992, Molecular analysis of the *Corynebacterium glutamicum gdh* gene encoding glutamate dehydrogenase. *Mol. Microbiol.* **6**:317–326.

Borodovsky, M., and McIninch, J. 1993, GENMARK: parallel gene recognition for both DNA strands, *Comp. Chem.* **17**:123–133.

Borodovsky, M., and McIninch, J., 1993, Recognition of genes in DNA sequence with ambiguities, *Biosystems* **30**:161–171.

Bowen, N.J., and McDonald, J.F., 1999, Genomic analysis of *Caenorhabditis elegans* reveals ancient families of retroviral-like elements, *Genome Res.* **9**(10):924–935.

Bowie, J.U., Luthy, R., and Eisenberg, D., 1991, A method to identify protein sequences that fold into a known three-dimensional structure, *Science* **253**:164–170.

Boyle, E.C., and Finlay, B.B., 2003, Bacterial pathogenesis: exploiting cellular adherence, *Curr. Opin. Cell. Biol.* **15**:633–639.

Bradley, P., Cowen, L., Menke, M., King, J., and Berger, B., 2001, BETAWRAP: successful prediction of parallel beta-helices from primary sequence reveals an association with many microbial pathogens, *Proc. Natl. Acad. Sci. USA* **98**:14819–14824.

Brants T., Chen F., and Tsochantradis, I., 2002, Topic-based document segmentation with probabilistic latent semantic analysis, in: *Proceedings of CIKM'02 ACM*; Nov 4–9 2002 McLean. Virginia (USA).

Brauch, H., Weirich, G., Brieger, J., Glavac, D., Rodl, H., Eichinger, M., Feurer, M., Weidt, E., Puranakanitstha, C., Neuhaus, C., Pomer, S., Brenner, W., Schirmacher, P., Storkel, S., Rotter, M., Masera, A., Gugeler, N., and Decker, H.J., 2000, VHL alterations in human clear cell renal cell carcinoma: association with advanced tumor stage and a novel hot spot mutation, *Cancer Res.* **60**:1942–1948.

Brendel, V., Bucher, P., Nourbakhsh, I.R., Edwin Blaisdell, B., and Karlin, S., 1992, Methods and algorithms for statistical analysis of protein sequences, *Proc. Natl. Acad. Sci. USA* **89**:2002–2006.

Brenner, S.E., Koehl, P., and Levitt, M., 2000, The ASTRAL compendium for protein structure and sequence analysis, *Nucleic Acids Res.* **28**:254–256.

Britten, R.J., 1995, Active gypsy/Ty3 retrotransposons or retroviruses in *Caenorhabditis elegans*, *Proc. Natl. Acad. Sci. USA* **92**(2):599–601.

Bucher, P., Karplus, K., Moeri, K., and Hofmann, N., 1996, A flexible search technique based on generalized profiles, *Comput. Chem.* **20**:3–24.

Buhlmann, P., and Wyner, A.J., 1999, Variable length Markov models, *Ann. Statistics* **27**:480–513.

Bulyk, M.L., McGuire, A.M., Masuda, N., and Church, G.M., 2004, A motif co-occurrence approach for genome-wide prediction of transcription-factor-binding sites in *Escherichia coli*, *Genome Res.* **14**(2):201–208.

Burge, C., and Karlin, S., 1997, Prediction of complete gene structures in human genomic DNA, *J. Mol. Biol.* **268**:78–94.

Burge, C.B., and Karlin, S., 1998, Finding the genes in genomic DNA, *Curr. Opin. Struct. Biol.* **8**:346–354.

Busygina, T.V, Ignatieva, E.V., and Osadchuk, A.V., 2003, Consensus sequence of transcription factor SF-1 binding site and putative binding site in the 5' flanking regions of genes encoding mouse steroidogenic enzymes 3betaHSDI and Cyp17, *Biochemistry (Mosk).* **68**:377–384.

Cachia, P.J., and Hodges, R.S., 2003, Synthetic peptide vaccine and antibody therapeutic development: Prevention and treatment of *Pseudomonas aeruginosa*, *Biopolymers* **71**:141–168.

Carter, R.J., Dubchak, I., and Holbrook, S.R., 2001, A computational approach to identify genes for functional RNAs in genomic sequences, *Nucleic Acids Res.* **29**:3928–3938.

Cartharius, K., Frech, K., Grote, K., Klocke, B., Haltmeier, M., Klingenhoff, A., Frisch, M., Bayerlein, M., and Werner, T., 2005, MatInspector and beyond: promoter analysis based on transcription factor binding sites, *Bioinformatics* Apr 28; [Epub ahead of print].

Cerdeno-Tarraga, A.M., Efstratiou, A., Dover, L.G., Holden, M.T., Pallen, M., Bentley, S.D., Besra, G.S., Churcher, C., James, K.D., De Zoysa, A., Chillingworth, T., Cronin, A., Dowd, L., Feltwell, T., Hamlin, N., Holroyd, S., Jagels, K., Moule, S., Quail, M.A., Rabbinowitsch, E., Rutherford, K.M., Thomson, N.R., Unwin, L., Whitehead, S.,

Barrell, B.G., and Parkhill, J. 2003, The complete genome sequence and analysis of *Corynebacterium diphtheriae* NCTC13129, *Nucleic Acids Res.* **31**:6516–6523.

Chaley, M.B., Korotkov, E.V., and Kudryashov, N.A., 2003, Latent periodicity of 21 bases typical for MCP II gene is widely present in various bacterial genes, *DNA Sequence* **14**(1):33–52.

Chaley, M.B., Korotkov, E.V., and Skryabin, K.G., 1999, Method revealing latent periodicity of the nucleotide sequences modified for a case of small samples, *DNA Research* **6**:153–163.

Chamary, J.V., Hurst, L.D., 2004, Similar rates but different modes of sequence evolution in introns and at exonic silent sites in rodents: evidence for selectively driven codon usage, *Mol. Biol. Evol.* **21**(6):1014–1023.

Chang, C.P., Chang, J.C., Chang, H.H., Tsai, W.J., and Lo, S.J., 2001, Positional importance of Pro53 adjacent to the Arg49-Gly50-Asp51 sequence of rhodostomin in binding to integrin alphaIIbbeta3, *Biochem. J.* **357**:57–64.

Chaouiya, C., Remy, E., Mossé, B., and Thieffry, D., 2003, Qualitative analysis of regulatory graphs: a computational tool based on a discrete formal framework, *LNCIS* **294**:119–126.

Chastonay, Y. de, Felder, H., Link, C., Aeby, P., Tobler, H., and Muller, F., 1992, Unusual features of the retroid element PAT from the nematode *Panagrellus redivivus*, *Nucleic Acids Res.* **20**(7):1623–1628.

Chechetkin, V.R., Lobzin, V.V., 1998, Study of correlations in segmented DNA sequences: application to structure coupling between exons and introns, *J. Theor. Biol.* **190**(1):69–83.

Chen, J.N., and Chang, J.S., 1998, Topical clustering of MRD sense based on information retrieval techniques, *Computational Linguistics* **24**(1):61–95.

Chen, Q.K., Hertz, G.Z., and Stormo, G.D., 1995, MATRIX SEARCH 1.0: a computer program that scans DNA sequences for transcriptional elements using a database of weight matrices, *Comput. Appl. Biosci.* **11**(5):563–566.

Chen, S., Lesnik, E.A., Hall, T.A., Sampath, R., Griffey, R.H., Ecker, D.J., and Blyn, L.B., 2001, A bioinformatics based approach to discover small RNA genes in the *Escherichia coli* genome, *BioSystems* **65**:157–177.

Chiu, S.W., Jakobsson, E., and Scott, H.L., 2001, Combined Monte Carlo and molecular dynamics simulation of hydrated lipid-cholesterol lipid bilayers at low cholesterol concentration, *Biophys. J.* **80**:1104–1114.

Chou, P.Y., and Fasman, G.D., 1974a, Conformational parameters for amino acids in helical, β-sheets and random coil regions calculated from proteins, *Biochemistry* **13**:211–221.

Chou, P.Y., and Fasman, G.D., 1974b, Prediction of protein conformation, *Biochemistry* **13**:222–244.

Chow, K.-S., Singh, D.P., Roper, J.M., and Smith, A.G., 1997, A single precursor protein for ferrochelatase-I from *Arabidopsis* is imported *in vitro* into both chloroplasts and mitochondria, *J. Biol. Chem.* **272**:27565–27571.

Christiaens, B., Symoens, S., Verheyden, S., Engelborghs, Y., Joliot, A., Prochiantz, A., Vandekerckhove, J., Rosseneu, M., and Vanloo, B., 2002, Tryptophan fluorescence study of the interaction of penetratin peptides with model membranes, *Eur. J. Biochem.* **269**:2918–2926.

Chugunov, A., Chavatte, P., and Efremov, R., 2005, Differences in the binding sites of two melatonin receptors help to explain their selectivity for some melatonin analogs: a molecular modeling study, *J. Med. Chem.* (in press).

Chui, H., and Rangarajan, A. 2000, A new algorithm for non-rigid point matching, *Proceeding of IEEE Conf. on Computer Vision and Pattern Recognition* **2**:44–51.

Chuzhanova, N.A., Anassis, E.J., Ball, E., Krawczak, M., and Cooper, D.N., 2003, Meta-analysis of indels causing human genetic disease: mechanisms of mutagenesis and the role of local DNA sequence complexity, *Hum. Mutation* **21**:28–44.

Claros, M.G., and Vincens, P., 1996, Computational method to predict mitochondrially imported proteins and their targeting sequences, *Eur. J. Biochem.* **241**:779–786.

Cliften, P., Sudarsanam, P., Desikan, A., Fulton, L., Fulton, B., Majors, J., Waterston, R., Cohen, B.A., and Johnston, M., 2003, Finding functional features in *Saccharomyces* genomes by phylogenetic footprinting, *Science* **301**:71–76.

Coggins, L.W., and O'Prey, M., 1989, DNA tertiary structures formed *in vitro* by misaligned hybridization of multiple tandem repeat sequences, *Nucleic Acids Res.* **17**:7417–7426.

Cohen, J.I., 1999, The biology of Epstein-Barr virus: lessons learned from the virus and the host, *Curr. Opin. Immunol.* **11**:365–370.

Cole, S.T., Brosch, R., Parkhill, J., Garnier, T., Churcher, C., Harris, D., Gordon, S.V., Eiglmeier, K., Gas, S., Barry, C.E.III, Tekaia, F., Badcock, K., Basham, D., Brown, D., Chillingworth, T., Connor, R., Davies, R., Devlin, K., Feltwell, T., Gentles, S., Hamlin, N., Holroyd, S., Hornsby, T., Jagels, K., Rutter, S., Seeger, K., Skelton, J., Squares, R., Squares, S., Sulston, J.E., Taylor, K., Whitehead, S., and Barrell, B.G., 1998, Deciphering the biology of *Mycobacterium tuberculosis* from the complete genome sequence, *Nature* **393**:537–544.

Cole, S.T., Eiglmeier, K., Parkhill, J., James, K.D., Thomson, N.R., Wheeler, P.R., Honore, N., Garnier, T., Churcher, C., Harris, D., Mungall, K., Basham, D., Brown, D., Chillingworth, T., Connor, R., Davies, R.M., Devlin, K., Duthoy, S., Feltwell, T., Fraser, A., Hamlin, N., Holroyd, S., Hornsby, T., Jagels, K., Lacroix, C., Maclean, J., Moule, S., Murphy, L., Oliver, K., Quail, M.A., Rajandream, M.A., Rutherford, K.M., Rutter, S., Seeger, K., Simon, S., Simmonds, M., Skelton, J., Squares, R., Squares, S., Stevens, K., Taylor, K., Whitehead, S., Woodward, J.R., and Barrell, B.G., 2001, Massive gene decay in the leprosy bacillus, *Nature* **409**:1007–1011.

Collado-Vides, J., and Hofestädt, R. 2002, *Gene Regulation and Metabolism – Post Genomic Computational Approaches*, MIT Press, Cambridge, MA.

Collatz, L., 1968, *Eigenvalues Problems*, Nauka, Moscow, pp. 504 (in Russian).

Conway, S., Canning, S.J., Barrett, P., Guardiola-Lamaitre, B., Delagrange, P., and Morgan, P.J., 1997, The roles of valine 208 and histidine 211 in ligand binding and receptor function of the ovine Mellab melatonin receptor, *Biochem. Biophys. Res. Commun.* **239**:418–423.

Conway, S., Mowat, E.S., Drew, J.E., Barrett, P., Delagrange, P., and Morgan, P.J., 2001, Serine residues 110 and 114 are required for agonist binding but not antagonist binding to the melatonin MT_1 receptor, *Biochem. Biophys. Res. Commun.* **282**:1229–1236.

Copeland, C.S., Brindley, P.J., Heyers, O., Michael, S.F., Johnston, D.A., Williams, D.L., Ivens, A.C., and Kalinna, B.H., 2003, Boudicca, a retrovirus-like long terminal repeat retrotransposon from the genome of the human blood fluke *Schistosoma mansoni*, *J. Virol.* **77**(11):6153–6166.

Cordier, M., Calender, A., Billaud, M., Zimber, U., Rousselet, G., Pavlish, O., Banchereau, J., Tursz, T., Bornkamm, and G., and Lenoir, G.M., 1990, Stable transfection of Epstein-Barr virus (EBV) nuclear antigen 2 in lymphoma cells containing the EBV P3HR1 genome induces expression of B-cell activation molecules CD21 and CD23, *J. Virol.* **64**(3):1002–1013.

Cornell, W.D., Cieplak, P., Bayly, C.I., Gould, I.R., Merz, K.M. Jr., Ferguson, D.M., Spellmeyer, D.C., Fox, T., Caldwell, J.W., and Kollman, P.A., 1995, A second generation force field for the simulation of proteins, nucleic acids and organic molecules, *J. Am. Chem. Soc.* **117**:5179–5197.

Cover, T.M., and Thomas, J.A., 1991, Elements of Information Theory. Wiley.

Covert, M., Schilling, W., and Famili, C.H., 2001, Metabolic modeling of microbial strains in silico, *Trends in Biochem. Sci.* **27**:179–186.

Coward, E., and Drablos, F., 1998, Detecting periodic patterns in biological sequences, *Bioinformatics* **14**:498–507.

Cox, R., and Mirkin, S.M., 1997, Characteristic enrichment of DNA repeats in different genomes, *Proc. Natl. Acad. Sci. USA* **94**:5237–5242.

Crow, J., and Kimura, M., 1971, *An Introduction to Population Genetics Theory*, Princeton Univ. Press, Princeton, New Jersey.

Currey, K.M., and Shapiro, B.A., 1997, Secondary structure computer prediction of the poliovirus 5' non-coding region is improved by a genetic algorithm, *Comput. Appl. Biosci.* **13**(1):1–12.

Czajlik, A., Mesko, E., Penke, B., and Perczel, A., 2002, Investigation of penetratin peptides. Part 1. The environment dependent conformational properties of penetratin and two of its derivatives, *J. Pept. Sci.* **8**:151–171.

Dandekar, T., Schuster, S., Snel B., et al., 1999, Pathway alignment: application to the comparative analysis of glycolytic enzymes, *Biochem. J.* **1**:115–124.

Daniell, H., Zheng, D., and Nielsen, B.L., 1995, Isolation and characterization of an in vitro DNA replication system from maize mitochondria, *Biochem. Biophys. Res. Commun.* **208**:287–294.

Dassow, G. von, Meir, E., Munro, E.M., and Odell, G.M., 2000, The segment polarity network is a robust developmental module, *Nature* **406**(6792):188–192.

deHaseth, P.L., and Helmann, J.D., 1995, Open complex formation by *Escherichia coli* RNA polymerase: the mechanism of polymerase-induced strand separation of double helical DNA, *Mol. Microbiol.* **16**:817–824.

Delcher, A.L., Harmon, D., Kasif, S., White, O., and Salzberg, S., 1999, Improved bacterial gene identification with Glimmer, *Nucleic Acids Res.* **27**:4636–4641.

DeMarco, R., Kowaltowski, A.T., Machado, A.A., Soares, M.B., Gargioni, C., Kawano, T., Rodrigues, V., Madeira, A.M., Wilson, R.A., Menck, C.F., Setubal, J.C., Dias-Neto, E., Leite, L. C., and Verjovski-Almeida, S., 2004, Saci-1, -2, and -3 and Perere, four novel retrotransposons with high transcriptional activities from the human parasite *Schistosoma mansoni*, *J. Virol.* **78**(6):2967–2978.

Dennis, M.S., Carter, P., and Lazarus, R.A., 1993, Binding interactions of kistrin with platelet glycoprotein IIb-IIIa: analysis by site-directed mutagenesis, *Proteins* **15**:312–321.

Derossi, D., Chassaing, G., and Prochiantz, A., 1998, Trojan peptides: the penetratin system for intracellular delivery, *Trends Cell Biol.* **8**:84–87.

Derweester, S., Dumais, S.T., Furnas, G.W., Landauer, T.K., and Harshman, R., 1990, Indexing by latent semantic analysis, *J. Amer. Assoc. Inform. Sci.* **41**:391–407.

Dieterich, C., Grossmann, S., Tanzer, A., Ropcke, S., Arndt, P.F., Stadler, P.F., and Vingron, M., 2005, Comparative promoter region analysis powered by CORG, *BMC Genomics* **6**(1):24.

Diez, J., Beguiristain, T., Le Tacon, F., Casacuberta, J.M., and Tagu, D., 2003, Identification of Ty1-copia retrotransposons in three ectomycorrhizal basidiomycetes: evolutionary relationships and use as molecular markers, *Curr. Genet.* **43**(1):34–44.

Dodin, G., Vandergheynst, P., Levoir, P., Cordier, C., and Marcourt, L., 2000, Fourier and wavelet transform analysis, a tool for visualizing regular patterns in DNA sequences, *J. Theor. Biol.* **206**:323–326.

Doerks, T., Bairoch, A., and Bork, P., 1998, Protein annotation: detective work for function prediction, *Trends Genet.* **14**:248–250.

Donelly, D., Overington, J.P., Ruffle, S.V., Nugent, J.H., and Blundell, T.L., 1993, Modeling α-helical transmembrane domains: the calculation and use of substitution tables for lipid-facing residues, *Protein Sci.* **2**:55–70.

Drew, H.R., and Travers, A.A., 1985, DNA bending and its relation to nucleosome positioning, *J. Mol. Biol.* **186**(4):773–790.

Drin, G., Mazel, M., Clair, P., Mathieu, D., Kaczorek, M., and Temsamani, J., 2001, Physico-chemical requirements for cellular uptake of pAntp peptide. Role of lipid-binding affinity, *Eur. J. Biochem.* **268**:1304–1314.

Du, P., and Alcotra, I., 1994, Sequence divergence analysis for the prediction of 7-helix membrane protein structures: I. Comparison with bacteriorhodopsin, *Prot. Engng.* **10**:1221–1229.

Dzhelyadin, T.R., Sorokin, A.A., Ivanova, N.N., Sivozhelezov, V.S., Kamzolova, S.G., and Polozov, R.V., 2001a, Characterization of electrostatic interaction between RNA polymerase from *E. coli* and T4 phage DNA promoters, *Biophysics (Mosk).* **46**:972–976.

Dzhelyadin, T.R., Sorokin, A.A., Sivozhelezov, V.S., Ivanova, N.N., Polozov, R.V., and Kamzolova, S.G., 2001b, New approaches in studying RNA polymerase- promoter recognition code, *J. Biomol. Struct. Dyn.* **18**:992–993.

Eddy, S.R., 1998, Profile hidden Markov models, *Bioinformatics* **14**(9):755–763.

Eddy, S.R., 1999, Noncoding RNA genes, *Curr. Opinion Genet. Dev.* **9**:695–699.

Edwards, R., and Glass, L., 2000, Combinatorial explosion in model gene networks, *Chaos* **10**:691–704.

Efremov, R.G., and Vergoten, G., 1995, Hydrophobic nature of membrane-spanning α-helical peptides as revealed by Monte Carlo simulations and molecular hydrophobicity potential analysis, *J. Phys. Chem.* **99**:10658–10666.

Efremov, R.G., Nolde, D.E., Konshina, A.G., Syrtcev, N.P., and Arseniev, A.S., 2004, Peptides and proteins in membranes: what can we learn via computer simulations? *Curr. Med. Chem.* **11**:2421–2442.

Ehret, G.B., Reichenbach, P., Schindler, U., et al., 2001, DNA binding specificity of different STAT proteins. Comparison of *in vitro* specificity with natural target sites, *J. Biol. Chem.* **276**:6675–6688.

Eliopoulos, A.G., Gallagher, N.J., Blake, S.M., Dawson, C.W., and Young, L.S., 1999, Activation of the p38 mitogen-activated protein kinase pathway by Epstein-Barr virus-encoded latent membrane protein 1 coregulates interleukin-6 and interleukin-8 production, *J. Biol. Chem.* **274**(23):16085–16096.

Elowitz, M.B., and Leibler, S., 2000, A synthetic oscillatory network of transcriptional regulators, *Nature* **403**:335–338.

Emanuelsson, O., Nielsen, H., Brunak, S., and Heijne, G. von, 2000, Predicting subcellular localization of proteins based on their N-terminal amino acid sequence, *J. Mol. Biol.* **300**:1005–1016.

Enukashvili, N.I., Kuznetsova, I.S., and Podgornaia, O.I., 2003, The mammalian centromere organization, *Tsitologiia* **45**(3):255–270.

Essmann, U., Perera, L., Berkowitz, M.L., Darden, T., Lee, H., and Pedersen, L.G., 1995, A smooth particle mesh Ewald method, *J. Chem. Phys.* **103**:8577–8593.

Eulenstein, O., Mirkin, B., and Vingron, M., 1998, Duplication-based measures of difference between gene and species, *J. Comput. Biol.* **5**:135–148.

Evans, M., Hastings, M., Peacock, B., 1993, *Statistical distributions*, Wiley and Sons, New York.

Evdokimov, E.V., 1999, *Problems of Regular Behavior and Determined Chaos in the Main Models of Population Dynamics: Theory and Experiment*, Doctoral (Biol.) Dissertation, Krasnoyarsk.

Faddeev, D.K., and Faddeeva, V.N., 1960, *Computational Methods of Linear Algebra* GIFML, Moscow, pp. 656 (in Russian).

Fadeev, S.I., 1985, On solution of a system of transcendental equations with parameter by Newton's method, Vychislitel'nye Sistemy (Computational Systems), Novosibirsk, **108**:78–93 (in Russian).

Fadeev, S.I., 1990, A program for numerical solution of nonlinear boundary value problems for systems of ordinary differential equations with parameter, in: *Computing Methods of Linear Algebra*, Nauka, Siberian Branch, Novosibirsk, pp. 104–198 (in Russian).

Fadeev, S.I., and Kogai, V.V., 2004, Using parameter continuation based on multiple shooting method for numerical research of nonlinear boundary value problems, *Int. J. Pure Appl. Math.* **14**:467–498 (in Russian).

Fadeev, S.I., and Likhoshvai, V.A., 2003, On hypothetical gene networks, *Sib. Zh. Industr. Matem.* **6**(3):134–153 (in Russian).

Fadeev, S.I., Pokrovskaya, S.A., Berezin, A.Yu., and Gainova, I.A., 1988, in: *Software Package STEP for Numerical Study of General Systems of Nonlinear Equations and Autonomous Systems*, Novosibirsk State University Press, Novosibirsk (in Russian).

Fariselli, P., and Casadio, R., 2000, Prediction of the number of residue contacts in proteins, *Proc. Int. Conf. Intell. Syst. Mol. Biol.* **8**:146–151.

Fariselli, P., Olmea, O., Valencia, A., and Casadio, R., 2001, Progress in predicting inter-residue contacts of proteins with neural networks and correlated mutations, *Proteins* **45** (Suppl. 5):157–162.

Fatemi, M., Hermann, A., Pradhan, S., and Jeltsch, A., 2001, The activity of the murine DNA methyltransferase Dnmt1 is controlled by interaction of the catalytic domain with the N-terminal part of the enzyme leading to an allosteric activation of the enzyme after binding to methylated DNA, *J. Mol. Biol.* **309**:1189–1199.

Felder, H., Herzceg, A., Chastonay, Y. de, Aeby, P., Tobler, H., and Muller, F., 1994, Tas, a retrotransposon from the parasitic nematode *Ascaris lumbricoides*, *Gene* **149**(2):219–225.

Felsenstein, J., 1985, Confidence limits on phylogenies: an approach using the bootstrap, *Evolution* **39**(4):783–791.

Festenstein, R., and Kioussis, D., 2000, Locus control regions and epigenetic chromatin modifiers, *Curr. Opin. Genet. Devel.* **10**:199–203.

Field, D., and Wills, C., 1996, Long, polymorphic microsatellites in simple organisms, *Proc. Royal Acad. London B* **263**:209–251.

Field, D., and Wills, C., 1998, Abundant microsatellite polymorphism in *Saccharomyces cerevisiae*, and different distributions of microsatellites in eight prokaryotes and *S. cerevisiae*, result from strong mutation pressures and variety of selective forces, *Proc. Natl. Acad. Sci.* **95**:1647–1652.

Finkelshtein, A.V., and Ptitsyn, O.B., 2002, *Physics of Protein: Lectures with Color and Stereoscopic Illustration*, Knizhniy dom 'Universitet', Moscow (in Russian).

Finlay, B.B., and Falkow, S., 1997, Common themes in microbial pathogenicity revisited, *Microbiol. Mol. Biol. Rev.* **61**:136–169.

Finney, A., and Hucka, M., 2003, Systems biology markup language: level 2 and beyond, *Biochem. Soc. Trans.* **31**:1472–1473.

Finney, A., Gor, V., Bornstein, B., and Mjolsness, E., 2003, *Systems Biology Markup Language (SBML) Level 3 Proposal: Array Features* http://www.sbml.org/wiki/arrays.

Finney, A., Hucka, M., Bornstein, B., Keating, S., Shapiro, B.E., Matthews, J., Kovitz, B., Schilstra, M., Funahashi, A., Doyle, J., and Kitano, H., 2005, Software infrastructure for effective communication and reuse of computational models, in: *System Modeling in*

Cellular Biology: From Concepts to Nuts and Bolts, S. Sultan, V. Perusal, J. Stalling, eds. (in press).

Finney, A., Hucka, M., Bornstein, B.J., Keating, S., Shapiro, B.E., Matthews, J., Kovitz, B., Funahashi, A., Schilstra, M., Doyle, J.C., and Kitano, H., 2004, Evolving a lingua franca and accompanying software infrastructure for computational systems biology: the Systems Biology Markup Language (SBML) Project, *IEE Systems Biology* **1**(1):41–53.

Fischer, D., Wolfson, H., Lin, L.S., and Nussinov, R., 1994, Three dimensional, sequence order-independent structural comparison of a serine protease against the crystallographic database reveals active site similarities: Potential implications to evolution and to protein folding, *Protein Sci.* **3**:769–778.

Fisher, D.H., 1987, Knowledge acquisition via incremental Conceptual Clustering, *Machine Learning* **2**:139–172.

Fitzgerald, D.J., and Anderson, J.N., 1999, DNA structural and sequence determinants for nucleosome positioning, *Gene Theor. Mol. Biol.* **4**:349–362.

Fitzgerald, D.J., Dryden, G.L., Bronson, E.C., Williams, J.S., and Anderson, J.N., 1994, Conserved patterns of bending in satellite and nucleosome positioning DNA, *J. Biol. Chem.* **269**(33):21303–21314.

Florea, L., McClelland, M., Riemer, C., Schwartz, S., and Miller, W., 2003, EnteriX 2003: Visualization tools for genome alignments of Enterobacteriaceae, *Nucleic Acids Res.* **31**:3527–3532.

Foe, V., and Alberts, B., 1983, Studies of nuclear and cytoplasmic behaviour during the five mitotic cycles that precede gastrulation in *Drosophila* embryogenesis, *J. of Cell Sci.* **61**:31–70.

Foissac, S., and Schiex, T., 2005, Integrating alternative splicing detection into gene prediction, *BMC Bioinformatics* **6**:25.

Forrest, L.R., and Sansom, M.S., 2000, Membrane simulations: bigger and better? *Curr. Opin. Struct. Biol.* **10**:174–181.

Forst, C.V., and Schulten, K., 1999, Evolution of metabolisms: a new method for the comparison of metabolic pathways using genomics information, *J. Comput. Biol.* **6**:343–360.

Forst, C.V., and Schulten, K., 2001, Phylogenetic analysis of metabolic pathways, *J. Mol. Evol.* **52**:471–489.

Frazer, K.A., Pachter, L., Poliakov, A., Rubin, E.M., and Dubchak, I., 2004, VISTA: computational tools for comparative genomics, *Nucleic Acids Res.* **32**:W273–W279.

Frishman, D., Mironov, A., Mewes, H.W., and Gelfand, M., 1998, Combining diverse evidence for gene recognition in completely sequenced bacterial genomes, *Nucleic Acids Res.* **26**:2941–2947.

Frisman, E.Ya., 1986, *Primary Genetic Divergence (Theoretical Analysis and Modeling)*, Dal'nevost. Nauch. Tsentr. Akad. Nauk SSSR, Vladivostok.

Frisman, E.Ya., and Zhdanova, O.L., 2003, An integral model of the dynamics of size and genetic composition of a Mendelian single-locus population of diploid organisms, *Tr. Dalnevost. Gos. T. Univ.* **133**:157–164 (in Russion).

Fudou, R., Jojima, Y., Seto, A., Yamada, K., Kimura, E., Nakamatsu, T., Hiraishi, A., and Yamanaka, S., 2002, *Corynebacterium efficiens* sp. nov., a glutamic-acid-producing species from soil and vegetables, *Int. J. Syst. Evol. Microbiol.* **52**:1127–1131.

Fujinaga, M., Chernaia, M.M., Tarasova, N.I., Mosimann, S.C., James, M.N.G., 1995, Crystal structure of human pepsin, *Protein Sci.* **4**:960–972.

Furet, P., Sele, A., and Cohen, N.C., 1988, 3D molecular lipophilicity potential profiles: a new tool in molecular modeling, *J. Mol. Graphics* **6**:182–200.

Garcia-Vallve, S., Guzman, E., Montero, M.A., Romeu, A., 2003, HGT-DB: a database of putative horizontally transferred genes in prokaryotes complete genomes, *Nucleic Acids Res.* **31**(1):187–189.

Gardner, T.S., Cantor, C.R., and Collins, J.J., 2000, Construction of a genetic toggle switch in *Escherichia coli*, *Nature* **403**:339–342.

Gelfand, M.S., 1989, Statistical analysis of mammalian pre-mRNA splicing sites, *Nucleic Acids Res.* **17**:6369–6382.

Gelfand, M.S., 1992, Statistical analysis and prediction of the exonic structure of human genes, *J. Mol. Evol.* **35**:239–252.

Gelfand, M.S., 1999, Recognition of regulatory sites by genomic comparison, *Research in Microbiology* **150**:755–771.

Gelfand, M.S., and Laikova, O.N., 2003, Prolegomena to the evolution of transcriptional regulation in bacterial genomes, in: *Functional Genomics Series, V. 3. Frontiers in Computational Genomics*, M.Y. Galperin, and E.V. Koonin, eds., Caister Academic Press, pp. 195–216.

Gelfand, M.S., Kozhukhin, C.G., and Pevzner, P.A., 1992, Extendable words in nucleotide sequences, *Comput. Appl. Biosci.* **8**:129–135.

Gelfand, M.S., Mironov, A.A., and Pevzner, P.A., 1996, Gene recognition via spliced sequence alignment, *Proc. Natl. Acad. Sci. USA* **93**:9061–9066.

Gelfand, M.S., Novichkov, P.S., Novichkova, E.S., and Mironov, A.A., 2000, Comparative analysis of regulatory patterns in bacterial genomes, *Briefings in Bioinformatics* **1**:357–371.

Gerasimova, A.V., Ravcheyev, D.A., Gelfand, M.S., and Rakhmaninova, A.B., 2004, Complex analysis of respiration switch in gamma-proteobacteria, in: *Proceedings of the Fourth Int. Conf. on Bioinformatics of Genome Regulation and Structure BGRS'2004*, V. 2, pp. 195–198.

Gerstein, M., 2000, Annotation of the human genome, *Science* **288**:1590.

Ghosh, D., 2000, Object-oriented transcription factors database (ooTFD), *Nucleic Acids Res.* **28**(1):308–310.

Ghosh, D., 2000, Object-oriented transcription factors database, (ooTFD), *Nucleic Acids Res.* **28**:308–310.

Gilbert, N., and Allan, J., 2001, Distinctive higher-order chromatin structure at mammalian centromeres, *Proc. Natl. Acad. Sci. USA* **98**:11949–11954.

Gilbert, N., Boyle, S., Fiegler., H., Woodfine, K., Carter, N.P., and Bickmore, W.A., 2004, Chromatin architecture of the human genome: gene-rich domains are enriched in open chromatin fibers, *Cell* **118**:555–566.

Gilson, M.K., 1993, Multiple-site titration and molecular modeling: two rapid methods for computing energies and forces for ionizable groups in proteins, *Proteins* **15**(3):266–282.

Glass, L., 1975, Classification of biological networks by their qualitative dynamics, *J. Theor. Biol.* **54**:85–107.

Glick, B.R., and Pasternak, J.J., 2002, *Molecular Biotechnology*, Mir, Moscow (Russian translation).

Godunov, S.K., 1994, *Ordinary Differential Equations with Constant Coefficients. V. 1. Boundary Value Problems,* Novosibirsk State University Press, Novosibirsk, pp. 264 (in Russian).

Godunov, S.K., 1997, *Modern Aspects of Linear Algebra*, Nauchnaya Kniga, Novosibirsk, pp. 390 (in Russian).

Godunov, S.K., Antonov, A.G., Kirilyuk, O.P., and Kostin, V.I., 1988, *Guaranteed Accuracy of Solutions of Systems of Linear Equations in Euclidian Spaces*, Nauka, Novosibirsk, Siberian Branch, pp. 456 (in Russian).

Goesmann, A., Haubrock, M., and Meyer, F., 2002, Pathfinder: reconstruction and dynamic visualization of metabolic pathways, *Bioinformatics* **18**:124–129.

Gold, S., and Rangarajan, A., 1996, Soft ax to Softassign: Neural network algorithms for combinatorial optimization, *J. Artificial Neural Networks* **2**(4):384–399.

Gold, S., Lu, C-P., Rangarajan, A., Papua, S., and Mjolsness, E., 1995, New algorithms for 2D and 3D point matching: pose estimation and correspondence, *Adv. Neural Inform. Processing Systems* **7**:957–964.

Goldman, N., 1993, Statistical tests of models of DNA substitution, *J. Mol. Evol.* **36**:182–198.

Golo, V.L., and Shaitan, K.V., 2002, Dynamic attractor in Berendsen thermostat and slow dynamics of biomacromolecules, *Biofizica* **47**:611–617 (in Russian).

Golo, V.L., Salnikov, Vl.N., and Shaitan, K.V., 2004, Harmonic oscillators in the Nose-Hoover environment, *Physical. Review E* **70**:046130.

Golubyatnikov, V.P., and Makarov, E.V., 2004, Closed trajectories in the Gene Networks, in: *Proceedings of the Fourth International conference BGRS-2004; 2004 July 25–30; Novosibirsk.* Institute of Cytology and Genetics SB RAS, Novosibirsk, pp. 42–45.

Golubyatnikov, V.P., Likhoshvai, V.A., Fadeev, S.I., Ratushny, A.V., Matushkin, Yu.G., and Kolchanov, N.A., 2003, Mathematical and Computer modeling of Genetic Networks, in: *Proceedings of the Sixth International Conference Human&Computers-2003; 2003 August 26-28; Aizu.* University of Aizu, Japan, pp. 200–205.

Golubyatnikov, V.P., Volokitin, E.P., Likhoshvai, V.A., and Osipov, A.F., 2004, Hopf bifurcation in the Gene Networks models, in: *Proceedings of the Fourth International conference BGRS-2004, 2004 July 25–30; Novosibirsk,* Institute of Cytology and Genetics SB RAS, Novosibirsk, pp. 46–48.

Goodman, M., Czelusniak, J., Moore, G.W., Romero-Herrera, A.E., and Matsuda, G., 1979, Fitting the gene lineage into its species lineage. A parsimony strategy illustrated by cladograms constructed from globulin sequences, *Syst. Zool.* **28**:132–163.

Goodwin, T.J., and Poulter, R.T., 2001, The diversity of retrotransposons in the yeast *Cryptococcus neoformans*, *Yeast.* **18**(9):865–880.

Gorban, A., Zinovyev, A., and Popova, T., 2003, Seven clusters in genomic triplet distributions. *In Silico Biol.* **3**:0039. (e-print: http://xxx.lanl.gov/abs/cond-mat/0305681).

Gorban, A., Zinovyev, A., and Popova, T., 2005, Four basic symmetry types in the universal 7-cluster structure of 143 complete bacterial genomic sequences, *In Silico Biol.* **5**:0025 http://www.bioinfo.de/isb/2005/05/0025/

Gorban', A.N., and Khlebopros, R.G., 1988, *Darwin's Demon: The Idea of Optimality and Natural Selection*, Nauka, Moscow.

Gorbunov, K.Yu., and Lyubetsky, V.A., 2005, Detection of ancestral genes introducing incongruence between gene and species trees, *Mol. Biol. (Mosk).* (in print).

Gordon, L., Chervonenkis, A., Gammerman, A., Shahmuradov, I.A., and Solovyev, V.V., 2003, Sequence alignment kernel for recognition of promoter regions, *Bioinformatics* **19**:1964–1971.

Goriely, S., Demonte, D., Nizet, S., De Wit, D., Willems, F., Goldman, M., and Lint, C. van, 2003, Human IL-12(p35) gene activation involves selective remodeling of a single nucleosome within a region of the promoter containing critical Sp1-binding sites, *Blood* **101**:4894–4902.

Gorski, K., Carneiro, M., and Schibler, U., 1986, Tissue-specific *in vitro* transcription from the mouse albumin promoter, *Cell* **47**:767–776.

Grabowski, B., and Kelman, Z., 2003, Archeal DNA replication: eukaryal proteins in a bacterial context, *Annu. Rev. Microbiol.* **57**:487–516.

Granovsky, A.A., 2004, PC GAMESS. Ref Type: Computer Program.

Gray, M.W., 1992, The endosymbiont hypothesis revisited, *Int. Rev. Cytol.* **141**:233–357.

Grol, C.J., and Jasen, J.M., 1996, The high affinity melatonin binding site probes with conformationally constrained ligands. II. Homology modeling of the receptor, *Bioorg. and Med. Chem.* **4**:1333–1339.

Grundy, F.J., and Henkin, T.M., 1998, The S-box regulon: a new global transcription termination control system for methionine and cysteine biosynthesis genes in gram-positive bacteria, *Mol. Microbiol.* **30**:737–749.

Gruschus, J.M., Tsao, D.H., Wang, L.H., Nirenberg, M., and Ferretti, J.A., 1997, Interactions of the vnd/NK-2 homeodomain with DNA by nuclear magnetic resonance spectroscopy: basis of binding specificity, *Biochemistry* **36**(18):5372–5380.

Guarino, N., 1995, Formal ontology, conceptual analysis and knowledge representation, *Intl. J. Human and Computer Studies* **45**(5, 6):625–640.

Guigo, R., Muchnik, I., and Smith, T., 1996, Reconstruction of ancient molecular phylogeny, *Mol. Phyl. Evol.* **6**:189–213.

Gultyaev, A.P., Batenburg, F.H.D. van, and Pleij, C.W.A., 1995, The computer simulation of RNA folding pathways using a genetic algorithm, *J. Mol. Biol.* **250**:37–51.

Gumucio, D.L., Heilstedt-Williamson, H., Gray, T.A., Tarle, S.A., Shelton, D.A., Tagle, D.A., Slightom, J.L., Goodman, M., and Collins, F.S., 1992, Phylogenetic footprinting reveals a nuclear protein which binds to silencer sequences in the human gamma and epsilon globin genes, *Mol. Cell. Biol.* **12**:4919–4929.

Gunsteren, W.F. van, and Berendsen, H.J.C., 1987, in: *Gromos-87 Manual.* Biomos BV. Nijenborgh 4, 9747 AG Groningen, the Netherlands.

Guo, F.B., Ou, H.Y., and Zhang, C.T., 2003, ZCURVE: a new system for recognizing protein-coding genes in bacterial and archaeal genomes, *Nucleic Acids Res.* **31**:1780–1789.

Gur-Arie, R., Cohen, T.J., Eaten, Y., et al., 2000, Simple sequence repeats in *Escherichia coli*: abundance, distribution, composition, and polymorphism, *Genome Res.* **10**:62–71.

Gusev, V.D., Nemytikova, L.A., and Chuzhanova, N.A., 1999, On the complexity measures of genetic sequences, *Bioinformatics* **15**:994–999.

Gutteridge, A., Bartlett, G.J., and Thornton, J.M., 2003, Using a neural network and spatial clustering to predict the location of active sites in enzymes, *J. Mol. Biol.* **330**:719–734.

Haeggstrom, J., 2000, Structure, function, and regulation of leukotriene A4 hydrolase, *Am. J. Respir. Crit. Care Med.* **161**:S25–S31.

Hahn, E., Wild, P., Hermanns, U., Sebbel, P., Glockshuber, R., Haner, M., Taschner, N., Burkhard, P., Aebi, U., and Muller, S.A., 2002, Exploring the 3D molecular architecture of *Escherichia coli* type 1 pili, *J. Mol. Biol.* **323**(5):845–857.

Halperin, A., Buhot, A., and Zhulina, E.B., 2004, Sensitivity, specificity, and the hybridization isotherms of DNA chips, *Biophys. J.* **86**(2):718–730.

Halperin, S.A., Scheifele, D., Mills, E., Guasparini, R., Humphreys, G., Barreto, L., and Smith, B., 2003, Nature, evolution, and appraisal of adverse events and antibody response associated with the fifth consecutive dose of a five-component acellular pertussis-based combination vaccine, *Vaccine* **21**:2298–2306.

Harari F., 1973, *The Graph Theory*, Mir, Moscow (Russian translation).

Hardenbol, P., Wang, J.C., and Van Dyke, M.W., 1997, Identification of preferred hTBP DNA binding sites by the combinatorial method REPSA, *Nucleic Acids Res.* **25**:3339–3344.

Harley, C.B., and Reynolds, R., 1987, Analysis of *Escherichia coli* promoter sequences, *Nucleic Acids Res.* **15**:2343–2361.

Hars, U., Horlacher, R., Boos, W., Welte, W., and Diederichs, K., 1998, Crystal structure of the effector-binding domain of the trehalose repressor of *Escherichia coli*, a member of the

LacI family, in its complexes with inducer trehalose-6-phosphate and noninducer trehalose, *Protein Sci.* **7**:2511–2521.

Havecker, E.R., Gao, X., and Voytas, D.F., 2004, The diversity of LTR retrotransposons, *Genome Biol.* **5**(6):225.

Hayes, W.S., and Borodovsky, M., 1998, How to interpret an anonymous bacterial genome: machine learning approach to gene identification, *Genome Res.* **8**:1154–1171.

Hedtke, B., Börner, T., and Weihe, A., 2000, One RNA polymerase serving two genomes, *EMBO Rep.* **1**:435–440.

Heinemeyer, T., Chen, X., Karas, H., Kel, A.E., Kel, O.V., Liebich, I., Meinhardt, T., Reuter, I., Schacherer, F., and Wingender, E., 1999, Expanding the TRANSFAC database towards an expert system of regulatory molecular mechanisms, *Nucleic Acids Res.* **27**:318–322.

Hekstra, D., Taussig, A.R., Magnasco, M., and Naef, F., 2003, Absolute mRNA concentrations from sequence-specific calibration of oligonucleotide arrays, *Nucleic Acids. Res.* **31**(7):1962–1968.

Held, G.A., Grinstein, G., and Tu, Y., 2003, Modeling of DNA microarray data by using physical properties of hybridization, *Proc. Natl. Acad. Sci. USA* **100**(13):7575–7580.

Helden, J. van, Naim, A., and Mancuso, R., 2000, Representing and analysing molecular and cellular function using the computer, *Biol. Chem.* **381**(9–10):921–935.

Hendlich, M., Bergner, A., Gunther, J., and Klebe, G., 2003, Relibase: design and development of a database for comprehensive analysis of protein-ligand interactions, *J. Mol. Biol.* **326**:607–620.

Henikoff, S., and Henikoff, J.G., 1993, Performance evaluation of amino acid substitution matrices, *Proteins* **17**:49–61.

Henkin, T.M., 1994, tRNA-directed transcription antitermination, *Mol. Microbiol.* **13**:381–387.

Hersberg, R., Altuvia, S., and Margalit, H., 2003, A survey of small RNA-encoding genes in *Escherichia coli*, *Nucleic Acids Res.* **31**:1813–1820.

Hertz, G.Z., and Stormo, G.D., 1996, *Escherichia coli* promoter sequences: Analysis and prediction, *Methods Enzymol.* **273**:30–42.

Hertz, G.Z., and Stormo, G.D., 1999, Identifying DNA and protein patterns with statistically significant alignments of multiple sequences, *Bioinformatics* **15**:563–577.

Herzel, H., and Große, I., 1997, Correlations in DNA sequences. The role of protein coding segments, *Physical Review E.* **55**:800–810.

Hetz, G.Z., and Stormo, G.D., 1996, *Escherichia coli* promoter sequences: analysis and prediction, *Meth. Enzymol.* **273**:30–42.

Hillis, D.M., and Huelsenbeck, J.P., 1992, Signal, noise, and reliability in molecular phylogenetic analyses, *J. Hered.* **83**:189–195.

Hivzer, J., Rozenberg, H., Frolow, F., Rabinovich, D., and Shakked, Z., 2001, DNA bending by an adenine-thymine tract and its role in gene regulation, *Proc. Natl. Acad. Sci. USA* **98**:8490–8495.

Hofacker, I.L., 2003, Vienna RNA secondary structure server, *Nucleic Acids Res.* **31**(13):3429–3431.

Holmes, I., and Durbin, R., 1998, Dynamic programming alignment accuracy, *J. Comput. Biol.* **5**:493–504.

Holodniok, M., Klic, A., Kubicek, M., and Marek, M., 1991, *Methods of Analysis of Nonlinear Dynamical Models*, Mir, Moscow, pp. 365 (in Russian).

Hooft, R.W.W., Sander, C., and Vriend, G., 1996, The PDBFINDER database: a summary of PDB, DSSP and HSSP information with added value, *CABIOS* **12**:525–529.

Hooser, A.A. van, Mancini, M.A., Allis, C.D., Sullivan, K.F., and Brinkley, B.R., 1999, The mammalian centromere: structural domains and the attenuation of chromatin modeling, *FASEB J.* **13:**Suppl. 2:S216–220.

Hoover, W.G., 2001, *Time Reversibility. Computer Simulations and Chaos*, World Scientific, Singapore.

Horton, P.B., and Kanehisa, M., 1992, An assessment of neural network and statistical approaches for prediction of *Escherichia coli* promoter sites, *Nucleic Acids Res.* **20**:4331–4338.

Hosid, S., Trigone I.N., and Bolton, A., 2004, Sequence periodicity of *Escherichia coli* is concentrated in intergenic regions, *BFC Mol. Biol.* **5**:14.

Hucka, M., Finney, A., Saburo, HIM., Balouris, H., Doyle, J.C., Kitano, H., Akin, I.P., Bornstein, B.J., Bray, D., Cornish-Bowden, A., Cuellar, A.A., Drano, S., Gilles, E.N.D., Inkle, M., Gor, V., Groaning, II., Hedley, T.J., Hodgeman, T.J., Homey, JTH., Hunter, P.J., Judy, S.N., Hansberger, J.P.L., Kremlin, A., Kumar, U., Le Novae, N., Lowe, L.S., Lucia, D., Mendes, P., Mjolsness, E.N.D., Nakayama, Y., Nelson, M.R., Nielsen, P.D.F., Sakurada, T., Schiff, J.C., Shapiro, B.E., Shimizu, S., Spence, H.A.D., Stalling, J., Takahashi, K., Tomita, M., Wagner, J., and Wang, J., 2003, The systems biology markup language (SBML): a medium for representation and exchange of biochemical network models, *Bioinformatics* **19**:513–523.

Huerta, A.M., and Collado-Vides, J., 2003, Sigma-70 promoters in *Escherichia coli*: specific transcription in dense regions of overlapping promoter-like signals, *J. Mol. Biol.* **333**:261–278.

Hughes, T.R., Robinson, M.D., Mitsakakis, N., and Johnston, M. 2004, The promise of functional genomics: completing the encyclopedia of a cell, *Curr. Opin. Microbiol.* **7**:546–554.

Hyde-DeRuyscher, R.P., Jennings, E., and Shenk, T., 1995, DNA binding sites for the transcriptional activator/repressor YY1, *Nucleic Acids Res.* **2**:4457–4465.

Hyvonen, M.T., Oorni, K., Kovanen, P.T., and Ala-Korpela, M., 2001, Changes in a phospholipid bilayer induced by the hydrolysis of a phospholipase A2 enzyme: a molecular dynamics simulation study, *Biophys. J.* **80**:565–578.

Ignatieva, E.V., Ananko, E.A., Podkolodnaya, O.A., Stepanenko, I.L., Khlebodarova, T.M., Merkulova, T.I., Pozdnyakov, M.A., Proscura, A.L., Grigorovich, D.A., Podkolodny, N.L., Naumochkin, A.N., Romashchenko, A.G., and Kolchanov, N.A., 2004, Transcription regulatory regions database (TRRD): description of transcription regulation and the main capabilities of the database, in: *Bioinformatics of Genome Regulation and Structure*, N.. Kolchanov, and R. Hofestaedt, eds., Kluwer Academic Publishers, Boston/Dordrecht/London, pp. 81–92.

Ikeda, M., and Nakagawa, S., 2003, The *Corynebacterium glutamicum* genome: features and impacts on biotechnological processes, *Appl. Microbiol. Biotechnol.* **62**:99–109.

Ingham, P.W., 1988, The molecular genetics of embryonic pattern formation in *Drosophila*, *Nature* **335**:25–34.

Ioshikhes, I., and Trifonov, E.N., 1993, Nucleosomal DNA sequence database, *Nucleic Acids Res.* **21**:4857–4859.

Ioshikhes, I., and Trifonov, E.N., 1993, Nucleosomal DNA sequence database, *Nucleic Acids Res.* **21**:4857–4859.

Ioshikhes, I., Bolshoy, A., Derenshteyn, K., Borodovsky, M., and Trifonov, E.N., 1996, Nucleosome DNA sequence pattern revealed by multiple alignment of experimentally mapped sequences, *J. Mol. Biol.* **262**:129–139.

Irizarry, R.A., Bolstad, B.M., Collin, F., Cope, L.M., Hobbs, B., and Speed, T.P., 2003, Summaries of affymetrix GeneChip probe level data, *Nucleic Acids. Res.* **31**(4):e15.

Ivanisenko, V.A., and Eroshkin, A.M., 1997, Search for sites with functionally important substitutions in sets of related or mutant protein, *Mol. Biol. (Mosk).* **31**:749–755.

Ivanisenko, V.A., Eroshkin, A.M., and Kolchanov, N.A., 2005b, WebProAnalyst: an interactive tool for analysis of quantitative structure-activity relationships in protein families, *Nucleic Acids Res.* Web server issue (in press).

Ivanisenko, V.A., Pintus, S.S., Grigorovich, D.A., and Kolchanov, N.A., 2005a, PDBSite: a database of the 3D structure of protein functional sites, *Nucleic Acids Res.* **33**:D183–D187.

Ivanisenko, V.A., Pintus, S.S., Grigorovich, D.A., and Kolchanov, N.A., 2004, PDBSiteScan: a program for searching for active, binding and posttranslational modification sites in the 3D structures of proteins, *Nucleic Acids Res.* **32**:W549–W554.

Ivanov, A.A., Voronkov, A.E., Baskin, I.I., Palyulin, V.A., and Zefirov, N.S., 2004, The study of the mechanism of binding of human MLlA melatonin receptor ligands using molecular modeling, *Dokl. Biochem. Biophys.* **394**:49–52.

Jacob, F., 1993, Du répresseur à I'agrégulat, Sciences de la vie/Life sciences, pp. 316, 331–333.

Jaeger, J., Blagov, M., Kosman, D., Kozlov, K.N., Manu, Myasnikova, E., Surkova, S., Vanario-Alonso, C.E., Samsonova, M., Sharp, D.H., and Reinitz, J., 2004, Dynamical analysis of regulatory interactions in the gap gene system of *Drosophila melanogaster*, *Genetics* **167**(4):1721–1737.

Jaeger, J.A., Turner, D.H., and Zuker, M., 1989, Improved predictions of secondary structures for RNA, *Proc. Natl. Acad. Sci. USA* **86**:7706.

Jain, A.K., Murty, M.N., and Flynn, P.J., 1999, Data clustering: a review, *ACM Computing Surveys* **31**(3):264–323.

Jakubke, H.-D., and Jeschkeit, H., 1985, *Aminosaeuren, Peptide, Proteine*, Mir, Moscow (Russian translation).

James, S.B., and Dewey, T.G., 1997, Multifractal analysis of solvent accessibilities in proteins, *Phys. Rev. E* **52**:880–887.

Janin, J., 1999, Wet and dry interfaces: the role of solvent in protein-protein and protein-DNA recognition, *Structure Fold Des.* **7**(12):R277–R279.

Jayaram, B., and Jain, T., 2004, The role of water in protein-DNA recognition, *Annu. Rev. Biophys. Biomol. Struct.* **33**:343–361.

Johnson, B.V., Bert, A.G., Ryan, G.R., Condina, A., and Cockerill, P.N., 2004, Granulocyte-macrophage colony-stimulating factor enhancer activation requires cooperation between NFAT and AP-1 elements and is associated with extensive nucleosome reorganization, *Mol. Cell Biol.* **18**:7914–7930.

Jones, B.K., Monks, B.R., Liebhaber, S.A., and Cooke, N.E., 1995, The human growth hormone gene is regulated by a multicomponent locus control region, *Mol. Cell. Biol.* **15**(12):7010–7021.

Jones, D.T., 1999, Protein secondary structure prediction based on position-specific scoring matrices, *J. Mol. Biol.* **292**:195–202.

Jones, S., Barker, J.A., Nobeli, I., and Thornton, J.M., 2003, Using structural motif templates to identify proteins with DNA binding function, *Nucleic Acids Res.* **31**:2811–2823.

Jönsson, H., Heisler, M., Reddy, G.V., Agrawal, V., Gor, V., Shapiro, B.E., Mjolsness, E., and Meyerowitz, E.M., 2005, Modeling the organization of the WUSCHEL expression domain in the shoot apical meristem, *Bioinformatics* (in press).

Jönsson, H., Shapiro, B.E., Meyerowitz, E.M., and Mjolsness E., 2003, Signaling in multicellular models of plant development, in: *On Growth, Form, and Computers*, P. Bentley and S. Kumar, eds., Academic Press, pp. 156–161.

Jorgensen, W.L., and Tirado-Rives, J., 1988, The OPLS potential functions for proteins. Energy minimization for crystals of cyclic peptides and crambin, *J. Am. Chem. Soc.* **110**:1657–1666.

Jorgensen, W.L., Chandrasekhar, J., Madura, J.D., Impey, R.W., and Klein, M.L., 1983, Comparison of simple potential functions for simulating liquid water, *J. Chem. Phys.* **79**:926–935.

Kabakcioglu, A., Kanter, I., Vendruscolo, M., and Domany, E., 2002, Statistical properties of contact vectors, *Phys. Rev. E. Soft. Matter. Phys.* **65**(4 Pt. 1):041904.

Kabsch, W., and Sander, C., 1983, Dictionary of protein secondary structure: pattern recognition of hydrogen-bonded and geometrical features, *Biopolymers* **22**(12):2577–2637.

Kajava, A.V., 2001, Proteins with repeated sequences – structural prediction and modeling, *J. Struct. Biol.* **134**:132–144.

Kalinina, O.V., Mironov, A.A., Gelfand, M.S., and Rakhmaninova, A.B., 2004, Automated selection of positions determining functional specificity of proteins by comparative analysis of orthologous groups in protein families, *Protein Science* **13**:443–456.

Kalinina, O.V., Novichkov, P.S., Mironov, A.A., Gelfand, M.S., and Rakhmaninova, A.B., 2004, SDPpred: a tool for prediction of amino acid residues that determine differences in functional specificity of homologous proteins, *Nucleic Acids Res.* **32**:W424–W428.

Kalinowski, J., Bathe, B., Bartels, D., Bischoff, N., Bott, M., Burkovski, A., Dusch, N., Eggeling, L., Eikmanns, B.J., Gaigalat, L., Goesmann, A., Hartmann, M., Huthmacher, K., Kramer, R., Linke, B., McHardy, A.C., Meyer, F., Mockel, B., Pfefferle, W., Puhler, A., Rey, D.A., Ruckert, C., Rupp, O., Sahm, H., Wendisch, V.F., Wiegrabe, I., and Tauch, A., 2003, The complete *Corynebacterium glutamicum* ATCC 13032 genome sequence and its impact on the production of L-aspartate-derived amino acids and vitamins, *J. Biotechnol.* **104**:5–25.

Kamzolova, S.G., and Postnikova, G.B., 1981, Spin-labeled nucleic acids, *Quart. Rev. Biophys.* **14**:223–288.

Kamzolova, S.G., Ivanova, N.N., and Kamzolov, S.S., 1999, Long-range interactions in T2 DNA during its complex formation with RNA polymerase from *E. coli*, *J. Biol. Phys.* **24**:157–161.

Kamzolova, S.G., Sivozhelezov, V.S., Sorokin, A.A., Dzhelyadin, T.R., Ivanova, N.N., and Polozov, R.V., 2000, RNA polymerase-promoter recognition. Specific features of electrostatic potential of 'early' T4 phage DNA promoters, *J. Biomol. Struct. Dyn.* **18**(3):325–334.

Kamzolova, S.G., Sorokin, A.A., Dzhelyadin, T.R., Osypov, A.A., and Beskaravainy, P.M., 2004, Analysis of oligonucleotide composition in DNA of *E. coli* genome and promoter sites, in: *Proceedings of the Fourth International Conference on Bioinformatics of Genome Regulation and Structure (BGRS'2004)*; July 25–30 2004; Novosibirsk, Russia. IC&G, Novosibirsk, V. 1, pp. 77–79.

Kananyan, G.Ch., et al., 1981, Enlarged model of lambda phage ontogenesis, *J. Theor. Biol.* **90**:301–315.

Kaplan, W., and Littlejohn, T.G., 2001, Swiss-PDB Viewer (Deep View), *Brief. Bioinform.* **2**(2):195–197.

Karlin, S., Dembo, A., and Kawabata, T., 1990, Statistical composition of high-scoring segments from molecular sequences, *Ann. Stat.* **18**:571–581.

Kashuba, E., Kashuba, V., Pokrovskaja, K., Klein, G., and Szekely, L., 2000, Epstein-Barr virus encoded nuclear protein EBNA-3 binds XAP-2, a protein associated with Hepatitis B virus X antigen, *Oncogene* **19**:1801–1806.

Kashuba, E., Pokrovskaja, K., Klein, G., and Szekely, L., 1999, Epstein-Barr virus-encoded nuclear protein EBNA-3 interacts with the epsilon-subunit of the T-complex protein 1 chaperonin complex, *J. Hum. Virol.* **2**:33–37.

Katti, M.V., Sami-Subbu, R., Ranjekar, P.K., and Gupta, V.S., 2000, Amino acid repeat patterns in protein sequences: their diversity and structural-functional implications, *Protein Sci.* **9**:1203–1209.

Katz, L., and Burge, COB. 2003, Widespread selection for local RNA secondary structure in coding regions of bacterial genes, *Genome Res.* **13**:2042–2051.

Kauffman, S.A., 1974, The large scale structure and dynamics of control circuits: an ensemble approach, *J. Theor. Biol.* **44**:167–190.

Kawahara, Y., Takahashi-Fuke, K., Shimizu, E., Nakamatsu, T., and Nakamori, S., 1997, Relationship between the glutamate production and the activity of 2-oxoglutarate dehydrogenase, *Brevibacterium lactofermentum, Biosci. Biotechnol. Biochem.* **61**:1109–1112.

Kellis, M., Patterson, N., Endrizzi, M., Birren, B., and Lander, E.S., 2003, Sequencing and analysis of yeast species to identify genes and regulatory elements, *Nature* **423**:241–254.

Kel-Margoulis, O.V., Kel, A.E., Reuter, I., Deineko, I.V., and Wingender, E., 2002, TRANSCompel: a database on composite regulatory elements in eukaryotic genes, *Nucleic Acids Res.* **30**:332–334.

Khorasanizadeh, S., 2004, The nucleosome: from genomic organization to genomic regulation, *Cell* **116**(2):259–272.

Kielman, M.F., Smits, R., Bernini, L.F., 1994, Localization and characterization of the mouse alpha-globin locus control region, *Genomics* **21**:431–433.

Kieser, A., Kilger, E., Gires, O., Ueffing, M., Kolch, W., and Hammerschmidt, W., 1997, Epstein-Barr virus latent membrane protein-1 triggers AP-1 activity via the c-Jun N-terminal kinase cascade, *EMBO J.* **16**(21):6478–6485.

Kikoin, I.K., 1976, *Tables of physical quantities. A Handbook*, Atomizdat, Moscow (in Russian).

Kim, S.J., and Lee, Y.S., 2002, In silico metabolic pathway modeling and analysis of Mycoplasma pneumoniae, *Genome Informatics* **12**:298–299.

Kimura, E., 2003, Metabolic engineering of glutamate production, in: *Advances in Biochemical Engineering Biotechnology. V. 79. Microbial Production of L-Amino Acids*, R. Faurie, and J. Thommel, eds., Springer, Berlin, Heidelberg, New York, pp. 37–57.

Kimura, M., 1983, *The Neutral Theory of Molecular Evolution*, Cambridge Univ. Press, Cambridge.

Kinzler, K.W., and Vogelstein, B., 1989, Whole genome PCR: application to the identification of sequences bound by gene regulatory proteins, *Nucleic Acids Res.* **17**:3645–3653.

Kirby, H., Rickinson, A., and Bell, A., 2000, The activity of the Epstein-Barr virus BamHI W promoter in B cells is dependent on the binding of CREB/ATF factors, *J. Gen. Virol.* **81**:1057–1066.

Kiyama, R., and Trifonov, E.N., 2002, What positions nucleosomes? A model, *FEBS Lett.* **523**:7–11.

Klingenhoff, A., Frech, K., Quandt, K., and Werner, T., 1999, Functional promoter modules can be detected by formal models independent of overall nucleotide sequence similarity, *Bioinformatics* **15**(3):180–186.

Knutson, J.C., 1990, The level of c-fgr RNA is increased by EBNA-2, an Epstein-Barr virus gene required for B-cell immortalization, *J. Virol.* **64**(6):2530–2536.

Ko, J., Murga, L.F., Andre, P., Yang, H., Ondrechen, M.J., Williams, R.J., Agunwamba, A., and Budil, D.E., 2005, Statistical criteria for the identification of protein active sites using theoretical microscopic titration curves, *Proteins: Structure Function Bioinformatics* (in press).

Kobrinsky, N.E., and Trakhtenbrot, B.A., 1962, Introduction into the Finite Automata Theory, PhysMathGiz, Moscow, 440 p. (in Russian).

Koehl, P., and Delarue, M., 1994, Polar and nonpolar atomic environments in the protein core: implications for folding and binding, *Proteins: Struct. Funct. Genet.* **20**:264–278.

Kogai, V.V., 2002, Application of parameter continuation for numerical study of periodic solutions of autonomous systems of ordinary differential equations, *Vestnik NGU* **4**:40–48 (in Russian).

Kogai, V.V., and Fadeev, S.I., 2001, Application of parameter continuation basing on multiple shooting method for numerical study nonlinear boundary value problems, *Sib. Zh. Industr. Matem.* **4**:83–101 (in Russian).

Kohonen, T., 2001, *Self-Organizing maps*, Springer-Verlag.

Kohonen, T., Kaski, S., Lagus, K., Salojärvi, J., Honkela, J., Paatero, A., and Saarela, A., 2000, Self organization of a massive document collection, *IEEE Transactions on Neural Networks* (Special Issue on Neural Networks for Data Mining and Knowledge Discovery) **11**(3):574–585.

Kolchanov N.A., et al., 1998, GeneNet: a gene network database and its automated visualization, *Bioinformatics* **14**(6):529–537.

Kolchanov N.A., et al., 2005, GeneNet in 2005, *Nucleic Acids Res.* **33** Database Issue:D425–D427.

Kolchanov, N.A., Ananko, E.A., Kolpakov, F.A., Podkolodnaya, O.A., Ignatieva, E.V., Goryachkovskaya, T.N., and Stepanenko, I.L., 2000, Gene Networks, *Mol. Biol. (Mosk).* **34**(4):533–544.

Kolchanov, N.A., Ananko, E.A., Likhoshvai, V.A., Podkolodnaya, O.A., Ignatieva, E.V., Ratushny, A.V., and Matushkin, Yu.G., 2002, Gene networks description and modeling in the GeneNet system, Chapter 7, in: *Gene Regulation and Metabolism*, J. Collado-Vides, and R. Hofestadt, eds., The MIT Press, Cambridge, Massachusetts. pp. 149–180.

Kolchanov, N.A., Ananko, E.A., Podkolodnaya, O.A., Ignatieva, E.V., Stepanenko, I.L., Kel-Margoulis, O.V., Kel, A.E., Merkulova, T.I., Goryachkovskaya, T.N., Busygina, T.V., Kolpakov, F.A., Podkolodny, N.L., Naumochkin, A.N., and Romashchenko, A.G., 1999, Transcription regulatory regions database (TRRD): its status in 1999, *Nucleic Acids Res.* **27**(1):303–306.

Kolchanov, N.A., Ignatieva, E.V., Ananko, E.A., Podkolodnaya, O.A., Stepanenko, I.L., Merkulova, T.I., Pozdnyakov, M.A., Podkolodny, N.L., Naumochkin, A.N., and Romashchenko, A.G., 2002, Transcription regulatory regions database (TRRD): its status in 2002, *Nucleic Acids Res.* **30**(1):312–317.

Kolchanov, N.A., Podkolodnaya, O.A., Ananko, E.A., Ignatieva, E.V., Stepanenko, I.L., Kel-Margoulis, O.V., Kel, A.E., Merkulova, T.I., Goryachkovskaya, T.N., Busygina, T.V., Kolpakov, F.A., Podkolodny, N.L., Naumochkin, A.N., Korostishevskaya, I.M., Romashchenko, A.G., and Overton, G.C., 2000, Transcription regulatory regions database (TRRD): its status in 2000, *Nucleic Acids Res.* **28**(1):298–301.

Kolchanov, N.A., Pozdnyakov, M.A., Orlov, Yu.L., Vishnevsky, O.V., Podkolodny, N.L., Vityaev, E.E., and Kovalerchuk, B., 2003, Computer System 'Gene Discovery' for Promoter Structure Analysis, in: *Artificial Intelligence and Heuristic Methods in Bioinformatics,* P. Frasconi and R. Shamir, eds., IOS Press, Amsterdam, pp. 173–192.

Kondrashov, A.S., 2003, Direct estimates of human per nucleotide mutation rates at 20 loci causing Mendelian diseases, *Hum. Mutat.* **21**(1):12–27.

Konstantinov, Y., Katyshev, A., Subota, I., and Tarasenko, V., 2003, Effects of redox conditions on DNA binding activity of mitochondrial topoisomerase I, *Maize Gen. Coop. Newslett.* **77**:37–38.

Kontopidis, G., Andrews, M.J., McInnes, C., Cowan, A., Powers, H., Innes, L., Plater, A., Griffiths, G., Paterson, D., Zheleva, D.I., Lane, D.P., Green, S., Walkinshaw, M.D., and

Fischer, P.M., 2003, Insights into cyclin groove recognition: complex crystal structures and inhibitor design through ligand exchange, *Structure (Camb)* **11**:1537–1546.

Koonin, E.V., Fedorova, N.D., Jackson, J.D., Jacobs, A.R., Krylov, D.M., Makarova, K.S., Mazumder, R., Mekhedov, S.L., Nikolskaya, A.N., Rao, B.S., Rogozin, I.B., Smirnov, S., Sorokin, A.V., Sverdlov, A.V., Vasudevan, S., Wolf, Y.I., Yin, J.J., and Natale, D.A., 2004, A comprehensive evolutionary classification of proteins encoded in complete eukaryotic genomes, *Genome Biol.* **5**:R7.1–R7.28 (genomebiology.com/2004/5/2/R7).

Korotkov, E.V., and Korotkova, M.A., 1995, DNA regions with latent periodicity in some human clones, *DNA Sequence* **5**:353–358.

Korotkov, E.V., and Kudryaschov, N., 2001, Latent periodicity of many genes, *Genome Informatics* **12**:437–39.

Korotkov, E.V., and Phoenix, D.A., 1997, Latent periodicity of DNA sequences of many genes, *Pac. Symp. Biocomput.* 222–231.

Korotkov, E.V., Korotkova, M.A., and Kudryashov, N.A., 2003, Information decomposition method for analysis of symbolical sequences, *Phys. Let. A* **312**:198–210.

Korotkova, M.A., Korotkov, E.V., and Rudenko, V.M., 1999, Latent periodicity in protein sequences, *J. Mol. Model.* **5**:103–115.

Koslowsky, T.J., and Yuille, ALL., 1994, The invisible hand algorithm: solving the assignment problem with statistics physics, *Neural Networks* **7**:477–490.

Kosman, D., Small, S., and Reinitz, J., 1998, Rapid preparation of a panel of polyclonal antibodies to *Drosophila* segmentation proteins, *Dev. Genes Evol.* **208**:290–294.

Kotelnikova, E.A., Makeev, V.Y., and Gelfand, M.S., 2005, Evolution of transcription factor DNA binding sites, *Gene* (in press).

Kraft, R., and Zien, J., 2004, Mining anchor terms for query refinement, in: *WWW2004 ACM*, May 17–22, 2004, New York, NY-USA.

Krogh, A., Mian, S., and Haussler, D., 1994, A hidden Markov model that finds genes in *E. coli* DNA, *Nucleic Acids Res.* **22**:4768–4778.

Krylov, V.I., Bobkov, V.V., and Monastyrny, P.I., 1972, *Computational Methods of Higher Mathematics*, Izd. Vysshaya Shkola, Minsk, pp. 584 (in Russian).

Kumar, S., Tamura, K., Jakobsen, I.B., and Nei, M., 2001, MEGA2: molecular evolutionary genetic analysis software, *Bioinformatics* **17**(12):1244–1245.

Kurland, C.G., and Andersson, S.G., 2000, Origin and evolution of the mitochondrial proteome, *Microbiol. Mol. Biol. Rev.* **64**:786–820.

Kurokawa, R., Yu, V.C., Naar, A., et al., 1993, Differential orientations of the DNA-binding domain and carboxy-terminal dimerization interface regulate binding site selection by nuclear receptor heterodimers, *Genes Dev.* **7**:1423–1435.

Kuznetsov, V.A., 2001, Distribution associated with stochastic processes of gene expression in a single eukaryotic cell, *EURASIP J. on Applied Signal Processing*, **4**:285–296.

Kuznetsov, V.A., 2002, Statistics of the numbers of transcripts and protein sequences encoded in the genome, in: *Computational and Statistical Methods to Genomics*, W. Zhang, I. Shmulevich, eds., Kluwer, Boston, pp. 125–171.

Kuznetsov, V.A., 2003a, Family of skewed distributions associated with the gene expression and proteome evolution, *Signal Processing* **83**(4):889–910.

Kuznetsov, V.A., 2003b, A stochastic model of evolution of conserved protein coding sequence in the archaeal, bacterial and eukaryotic proteomes, *Fluctuation and Noise Letters* **3**(3):L295–L324.

Kuznetsov, V.A, Pickalov, V.V., Senko, O.V., Knott, G.D., 2002, Analysis of evolving proteomes: Predictions of the number of protein domains in nature and the number of genes in eukaryotic organisms, *J. Biol. Systems* **10**(4):381–407.

Kuznetzov, Yu.A., 1995, *Elements of Applied Bifurcation Theory*, Springer-Verlag, New York, pp. 1–591.

Labeots, L.A., and Weiss, M.A., 1997, Electrostatic and hydration at the homeodomain-DNA interface: chemical probes of an interfacial water cavity, *J. Mol. Biol.* **269**(1): 113–128.

LaLonde, J.M., Levenson, M.A., Roe, J.J., Bernlohr, D.A., and Banaszak, L.J., 1994, Adipocyte lipid-binding protein complexed with arachidonic acid. Titration calorimetry and x-ray crystallographic studies, *J. Biol. Chem.* **269**:25339–25347.

Landau, G.M., Schmidt, J.P, and Sokol, D., 2001, An algorithm for approximate tandem repeats, *J. Comp. Biol.* **8**:1–18.

Landau, L.D., and Lifshits, E.M., 1986, *Statistical Mechanics*, Nauka, Moscow (in Russian).

Landolt-Bornstein, 1989, in: *Numerical Data and Functional Relationships in Science and Technology*, W. Saenger, ed., New Series VII/1b, Springer-Verlag, Berlin.

Langermann, S., Mollby, R., Burlein, J.E., Palaszynski, S.R., Auguste, C.G., DeFusco, A., Strouse, R., Schenerman, M.A., Hultgren, S.J., Pinkner, J.S., Winberg, J., Guldevall, L., Soderhall, M., Ishikawa, K., Normark, S., and Koenig, S., 2000, Vaccination with FimH adhesin protects cynomolgus monkeys from colonization and infection by uropathogenic *Escherichia coli*, *J. Infect. Dis.* **181**:774–778.

Laskin, A.A., Korotkov, E.V., and Kudryashov, N.A., 2004, Latent periodicity of many domains in protein sequences reflects their structure, function and evolution, in: *Bioinformatics of Genome Regulation and Structure*, N. Kolchanov, R. Hofestaedt, eds., Kluwer Acad. Publ., New York, pp. 135–144.

Laskin, A.A., Korotkov, E.V., Chaley, M.B., and Kudryashov, N.A., 2003, The locally optimal method of cyclic alignment to reveal latent periodicities in genetic texts: the NAD-binding protein sites, *Mol. Biol.* **37**:561–570.

Laskowski, R.A., Luscombe, N.M., Swindells, M.B., Thornton, J.M., 1996, Protein clefts in molecular recognition and function, *Protein Sci.* **5**:2438–2452.

Ledneva, R.K., Alekseevskii, A.V., Vasil'ev, S.A., Spirin, S.A., and Kariagina, A.S., 2001, Structural aspects of homeodomain interactions with DNA, *Mol. Biol. (Mosk).* **35**(5):764–777 (in Russian).

Lee, B., and Richards, F.M., 1971, The interpretation of protein structures: estimation of static accessibility, *J. Mol. Biol.* **55**:379–400.

Lehninger, A.L., Nelson, D.L., 1993, *Principles of Biochemistry*, 2nd ed., Worth Publishers, New York.

Leirmo, S., and Gourse, P., 1991, Factor-independent activation of *Escherichia coli* rRNA transcription. 1. Kinetic analysis of the roles of the upstream activation region and super coiling on transcription of the rrnBP1 promoter *in vitro*, *J. Mol. Biol.* **220**:555–568.

Lemak, A.S., and Balabaev, N.K., 1994, On the Berendsen thermostat, *Mol. Simulation* **13**:177–187.

Lescot, M., Dehais, P., Thijs, G., Marchal, K., Moreau, Y., Van de Peer, Y., Rouze, P., and Rombauts, S., 2002, PlantCARE, a database of plant cis-acting regulatory elements and a portal to tools for in silico analysis of promoter sequences, *Nucleic Acids Res.* **30**(1):325–327.

Letoha, T., Gaal, S., Somlai, C., Czajlik, A., Perczel, A., and Penke, B., 2003, Membrane translocation of penetratin and its derivatives in different cell lines, *J. Mol. Recognit.* **16**:272–279.

Levitsky, V.G., 2004, RECON: a program for prediction of nucleosome formation potential, *Nucleic Acids. Res.* **32**:W346–W349.

Levitsky, V.G., and Katokhin, A.V., 2003, Recognition of eukaryotic promoters using a genetic algorithm based on iterative discriminant analysis, *In Silico Biol.* **3**:81–87.

Levitsky, V.G., Ignatieva, E.V., Vasiliev, G.V., Klimova, N.V., Busygina, T.V., Merkulova, T.I., and Kolchanov, N.A., 2005, The SiteGA tool for transcription factor binding sites recognition and context analysis: significant dinucleotide features besides the canonical consensus exemplified by SF-1 binding site. *This issue.*

Levitsky, V.G., Katokhin, A.V., Podkolodnaya, O.A., Furman, D.P., and Kolchanov, N.A., 2005, NPRD: nucleosome positioning region database, *Nucleic Acids Res.* **33** Database Issue:D67–D70.

Levitsky, V.G., Podkolodnaya, O.A., Kolchanov, N.A., and Podkolodny, N.L., 2001a, Nucleosome formation potential of eukaryotic DNA: tools for calculation and promoters analysis, *Bioinformatics* **17**:998–1010.

Levitsky, V.G., Podkolodnaya, O.A., Kolchanov, N.A., and Podkolodny, N.L., 2001b, Nucleosome formation potential of exons, introns, and Alu repeats, *Bioinformatics* **17**:1062–1064.

Levitsky, V.G., Ponomarenko, M.P., Ponomarenko, J.V., Frolov, A.S., and Kolchanov, N.A., 1999, Nucleosomal DNA property database, *Bioinformatics* **15**(7–8):582–592.

Lewontin, R.C., 1974, *The Genetic Basis of Evolutionary Change*, Columbia Univ. Press, New York.

Li, W., 1997, The complexity of DNA, *Complexity* **3**:33–37.

Liang, J., Edelsbrunner, H., and Woodward, C., 1998, Anatomy of protein pockets and cavities: measurement of binding site geometry and implications for ligand design, *Protein Sci.* **7**:1884–1897.

Liao, L., Kim, S., and Tomb, J.F., 2002, Genome comparisons based on profiles of metabolic pathways, in: *Proceedings of the Sixth International Conference on Knowledge-Based Intelligent Information and Engineering Systems (KES 2002)*, Crema, Italy, pp. 469–476.

Likhoshvai, V.A., Fadeev, S.I., and Matushkin, Yu.G., 2004, The global operation modes of gene networks determined by the structure of negative feedbacks, in: *Bioinformatics of Genome Regulation and Structure*, N. Kolchanov, and R. Hofestaedt, eds., Kluwer Academic Publishers, Boston/Dordrecht/London, pp. 319–330.

Likhoshvai, V.A., Matushkin, Yu.G., and Fadeev, S.I., 2001, Relationship between a gene network graph and qualitative model of its functioning, *Mol. Biol.* **35**(6):926–932.

Likhoshvai, V.A., Matushkin, Yu.G., and Fadeev, S.I., 2003, Problems of the theory of gene network operation, *Sib. Zh. Industr. Matem.* **4**:64–80 (in Russian).

Likhoshvai, V.A., Matushkin, Yu.G., Ratushny, A.V., Ananko, E.A., Ignatieva, E.V., and Podkolodnaia, O.A., 2001, A generalized chemical-kinetic method for modeling gene networks, *Mol. Biol. (Mosk).* **35**:1072–1079.

Lim, A., and Zhang, L. 1999, WebPHYLIP: A Web Interface to PHYLIP, *Bioinformatics* **15**:1068–1069.

Ling, P.D., Hsieh, J.J., Ruf, I.K., Rawlins, D.R., and Hayward, S.D., 1994, EBNA-2 upregulation of Epstein-Barr virus latency promoters and the cellular CD23 promoter utilizes a common targeting intermediate, CBF1, *J. Virol.* **68**(9):5375–5383.

Lipman, D.J., Wilbur, W.J., 1984, Interaction of silent and replacement changes in eukaryotic coding sequences, *J. Mol. Evol.* **21**(2):161–167.

Liu, H., and Wong, L., 2003, Data mining tools for biological sequences, *J. Bioinformatics Comput. Biol.* **1**(1):139–167.

Livingstone, C.D., and Barton, G.J., 1993, Protein sequence alignments: a strategy for the hierarchical analysis of residue conservation, *Comput. Appl. Biosci.* **9**:745–756.

Lobov, I.B., Tsutsui, K., Mitchell, A.R., and Podgornaya, O.I., 2001, Specificity of SAF-A and lamin B binding *in vitro* correlates with the satellite DNA bending state, *J. Cell. Biochem.* **83**(2):218–229.

Lobry, J., 1997, Influence of genomic G+C content on average amino-acid composition of proteins from 59 bacterial species, *Gene* **205**(1–2):309–316.

Lobry, J.R., and Sueoka, N., 2002, Asymmetric directional mutation pressures in bacteria, *Genome Biol.* **3**(10):0058.

Lohmann, V., Korner, F., Herian, U., and Bartenschlager, R., 1997, Biochemical properties of hepatitis C virus NS5B RNA-dependent RNA polymerase and identification of amino acid sequence motifs essential for enzymatic activity, *J. Virol.* **71**:8416–8428.

Long, F., Liu, H., Hahn, C., Sumazin, P., Zhang, M.Q., and Zilberstein, A., 2004, Genome-wide prediction and analysis of function-specific transcription factor binding sites, *In Silico Biol.* **4**:0033.

Long, J., and Barton, M., 2000, Initiation of axillary and floral meristems in *Arabidopsis*, *Dev. Biol.* **218**(2):341–353.

Lopez Cascales, J.J., Torre, J.G. de la, Marrink, S.J., and Berendsen, H.J. C., 1996, Molecular dynamics simulation of a charged biological membrane, *J. Chem. Phys.* **104**:2713–2720.

Lowary, P.T., and Widom, J., 1997, Nucleosome packaging and nucleosome positioning of genomic DNA, *Proc. Natl. Acad. Sci. USA* **94**:1183–1188.

Luan, D.D., Korman, M.H., Jakubczak, J.L., and Eickbush, T.H., 1993, Reverse transcription of R2Bm RNA is primed by a nick at the chromosomal target site: a mechanism for non-LTR retrotransposition, *Cell* **72**(4):595–605.

Lukashin, I.N., and Borodovsky, M., 1998, GeneMark.hmm: a new solutions for gene finding, *Nucleic Acids Res.* **26**:1107–1115.

Lung, S., Mulish, C., and Robertson, D., 2001, Basic gene grammars and DNA-Chart Parser for language processing of *Escherichia coli* promoter DNA sequences, *Bioinformatics* **17**:226–236.

Lyubetsky, V.A., and V'yugin, V.V., 2003a, Methods of horizontal gene transfer determination using phylogenetic data, *In Silico Biology* **3**:17–31.

Lyubetsky, V.A., and V'yugin, V.V., 2003b, Reconstruction of evolutionary events basing on comparison of gene and species trees, *Biophysics (Mosk).* **48** (Suppl. 1):97–106.

Ma, B., Elkayam, T., Wolfson, H., and Nussinov, R., 2001, Protein-protein interactions: Structurally conserved residues distinguish between binding sites and exposed protein surfaces, *Comput. Chem.* **26**:79–84.

Mackey, M., and Glass, L., 1997, Oscillation and chaos in physiological control systems, *Science* **15**:287–289.

Madura, J.D., Briggs, J.M., Wade, R.C., Davis, M.E., Luty, B.A., Ilin, A., Antosiewicz, J., Gilson, M.K., Bagheri, B., Scott, L.R., and McCammon, J.A., 1995, Electrostatics and diffusion of molecules in solution – simulations with the University of Houston Brownian Dynamics program, *Comp. Phys. Commun.* **91**:57–95.

Magzoub, M., Eriksson, L. E, and Graslund, A., 2002, Conformational states of the cell-penetrating peptide penetratin when interacting with phospholipid vesicles: effects of surface charge and peptide concentration, *Biochim. Biophys. Acta* **1563**:53–63.

Magzoub, M., Eriksson, L.E., and Graslund, A., 2003, Comparison of the interaction, positioning, structure induction and membrane perturbation of cell-penetrating peptides and non-translocating variants with phospholipid vesicles, *Biophys. Chem.* **103**:271–288.

Mahadevan, I., and Ghosh, I., 1994, Analysis of *E. coli* promoter structures using neural networks, *Nucleic Acids Res.* **22**:2158–2165.

Mahalanobis, P.C., 1936, On the generalised distance in statistics, *Proc. Natl. Inst. Sci. India* **12**:49–55.

Makarova, K.S., Mironov, A.A., and Gelfand, M.S., 2001, Conservation of the binding site for the arginine repressor in all bacterial lineages, *Genome Biol.* **2**: RESEARCH0013.

Makeev, V.Ju., and Tumanyan, V.G., 1996, Search of periodicities in primary structure of biopolymers: a general Fourier approach, *Comput. Appl. Biosci.* **12**:49–54.

Malik, H.S., and Eickbush, T.H., 1999, Modular evolution of the integrase domain in the Ty3/gypsy class of LTR retrotransposons, *J. Virol.* **73**(6):5186–5190.

Malik, H.S., and Eickbush, T.H., 2001, Phylogenetic analysis of ribonuclease H domains suggests a late, chimeric origin of LTR retrotransposable elements and retroviruses, *Genome Res.* **11**(7):1187–1197.

Mantegna, R.N., Buldyrev, S.V., Goldberger, A.L., Havlin, S., Peng, C.K., Simons, M., and Stanley, H., 1994, Linguistic features of non-coding DNA sequences, *Physical Review Letters* **73**:3169–3172.

Marchler-Bauer, A., Panchenko, A.R., Shoemaker, B.A., Thiessen, P.A., Geer, L.Y., and Bryant, S.H., 2002, CDD: a database of conserved domain alignments with links to domain three-dimensional structure, *Nucleic Acids Res.* **30**:281–283.

Margalit, H., Shapiro, B.A., Nussinov, R., Owens, J., and Jernigan, R.L., 1988, Helix stability in prokaryotic promoter regions, *Biochemistry* **27**:5179–5188.

Markstein, M., Markstein, P., Markstein, V., and Levine, M., 2002, Genome-wide analysis of clustered Dorsal binding sites identifies putative target genes in the Drosophila embryo, *Proc. Natl. Acad. Sci. USA* **99**(2):763–768.

Marti-Renom, M.A., Stuart, A., Fiser, A., Sanchez, R., Melo, F., and Sali, A., 2000, Comparative protein structure modeling of genes and genomes, *Annu. Rev. Biophys. Biomol. Struct.* **29**:291–325.

Masulis, I., Buckin, V., and Ozoline, O.N., 2002, Flexible elements in the structure of promoter DNA as probed by cationic surfactant binding, *J. Biomol. Struct. and Dynamics* **19**:919–927.

Masumoto, H., Nakano, M., and Ohzeki, J., 2004, The role of CENP-B and alpha-satellite DNA: de novo assembly and epigenetic maintenance of human centromeres, *Chromosome Res.* **12**(6):543–556.

Mathe, C., Sagot, M.F., Schiex, T., and Rouze, P., 2002, Current methods of gene prediction, their strengths and weaknesses, *Nucleic Acids Res.* **30**(19):4103–4117.

Matthews, B.W., 1975, Comparison of the predicted and observed secondary structure of T4 phage lysozyme, *Biochim. Biophys. Acta* **405**:442–451.

Matys, V., Fricke, E., Geffers, R., Gossling, E., Haubrock, M., Hehl, R., Hornischer, K., Karas, D., Kel, A.E., Kel-Margoulis, O.V., Kloos, D.U., Land, S., Lewicki-Potapov, B., Michael, H., Munch, R., Reuter, I., Rotert, S., Saxel, H., Scheer, M., Thiele, S., and Wingender, E., 2003, TRANSFAC: transcriptional regulation, from patterns to profiles, *Nucleic Acids Res.* **31**(1):374–378.

Mavrovouniotis, M.L., 1995, Computational methods for complex metabolic systems: representation of multiple levels of detail, in: *Bioinformatics and Genome Research*, H.A. Lim, and C.R. Cantor, eds., World Scientific, pp. 265–273.

McAdams, H.H., and Shapiro L., 1995, Circuit simulation of genetic networks, *Science* **269**(5224):650–656.

McCray, A.T., and Miller, R.A., 1998, Making the conceptual connections: the unified medical language system (UMLS) after a decade of research and development, *J. Am. Med. Inf. Assoc.* **4**(6):484–500.

McCulloch, W.S., 1945, A heterarchy of values determined by the topology of nervous nets, *Bull. Mathem. Biophys.* **7**:68–72.

McLachlan, A.D., 1993, Multichannel Fourier analysis of patterns in protein sequences, *J. Phys. Chem.* **97**:3000–3006.

Medford, J., 1992, Vegetative apical meristems, *Plant Cell* **4**(9):1029–1039.

Meeta, R., Mitra, C.K., Cserzo, M., and Simon, I., 1995, Proteins as special subsets of polypeptides, *J. Biosci.* **20**:579–590.

Mei, R., Hubbell, E., Bekiranov, S., Mittmann, M., Christians, F.C., Shen, M.-M., Lu, G., Fang, J., Liu, W.-M., Ryder, T., Kaplan, P., Kulp, D., and Webster, T.A., 2003, Probe selection for high-density oligonucleotide arrays, *Proc. Natl. Acad. Sci. USA* **100**(20):11237–11242.

Merrill, P., Sweeton, D., and Wieschaus, E., 1988, Requirements for autosomal gene activity during precellular stages of *Drosophila melanogaster*, *Development* **104**:495–509.

Metzgar, D., Thomas, E., Davis, C., Field, D., and Wills, C., 2001, The microsatellites of *Escherichia coli*: rapidly evolving repetitive DNAs in a non-pathogenic prokaryote, *Mol. Microbiol.* **39**(1):183–190.

Meyer, I.M., and Durbin, R., 2004, Gene structure conservation aids similarity based gene prediction, *Nucleic Acids Res.* **32**:776–783.

Michalski, R., and Stepp, R., 1985, Automated construction of classification: conceptual clustering versus numerical taxonomy, *IEEE Trans Patt. Analysis and Machine Intelligence* **5**:396–409.

Mihalek, I., Res, I., and Lichtarge, O., 2004, A family of evolution-entropy hybrid methods for ranking protein residues by importance, *J. Mol. Biol.* **336**:1265–1282.

Miller, K., Lynch, C., Martin, J., Herniou, E., and Tristem, M., 1999, Identification of multiple Gypsy LTR-retrotransposon lineages in vertebrate genomes, *J. Mol. Evol.* **49**(3):358–366.

Mills, M., Lacroix, L., Arimondo, P.B., Leroy, J.L., Francois, J.C., Klump, H., and Mergny, J.L., 2002, Unusual DNA conformations: implications for telomeres, *Curr. Med. Chem. Anti-Canc. Agents* **2**:627–644.

Mineta, K., Nakazawa, M., Cebria, F., Ikeo, K., Agata, K., and Gojobori, T., 2003, Origin and evolutionary process of the CNS elucidated by comparative genomics analysis of planarian ESTs, *Proc. Natl. Acad. Sci. USA* **100**:7666–7671.

Mirny, L., and Gelfand, M.S., 2002, Structural analysis of conserved base-pairs in protein-DNA complexes, *Nucleic Acids Res.* **30**:1704–1711.

Misra, V.K., Hecht, J.L., Sharp, K.A., Friedman, R.A., and Honig, B., 1994, Salt effects on protein-DNA interactions. The lambda cI repressor and EcoRI endonuclease, *J. Mol. Biol.* **238**(2):164–180.

Mitchell, A.R., 1996, The mammalian centromere: its molecular architecture, *Mutat. Res.* **372**(2):153–162.

Mitra, C.K., and Sen, A., 2001, Towards a dynamical systems approach to protein sequence structure, *Calcutta Statistical Association Bull.* **51**:203–204.

Moor, E.F., 1962, Mathematical models of self-reproduction, in: *Proceedings of Symposia in Applied Mathematics, XIV*, American Mathematical Society Press, Providence, pp. 36–62.

Moser, J., Gerstel, B., Meyer, J.E.W., Chakraborty, T., Wehland, J., Heinz, D.W., 1997, Crystal structure of the phosphatidylinositol-specific phospholipase C from the human pathogen *Listeria* monocytogenes 1, *J. Mol. Biol.* **273**(1):269–282.

Mouse Genome Sequencing Consortium, 2002, Initial sequencing and comparative analysis of the mouse genome, *Nature* **420**(6915):520–562.

Moxon, E.R., and Wills, C., 1999, DNA microsatellites: Agents of evolution? *Sci. Am.* **280**:94–99.

Mrazek, J., and Karlin, S., 1998, Strand compositional asymmetry in bacterial and large viral genomes, *Proc. Natl. Acad. Sci. USA* **95**:3720–3725.

Mukhopadhyay, P., Monticelli, L., and Tieleman D.P., 2004, Molecular dynamics simulation of a palmitoyl-oleoyl phosphatidylserine bilayer with Na+ counterions and NaCl, *Biophys. J.* **86**:1601–1609.

Mulder, N.J., Apweiler, R, Attwood, T.K., et al., 2003, The InterPro Database, 2003 brings increased coverage and new features, *Nucleic Acids Res.* **31**(1):315–318.

Murga, L.F., Wei, Y., Andre, P., Clifton, J.G., Ringe, D., and Ondrechen, M.J., 2004, Physicochemical methods for prediction of functional information for proteins, *Israel J. Chem.* **44**:299–308.

Mushegian, R.A., and Koonin, V.E., 1996, A minimal gene set for cellular life derived by comparison of complete bacterial genomes, *Proc. Natl. Acad. Sci. USA* **93**(19):10268–10273.

Myasnikova, E., Kosman, D., Samsonova, M., and Reinitz, J., 2005, Removal of background signal from *in situ* data on the expression of segmentation genes in *Drosophila, Dev. Genes and Evol.* DOI: 10.1007/s00427-005-0472-2, http://www.springerlink.com/index/10.1007/s00427-005-0472-2.

Myasnikova, E., Samsonova, A., Kozlov, K., Samsonova, M., and Reinitz, J., 2001, Registration of the expression patterns of *Drosophila* segmentation genes by two independent methods, *Bioinformatics* **17**:3–12.

Myasnikova, E., Samsonova, A., Samsonova, M., and Reinitz, J., 2002, Support vector regression applied to the determination of the developmental age of a *Drosophila* embryo from its segmentation gene expression patterns, *Bioinformatics* **18**(1):S87–S95.

Mor, M., Plazzi, P.V., Spadoni, G., and Tarzia, G., 1999, Melatonin, *Curr. Med. Chem.* **6**:501–518.

Naef, F., and Magnasco, M.O., 2003, Solving the riddle of the bright mismatches: hybridization in oligonucleotide arrays, *Phys. Rev. E* **68**:11906–11910.

Naef, F., Lim, D.A., Patil, N., and Magnasco, M., 2002, DNA hybridization to mismatched templates: A chip study, *Phys. Rev. E* **65**:4092–4096.

Nakai, K., and Kanehisa, M., 1991, Expert system for predicting protein localization sites in gram-negative bacteria, *Proteins* **11**:95–110.

Nakamura, Y., Nishio, Y., Ikeo, K., and Gojobori, T., 2003, The genome stability in *Corynebacterium* species due to lack of the recombinational repair system, *Gene* **317**:149–155.

Nakao, M., Bono, H., and Kawashima, S., 1999, Genome-scale gene expression analysis and pathway reconstruction in KEGG, *Genome Informatics* **10**:94–103.

Navajas, C., Kokkola, T., Poso, A., Honka, N., Gynther, J., and Laitinen, J.T., 1996, A rhodopsin-based model for melatonin recognition at its G-protein coupled receptor, *Eur. J. Pharmacol.* **304**:173–183.

Nazina, A., and Papatsenko, D., 2003, Statistical extraction of Drosophila cis-regulatory modules using exhaustive assessment of local word frequency, *BMC Bioinformatics* **4**:65–78.

Nei, M., and Sudhir, K. 2000, *Molecular Evolution and Phylogenetics*, Oxford University Press, New York, pp. 17–31.

Nenadic, G., et al., 2002, Terminology-driven literature and knowledge acquisition in biomedicine, *Intl. J. of Medical Informatics* **67**:33–48.

Neuveglise, C., Sarfati, J., Latge, J.-P., and Paris, S., 1996, *Afut1*, a retrotransposon-like element from *Aspergillus fumigatus*, *Nucleic Acids Res.* **24**(8):1428–1434.

Newcomer, M., Gilliland, G., and Quiocho, F.A., 1981, *L*-Arabinose binding protein – Sugar complex at 2.4 A resolution, *J. Biol. Chem.* **256**:13213–13217.

Nicolas, P., Bize, L., Muri, F., Hoebeke, M., Rodolphe, F., Ehrlich, S.D., Prum, B., and Bessieres, P., 2002, Mining Bacillus *subtilis* chromosome heterogeneities using hidden Markov models, *Nucleic Acids Res.* **30**(6):1418–1426.

Nishio, Y., Nakamura, Y., Kawarabayasi, Y., Usuda, Y., Kimura, E., Sugimoto, S., Matsui, K., Yamagishi, A., Kikuchi, H., Ikeo, K., and Gojobori, T., 2003, Comparative complete genome sequence analysis of the amino acid replacements responsible for the thermostability of *Corynebacterium efficiens*, *Genome Res.* **13**:1572–1579.

Nishio, Y., Nakamura, Y., Usuda, Y., Sugimoto, S., Matsui, K., Kawarabayasi, Y., Kikuchi, H., Gojobori, T., and Ikeo, K., 2004, Evolutionary process of amino acid biosynthesis in *Corynebacterium* at the whole genome level, *Mol. Biol. Evol.* **21**:1683–1691.

Notredame, C., and Abergel, C., 2003, Using multiple alignment methods to assess the quality of genomic data analysis, in: *Bioinformatics and Genomes: Current Perspectives*, M. Andare, ed., Horizon Scientific Press, Wymondham, UK, pp. 30–55.

Ofek, I., Hasty, D.L., and Sharon, N., 2003, Anti-adhesion therapy of bacterial diseases: prospects and problems, *FEMS Immunol. Med. Microbiol.* **38**:181–191.

Ogata, H., Goto, S., Fujibuchi, W., et al., 1998, Computation with the KEGG pathway database, *BioSystems* **47**:119–128.

Ogata, K., and Wodak, S.J., 2002, Conserved water molecules in MHC class-I molecules and their putative structural and functional roles, *Protein Eng.* **15**(8):697–705.

Ohler, U., and Niemann, H., 2001, Identification and analysis of eukaryotic promoters: recent computational approaches, *Trends Genet.* **17**(2):56–60.

Ohno, S., 1970, *Evolution by Gene Duplication*, Springer-Verlag, Berlin.

Ohno, S., 1984, Repeats of base oligomers as the primordial coding sequences of the primeval earth and their vestiges in modern genes, *J. Mol. Evol.* **20**:313–321.

Ohno, S., and Epplen, J.T., 1983, The primitive code and repeats of base oligomers as the primordial protein-encoding sequence, *Proc. Natl. Acad. Sci. USA* **80**:3391–3395.

Ohno, M., Fukagawa, T., Lee, J.S., and Ikemura, T., 2002, Triplex-forming DNAs in the human interphase nucleus visualized in situ by polypurine/polypyrimidine DNA probes and antitriplex antibodies, *Chromosoma* **111**:201–213.

Ondrechen, M.J., 2002, THEMATICS as a tool for functional genomics, *Genome Informatics* **13**:563–564.

Ondrechen, M.J., Clifton, J.G., and Ringe D., 2001, THEMATICS: a simple computational predictor of enzyme function from structure, *Proc. Natl. Acad. Sci. USA* **98**:12473–12478.

Ondrechen, M.J., Murga, L.F., Clifton, J.G., and Ringe, D., 2003, Prediction of protein function with THEMATICS, *Currents in Comp. Mol. Biol.* :21–22.

Onufriev, A., Case, D.A., and Ullmann, G.M., 2001, A novel view of pH titration in biomolecules, *Biochemistry* **40**:3413–3419.

Oppenheim, A., and Lim, S., 1981, The importance of phase in signals, *Proc. IEEE* **69**:529–541.

Orlov, Y.L., and Potapov, V.N., 2004, Complexity: an internet resource for analysis of DNA sequence complexity, *Nucleic Acids Res.* **32**(Web Server issue):W628–W633.

Orlov, Yu.L., Filippov, V.P., Potapov, V.N., and Kolchanov, N.A., 2002, Construction of stochastic context trees for genetic texts, *In Silico Biology* **2**(3):233–247.

Oshchepkov, D.Yu., Vityaev, E.E., Grigorovich, D.A., Ignatieva, E.V., and Khlebodarova, T.M., 2004a, SITECON: a tool for detecting conservative conformational and physicochemical properties in transcription factor binding site alignments and for site recognition, *Nucleic Acids Res.* **32**(Web Server issue):W208–W212.

Oshchepkov, D.Yu., Turnaev, I.I., Pozdnyakov, M.A., Milanesi, L., Vityaev, E.E., and Kolchanov, N.A., 2004b, SITECON – a tool for analysis of DNA physicochemical and conformational properties: E2F/DP transcription factor binding site analysis and recognition, in: *Bioinformatics of Genome Regulation and Structure*, N. Kolchanov, R. Hofestaedt, eds, Kluwer Academic Publishers, Boston/Dordrecht/London, pp. 93–102.

Otu, H., and Sayood, K., 2003, A new sequence distance measure for phylogenetic tree construction, *Bioinformatics* **19**(16):2122–2130.

Otwinowski, Z., Schevitz, R.W., Zhang, R,G., Lawson, C.L., Joachimiak, A., Marmorstein, R.Q., Luisi, B.F., and Sigler, P.B., 1988, Crystal structure of trp repressor/operator complex at atomic resolution, *Nature* **335**(6188):321–329.

Ou, H.Y., Guo, F.B., and Zhang, C.T., 2003, Analysis of nucleotide distribution in the genome of Streptomyces coelicolor A3(2) using the Z curve method, *FEBS Lett.* **540**(1–3):188–194.

Ovchinnikov, Yu.A., 1987, *Bioorganic Chemistry* Prosveshchenie, Moscow.

Overbeek, R., Larsen, N., and Smith, W., 1997, Representation of function: the next step, *Gene* **191**:GC1–9.

Ozoline, O.N., Deev, A.A., and Trifonov, E.N., 1999a, DNA bendability – a novel feature in *E. coli* promoter recognition, *J. Biomol. Struct. and Dynamics* **16**:825–831.

Ozoline, O.N., Deev, A.A., Arkhipova, M.V., Chasov, V., and Travers, A., 1999b, Proximal transcribed regions of bacterial promoters have non-random distribution of A/T-tracts, *Nucleic Acids Res.* **27**:4768–4774.

Paar, V., Pavin, N., Rosandic, M., Gluncic, M., Basar, I., Pezer, R., and Zinic, S.D., 2005, ColorHOR–novel graphical algorithm for fast scan of alpha satellite higher-order repeats and HOR annotation for GenBank sequence of human genome, *Bioinformatics* **21**(7):846–852.

Page, R.D.M., and Charlstone, M.A., 1997, From gene to organismal phylogeny: reconciled trees and gene tree/species tree problem, *Mol. Phylogenet. Evol.* **7**:231–240.

Palczewski, K., Kumasaka, T., Hori, T., Behnke, C.A., Motoshima, H., Fox, B.A., Le Trong, I., Teller, D.C., Okada, T., Stenkamp, R.E., Yamamoto, M., and Miyano, M., 2000, Crystal structure of rhodopsin: a G protein-coupled receptor, *Science* **289**:739–745.

Pandit, S.A., and Berkowitz, M.L., 2002, Molecular dynamics simulation of dipalmitoylphosphatidylserine bilayer with Na+ counterions, *Biophys. J.* **82**:1818–1827.

Panina, E.M., Mironov, A.A., and Gelfand, M.S., 2001, Comparative analysis of FUR regulons in gamma-proteobacteria, *Nucleic Acids Res.* **29**:5195–5206.

Panina, E.M., Mironov, A.A., and Gelfand, M.S., 2003a, Comparative genomics of bacterial zinc regulons: enhanced ion transport, pathogenesis, and rearrangement of ribosomal proteins, *Proc. Natl. Acad. Sci. USA* **100**:9912–9917.

Panina, E.M., Vitreschak, A.G., Mironov, A.A., and Gelfand, M.S., 2003b, Regulation of biosynthesis and transport of aromatic amino acids in low-GC Gram-positive bacteria, *FEMS Microbiol. Lett.* **222**:211–220.

Papatsenko, D., Makeev, V., Lifanov, A., Regnier, M., Nazina, A., and Desplan, C., 2002, Extraction of functional binding sites from unique regulatory regions: the Drosophila early developmental enhancers, *Genome Res.* **12**(1):470–481.

Paris, S., and Latge, J.P., 2001, Afut2, a new family of degenerate gypsy-like retrotransposon from *Aspergillus fumigatus*, *Med. Mycol.* **39**(2):195–198.

Patra, M., Karttunen, M., Hyvonen, M.T., Falck, E., Lindqvist, P., and Vattulainen, I., 2003, Molecular dynamics simulations of lipid bilayers: major artifacts due to truncating electrostatic interactions, *Biophys. J.* **84**:3636–3645.

Pearlman, D.A., Case, D.A., Caldwell, J.W., Seibel, G.L., Singh, U.C., Weiner, P., and Kollman, P.A., 1991, *AMBER 4.0.* University of California, San Francisco.

Pearson, W.R., 2000, Flexible sequence similarity searching with the FASTA3 program package, *Methods Mol. Biol.* **132**:185–219.

Pearson, W.R., and Lipman, D.J., 1988, Improved tools for biological sequence comparison, *Proc. Natl. Acad. Sci. USA* **85**:2444–2448.

Pedersen, A.G., and Engelbrecht, J., 1995, Investigations of *Escherichia coli* promoter sequences with artificial neural networks: new signals discovered upstream of the transcriptional start point, *Proceedings of Int. Conf. Intell. Syst. Mol. Biol.* **3**:292–299.

Pedersen, J.S., Meyer, I.M., Forsberg, R., Simminds, P., and Hein, J. 2004, A comparative method for finding and folding RNA secondary structures within protein coding regions, *Nucleic Acids Res.* **32**:4925–4936.

Peelman, F., Beneden, K. van, Zabeau, L., Iserentant, H., Ulrichts, P., Defeau, D., Verhee, A., Catteeuw, D., Elewaut, D., and Tavernier, J., 2004, Mapping of the leptin binding sites and design of a leptin antagonist, *J. Biol. Chem.* **279**:41038–41046.

Peng, C.K., Buldyrev S.V., Havlin, S., Simons, M., Stanley, H.E., and Goldberger, A., 1994, Mosaic organization of nucleotides, *Physical Rev. E.* **1**:1685–1689.

Perez-Martin, J., Rojo, F., and de Lorenzo, V., 1994, Promoters responsive to DNA bending: a common theme in prokaryotic gene expression, *Microbiol. Rev.* **58**:268–290.

Permina, E.A., and Gelfand, M.S., 2003, Heat shock (sigma32 and HrcA/CIRCE) regulons in beta-, gamma- and epsilon-proteobacteria, *J. Mol. Microbiol. Biotechnol.* **6**:174–181.

Petrache, H.I., Tristram-Nagle, S., Gawrisch, K., Harries, D., Parsegian, V.A., and Nagle, J.F., 2004, Structure and fluctuations of charged phosphatidylserine bilayers in the absence of salt, *Biophys. J.* **86**:1574–1586.

Pevzner, P.A., 2000, *Computational molecular biology: an algorithmic approach*, The MIT Press, Cambridge, Massachusetts.

Pevzner, P.A., Borodovsky, M., and Mironov, A.A., 1989, Linguistics of nucleotide sequences. I: The significance of deviations from mean statistical characteristics and prediction of the frequencies of occurrence of words, *J. Biomol. Struct. Dyn.* **6**:1013–1026.

Pickering, S.J., Pulpitt, A.J., Efford, N., Gold, N.D., and Westhead, D.R., 2001, AI-based algorithms for protein surface comparisons, *Comput. Chem.* **26**:79–84.

Politi, V., Perini, G., Trazzi, S., Pliss, A., Raska, I., Earnshaw, W.C., and Della Valle G., 2002, CENP-C binds the alpha-satellite DNA *in vivo* at specific centromere domains, *J. Cell. Sci.* **115**(Pt 11):2317–2327.

Pollastri, G., Baldi, P., Fariselli, P., and Casadio, R., 2001, Improved prediction of the number of residue contacts in proteins by recurrent neural networks, *Bioinformatics* **17** (Suppl. 1) S234–242.

Pollastri, G., Baldi, P., Fariselli, P., and Casadio, R., 2002, Prediction of coordination number and relative solvent accessibility in proteins, *Proteins* **47**:142–153.

Pollock, R., and Treisman, R., 1990, A sensitive method for the determination of protein-DNA binding specificities, *Nucleic Acids Res.* **18**:6197–6204.

Polozov, R.V., Dzhelyadin, T.R., Sorokin, A.A., Ivanova, N.N., Sivozhelezov, V.S., and Kamzolova, S.G., 1999, Electrostatic potentials of DNA. Comparative analysis of promoter and nonpromoter nucleotide sequences, *J. Biomol. Struct. Dyn.* **16**(6):1135–1143.

Pommier, Y., 1998, Diversity of DNA topoisomerases I and inhibitors, *Biochimie* **80**:255–270.

Ponomarenko, J.V., Orlova, G.V., Ponomarenko, M.P., et al., 2000, SELEX_DB: an activated database on selected randomized DNA/RNA sequences addressed to genomic sequence annotation, *Nucleic Acids Res.* **28**:205–208.

Ponomarenko, M.P., Ponomarenko, Iu.V., Kel', A.E., Kolchanov, N.A., Karas, H., Wingender, E., and Sklenar, H., 1997, Computer analysis of conformational features of the eukaryotic TATA-box DNA promoters, *Mol. Biol. (Mosk).* **31**(4):733–740.

Pontryagin, L.S., 1961, *Ordinary Differential Equations* GIFML, Moscow, pp. 312 (in Russian).

Popov, E.M., 1997, *Problem of Protein*, V. 3. *Structural organization of protein*, Nauka, Moscow (in Russian).

Porter, C.M., Havens, M.A., and Clipstone, N.A., 2000, Identification of amino acid residues and protein kinases involved in the regulation of NFATc subcellular localization, *J. Biol. Chem.* **275**:3543–3551.

Porter, M., 1980, An algorithm for suffix stripping, *Program.* **14**(3):130–137.

Pozdniakov, M.A., Vitiaev, E.E., Ananko, E.A., Ignatieva, E.V., Podkolodnaia, O.A., Podkolodnyi, N.L., Lavriushev, S.V., and Kolchanov, N.A., 2001, Comparative analysis of methods for recognizing potential transcription factor binding sites, *Mol. Biol. (Mosk).* **35**(6):961–969 (in Russian).

Prabu-Jeyabalan, M., Nalivaika, E., Schiffer, C.A., 2000, How does a symmetric dimer recognize an asymmetric substrate? A substrate complex of HIV-1 protease, *J. Mol. Biol.* **301**:1207–1220.

Press, W., Teukolsky, S., Vetterling, W., and Flannery B., 1992, *Numerical Recipes in Fortran: the Art of Scientific Computing*, 2nd ed., Cambridge University Press, New York.

Prinz, C., Hafsi, N., and Volant, P., 2003, *Helicobacter pylori* virulence factors and the host immune response: implications for therapeutic vaccination, *Trends in Microbiol.* **11**:134–138.

Pullner, A., Mautner, J., Albert, T., and Eick, D., 1996, Nucleosomal structure of active and inactive c-myc genes, *J. Biol. Chem.* **271**(49):31452–31457.

Qiu, P., 2003, Recent advances in computational promoter analysis in understanding the transcriptional regulatory network, *Biochem. Biophys. Res. Commun.* **309**(3):495–501.

Quandt, K., Frech, K., Karas, H., Wingender, E., and Werner, T., 1995, MatInd and MatInspector: new fast and versatile tools for detection of consensus matches in nucleotide sequence data, *Nucleic Acids Res.* **23**:4878–4884.

Rackovsky, S., 1998, Hidden sequence periodicities and protein architecture, *Proc. Natl. Acad. Sci. USA* **95**:8580–8584.

Radic, M.Z., Lundgren, K., and Hamkalo, B.A., 1987, Curvature of mouse satellite DNA and condensation of heterochromatin, *Cell* **50**(7):1101–1108.

Ramseier, T.M., Bledig, S., Michotey, V., Feghali, R., and Saier, M.H. Jr., 1995, The global regulatory protein FruR modulates the direction of carbon flow in *Escherichia coli*, *Mol. Microbiol.* **16**:1157–1169.

Rapola, S., Jaunty, V., Areola, M., Helena Macula, P., Kathy, H., and Kelpie, T., 2003, Anti-PsaA and the risk of pneumococcal AIM and carriage, *Vaccine* **21**:3608–3613.

Rat Genome Sequencing Project Consortium, 2004, Genome sequence of the Brown Norway rat yields insights into mammalian evolution, *Nature* **428**(6982):493–521.

Ratner, V.A., 1977, *Mathematical Population Genetics (An Elementary Course)*, Nauka, Novosibirsk.

Ratner, V.A., and Tchuraev, R.N., 1978, Simplest genetic systems controlling ontogenesis: organization principle and models of their function, in: *Progress in Theoretical Biology*, R. Rosen, F.M. Snell, eds., Acad. Press, New York/San Francisco/London **5**:81–127.

Ravcheev, D.A., Gelfand, M.S., Mironov, A.A., and Rakhmaninova, A.B., 2002, Purine regulon of gamma-proteobacteria, *Genetika* **38**:1203–1214 (in Russian).

Raychauduri, S., Chang, J.T., Imam, F. and Altman, R., 2003, The computational analysis of scientific literature to define and recognize gene expression clusters, *Nucleic Acids. Res.* **31**(15):4553–4560.

Ren, P., and Ponder, J.W., 2003, Polarizable atomic multipole water model for molecular mechanics simulation, *J. Phys. Chem.* **107**:5933–5947.

Rhee, S., Beavis, W., Berardini, T.Z., Chen, G., Dixon, D., Doyle, A., Garcia-Hernandez, M., Huala, E., Lander, G., Montoya, M., Miller, N., Mueller, L.A., Mundodi, S., Reiser, L., Tacklind, J., Weems, D.C, Wu, Y., Xu, I., Yoo, D., Yoon, J., and Zhang, P., 2003, The *Arabidopsis* information resource (TAIR): a model organism database providing a centralized, curated gateway to *Arabidopsis* biology, research materials and community, *Nucleic Acids Res.* **31**(1):224–228.

Richardson, C.J., and Barlow, D.J., 1999, The bottom line for prediction of residue solvent accessibility, *Protein Eng.* **12**:1051–1054.

Rideout, W.M.I., Coetzee, G.A., Olumi, A.F., and Jones, P.A., 1990, 5-Methylcytosine as an endogenous mutagen in the human LDL receptor and p53 genes, *Science* **249**:1288–1290.

Ringe, D., Wei, Y., Boino, K.R., and Ondrechen, M.J., 2004, Protein structure to function: insights from computation, *Cellular Mol. Life Sci.* **61**:387–392.

Rissanen, J., 1986, Complexity of strings in the class of Markov sources, *IEEE Trans. Inform. Theory* IT-32:526–532.

Rivas, E., and Eddy, S.R., 2000, Secondary structure alone in generally not statistically significant for the detection of noncoding RNAs, *Bioinformatics* **16**(7):583.

Rivas, E., Klein, R.J., Jones, T.A., and Eddy, S.R, 2001, Computational identification of noncoding RNAs in *E. coli* by comparative genomics, *Curr. Biol.* **11**:1369–1373.

Rivera-Pomar, R., and Jäckle, H., 1996, From gradients to stripes in *Drosophila* embryogenesis: filling in the gaps, *Trends Genet.* **12**:478–483.

Robison, K., McGuire, A.M., and Church, G.M., 1998, A comprehensive library of DNA-binding site matrices for 55 proteins applied to the complete *Escherichia coli* K-12 genome, *J. Mol. Biol.* **284**:241–254.

Rodionov, D.A., Mironov, A.A., and Gelfand, M.S., 2002a, Conservation of the biotin regulon and the BirA regulatory signal in Eubacteria and Archaea, *Genome Res.* **12**:1507–1516.

Rodionov, D.A., Mironov, A.A., Rakhmaninova, A.B., and Gelfand, M.S., 2000, Transcriptional regulation of transport and utilization systems for hexuronides, hexuronates and hexonates in gamma purple bacteria, *Mol. Microbiol.* **38**:673–683.

Rodionov, D.A., Vitreschak, A.G., Mironov, A.A., and Gelfand, M.S., 2002b, Comparative genomics of thiamin biosynthesis in prokaryotes. New genes and regulatory mechanisms, *J. Biol. Chem.* **277**:48949–48959.

Rodionov, D.A., Vitreschak, A.G., Mironov, A.A., and Gelfand, M.S., 2004, Comparative genomics of the methionine metabolism in Gram-positive bacteria: a variety of regulatory systems, *Nucleic Acids Res.* **32**:3340–3345.

Rodionov, M.A., Galaktionov, S.G., and Akhrem, A.A., 1981, Prediction of the degree of exposure of amino acid residues in globular proteins, *Dokl. Akad. Nauk SSSR* **261**:756–759.

Rog, T., Murzyn, K., and Pasenkiewicz-Gierula, M., 2003, Molecular dynamics simulations of charged and neutral lipid bilayers: treatment of electrostatic interactions, *Acta Biochim. Pol.* **50**:789–798.

Røgen, P., and Fain, B., 2003, Automatic classification of protein structure by using Gauss integrals, *Proc. Natl. Acad. Sci. USA* **100**:119–124.

Ron, D., Singer, Y., and Tishby, N., 1996, The power of amnesia: learning probabilistic automata with variable memory length, *Machine Learning* **25**:117–149.

Rost, B., and Sander, C., 1993, Improved prediction of protein secondary structure by use of sequence profiles and neural networks, *Proc. Natl. Acad. Sci. USA* **90**:7558–7562.

Rost, B., and Sander, C., 1994a, Combining evolutionary information and neural networks to predict protein secondary structure, *Proteins* **19**:55–72.

Rost, B., and Sander, C., 1994b, Conservation and prediction of solvent accessibility in protein families, *Proteins* **20**:216–226.

Roulet, E., Bucher, P., et al., 2000, Experimental analysis and computer prediction of CTF/NFI transcription factor DNA binding sites, *J. Mol. Biol.* **297**:833–848.

Roulet, E., Busso, S., Camargo, A.A., Simpson, A.J., Mermod, N., and Bucher, P., 2002, High-throughput SELEX SAGE method for quantitative modeling of transcription-factor binding sites, *Nat. Biotechnol.* **20**:831–835.

Rudberg, P., Tholander, F., Thunnissen, M.M., Samuelsson, B., and Haeggstrom, J., 2002, Leukotriene a4 hydrolase: Selective abrogation of leukotriene b4 formation by mutation of aspartic acid 375, *Proc. Natl. Acad. Sci. USA* **99**:4215–4220.

Rumelhart, D.E., Hinton, G.E., and Williams, R.J., 1986, Learning internal representations by error propagation, in: *Parallel Distributed Processing*, D.E. Rumelhart, J.L. McClelland, eds., V. 1, MIT Press, Cambridge, MA, pp. 318–362.

Sable, J.H., Carlin, B.G., and Sievert, M.C., 2000, Creating local bibliographic databases: new tools for evidence-based health care, *Bull. Med. Libr. Assoc.* **88**(2):139–144.

Sachdeva, G., Kumar, K., Jain, P., and Ramachandran, S., 2005, SPAAN: a software program for prediction of adhesins and adhesin-like proteins using neural networks, *Bioinformatics* **21**:483–491.

Saitou, N., and Nei, M., 1987, The neighbor-joining method: a new method for reconstructing phylogenetic trees, *Mol. Biol. Evol.* **4**:406–425.

Salton, G., and McGill, M.J., 1983, *Introduction to Modern Information Retrieval*, McGraw-Hill, Inc. New York.

Salzberg, S.L., Delcher, A.L., Kasif, S., and White, O., 1998, Microbial gene identification using interpolated Markov Models, *Nucleic Acids Res.* **26**(2):544–548.

Sánchez, L., and Thieffry, D., 2001, A logical analysis of the gap gene system. *J. Theor. Biol.* **211**(2):115–141.

Sánchez, L., and Thieffry, D., 2003, Segmenting the fly embryo: a logical analysis of the pair-rule cross-regulatory module, *J. Theor. Biol.* **224**(4):517–537.

Sandberg, M., Eriksson, L., Jonsson, J., Sjostrom, M., and Wold, S., 1998, New chemical descriptors relevant for the design of biologically active peptides, a multivariate characterization of 87 amino acids, *J. Med. Chem.* **41**:2481–2491.

Sandelin, A., Alkema, W., et al., 2004, JASPAR: an open-access database for eukaryotic transcription factor binding profiles, *Nucleic Acids Res.* **32**:D91–94.

SAS Institute Inc., 1989, *SAS/STAT User's guide. Version 6*, Vol. 1. SAS Institute Inc., Cary, NC.

Saxonov, S., Daizadeh, I., Fedorov, A., and Gilbert, W., 2000, EID: the Exon-Intron Database-an exhaustive database of protein-coding intron-containing genes, *Nucleic Acids Res.* **28**(1):185–90.

Sayle, R., and Milner-White, E.J, 1995, RasMol: Biomolecular graphics for all, *Trends in Biochem. Sci. (TIBS)* **20**(9):374.

Schaffer, A.A., Aravind, L., Madden, T.L., Shavirin, S., Spouge, J.L., Wolf, Y.I., Koonin, E.V., and Altschul, S.F., 2001, Improving the accuracy of PSI-BLAST protein database searches with composition-based statistics and other refinements, *Nucleic Acids Res.* **29**:2994–3005.

Schmid, C.D., Praz, V., Delorenzi, M., Perier, R., and Bucher, P., 2004, The eukaryotic promoter database EPD: the impact of in silico primer extension, *Nucleic Acids Res.* **32**(1):D82–D85.

Schmidt, M.W., Baldridge, K.K, Boatz, J.A., Elbert, S.T., Gordon, M.S., Jensen, J.J., Koseki, S., Matsunaga, N., Nguyen, K.A., Su, S., Windus, T.L., Dupuis, M., and Montgomery, J.A., 1993, General atomic and molecular electronic structure system GAMESS, *J. Comput. Chem.* **14**:1347–1363.

Schneider, G., 1999, How many potentially secreted proteins are contained in a bacterial genome? *Gene* **237**:113–121.

Schneidman-Duhovny, D., Nussinov, R., and Wolfson, H.J., 2004, Predicting molecular interactions in silico: II. Protein-protein and protein-drug docking, *Curr. Med. Chem.* **11**:91–107.

Schölkopf, B., and Smola, A.J., 2002, *Learning with Kernels*, MIT Press, Cambridge, MA.

Schoof, H., Ernst, R., Nazarov, V., Pfeifer, L., Mewes, H., and Mayer, K.F.X., 2004, MIPS *Arabidopsis thaliana* database (MAtDB): an integrated biological knowledge resource for plant genomics, *Nucleic Acids Res.* **32**(1):373–376.

Schuster, H.G., 1984, *Deterministic Chaos*, Physik-Verlag, Weinheim, pp. 1–240.

Schwabe, J.W., 1997, The role of water in protein-DNA interactions, *Curr. Opin. Struct. Biol.* **7**(1):126–134.

Schwarz, G., 1978, Estimating the dimension of a model, *Ann. Statistics* **6**:461–464.

Seledtsov, I.A., Solovyev, V.V., and Merkulova, T.I., 1991, New elements of glucocorticoid-receptor binding sites of hormone-regulated genes, *Biochim. Biophys. Acta* **1089**(3):367–376.

Selinger, D.W., Cheung, K.J., Mei, R., Johansson, E.M., Richmond, C.S., Blattner, F.R., Lockhart, D.J., and Church, G.M., 2000, RNA expression analysis using a 30 base pair resolution *Escherichia coli* genome arrays, *Nature Biotechnol.* **18**:1262–1268.

Selkov, E., Maltsev, N., and Olsen, G., 1997, A reconstruction of the metabolism of *Methanococcus jannaschii* from sequence data, *Gene* **197**:GC11–26.

Semendyaev, K.A., 1943, On finding eigenvalues and invariant manifolds for matrices by iterations, *Prikl. Matem. Mekhan.* **7**:193–222 (in Russian).

Shaitan, K.V., and Saraikin, S.S., 2002, The calibrated virtual medium for diffusion constant calculations by molecular dynamic simulations, *J. Phys. Chem.* **76**:1091–1096 (in Russian).

Shaitan, K.V., Balabaev, N.K., Lemak, A.S., Ermolaeva, M.D., Ivaikina, A.G., Kislyuk, O.S., Orlov, M.V., and Gelfand, E.V., 1997a, Molecular dynamics of oligopeptides 1. Use of the long trajectories and high temperatures for definition of a statistical weight of conformational sub-states, *Biophysics* **42**:47–53 (in Russian).

Shaitan, K.V., Ermolaeva, M.D., and Saraikin, S.S., 2000, Nonlinear dynamics of the molecular systems and the correlations of internal motions in the oligopeptides, *Ferroelectrics* **220**:205–220.

Shaitan, K.V., Ermolaeva, M.D., Balabaev, N.K., Lemak, A.S., and Orlov, M.V., 1997b, Molecular dynamics of oligopeptides. 2. Correlation functions of intrinsic degrees of freedom of the modified dipeptides, *Biophysics* **42**:558–565 (in Russian).

Shakked, Z., Guzikevich-Guerstein G., Frolow F., Rabinovich, D., Joachimiak, A., and Sigler, P.B., 1994, Determinants of repressor/operator recognition from the structure of the trp operator binding site, *Nature* **368**:469–473.

Shapiro, D.J., Sharp, P.A., Wahli, W.W., and Keller, M.J., 1988, A high-efficiency HeLa cell nuclear transcription extract, *DNA* **7**:47–55.

Sharp, P.M., and Li, W.H., 1987, The codon adaptation index – a measure of directional synonymous codon usage bias, and its potential applications, *Nucleic Acids Res.* **15**:1281–1295.

Shehadi, I.A., Abyzov, A., Uzun, A., Wei, Y., Murga, L.F., Ilyin, V., and Ondrechen, M.J., 2004, Active site prediction for comparative model structures with THEMATICS, *J. Bioinformatics and Computational Biology* (in press).

Shehadi, I.A., Yang, H., and Ondrechen, M.J., 2002, Future directions in protein function prediction, *Mol. Biol. Rep.* **29**:329–335.

Shimizu, H., Tanaka, T., Nakato, A., Nagahisa, K., Kimura, E. and Shioya, S., 2003, Effects of the changes in enzyme activities on metabolic flux redistribution around the 2-oxoglutarate branch in glutamate production by *Corynebacterium glutamicum*, *Bioprocess Biosyst. Eng.* **25**:291–298.

Shultzaberger, R.K., and Schneider, T.D., 1999, Using sequence logos and information analysis of Lrp DNA binding sites to investigate discrepancies between natural selection and SELEX, *Nucleic Acids Res.* **27**:882–887.

Simmen, M.W., and Bird, A., 2000, Sequence analysis of transposable elements in the sea squirt, *Ciona intestinalis, Mol. Biol. Evol.* **17**(11):1685–1694.

Smith, N.G., Hurst, L.D., 1999, The effect of tandem substitutions on the correlation between synonymous and nonsynonymous rates in rodents, *Genetics* **153**(3):1395–1402.

Smith, P., and Goodman, R.M., 1992, Information theoretic approach to rule induction from databases, *IEEE Trans Knowledge and data Eng.* **4**:301–316.

Sorokin, A.A., 2001, Functional analysis of *E. coli* promoter sequences. New promoter determinants. Ph.D. Thesis. Pushchino, Institute of Theoretical and Experimental Biophysics RAS.

Sorokin, A.A., Dzhelyadin, T.R., Ivanova, N.N., Polozov, R.V., and Kamzolova, S.G., 2001, The quest for new forms of promoter determinants. Relationship of promoter nucleotide sequences to their electrostatic potential distribution, *J. Biomol. Struct. Dyn.* **18**:1020.

Spingola, M., Grate, L., Haussler, D., and Ares, M. Jr., 1999, Genome-wide bioinformatic and molecular analysis of introns in Saccharomyces cerevisiae, *RNA* **5**(2):221–234.

St. Game, J.W. III., 1996, Progress towards a vaccine for nontypable *Haemophilus influenzae. The Finnish Medical Society DUODECIM. Ann. Med.* **28**:31–37.

Stern, L., Allison, L., Coppel, R.L., and Dix, T.I., 2001, Discovering patterns in Plasmodium falciparum genomic DNA, *Mol. Biochem. Parasitol.* **118**:175–186.

Stewart, M., 2000, Insights into the molecular mechanism of nuclear trafficking using nuclear transport factor 2 (NTF2), *Cell Struct. Funct.* **25**:217–225.

Stormo, G.D., 2000, DNA binding sites: representation and discovery, *Bioinformatics* **16**:16–23.

Strauss-Soukup, J.K., and Maher, L.J., 3rd., 1998, Electrostatic effects in DNA bending by GCN4 mutants, *Biochemistry* **37**(4):1060–1066.

Strimmer, K., Haeseler, A. von, 1997, Likelihood-mapping: a simple method to visualize phylogenetic content of a sequence alignment, *Proc. Natl. Acad. Sci. USA* **94**:6815–6819.

Strom, M.S., and Lory, S., 1993, Structure-function and biogenesis of the type IV pili, *Annu. Rev. Microbiol.* **47**:565–596.

Stull, D., Vestram, A., and Zincke, H., 1971, *Chemical Thermodynamics of Organic Compounds,* Mir, Moscow (Russian translation).

Sueoka, N., 1999, Two aspects of DNA base composition: G+C content and translation-coupled deviation from intra-strand rule of A = T and G = C, *J. Mol. Evol.* **49**:49–62.

Sugano, N., Chen, W., Roberts, M.L., and Cooper N.R., 1997, Epstein-Barr virus binding to CD21 activates the initial viral promoter via NF-kappaB induction, *J. Exp. Med.* **186**(5):731–737.

Sugden, D., Chong, N.W.S., and Lewis, D.F.W., 1995, Structural requirements at the melatonin receptor, *Br. J. Pharmacol.* **114**:618–623.

Sun, H., Palaniswamy, S.K., and Davuluri, R.V., 2003, MPromDb – a database of experimentally supported cis-regulatory elements in mammalian genomes, in: *Systems Biology: Genome Approaches to Transcriptional Regulation, 2003 March 6 - March 9; Cold Spring Harbor*, NY, USA.

Suzuki, Y., Yamashita, R., Sugano, S., and Nakai, K., 2004, DBTSS, DataBase of transcriptional start sites: progress report 2004, *Nucleic Acids Res.* **32**(1):D78–D81.

Sved, J., and Bird, A., 1990, The expected equilibrium of the CpG dinucleotide in vertebrate genomes under a mutation model, *Proc. Natl. Acad. Sci. USA* **87**:4692–4696.

Svirezhev, Yu.M., and Pasekov, V.P., 1982, *Fundamentals of Mathematical Genetics*, Nauka, Moscow.

Tanaka, Y., Nureki, O., Kurumizaka, H., Fukai, S., Kawaguchi, S., Ikuta, M., Iwahara, J., Okazaki, T., and Yokoyama, S., 2001, Crystal structure of the CENP-B protein-DNA complex: the DNA-binding domains of CENP-B induce kinks in the CENP-B box DNA, *EMBO J.* **20**:6612–6618.

Tao, Y., Kassatly, R.F., Cress, W.D., and Horowitz, J.M., 1997, Subunit composition determines E2F DNA-binding site specificity, *Mol. Cell. Biol.* **17**:6994–7007.

Tatusov, R.L., Koonin, E.V., and Lipman, D.J., 1997, A genomic perspective on protein families, *Science* **278**:631–637.

Tatusov, R.L., Natale, D.A., Garcavtsev, I.V., Tatusova, T.A., Shankavaram, U.T., Rao, B.S., Kiryutin, B., Galperin, M.Y., Fedorova, N.D., and Koonin, E.V., 2001, The COG database: new developments in phylogenetic classification of protein from complete genomes, *Nucleic Acids Res.* **29**:22–38.

Tautz, D., 1989, Hypervariability of simple sequences as a general source for polymorphic DNA markers, *Nucleic Acids Res.* **17**:6463–6471.

Tchuraev, R.N., 1975, Mathematic-logical models for molecular control systems, in: *Investigations on Mathematical Genetic,* V.A. Ratner, ed., ICG Press, Novosibirsk, pp. 67–76 (in Russian).

Tchuraev, R.N., 1980, On a stochastic model of a molecular genetic system capable of differentiation and reproduction of the initial state, *J. Biom.* **22**(2):189–194.

Tchuraev, R.N., 1991, A new method for the analysis of the dynamics of the molecular genetic control systems. I. Description of the method of generalized threshold models, *J. Theor. Biol.* **151**:71–87.

Tchuraev, R.N., 1993, *The Method of Generalized Threshold Models for Analyzing the Dynamics of Control Eukaryotic Molecular-Genetic Systems*, RAS Ufa Research Center, Ufa, 32 p. (in Russian).

Tchuraev, R.N., 1997, The Ėpigene Hypothesis, *Biopol. And Cell* **12**:76–82.

Tchuraev, R.N., 1998, The equations of dynamics of genes activities in a general view, in: *Proceeding of the First International Conference on Cenome Regulation and Structure (BGRS'1998)*, Novosibirsk, pp. 128–131.

Tchuraev, R.N., 2000, On storing, coding, passing and processing the hereditary information in living system, *Computational Technologies* **5**(2):100–111.

Tchuraev, R.N., 2005, The frame of non-canonical theory of heredity: from genes to epigenes, *J. General Biol.* **66**(2):99–122 (in Russian).

Tchuraev, R.N., and Galimzyanov, A.V., 2003, Parametric stability evaluation in computer experiments on the mathematical model of *Drosophila* control gene subnetwork, *In Silico Biol.* **3**:101–115.

Tchuraev, R.N., and Ratner, V.A., 1973, On modeling of molecular-genetic control systems in terms of automata theory, *J. Genetics* **9**(2):173–175 (in Russian).

Tchuraev, R.N., et al., 2000, Epigenes: design and construction of new hereditary units, *FEBS Lett.* **486**(3):200–202.

Tech, M., and Merkl, R., 2003, YACOP: Enhanced gene prediction obtained by a combination of existing methods, *In Silico Biol.* **3**:441–451.

Teller, D.C., Okada, T., Behnke, C.A., Palczewski, K., and Stenkamp, R.E., 2001, Advances in determination of a high-resolution three-dimensional structure of rhodopsin, a model of G-protein-coupled receptors (GPCRs), *Biochemistry* **40**:7761–7772.

Tenney, A.E., Brown, R.H., Vaske, C., Lodge, J.K., Doering, T.L., and Brent, M.R., 2004, Gene prediction and verification in a compact genome with numerous small introns, *Genome Res.* **14**:2330–2335.

Terai, G., Takagi, T., and Nakai, K., 2001, Prediction of co-regulated genes in *Bacillus subtilis* on the basis of upstream elements conserved across three closely related species, *Genome Biol.* 2: RESEARCH0048.

Thiesen, H.J., and Bach, C., 1990, Target detection assay TDA: a versatile procedure to determine DNA binding sites as demonstrated on SP1 protein, *Nucleic Acids Res.* **18**:3203–3209.

Thijs, G., Marchal, K., Lescot, M., Rombauts, S., De Moor, B., Rouze, P., and Moreau, Y., 2002, A Gibbs sampling method to detect overrepresented motifs in the upstream regions of coexpressed genes, *J. Comput. Biol.* **9**:447–464.

Thomas, J.W., Touchman, J.W., Blakesley, R.W., Bouffard, G.G., Beckstrom-Sternberg, S.M., Margulies, E.H., Blanchette, M., Siepel, A.C., Thomas, P.J., McDowell, J.C., Maskeri, B., Hansen, N.F., Schwartz, M.S., Weber, R.J., Kent, W.J., Karolchik, D., Bruen, T.C., Bevan, R., Cutler, D.J., Schwartz, S., Elnitski, L., Idol, J.R., Prasad, A.B., Lee-Lin S.Q., Maduro, V.V., Summers, T.J., Portnoy, M.E., Dietrich, N.L., Akhter, N., Ayele, K., Benjamin, B., Cariaga, K., Brinkley, C.P., Brooks, S.Y., Granite, S., Guan, X., Gupta, J., Haghighi, P., Ho, S.L., Huang, M.C., Karlins, E., Laric, P.L., Legaspi, R., Lim, M.J., Maduro, Q.L., Masiello, C.A., Mastrian, S.D., McCloskey, J.C., Pearson, R., Stantripop, S., Tiongson, E.E., Tran, J.T., Tsurgeon, C., Vogt, J.L., Walker, M.A., Wetherby, K.D., Wiggins, L.S., Young, A.C., Zhang, L.H., Osoegawa, K., Zhu, B., Zhao, B., Shu, C.L., De Jong, P.J., Lawrence, C.E., Smit, A.F., Chakravarti, A., Haussler, D., Green, P., Miller, W., and Green, E.D., 2003, Comparative analyses of multi-species sequences from targeted genomic regions, *Nature* **424**:788–793.

Thomas, R., 1991, Regulatory networks seen as asynchronous automata: a logical description, *J. Theor. Biol.* **153**:1–23.

Thomas, R., Thieffry, D., and Kaufman, M., 1995, Dynamical behaviour of biological regulatory networks. I. Biological role of feedback loops and practical use of the concept of the loop-characteristic state, *Bull. Math. Biol.* **57**(2):247–276.

Thompson, J.D., Gibson, T.J., Plewniak, F., Jeanmougin, F., and Higgins, D.G., 1997, The CLUSTAL_X windows interface: flexible strategies for multiple sequence alignment aided by quality analysis tools, *Nucleic Acids Res.* **25**:4876–4882.

Thompson, J.D., Higgins, D.G., Gibson, T.J., 1994, CLUSTAL W: improving the sensitivity of progressive multiple sequence alignment through sequence weighting, position-specific gap penalties and weight matrix choice, *Nucleic Acids Res.* **22**(22):4673–4680.

Thunnissen, M.M., Nordlund, P., and Haeggstrom, J., 2001, Crystal structure of human leukotriene a4 hydrolase, a bifunctional enzyme in inflammation, *Nat. Struct. Biol.* **8**:131–135.

Tierney, R., Kirby, H., Nagra, J., Rickinson, A., and Bell, A., 2000, The Epstein-Barr virus promoter initiating B-cell transformation is activated by RFX proteins and the B-cell-specific activator protein BSAP/Pax5, *J. Virol.* **74**:10458–10467.

Tikhonov, A.N., 1952, Systems of differential equations containing small parameters, *Mat. Sbornik* **31**:575–586 (in Russian).

Titov, I.I., and Palyanov, A.Yu., 2004, A genetic algorithm for inverse folding of RNA, in: *Bioinformatics of Genome Regulation and Structure*. N.A. Kolchanov and R. Hofestaedt, eds., Kluwer Academic Publishers, Boston/Dordrecht/London, pp. 193–202.

Titov, I.I., Ivanisenko, V.A., and Kolchanov, N.A., 2000a, FITNESS—A WWW-resource for RNA folding simulation based on genetic algorithm with local optimization, *Comput. Technol.* 5:48–56.

Titov, I.I., Vorobiev, D.G., and Kolchanov, N.A., 2000b, Mass analysis of RNA secondary structures using a genetic algorithm, in: *Proceedings of the Second International*

Conference on Bioinformatics of Genome Regulation and Structure (BGRS'2000), V. 2, pp. 139–143.

Titov, I.I., Vorobiev, D.G., Ivanisenko, V.A., and Kolchanov, N.A., 2002, A fast genetic algorithm for RNA secondary structure analysis, *Russ. Chem. Bull.* **51**(7):1135–1144.

Tjaden, B., Saxena, R.M., Stolyar, S., Haynor, D.R., Kolker, E., and Rosenow, C., 2002, Transcriptome analysis of *Escherichia coli* using high-density oligonucleotide probe arrays, *Nucleic Acids Res.* **30**:3732–3738.

Tohsato, Y., Matsuda, H., and Hashimot, A., 2000, A multiple alignment algorithm for metabolic pathway analysis using enzyme hierarchy, in: *Proceedings of the Eighth International Conference on Intelligent Systems for Molecular Biology (ISMB 2000)*, pp. 376–383.

Tomatsu, S., Orii, K.O., Bi, Y., Gutierrez, M.A., Nishioka, T., Yamaguchi, S., Kondo, N., Orii, T., Noguchi, A., and Sly, W.S., 2004, General implications for CpG hot spot mutations: methylation patterns of the human iduronate-2-sulfatase gene locus, *Hum. Mutat.* **23**:590–598.

Travers, A.A., 1989, DNA conformation and protein binding, *Annu. Rev. Biochem.* **58**:427–453.

Trifonov, E.N., 1997, Genetic level of DNA sequences is determined by superposition of many codes, *Mol. Biol. (Mosk).* **4**:759–767.

Trifonov, E.N., and Sussman, J.L., 1980, The pitch of chromatin DNA is reflected in its nucleotide sequence, *Proc. Natl. Acad. Sci. USA* **77**:3816–3820.

Tse-Dinh, Y.C., 1998, Bacterial and archeal type I topoisomerases, *Biochim. Biophys. Acta* **1400**:19–27.

Tu, Z., and Orphanidis, S.P., 2001, Microuli, a family of miniature subterminal inverted-repeat transposable elements (MSITEs): transposition without terminal inverted repeats, *Mol. Biol. Evol.* **18**(5):893–895.

Tucker-Kellogg, L., Rould, M.A., Chambers, K.A., Ades, S.E., Sauer, R.T., and Pabo, C.O., 1997, Engrailed (Gln50 → Lys) homeodomain-DNA complex at 1.9 A resolution: structural basis for enhanced affinity and altered specificity, *Structure* **5**(8):1047–1054.

Tuzinkevich, A.V., 1989, *Integral Models of Spatial and Temporal Dynamics of Ecosystems*, Dal'nevost. Nauch. Tsentr. Akad. Nauk SSSR, Vladivostok.

Ulanovsky, L.E., and Trifonov, E.N., 1987, Estimation of wedge components in curved DNA, *Nature* **326**:720–722.

Usuda, Y., Tujimoto, N., Abe, C., Asakura, Y., Kimura, E., Kawahara, Y., Kurahashi, O., and Matsui, H., 1996, Molecular cloning of the *Corynebacterium glutamicum* (*'Brevibacterium lactofermentum*' AJ12036) *odhA* gene encoding a novel type of 2-oxoglutarate dehydrogenase, *Microbiology* **142**:3347–3354.

V'yugin, V.V., Gelfand, M.S., and Lyubetsky, V.A., 2002, Trees reconciliation: reconstruction of species phylogeny by phylogenetic gene trees, *Mol. Biol. (Mosk).* **36**(5):650–658.

V'yugin, V.V., Gelfand, M.S., and Lyubetsky, V.A., 2003, Identification of horizontal gene transfer from phylogenetic gene trees, *Mol. Biol. (Mosk).* **37**(4):571–584.

Vainrub, A., and Pettitt, B.M., 2002, Coulomb blockage of hybridization in two-dimensional DNA arrays, *Phys. Rev. E* **66**:art. No. 041905.

Vakser, I.A., Matar, O.G., and Lam, C.F., 1999, A systematic study of low resolution recognition in protein-protein complexes, *Proc. Natl. Acad. Sci. USA* **96**:8477–8482.

Val, P., Lefrancois-Martinez, A.M., Veyssiere, G., and Martinez, A., 2003, SF-1 a key player in the development and differentiation of steroidogenic tissues, *Nuclear. Recept.* **1**:8–45.

Vanet, A., Marsan, L., and Sagot, M.F., 1999, Promoter sequences and algorithmical methods for identifying them, *Res. Microbiol.* **150**:779–799.

Vapnik, V., 1995, *The Nature of Statistical Learning Theory*, Springer, N.Y.

Vassylyev, D.G., Tomitori, H., Kashiwagi, K., Morikawa, K., and Igarashi, K., 1998, Crystal structure and mutational analysis of the Escherichia coli putrescine receptor. Structural basis for substrate specificity, *J. Biol. Chem.* **273**:17604–17609.

Via, A., Ferre, F., Brannetti, B., and Helmer-Citterich, M., 2000, Protein surface similarities: a survey of methods to describe and compare protein surfaces, *Cell. Mol. Life Sci.* **57**:1970–1977.

Vieille, C., and Zeikus, G.J., 2001, Hyperthermophilic enzymes: sources, uses, and molecular mechanisms for thermostability, *Microbiol. Mol. Biol. Rev.* **65**:1–43.

Viswanadhan, V.N., Ghose, A.K., Revankar, R.R., and Robins, R.K., 1989, Atomic physicochemical parameters for three dimensional structure directed quantitative structure-activity relationships. 4. Additional parameters for hydrophobic and dispersive interactions and their application for an automated superposition of certain naturally occurring nucleoside antibiotics, *J. Chem. Inf. Comput. Sci.* **29**:163–172.

Vityaev, E.E., Orlov, Yu.L., Pozdnyakov, M.A., Vishnevsky, O.V., Kolchanov, N.A., and Kovalerchuk, B.K., 2002, Knowledge Discovery for Promoter Structure Analysis, in: *Proceeding of the International Conference on Imaging Science, Systems, and Technology, Las Vegas*, Nevada, USA, June 24–27, 2002, Hamid R. Arabnia, Youngsong Mun, eds., CSREA Press, V. 1, pp. 122–128.

Vityaev, E.E., Orlov, Yu.L., Vishnevsky, O.V., Belenok, A.S., and Kolchanov, N.A., 2001, Computer system Gene Discovery for regularities search in eukaryotic regulatory regions, *Mol. Biol. (Mosk).* **35**(6):952–960 (in Russian).

Vogel, J., Bartels, V., Tang, T.H., Churakov, G., Slagter-Jager, J.G., Huttenhofer, A.,J., and Wagner, E.G.H., 2003, RNomics in *Escherichia coli* detects new sRNA species and indicates parallel transcriptional output in bacteria, *Nucleic Acids Res.* **31**:6435–6443.

Vogt, P., 1990, Potential genetic functions of tandemly repeated DNA sequence blocks in the human genome are based on a highly conserved 'chromatin folding code', *Hum. Genet.* **84**:301–336.

Volff, J.-N., Korting, C., Altschmied, J., Duschl, J., Sweeney, K., Wichert, K., Froschauer, A., and Schartl, M., 2001, Jule from the fish *Xiphophorus* is the first complete vertebrate Ty3/Gypsy retrotransposon from the Mag family, *Mol. Biol. Evol.* **18**(2):101–111.

Volokitin, E.P., 2004, On limit cycles in elemental model of hypothetical Gene Networks, *Sib. J. Industr. Mathem.* **7**(3):57–65.

Voss, R., 1992, Evolution of long-range fractal correlations and 1/f noise in DNA base sequences, *Physical Review Letters* **68**:3805–3808.

Wada, T., Yamazaki, T., and Kyogoku, Y., 2000, The structure and the characteristic DNA binding property of the C-terminal domain of the RNA polymerase α subunit from *Thermus thermophilus*, *J. Biol. Chem.* **275**:16057–16063.

Wagner, E.G., Altuvia, S., and Romby, P., 2002, Antisense RNAs in bacteria and their genetic elements, *Adv. Genet.* **46**:361–398.

Walker, M., Pavlovic, V., and Kasif, S., 2002, A comparative genomic method for computational identification of prokaryotic translation initiation sites, *Nucleic Acids Res.* **30**:3181–3191.

Wallace, A.C., Borkakoti, N., and Thornton, J.M., 1997, TESS: a geometric hashing algorithm for deriving 3D coordinate templates for searching structural databases, Application to enzyme active sites, *Protein Sci.* **6**:2308–2323.

Wan, H., Li, L., Federhen, S., and Wootton, J.C., 2003, Discovering simple regions in biological sequences associated with scoring schemes, *J. Comput. Biol.* **10**(2):171–185.

Wan, X.F., Xu, D., Kleinhofs, A., and Zhou, J., 2004, Quantitative relationship between synonymous codon usage bias and GC composition across unicellular genomes, *BMC Evol. Biol.* **4**(1):19.

Wassarman, K.M., and Storz, G., 2000, 6S RNA regulates *E. coli* RNA polymerase activity, *Cell* **101**:613–623.

Wassarman, K.M., Repoila, F., Rosenow, C., and Storz, G., 2001, Identification of novel small RNAs using comparative genomics and microarrays, *Genes and Dev.* **15**:1637–1651.

Webb, E.C., 1992. *Enzyme Nomenclature 1992: Recommendations of the Nomenclature Committee of the International Union of Biochemistry and Molecular Biology on the Nomenclature and Classification of Enzymes.* Academic Press, New York, NJ.

Weber, J.L., 1990, Informativeness of human poly$(GT)_n$ polymorphisms, *Genomics* **7**:524–530.

Weeber, M., et al., 2003, Generating hypothesis by discovering implicit associations in the literature: a case report of a search for new potential therapeutic uses of thalidomide, *J. Am. Med. Inform. Assoc.* **10**:252–259.

Weiner, P., and Kollman, P.A., 1981, AMBER: assisted aodel auilding with anergy aefinement. A aeneral arogram for aodeling aolecules and aheir anteractions, *J. Comp. Chem.* **2**:287–303.

Weiner, S.J., Kollman, P.A., Case, D.A., Singh, U.C., Ghio, C., Alagona, G., Profeta, S., and Weiner, P.K., 1984, A new force field for molecular mechanical simulation of nucleic acids and proteins, *J. Am. Chem. Soc.* **106**:765–784.

Weiner, S.J., Kollman, P.A., Nguyen, D.T., and Case, D.A., 1986, An all atom force field for simulations of proteins and nucleic acids, *J. Comp. Chem.* **7**:230–252.

Westbrook, J., Feng, Z., Chen, L., Yang, H., and Berman, H.M., 2003, The protein data bank and structural genomics, *Nucleic Acids Res.* **31**:489–491.

Weston, S.A., Lahm, A., and Suck, D., 1992, X-ray structure of the DNase-d2 complex at 2.3 resolution, *J. Mol. Biol.* **226**:1237–1256.

Wilson, D.S., Guenther, B., Desplan, C., and Kuriyan, J., 1995, High resolution crystal structure of a paired (Pax) class cooperative homeodomain dimer on DNA, *Cell* **82**(5):709–719.

Wingender, E., 1997, Classification of eukaryotic transcription factors, *Mol. Biol. (Mosk).* **31**:584–600.

Wizemann, T.M., Adamou, J.E., and Langermann, S., 1999, Adhesins as targets for vaccine development, *Emerg. Infect. Dis.* **5**:395–403.

Woda, J., Schneider, B., Patel, K., Mistry, K., and Berman H.M., 1998, An analysis of the Relationship between hydration and Protein-DNA Interactions, *Biophys. J.* **75**:2170–2177.

Wolf, Y., Rogozin, I., Grishin, N., Tatusov, R., and Koonin, E., 2001, Genome trees constructed using five different approaches suggest new major bacterial clades, *BMC Evol. Biol.* **1**:8.

Woolfe, A., Goodson, M., Goode, D., Snell, P., Smith, S., Vavouri, T., McEwen, G., Gilks, W., Walter, K., Abnizova, I., Edwards, Y., and Elgar, G., 2005, Highly conserved non-coding sequences are associated with vertebrate development, *PloS Biology* **3,** Issue 1 on line.

Wright, W.E., Binder, M., and Funk, W., 1991, Cyclic amplification and selection of targets (CASTing) for the myogenin consensus binding site, *Mol. Cell. Biol.* **11**:4104–4110.

Xia, X., 1998, The rate heterogeneity of nonsynonymous substitutions in mammalian mitochondrial genes, *Mol. Biol. Evol.* **15**:336–344.

Xia, X., 2001, *Data Analysis in Molecular Biology and Evolution*, Kluwer Academic Publishers, Boston.

Xia, X., 2004, A peculiar codon usage pattern revealed after removing the effect of DNA methylation, in: *Proceeding of the Fourth International Conference on Bioinformatics of*

Genome Regulation and Structure (BGRS'2004), Novosibirsk, Russia: IC&G, Novosibirsk, V. 1, pp. 216–220.

Xia, X., and Li, W.H., 1998, What amino acid properties affect protein evolution? *J. Mol. Evol.* **47**:557–564.

Xia, X., and Xie, Z., 2001, DAMBE: Software package for data analysis in molecular biology and evolution, *J. Hered.* **92**:371–373.

Xia, X., Hafner, M.S., and Sudman, P.D., 1996, On transition bias in mitochondrial genes of pocket gophers, *J. Mol. Evol.* **43**:32–40.

Xia, X.H., 2003, DNA methylation and mycoplasma genomes, *J. Mol. Evol.* **57**:S21–S28.

Xiong, Y., and Eickbush, T.H., 1990, Origin and evolution of retroelements based upon their reverse transcriptase sequences, *EMBO J.* **9**(10):3353–3362.

Yada, T., Nakao, M., Totoki, Y., and Nakai, K., 1999, Modeling and predicting transcriptional units of *Escherichia coli* genes using hidden Markov models, *Bioinformatics* **15**:987–993.

Yamada, K., and Seto, A., 1987, JP 63-240779, 6 October.

Ye, J., and Berg, B. van den, 2004, Crystal structure of the bacterial nucleoside transporter tsx, *EMBO J.* **23**:3187–3195.

Yoda, K., Ando, S., Okuda, A., Kikuchi, A., and Okazaki, T., 1998, In vitro assembly of the CENP-B/alpha-satellite DNA/core histone complex: CENP-B causes nucleosome positioning, *Genes Cells* **3**:533–548.

Yu, H., Yoo, A.S., and Greenwald, I., 2004, Cluster Analyzer for Transcription Sites (CATS): a C++-based program for identifying clustered transcription factor binding sites, *Bioinformatics* **20**(7):1198–1200.

Zaslaver, A., Mayo, A.E., Rosenberg, R., Bashkin, P., Sberro, H., Tsalyuk, M., Surette, M.G., and Alon, U., 2004, Just-in-time transcription program in metabolic pathways, *Nat. Genet.* **36**:486–491.

Zhang, A., Wassarman, K. M., Rosenow, C., Tjaden, B.C., Storz, G., and Gottesman, S., 2003, Global analysis of small RNA and mRNA targets of Hfq, *Mol. Microbiol.* **50**:1111–1124.

Zhang, L., Miles, M.F., and Aldape, K.D., 2003, A model of molecular interactions on short oligonucleotide microarrays, *Nat. Biotechnol.* **21**:818–828.

Zhang, M.Q., 1998, Identification of human gene core-promoters in silico, *Genome Res.* **8**:319–326.

Zheng, J, Wu, J, and Sun, Z., 2003, An approach to identify over-represented cis-elements in related sequences, *Nucleic Acids Res.* **31**(7):1995–2005.

Zhurkin, V.B., Poltev, V.I., and Florent'ev, V.L., 1980, Atom-atomic potential functions for conformational calculations of nucleic acids, *Mol. Biol. (Mosk).* **14**(5):1116–1130.

Zinovyev A., 2002, Cluster structures in genomic word frequency distributions. Web-site with supplementary materials. http://www.ihes.fr/~zinovyev/7clusters/index.htm

Zinovyev A., Gorban A., and Popova T., 2003, Self-organizing approach for automated gene identification, 2003, *Open. Sys. and Inf. Dyn.* **10**(4):321–333.

Zuegge, J., Ralph, S., Schmuker, M., McFadden, G.I., and Schneider, G., 2001, Deciphering apicoplast targeting signals–feature extraction from nuclear-encoded precursors of *Plasmodium falciparum* apicoplast proteins, *Gene* **280**:19–26.

Zuker, M., 2003, Mfold web-server for nucleic acid folding and hybridization prediction, *Nucleic Acids Res.* **31**(13):3406–3415.

Index